T0182098

Undergraduate Texts in Mathematics

Undergraduate Texts in Mathematics

Undergraduate Texts in Mathematics are generally aimed at third- and fourth-year undergraduate mathematics students at North American universities. These texts strive to provide students and teachers with new perspectives and novel approaches. The books include motivation that guides the reader to an appreciation of interrelations among different aspects of the subject. They feature examples that illustrate key concepts as well as exercises that strengthen understanding.

More information about this series at http://www.springer.com/series/666

Nakhlé H. Asmar · Loukas Grafakos

Complex Analysis
with Applications

 Springer

Nakhlé H. Asmar
Department of Mathematics
University of Missouri
Columbia, MO, USA

Loukas Grafakos
Department of Mathematics
University of Missouri
Columbia, MO, USA

ISSN 0172-6056   ISSN 2197-5604   (electronic)
Undergraduate Texts in Mathematics
ISBN 978-3-030-06788-5   ISBN 978-3-319-94063-2   (eBook)
https://doi.org/10.1007/978-3-319-94063-2

Mathematics Subject Classification (2010): 97-XX, 97I80

This Springer imprint is published by the registered company Springer Nature Switzerland AG
The registered company address is: Gewerbestrasse 11, 6330 Cham, Switzerland

Preface

Our goal in writing this book was to present a rigorous and self-contained introduction to complex variables and their applications. The book is based on notes from an undergraduate course in complex variables that we taught at the University of Missouri and on the book Applied Complex Analysis with Partial Differential Equations by N. Asmar (with the assistance of Gregory C. Jones), published by Prentice Hall in 2002.

A course in complex variables must serve students with different mathematical backgrounds from engineering, physics, and mathematics. The challenge in teaching such a course is to find a balance between rigorous mathematical proofs and applications. While recognizing the importance of developing proof-writing skills, we have tried not to let this process hinder a student's ability to understand and appreciate the applications of the theory. This book has been written so that the instructor has the flexibility to choose the level of proofs to present to the class. We have included complete proofs of most results. Some proofs are very basic (e.g., those found in the early sections of each chapter); others require a deeper understanding of calculus (e.g., use of differentiability in Sections 2.4, 2.5); and yet others propel the students to the graduate level of mathematics. The latter are found in optional sections, such as Section 3.5.

The core material for a one-semester course is contained in the first five chapters of the book. Aiming for a flexible exposition, we have given at least two versions of Cauchy's theorem, which is the most fundamental result contained in this book. In Section 3.4 we provide a quick proof of Cauchy's theorem as a consequence of Green's theorem which covers practically most applications. Then in Section 3.4 we discuss a more theoretical version of Cauchy's theorem for arbitrary homotopic curves; this approach may be skipped without altering the flow of the presentation. The book contains classical applications of complex variables to the computation of definite integrals and infinite series. Further applications are given related to conformal mappings and to Dirichlet and Neumann problems; these boundary value problems motivate the introduction to Fourier series, which are briefly discussed in Section 6.4.

The importance that we attribute to the exercises and examples is clear from the space they occupy in the book. We have included far more examples and exercises than can be covered in one course. The examples are presented in full detail. As with the proofs, the objective is to give the instructor the option to choose the examples that are suitable to the class, while providing the students many more illustrations to assist them with the homework problems.

The exercises vary in difficulty from straightforward ones to more involved project problems. Hints are provided in many cases. Solutions to the exercises can be provided upon request free-of-charge to instructors who use the text. Complimentary solutions to every-other-odd exercise and other material related to the book, such as errata and improvements, can be found at the Web site:

https://www.springer.com/us/book/9783319940625

We wish to thank Professors Tanya Christiansen and Stephen Montgomery-Smith who have used the text in the classroom and have provided us with valuable comments. We also wish to thank the following individuals who provided us with assistance and corrections: Dimitrios Betsakos, Suprajo Das, Hakan Delibas, Haochen Ding, Michael Dotzel, Nikolaos Georgakopoulos, Rebecca Heinen, Max Highsmith, Jeremy Hunn, Dillon Lisk, Caleb Mayfield, Vassilis Nestoridis, Georgios Ntosidis, Adisak Seesanea, Yiorgos-Sokratis Smyrlis, Suzanne Tourville, Yanni Wu, and Run Yan.

We are thankful to all of Springer's excellent staff, but especially to Elizabeth Loew for being so helpful during the preparation of the book. Finally, we are especially grateful for the support and encouragement that we have received from our families. This book is dedicated to them.

Columbia, Missouri, USA   Nakhlé H. Asmar
                                                                            Loukas Grafakos

Contents

Chapter 1
Complex Numbers and Functions

> *Dismissing mental tortures, and multiplying* $5 + \sqrt{-15}$
> *by* $5 - \sqrt{-15}$, *we obtain* $25 - (-15)$. *Therefore the*
> *product is* 40. ... *and thus far does arithmetical sub-*
> *tlety go, of which this, the extreme, is, as I have said,*
> *so subtle that it is useless.*
> -Girolamo Cardano (or Cardan) (1501–1576)
> [First explicit use of complex numbers, which ap-
> peared around 1545 in Cardan's solution of the prob-
> lem of finding two numbers whose sum is 10 and
> whose product is 40.]

This chapter starts with the early discovery of complex numbers and their role in solving algebraic equations. Complex numbers have the algebraic form $x + iy$, where x, y are real numbers, but they can also be geometrically represented as vectors (x, y) in the plane. Both representations have important advantages; the first one is well-suited for algebraic manipulations while the second provides significant geometric intuition. There is also a natural notion of distance between complex numbers that satisfies the familiar triangle inequality. Complex numbers also have a polar form (r, θ) based on their distance r to the origin and angle θ from the positive real semi-axis. This alternative representation provides additional insight, both algebraic and geometric, and this is explicitly manifested even in simple operations, such as multiplication and division.

Complex analysis is in part concerned with the study of complex-valued functions of a complex variable. The most important of these functions is the complex exponential e^z which is used in the definition of the trigonometric and logarithmic functions. Since we cannot plot the graphs of complex-valued functions of a complex variable (this would require four dimensions), we visualize these functions as mappings from one complex plane, the z-plane, into another plane, the w-plane. Complex-valued functions of a complex variable and their mapping properties are explored in this chapter.

Complex numbers, like many other ideas in mathematics, have significant applications in the sciences and can be used to solve real-world problems. Some of these applications are discussed in the last two chapters. In this chapter the applications are limited to finding roots of certain polynomial, algebraic, and trigonometric equations.

© Springer International Publishing AG, part of Springer Nature 2018
N. H. Asmar and L. Grafakos, *Complex Analysis with Applications*,
Undergraduate Texts in Mathematics, https://doi.org/10.1007/978-3-319-94063-2_1

1.1 Complex Numbers

Complex numbers were discovered in the sixteenth century for the purpose of solving algebraic equations that do not have real solutions. As you know, the equation

$$x^2 + 1 = 0$$

has no real roots, because there is no real number x such that $x^2 = -1$; or equivalently, we cannot take the square root of -1. The Italian mathematician Girolamo Cardano (1501–1576), better known as Cardan, stumbled upon the square roots of negative numbers and used them in his work. While Cardan was reluctant to accept these "imaginary" numbers, he did realize their role in solving algebraic equations.

Two centuries later, the Swiss mathematician Leonhard Euler (1707–1783) introduced the symbol i by setting

$$i = \sqrt{-1}, \text{ or equivalently, } i^2 = -1.$$

Although Euler used numbers of the form $a + ib$ routinely in computations, he was skeptical about their meaning and referred to them as imaginary numbers. It took the authority of the great German mathematician Karl Friedrich Gauss (1777–1855) to definitively recognize the importance of these numbers, introducing for the first time the term complex numbers that we now widely use.

Definition 1.1.1. A **complex number** z is an ordered pair (a, b) of real numbers. The set of all complex numbers is denoted by $\mathbb{C}$. We think of $\mathbb{C}$ as a vector space over the real numbers, and we define $1 \equiv (1, 0)$ and $i \equiv (0, 1)$. Then the set of real numbers $\mathbb{R}$ is contained in $\mathbb{C}$, and for a, b real we have the identification

$$(a, b) = a(1, 0) + b(0, 1) \equiv a1 + bi = a + bi.$$

If $z = a + bi$ is a complex number, then the real number a is called the **real part** of z and is denoted $\operatorname{Re} z$. The real number b is called the **imaginary part** of z and is denoted $\operatorname{Im} z$. For example, if $z = 3 + i$, then

$$\operatorname{Re} z = \operatorname{Re}(3 + i) = 3 \quad \text{and} \quad \operatorname{Im} z = \operatorname{Im}(3 + i) = 1.$$

Note that the imaginary part of a complex number is itself a real number. The imaginary part of $a + bi$ is just b, not bi, when a, b are real.

Two complex numbers are equal if they have the same real and imaginary parts. That is, $z_1 = z_2$ if and only if $\operatorname{Re} z_1 = \operatorname{Re} z_2$ and $\operatorname{Im} z_1 = \operatorname{Im} z_2$.

We do not distinguish between the forms $a + ib$ and $a + bi$; for example, $-2 + i4$ and $-2 + 4i$ are the same complex numbers. When a complex number has a zero imaginary part like $a + 0i$ we simply write it as a. These are known as **purely real** numbers and are just new interpretations of real numbers. Sometimes, via a minor abuse of language, we say that z is "real" when z has a zero imaginary part. When a complex number has a zero real part like $0 + bi$ we simply write it as bi. These numbers are called **purely imaginary**. For example, i, $2i$, πi, $-\frac{2}{3}i$ are all purely imaginary numbers. The unique complex number with zero real and imaginary parts is denoted as 0, instead of $0 + 0i$.

Algebraic Properties of Complex Numbers

We define **addition** among complex numbers as if i obeyed the same basic algebraic relations that real numbers do. To add complex numbers, we add their real and imaginary parts:

$$(a + bi) + (c + di) = (a + c) + (b + d)i.$$

For example, $(3 + 2i) + (-1 - 4i) = (3 - 1) + i(2 - 4) = 2 - 2i$.

The following properties of addition are straightforward to check: If z_1, z_2, and z_3 are complex numbers, then

$$z_1 + z_2 = z_2 + z_1 \qquad \text{(Commutative property)}$$
$$z_1 + (z_2 + z_3) = (z_1 + z_2) + z_3 \qquad \text{(Associative property)}$$

The complex number 0 is the **additive identity**: $0 + z = z + 0 = z$ for all complex numbers z. The **additive inverse** of $z = x + yi$ is the complex number $-z = -x - yi$; since $z + (-z) = 0$.

We define **subtraction** the same way:

$$(a + bi) - (c + di) = (a - c) + (b - d)i.$$

For example, $(3 + 2i) - (-1 + 4i) = (3 - (-1)) + (2 - 4)i = 4 - 2i$.

Multiplication of two complex numbers is defined as follows:

$$(a + bi)(c + di) = (ac - bd) + (ad + bc)i. \qquad (1.1.1)$$

Taking $a = c = 0$ and $b = d = 1$ in (1.1.1) we obtain that $i^2 = -1$. Then observe that (1.1.1) is obtained as a product of two binomial expressions using $i^2 = -1$. Indeed,

$$(a + bi)(c + di) = ac + a(di) + (bi)c + (bi)(di) = (ac - bd) + (ad + bc)i.$$

For example, we have $(-1 + i)(2 + i) = -2 - i + 2i + i^2 = -3 + i$. Also, $-i(4 + 4i) = -4i - 4(i)^2 = -4i + 4 = 4 - 4i$. The product of z_1 and z_2 is denoted by $z_1 z_2$. Multiplication satisfies the following properties:

$$z_1 z_2 = z_2 z_1 \qquad \text{(Commutative property)}$$
$$(z_1 z_2)z_3 = z_1(z_2 z_3) \qquad \text{(Associative property)}$$
$$z_1(z_2 + z_3) = z_1 z_2 + z_1 z_3 \qquad \text{(Distributive property)}$$

The **multiplicative identity** is the number 1; since $z1 = 1z = z$ for all complex numbers z. We show in this section that every nonzero complex number has a multiplicative inverse. For this purpose, it will be convenient to introduce another important operation on complex numbers.

For $z = a + bi$ we define the **complex conjugate** $\bar{z}$ of z by

$$\bar{z} = \overline{a + bi} = a - bi.$$

Conjugation changes the sign of the imaginary part of a complex number but leaves the real part unaltered. Thus

$$\operatorname{Re}\bar{z} = \operatorname{Re}z \quad \text{and} \quad \operatorname{Im}\bar{z} = -\operatorname{Im}z.$$

Example 1.1.2. (Basic operations) Write the expressions in the form $a + bi$, where a and b are real numbers.

(a) $(2 - 7i) + \overline{(2 - 7i)}$ (b) $(2 - 7i) - \overline{(2 - 7i)}$

(c) $(2 - 7i)(2 - 7i)$ (d) $(2 - 7i)\overline{(2 - 7i)}$.

Solution. (a) We have $\overline{2 - 7i} = 2 + 7i$, and so

$$(2 - 7i) + \overline{(2 - 7i)} = (2 - 7i) + (2 + 7i) = 4.$$

(b) Similarly,

$$(2 - 7i) - \overline{(2 - 7i)} = (2 - 7i) - (2 + 7i) = -14i.$$

(c) Multiplying $2 - 7i$ by itself we find

$$(2 - 7i)(2 - 7i) = 4 - 14i - 14i + 49i^2 = 4 - 49 - 28i = -45 - 28i.$$

(d) Taking the conjugate of $2 - 7i$ and then performing the multiplication, we find

$$(2 - 7i)\overline{(2 - 7i)} = (2 - 7i)(2 + 7i) = 4 + 14i - 14i - 49i^2 = 4 + 49 = 53. \quad \square$$

Recall that for a real number x, we have $x^2 \geq 0$. Example 1.1.2(c) shows that this statement is no longer true for complex numbers: $(2 - 7i)^2 = -45 - 28i$, which is not even a real number. What should we multiply z by to get a nonnegative real number? Example 1.1.2(d) gives us a hint: For a complex number $z = x + iy$ (with x, y real) we have

$$z\bar{z} = (x + iy)(x - iy) = x^2 + y^2, \tag{1.1.2}$$

which is *always* a nonnegative real number. Identity (1.1.2) is very important. It can be used for instance to obtain multiplicative reciprocals. Note that if $z = x + iy \neq 0$, then $z\bar{z} = x^2 + y^2 > 0$; and so

$$\frac{1}{z\bar{z}} = \frac{1}{x^2 + y^2}$$

is also a positive *real* number.

Proposition 1.1.3. (Multiplicative Inverse) *Let x, y be real numbers and $z = x + iy$ be a nonzero complex number. Then the multiplicative inverse of z, denoted $\frac{1}{z}$ or $1/z$ or z^{-1}, is*

$$\frac{1}{z\bar{z}}\bar{z} = \frac{x}{x^2 + y^2} - i\frac{y}{x^2 + y^2}. \tag{1.1.3}$$

Proof. Note that if α is a real number, then $\alpha(x - iy) = \alpha x - i\alpha y$. So, taking $\alpha = \frac{1}{z\bar{z}}$ we obtain

$$\frac{1}{z\bar{z}}\bar{z} = \frac{1}{x^2 + y^2}(x - iy) = \frac{x}{x^2 + y^2} - i\frac{y}{x^2 + y^2},$$

which establishes the second equality in (1.1.3). To prove the proposition, it suffices to show that z times the multiplicative inverse $1/z$ is equal to 1. Indeed, using the associativity and commutativity of multiplication, we obtain

$$z \cdot \frac{1}{z\bar{z}}\bar{z} = \frac{1}{z\bar{z}}z\bar{z} = 1,$$

because we are multiplying and dividing by the same nonzero real number $z\bar{z}$. ∎

We now define **division** by a complex number $z \neq 0$ to be multiplication by $1/z$. So if $c + di \neq 0$, then, using (1.1.3), we find

$$\begin{aligned}
\frac{a + bi}{c + di} &= (a + bi)\frac{1}{c + di} \\
&= (a + bi)\left(\frac{c}{c^2 + d^2} - \frac{d}{c^2 + d^2}i\right) \\
&= \frac{ac + bd}{c^2 + d^2} + \frac{bc - ad}{c^2 + d^2}i,
\end{aligned}$$

where in the last step we used (1.1.1).

An alternative way to compute the ratio of two complex numbers is by multiplying and dividing by the complex conjugate of the denominator, i.e.,

$$\begin{aligned}
\frac{a + bi}{c + di} &= \frac{a + bi}{c + di}\frac{c - di}{c - di} \\
&= \frac{(ac + bd) + (bc - ad)i}{c^2 + d^2} \\
&= \frac{ac + bd}{c^2 + d^2} + \frac{bc - ad}{c^2 + d^2}i.
\end{aligned} \tag{1.1.4}$$

It is not necessary to memorize formula (1.1.4) but suffices to recall that it is obtained by multiplying and dividing the fraction by the complex conjugate of the denominator. Here are several illustrations.

Example 1.1.4. (Inverses and quotients) Express the complex numbers in the form $a + bi$, where a and b are real numbers.

(a) $(1+i)^{-1}$ (b) $\dfrac{1}{1-i}$ (c) $\dfrac{2+i}{3-i}$ (d) $\dfrac{i}{i-1}$ (e) $\dfrac{1}{i}$ (f) $\dfrac{3+5i}{-i}$

Solution. (a) To write an equivalent expression without a complex number in a denominator, multiply and divide by the conjugate number of the denominator and use (1.1.2). So we have

$$(1+i)^{-1} = \frac{1}{1+i} = \frac{1-i}{(1+i)(1-i)} = \frac{1-i}{1^2+1^2} = \frac{1}{2} - \frac{1}{2}i.$$

(b) Similarly,

$$\frac{1}{1-i} = \frac{1+i}{(1-i)(1+i)} = \frac{1+i}{2} = \frac{1}{2} + \frac{1}{2}i.$$

(c) Here we multiply and divide by $3+i$, the conjugate of $3-i$:

$$\frac{2+i}{3-i} = \frac{2+i}{3-i}\frac{3+i}{3+i} = \frac{(2+i)(3+i)}{3^2+(-1)^2} = \frac{5+5i}{10} = \frac{1}{2} + \frac{1}{2}i.$$

(d) Start by writing the denominator in the form $a + bi$. The multiply and divide by $a - bi$:

$$\frac{i}{i-1} = \frac{i}{-1+i} = \frac{i}{-1+i}\frac{-1-i}{-1-i} = \frac{i(-1-i)}{(-1)^2+1^2} = \frac{1}{2} - \frac{1}{2}i.$$

(e) This leads to an interesting fact:

$$\frac{1}{i} = \frac{1}{i}\cdot\frac{-i}{-i} = -i.$$

Thus the multiplicative inverse of i is equal to the additive inverse of i.

(f) Using (e), we find

$$\frac{3+5i}{-i} = i(3+5i) = -5+3i. \qquad\qquad \square$$

If z is a complex number, then $z^1 = z$, $z^2 = z \cdot z$, and for a positive integer n,

$$z^n = \overbrace{z \cdot z \cdots z}^{n \text{ terms}}.$$

If $z \neq 0$, z^0 is defined to be 1. Also, $z^{-n} = \frac{1}{z^n}$. As a consequence of the definition, we have the familiar results for exponents such as $z^m z^n = z^{m+n}$, $(z^m)^n = z^{mn}$, and $(zw)^m = z^m w^m$. Using the identities

$$i^0 = 1, \quad i^1 = i, \quad i^2 = -1, \quad i^3 = -i, \quad i^4 = 1, \quad i^5 = i, \quad \ldots,$$

we conclude that for $n = 0, 1, 2, \ldots$

$$i^n = \begin{cases} 1 & \text{if } n = 4k, \\ i & \text{if } n = 4k+1, \\ -1 & \text{if } n = 4k+2, \\ -i & \text{if } n = 4k+3, \end{cases}$$

where $k = 0, 1, 2, \ldots$. We say that the sequence of complex numbers $\{i^n\}_{n=0}^{\infty}$ is **periodic** with period 4, since it repeats every four terms. Since $1/i = -i$, we also have, for $n = 0, 1, 2, \ldots$,

$$\frac{1}{i^n} = \left(\frac{1}{i}\right)^n = (-i)^n = (-1)^n i^n = \begin{cases} 1 & \text{if } n = 4k, \\ -i & \text{if } n = 4k+1, \\ -1 & \text{if } n = 4k+2, \\ i & \text{if } n = 4k+3. \end{cases}$$

Proposition 1.1.5. *Let z, z_1, and z_2 be complex numbers. Then the following properties are valid:*

(1) $\overline{z_1 + z_2} = \overline{z_1} + \overline{z_2}$
(2) $\overline{z_1 - z_2} = \overline{z_1} - \overline{z_2}$

(3) $\overline{z_1 z_2} = \overline{z_1}\,\overline{z_2}$
(4) $\overline{\left(\frac{z_1}{z_2}\right)} = \frac{\overline{z_1}}{\overline{z_2}}$ $(z_2 \neq 0)$

(5) $\overline{(z^n)} = (\overline{z})^n$, $n = 1, 2, \ldots$
(6) $\overline{\overline{z}} = z$

(7) $z + \overline{z} = 2\operatorname{Re} z$
(8) $z - \overline{z} = 2i\operatorname{Im} z$.

Proof. Identities (1) and (2) are left to the reader. To prove (3) we set $z_1 = a + ib$, $z_2 = c + id$ with a, b, c, d being real. Then $z_1 z_2 = ac - bd + i(ad + bc)$ and $\overline{z_1 z_2} = ac - bd - i(ad + bc)$. But $\overline{z_1}\,\overline{z_2} = (a - ib)(c - id) = ac - bd - i(ad + bc)$, hence (3) holds. Setting $z = z_1 = z_2$ we obtain (5) for $n = 2$; the case of general n follows from the case $n = 2$ by induction. To prove (4), we notice that in identity (1.1.4), if we replace b by $-b$ and d by $-d$, the real part of the outcome remains unchanged but the imaginary part is changed by a minus sign. This proves (4). Identity (6) is saying that the conjugate of the conjugate of a complex number is itself. It is proved as follows: If $z = a + ib$, where a, b are real, we have $\overline{z} = a - ib$ and $\overline{\overline{z}} = a - ib = a + ib = z$. Finally (7) and (8) are left to the reader. ∎

Example 1.1.6. (A linear equation) Solve the equation $(2 + i)\overline{z} - i = 3 + 2i$.

Solution. Add i to both sides and then divide by $2 + i$:

$$\overline{z} = \frac{1}{2+i}(3 + 3i) = \frac{(3 + 3i)(2 - i)}{(2 + i)(2 - i)} = \frac{9 + 3i}{2^2 + 1^2} = \frac{9}{5} + \frac{3}{5}i.$$

To find z we only need to change the sign of the imaginary part of $\overline{z}$. This gives $z = \frac{9}{5} - \frac{3}{5}i$. □

Example 1.1.7. (A system of two equations) Find the values of z_1 and z_2 that solve the system

$$\begin{cases} z_1 + \overline{z_2} = 3 + 2i \\ i\overline{z_1} + z_2 = 3. \end{cases}$$

Solution. Conjugate both sides of the first equation. Since $\overline{\overline{z_2}} = z_2$ and $\overline{3 + 2i} = 3 - 2i$, we obtain $\overline{z_1} + z_2 = 3 - 2i$. Then subtract $\overline{z_1} + z_2 = 3 - 2i$ from the second equation in the system and solve for z_1. We obtain

$$(-1 + i)\overline{z_1} = 2i$$

$$\overline{z_1} = \frac{2i}{-1 + i} = \frac{2i(-1 - i)}{(-1)^2 + (-1)^2} = 1 - i$$

$$z_1 = 1 + i.$$

Replace this value of z_1 into the first equation of the system and solve for z_2:

$$(1 + i) + \overline{z_2} = 3 + 2i \quad \Longrightarrow \quad \overline{z_2} = 2 + i \quad \Longrightarrow \quad z_2 = 2 - i.$$

Thus the solutions of the system are $z_1 = 1 + i$ and $z_2 = 2 - i$. □

The complex number system, endowed with the binary operations of addition, subtraction, multiplication, and division, satisfies the same basic algebraic properties as the real number system. These include the commutative and associative properties of addition and multiplication; the distributive property; the existence of additive inverses; and the existence of multiplicative inverses for nonzero complex numbers; see Exercise 21. Consequently, all the algebraic identities that are true for real numbers remain true for complex numbers. For example,

$$(z_1 + z_2)^2 = z_1^2 + 2z_1 z_2 + z_2^2$$
$$(z_1 - z_2)^2 = z_1^2 - 2z_1 z_2 + z_2^2$$
$$z_1^2 - z_2^2 = (z_1 - z_2)(z_1 + z_2)$$

for arbitrary complex numbers z_1 and z_2.

Square Roots of Negative Numbers and Quadratic Equations

We end this section by revisiting the quadratic equation. As you may suspect, solving such an equation involves computing square roots. In the complex number system, a negative number has two square roots. For example, the square roots of -7 are $i\sqrt{7}$ and $-i\sqrt{7}$, since $(i\sqrt{7})^2 = i^2 7 = -7$ and $(-i\sqrt{7})^2 = i^2 7 = -7$. We need a convention to distinguish between the two roots. For a positive real number r, we call $i\sqrt{r}$ the **principal value** of the square root of $-r$. The second root of $-r$ is then $-i\sqrt{r}$. Thus, if A is real, the square roots of $-A^2$ are iA and $-iA$, and the principal value of the square root is $i|A|$.

Via the use of square roots of negative numbers, we are able to factor quadratic expressions of the form $x^2 + A^2$, where x, A are real numbers, as follows:

$$x^2 + A^2 = x^2 - (-A^2) = x^2 - (iA)^2 = (x + iA)(x - iA). \tag{1.1.5}$$

Let $a \neq 0$, b, and c be real numbers. As an application of (1.1.5) we factor the quadratic expression $ax^2 + bx + c$ when its discriminant $b^2 - 4ac$ is a negative number. We write

$$\begin{aligned}
ax^2 + bx + c &= a\left(x^2 + 2x\frac{b}{2a} + \frac{c}{a}\right) \\
&= a\left(x^2 + 2x\frac{b}{2a} + \frac{b^2}{4a^2} - \frac{b^2}{4a^2} + \frac{c}{a}\right) \\
&= a\left(\left(x + \frac{b}{2a}\right)^2 + \frac{4ac - b^2}{4a^2}\right) \\
&= a\left(\left(x + \frac{b}{2a}\right)^2 + \left(\frac{\sqrt{4ac - b^2}}{2a}\right)^2\right) \\
&= a\left(x + \frac{b}{2a} + i\frac{\sqrt{4ac - b^2}}{2a}\right)\left(x + \frac{b}{2a} - i\frac{\sqrt{4ac - b^2}}{2a}\right),
\end{aligned} \tag{1.1.6}$$

where in the last step we made use of the identity in (1.1.5).

Now consider the quadratic equation $ax^2 + bx + c = 0$. If $b^2 - 4ac \geq 0$, the well-known **quadratic formula** gives the solutions

$$x_1 = \frac{-b + \sqrt{b^2 - 4ac}}{2a} \quad \text{and} \quad x_2 = \frac{-b - \sqrt{b^2 - 4ac}}{2a}.$$

If $b^2 - 4ac > 0$, then x_1, x_2 are two distinct real solutions, while if $b^2 - 4ac = 0$ then we have one double root. If $b^2 - 4ac < 0$, then the solutions are

$$x_1 = \frac{-b + i\sqrt{4ac - b^2}}{2a} \quad \text{and} \quad x_2 = \frac{-b - i\sqrt{4ac - b^2}}{2a}, \tag{1.1.7}$$

as it follows from the factorization in (1.1.6). Note that the solutions are mutually conjugate. That is, $x_1 = \overline{x_2}$.

Example 1.1.8. (The quadratic formula) Find the roots of $x^2 + x + 1 = 0$.

Solution. Since the discriminant is negative (it is equal to $1^2 - 4 \cdot 1 \cdot 1 = -3$), we expect two distinct complex conjugate roots. In view of (1.1.7) these are

$$x_1 = \frac{-1 + i\sqrt{3}}{2}, \quad x_2 = \frac{-1 - i\sqrt{3}}{2}. \qquad \square$$

We have discussed only square roots of negative numbers for the purpose of solving quadratic equations with real coefficients. You may wonder about the square roots of arbitrary complex numbers. Indeed, all complex numbers have two square roots; see Exercises 47–49 and Section 1.3.

Exercises 1.1

In Exercises 1–20, write the complex expressions in the form $a + bi$, where a and b are real numbers.

1. $\dfrac{1-i}{2}$ 2. $\dfrac{5+i}{3}$ 3. $i\dfrac{3+i}{3}$

4. $\overline{4i}(2-i)^2$ 5. $(\overline{2-i})^2$ 6. $(3+2i)-(i-\pi)$

7. $(x+iy)^2$ 8. $i\overline{(2+i)^2}$ 9. $(\frac{1}{2}+\frac{i}{7})(\frac{3}{2}-i)$

10. $\left(\dfrac{1}{2}+\dfrac{\sqrt{3}}{2}i\right)^3$ 11. $(2i)^5$ 12. $i^{12}+i^{25}-7i^{111}$

13. $\dfrac{14+13i}{2-i}$ 14. $\dfrac{1+i}{2i}$ 15. $\dfrac{(1-i)^2}{3+i}i$

16. $\dfrac{7i}{2-i}$ 17. $\dfrac{x+iy}{x-iy}$ 18. $\dfrac{101+i}{100+i}$

19. $\dfrac{i-\pi}{i+\pi}$ 20. $\dfrac{(2-i)(3+i)(4+i)^2}{1+i}$

21. Let z_1, z_2, z_3 be complex numbers. Prove that
(a) $z_1 + z_2 = z_2 + z_1$ (commutativity of addition)
(b) $z_1 z_2 = z_2 z_1$ (commutativity of multiplication)
(c) $(z_1 + z_2) + z_3 = z_1 + (z_2 + z_3)$ (associativity of addition)
(d) $(z_1 z_2)z_3 = z_1(z_2 z_3)$ (associativity of multiplication)
(e) $z_1(z_2 + z_3) = z_1 z_2 + z_1 z_3$ (distributive property)

22. (a) Show that $\text{Re}(z_1 \pm z_2) = \text{Re}(z_1) \pm \text{Re}(z_2)$.
(b) Give an example to show that, in general, $\text{Re}(z_1 z_2) \neq \text{Re}(z_1)\,\text{Re}(z_2)$.
(c) Show that $\text{Re}(z_1 z_2) = \text{Re}(z_1)\,\text{Re}(z_2)$ if and only if either z_1 or z_2 is a real number.

23. (a) Show that $i^{-n} = i^n$ for n even, and $i^{-n} = -i^n$ for n odd.
(b) Show that $\overline{i^n} = i^{-n}$ for all n.

In Exercises 24–31, solve for z.

24. $\dfrac{z}{2i} - 3 + i = 7 + 2i$ 25. $(2+3i)z = (2-i)z - i$ 26. $(1-i)\bar{z} = 6 + 3i$

27. $\overline{z+2+i} = 6i$ 28. $\bar{z}+i = 1-i$ 29. $\overline{iz+2i} = 4$

30. $\dfrac{z}{1+i} = z - 2$ 31. $\dfrac{1-z}{1+z} = 2i$

In Exercises 32–35, solve the systems for z_1 and z_2.

32. $\begin{cases} z_1 + 2z_2 = 12 - 3i \\ 3z_1 + z_2 = 16 + 6i \end{cases}$ 　　　**33.** $\begin{cases} (1-i)z_1 + z_2 = 3 + 2i \\ z_1 + (2-i)z_2 = 2 + i \end{cases}$

34. $\begin{cases} z_1 + z_2 = \frac{7}{2} - 6i \\ 2\overline{z_1} + 3iz_2 = 22 + 7i \end{cases}$ 　　　**35.** $\begin{cases} z_1 + 3\overline{z_2} = 6 + 3i \\ \overline{z_1} + (1+i)z_2 = 5 \end{cases}$

In Exercises 36–39, solve the quadratic equations. Express your answers in the form $a + bi$, where a and b are real.

36. $x^2 + 6 = 0$ 　　　　**37.** $x^2 + 4x + 5 = 0$

38. $2x^2 + x + 1 = 0$ 　　　**39.** $3x^2 + x = -2$

40. Find two numbers whose sum is 10 and whose product is 40. (This problem is of some historical value. It is said that Girolamo Cardan (1501–1576) first stumbled upon complex numbers while solving it.)

41. Let n be a positive integer and let $a_0, a_1, \ldots, a_n, z$ be complex numbers. If $a_n \neq 0$ and z varies, the expression $a_n z^n + a_{n-1} z^{n-1} + \cdots + a_1 z + a_0$ is called a **polynomial** of degree n in z. The polynomial is said to have real coefficients if the numbers a_j are real. For such a polynomial, show that

$$\overline{a_n z^n + a_{n-1} z^{n-1} + \cdots + a_1 z + a_0} = a_n (\overline{z})^n + a_{n-1} (\overline{z})^{n-1} + \cdots + a_1 \overline{z} + a_0.$$

42. Recall that z_0 is a **root** of a polynomial $p(z)$ if $p(z_0) = 0$. Show that if z_0 is a root of a polynomial with real coefficients, then $\overline{z_0}$ is also a root. Thus the nonreal complex roots of polynomials with real coefficients always appear in conjugate pairs.

In Exercises 43–46, use the given roots to find other roots with the help of Exercise 42. Then factor the polynomial and find all its roots.

43. $z^3 + z^2 + z + 1 = 0$, 　$z = i$

44. $z^3 + 10z^2 + 29z + 30 = 0$, 　$z = -2 + i$

45. $z^4 + 4 = 0$, 　$z_1 = 1 + i$

46. $z^4 - 6z^3 + 15z^2 - 18z + 10 = 0$, 　$z_1 = 1 + i$, $z_2 = 2 + i$

47. Project Problem: Computing square roots. The problem of finding nth roots of complex numbers will be discussed later in this chapter. The case of square roots is particularly interesting and can be solved by reducing to two equations in two unknowns. In this exercise, you are asked to compute $\sqrt{1 + i}$ to illustrate the process.
(a) Finding $\sqrt{1 + i}$ is equivalent to solving $z^2 = 1 + i$. Let $z = x + iy$ and obtain

$$\begin{cases} x^2 - y^2 = 1, \\ 2xy = 1. \end{cases}$$

(b) Derive the following equation in x: $4x^4 - 4x^2 - 1 = 0$ (a quadratic in x^2).
(c) Keep in mind that x^2 is nonnegative and obtain that $x^2 = \frac{1 + \sqrt{2}}{2}$. Thus

$$x = \pm \sqrt{\frac{1 + \sqrt{2}}{2}} \quad \text{and} \quad y = \pm \frac{1}{\sqrt{2 + 2\sqrt{2}}}.$$

(d) Conclude that the square roots of $1 + i$ are

$$\sqrt{\frac{1 + \sqrt{2}}{2}} + \frac{i}{\sqrt{2 + 2\sqrt{2}}} \quad \text{and} \quad -\sqrt{\frac{1 + \sqrt{2}}{2}} - \frac{i}{\sqrt{2 + 2\sqrt{2}}}.$$

48. Find the two square roots of i.

49. Find the two square roots of $-3 + 4i$.

50. Project Problem: The cubic equation. We derive the solution of the cubic equation

$$x^3 + ax^2 + bx + c = 0, \tag{1.1.8}$$

where a, b, and c are real numbers.

(a) Use the change of variables $x = y - \frac{a}{3}$ to transform the equation to the following reduced form

$$y^3 + py + q = 0, \tag{1.1.9}$$

which does not contain a quadratic term in y, where $p = b - \frac{a^2}{3}$ and $q = \frac{2a^3}{27} - \frac{ab}{3} + c$. (This trick is due to the Italian mathematician Niccolò Tartaglia (1500–1557).)

(b) Let $y = u + v$, and show that $u^3 + v^3 + (3uv + p)(u + v) + q = 0$.

(c) Require that $3uv + p = 0$; then directly we have $u^3v^3 = -\frac{p^3}{27}$, and from the equation in part (b) we have $u^3 + v^3 = -q$.

(d) Suppose that U and V are numbers satisfying $U + V = -\beta$ and $UV = \gamma$. Show that U and V are solutions of the quadratic equation $X^2 + \beta X + \gamma = 0$.

(e) Use (c) and (d) to conclude that u^3 and v^3 are solutions of the quadratic equation $X^2 + qX - \frac{p^3}{27} = 0$. Thus,

$$u = \sqrt[3]{-\frac{q}{2} + \sqrt{\left(\frac{q}{2}\right)^2 + \left(\frac{p}{3}\right)^3}} \quad \text{and} \quad v = \sqrt[3]{-\frac{q}{2} - \sqrt{\left(\frac{q}{2}\right)^2 + \left(\frac{p}{3}\right)^3}}.$$

(f) Derive a solution of (1.1.8),

$$x = \sqrt[3]{-\frac{q}{2} + \sqrt{\left(\frac{q}{2}\right)^2 + \left(\frac{p}{3}\right)^3}} + \sqrt[3]{-\frac{q}{2} - \sqrt{\left(\frac{q}{2}\right)^2 + \left(\frac{p}{3}\right)^3}} - \frac{a}{3}.$$

This is **Cardan's formula**, named after him because he was the first one to publish it. In the case $\left(\frac{q}{2}\right)^2 + \left(\frac{p}{3}\right)^3 \geq 0$, the formula clearly yields one real root of (1.1.8). You can use this root to factor (1.1.8) down into a quadratic equation, which you can solve to find all the roots of (1.1.8). The case $\left(\frac{q}{2}\right)^2 + \left(\frac{p}{3}\right)^3 < 0$ baffled the mathematicians of the sixteenth century. They knew that the cubic equation (1.1.8) must have at least one real root, yet the solution in this case involves square roots of negative numbers, which are imaginary numbers. It turns out in this case that u and v are complex conjugate numbers, hence their sum is a real number and the solution x is real! This was discovered by the Italian mathematician Rafael Bombelli (1527–1572) (see Exercise 57). Not only was Bombelli bold enough to work with complex numbers; by using them to generate real solutions, he demonstrated that complex numbers were not merely the product of our imagination but tools that are essential to derive real solutions. This theme will occur over and over again in this book when we will appeal to complex-variable techniques to solve real-life problems calling for real-valued solutions.

For an interesting account of the history of complex numbers, we refer to the book *The History of Mathematics, An Introduction*, 3rd edition, by David M. Burton (McGraw-Hill, 1997).

51. (Bombelli's equation) An equation of historical interest is $x^3 - 15x - 4 = 0$, which was investigated by Bombelli.

(a) Use Cardan's formula to derive the solution

$$x = u + v = \sqrt[3]{2 + 11i} + \sqrt[3]{2 - 11i},$$

where u is the first cube root and v is the second.

(b) Bombelli had the incredible insight that u and v have to be conjugate for $u + v$ to be real. Set $u = a + ib$ and $v = a - ib$, where a and b are to be determined. Cube both sides of the equations and note that $a = 2$, $b = 1$ will work for both equations.

(c) What is the real solution, x, of Bombelli's equation? What are the other two solutions?

1.2 The Complex Plane

A useful way of visualizing complex numbers is to plot them as points in a plane. To do this, we associate to each complex number $z = x + iy$ the ordered pair (x, y) and then plot the point $P = (x, y)$ in the Cartesian xy-plane. Since x and y uniquely determine z, we thus obtain a one-to-one correspondence between complex numbers $z = x + iy$ and points (x, y) in the Cartesian plane. The horizontal axis is called the **real axis**, since the abscissa of a complex number is its real part; and complex numbers lying on the horizontal axis are purely real. The vertical axis is called the **imaginary axis**, since the ordinate of a complex number is its imaginary part; and complex numbers lying on the vertical axis are purely imaginary. The Cartesian plane is referred to as the **complex plane**, also commonly called the z-**plane**. It is not unusual to denote a point (x, y) in the complex plane by the corresponding complex number $x + iy$ (see Figure 1.1). We can also think of a complex number $z = x + iy$ as a two-dimensional vector in the complex plane, with its tail at the origin and its head at $P = (x, y)$.

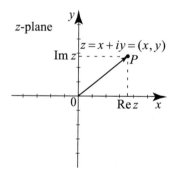

Fig. 1.1 The complex plane.

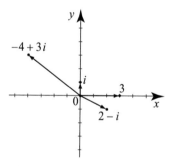

Fig. 1.2 Complex numbers as points and vectors.

Historically, the geometric representation of complex numbers is due to Gauss and two lesser known mathematicians: the Frenchman Jean-Robert Argand (1768–1822) and a Norwegian-Danish surveyor Caspar Wessel (1745–1818). This relatively simple idea dispelled the mystery and skepticism surrounding complex numbers. Much as real numbers are represented as points on a line, complex numbers are represented as points in the plane. Our ability to visualize complex numbers greatly enhances our understanding of their properties and provides significant intuition.

Example 1.2.1. (Points and vectors in the plane) Label the following points in the complex plane: $3, 0, i, 2 - i, -4 + 3i$. Draw their associated vectors, emanating from the origin.

Solution. The points and the vectors are depicted in Figure 1.2. The complex number 3, being purely real, lies on the horizontal axis; while i, being purely imaginary, lies on the vertical axis. Note that one cannot draw a vector to represent 0.   □

Geometric Interpretation of Algebraic Rules

The vector representation provides a nice geometric interpretation of addition of complex numbers via the usual **head-to-tail** or **parallelogram method** (see Figure 1.3).

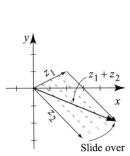 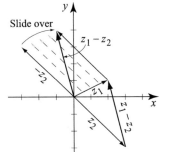 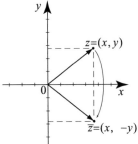

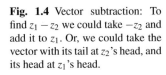

Fig. 1.3 Vector addition: Slide z_2 over maintaining its direction so its tail lies atop z_1's head. The resulting vector is $z_1 + z_2$.

Fig. 1.4 Vector subtraction: To find $z_1 - z_2$ we could take $-z_2$ and add it to z_1. Or, we could take the vector with its tail at z_2's head, and its head at z_1's head.

Fig. 1.5 Complex conjugation reflects a point $z = (x, y)$ about the horizontal axis, resulting the point $\bar{z} = (x, -y)$.

Multiplying a complex number by -1 has the effect of reflecting it about the origin. If $z = x + iy = (x, y)$, then $-z = -x - iy = (-x, -y)$. This allows us to subtract complex numbers. The complex subtraction $z_1 - z_2$ can be performed by first multiplying the z_2 vector by -1, then adding the resultant to z_1. Alternatively, if we draw both vectors with their tails at the origin, then the difference $z_1 - z_2$ is the vector that points from the head of z_2 to the head of z_1 (see Figure 1.4).

Conjugation has an interesting interpretation in the complex plane. Since the conjugate of $z = x + iy = (x, y)$ is $\bar{z} = x - iy = (x, -y)$, conjugates are reflections of each other across the real axis (see Figure 1.5).

Example 1.2.2. (Vector addition and subtraction of complex numbers) Let $z_1 = 2 + i$ and $z_2 = 3 - 3i$. Find graphically $z_1 + z_2$ and $z_1 - z_2$.

Solution. First draw the points in the plane and their associated vectors (see Figure 1.3). Then take the vector representing z_2 and slide it so that its tail lies on z_1's head. This gives $z_1 + z_2 = 5 - 2i$.

Reflect z_2 about the origin to obtain $-z_2$. Now add $-z_2$ to z_1 as vectors to get $z_1 - z_2$. The result is $z_1 - z_2 = -1 + 4i$ (see Figure 1.4). □

The Absolute Value

For a complex number $z = x + iy$, we define the **norm** or **modulus** or **absolute value** of z by

$$|z| = \sqrt{x^2 + y^2}. \tag{1.2.1}$$

If $z = x$ is real, then $|z| = \sqrt{x^2} = |x|$. Thus the absolute value of a complex number z reduces to the familiar absolute value when z is real. Just as the absolute value $|x|$ of a real number x represents the distance from x to the origin on the real line, the absolute value $|z|$ of a complex number z represents the *distance* from the point z to the origin in the complex plane (see Figure 1.6).

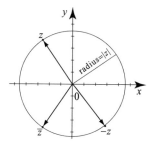

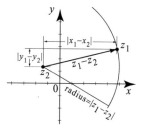

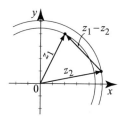

Fig. 1.6 The distance from z to the origin is $|z|$.

Fig. 1.7 The coordinates of $z_1 - z_2$ have absolute values $|x_1 - x_2|$ and $|y_1 - y_2|$.

Fig. 1.8 The distance from z_1 to z_2 is the modulus of $z_1 - z_2$.

It is easy to see from (1.2.1) or Figure 1.6 that

$$|z| = 0 \Leftrightarrow z = 0; \tag{1.2.2}$$

and

$$|-z| = |z|. \tag{1.2.3}$$

If $z_1 = x_1 + iy_1$ and $z_2 = x_2 + iy_2$, applying (1.2.1) to the complex number $z_1 - z_2 = (x_1 - x_2) + i(y_1 - y_2)$, we obtain

$$|z_1 - z_2| = \sqrt{(x_1 - x_2)^2 + (y_1 - y_2)^2}, \tag{1.2.4}$$

which is the familiar formula for distance between the points (x_1, y_1) and (x_2, y_2). Thus $|z_1 - z_2|$ has a concrete geometric interpretation as the **distance** between (the points) z_1 and z_2 (see Figure 1.7).

Example 1.2.3. (The absolute value as a distance)
(a) For $z_1 = 2 + 4i$ and $z_2 = 5 + i$, we have

$$|z_1| = \sqrt{2^2 + 4^2} = 2\sqrt{5} \approx 4.472, \qquad |z_2| = \sqrt{5^2 + 1^2} = \sqrt{26} \approx 5.099.$$

Thus we see that $|z_1| < |z_2|$. Geometrically, this means that z_2 lies farther from the origin in the complex plane (see Figure 1.8).
(b) The distance between z_1 and z_2 is

$$|z_1 - z_2| = \sqrt{(2 - 5)^2 + (4 - 1)^2} = \sqrt{18} = 3\sqrt{2} \approx 4.24.$$

This distance is also the length of the vector $z_1 - z_2$; see Figures 1.7 and 1.8.   □

Note that in Example 1.2.3(a) we compared the sizes of z_1 and z_2 and *not* the numbers themselves. In general, it does not make sense to write an inequality such as $z_1 \leq z_2$ or $z_2 \leq z_1$, unless z_1 and z_2 are real. This is because the complex numbers do not have a linear ordering like the real numbers.

We can use the geometric interpretation of the absolute value in describing subsets of the complex numbers as subsets of the plane.

Example 1.2.4. (Circles, disks, and ellipses)
(a) Find and plot all complex numbers z satisfying

$$|z + 4 - i| = 2. \tag{1.2.5}$$

(b) Find and plot the points z in the complex plane satisfying

$$|z + 4 - i| \leq 2. \tag{1.2.6}$$

(c) Find and plot all complex numbers z such that

$$|z + 2 + 2i| + |z + 1 + i| = 3\sqrt{2}. \tag{1.2.7}$$

Solution. (a) In these questions, when we write an absolute value of the form $|z - z_0|$ we interpret it as a distance between z and z_0. Thus (1.2.5) is equivalent to

$$|z - (-4 + i)| = 2.$$

Reading the absolute value as a distance, the question becomes: What are the points z whose distance to $-4 + i$ is 2? Now the answer is obvious:

$|z-(-4+i)|=2 \Leftrightarrow z$ lies on the circle centered at $-4+i$, with radius 2.

(See Figure 1.9.) The Cartesian equation of the circle is $(x+4)^2+(y-1)^2=4$. To derive this equation, write $z=x+iy$ and use (1.2.4). Thus

$$|z-(-4+i)|=2 \Leftrightarrow |(x+4)+i(y-1)|=2 \Leftrightarrow \sqrt{(x+4)^2+(y-1)^2}=2,$$

and the Cartesian equation of the circle follows upon squaring both sides.
(b) Reading the absolute value as a distance, we ask: What are the points z whose distance to $-4+i$ is less than or equal to 2? The answer is clear:

$$|z-(-4+i)| \le 2 \Leftrightarrow z \text{ lies inside or on the circle centered at } -4+i, \text{with radius 2.}$$

In other words, z lies in the disk centered at $-4+i$, with radius 2 (Figure 1.9).

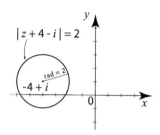

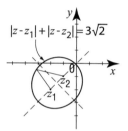

Fig. 1.9 The circle in Example 1.2.4(a).

Fig. 1.10 The ellipse in Example 1.2.4(c). Here $z_1 = -2 - 2i$ and $z_2 = -1 - i$.

(c) Write (1.2.7) in the form

$$|z-(-2-2i)|+|z-(-1-i)|=3\sqrt{2}.$$

This time we are looking for all points z the sum of whose distances to the two points $z_1 = -2-2i$ and $z_2 = -1-i$ is constant and equals $3\sqrt{2}$. From elementary geometry, we know this is the ellipse with foci[1] located at z_1 and z_2, and major axis $3\sqrt{2}$. (See Figure 1.10.) Notice that this ellipse passes through the origin. □

The absolute value of complex numbers satisfies many interesting properties that include and extend all those of the absolute value of real numbers.

Proposition 1.2.5. (Absolute Value Identities) *Let $z, z_1, z_2,\ldots$ be complex numbers. We have*

$$|z| = \sqrt{z\bar{z}} \quad or \quad |z|^2 = z\bar{z}. \tag{1.2.8}$$

Furthermore, we have

---

[1] plural of focus

$$|z| = |\bar{z}| \tag{1.2.9}$$

$$|z_1 z_2| = |z_1| |z_2| \tag{1.2.10}$$

$$|z_1 z_2 \cdots z_n| = |z_1| |z_2| \cdots |z_n| \tag{1.2.11}$$

$$|z^n| = |z|^n \quad (n = 1, 2, \ldots). \tag{1.2.12}$$

Moreover, a quotient satisfies

$$\left| \frac{z_1}{z_2} \right| = \frac{|z_1|}{|z_2|} \quad (z_2 \neq 0). \tag{1.2.13}$$

Remark 1.2.6. If z complex number, notice that $z\bar{z}$ is always a nonnegative real number, so there is no problem in taking the square root in (1.2.8) and the two identities in (1.2.8) are equivalent.

Proof. Write $z = x + iy$. Squaring both sides of (1.2.1) we obtain

$$|z|^2 = x^2 + y^2 = \underbrace{(x+iy)}_{z} \underbrace{(x-iy)}_{\bar{z}} = z\bar{z},$$

and (1.2.8) follows. Using (1.2.1) with $\bar{z} = x - iy$, we find

$$|\bar{z}|^2 = x^2 + (-y)^2 = x^2 + y^2 = |z|^2.$$

Using (1.2.8) to compute $|z_1 z_2|$, we find

$$|z_1 z_2|^2 = (z_1 z_2)\overline{(z_1 z_2)} = (z_1 z_2)(\overline{z_1}\, \overline{z_2}) = (z_1\overline{z_1})(z_2\overline{z_2}) = |z_1|^2 |z_2|^2$$

and (1.2.10) follows by taking square roots. The proof of (1.2.11) follows by induction from the case $n = 2$, while (1.2.12) is a special case of (1.2.11). Replacing z_1 by $\frac{z_1}{z_2}$ (with $z_2 \neq 0$) in (1.2.10), we obtain

$$|z_1| = \left| \frac{z_1}{z_2} z_2 \right| = \left| \frac{z_1}{z_2} \right| |z_2| \quad \Rightarrow \quad |z_1| = \left| \frac{z_1}{z_2} \right| |z_2|,$$

and (1.2.13) follows upon dividing by $|z_2| \neq 0$. ∎

Example 1.2.7. (Moduli of products and quotients) Compute the absolute values. (Take n to be a positive integer.)

(a) $|(1+i)\overline{(2-i)}|$ (b) $|(1-i)^4(1+2i)(1+\sqrt{2}i)|$ (c) $|i^n|$

(d) $\left| \dfrac{1+2i}{1-i} \right|$ (e) $\left| \dfrac{(3+4i)^2(3-i)^{10}}{(3+i)^9} \right|$

Solution. We will use as much as possible the properties of the absolute value to avoid excessive computations.

(a) Using (1.2.10) and (1.2.9), we have

$$|(1+i)\overline{(2-i)}| = |1+i|\,|\overline{(2-i)}| = |1+i|\,|2-i| = \sqrt{2}\sqrt{5} = \sqrt{10}.$$

(b) Using (1.2.11) and (1.2.12), we have

$$|(1-i)^4(1+2i)(1+\sqrt{2}i)| = |(1-i)^4|\,\overbrace{|1+2i|}^{\sqrt{5}}\,\overbrace{|1+\sqrt{2}i|}^{\sqrt{3}} = \overbrace{|1-i|^4}^{(\sqrt{2})^4}\sqrt{15} = 4\sqrt{15}.$$

(c) From (1.2.12) and the fact that $|i| = 1$, we have

$$|i^n| = |i|^n = 1.$$

(d) Using (1.2.13), we have

$$\left|\frac{1+2i}{1-i}\right| = \frac{|1+2i|}{|1-i|} = \sqrt{\frac{5}{2}}.$$

(e) We have

$$\left|\frac{(3+4i)^2(3-i)^{10}}{(3+i)^9}\right| = \frac{\overbrace{|3+4i|^2}^{25}|3-i|^{10}}{|3+i|^9} = 25|3-i| = 25\sqrt{10},$$

because $|3+i| = |3-i|$, in view of (1.2.9). $\qquad\square$

In addition to the identities that we just proved, the absolute value satisfies fundamental inequalities, which are in some cases immediate consequences of elementary facts from geometry. We recall three facts from geometry:

- In a right triangle, the hypotenuse is equal to the square root of the sum of the squares of the two other sides, hence it is larger than either one of the other sides.

- In a triangle, each side is smaller than the sum of the two other sides.

- In a triangle, the length of one side is larger than the difference of the two other sides.

Proposition 1.2.8. (Absolute Value Inequalities) *Let $z, z_1, z_2, \ldots$ be complex numbers. We have*

$$|\operatorname{Re} z| \le |z|, \qquad |\operatorname{Im} z| \le |z|; \tag{1.2.14}$$
$$|z| \le |\operatorname{Re} z| + |\operatorname{Im} z|. \tag{1.2.15}$$

The absolute value of the sum $z_1 + z_2$ satisfies the fundamental inequality

$$|z_1 + z_2| \le |z_1| + |z_2|, \tag{1.2.16}$$

known as the **triangle inequality**. *More generally, we have*

$$|z_1 + z_2 + \cdots + z_n| \leq |z_1| + |z_2| + \cdots + |z_n|. \tag{1.2.17}$$

The absolute value of the difference $z_1 - z_2$ satisfies

$$|z_1 - z_2| \leq |z_1| + |z_2|. \tag{1.2.18}$$

Moreover, we have the lower estimates

$$|z_1 + z_2| \geq ||z_1| - |z_2||, \tag{1.2.19}$$

and

$$|z_1 - z_2| \geq ||z_1| - |z_2||. \tag{1.2.20}$$

Proof. Consider a nondegenerate right triangle with vertices at 0, $z = (x, y)$, and $\operatorname{Re} z = x$, as shown in Figure 1.11. The sides of the triangle are $|\operatorname{Re} z| = |x|$, $|\operatorname{Im} z| = |y|$, and the hypotenuse is $|z|$. Since the hypothenuse is larger than either of the other two sides, we obtain (1.2.14). Since the sum of two sides in a triangle is larger than the third side, we obtain (1.2.15). Of course, (1.2.14) and (1.2.15) are also consequences of the inequalities

$$|x| \leq \sqrt{x^2 + y^2}, \quad |y| \leq \sqrt{x^2 + y^2}, \text{ and } \quad \sqrt{x^2 + y^2} \leq |x| + |y|,$$

which are straightforward to prove.

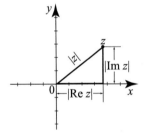

Fig. 1.11 Related to inequalities (1.2.14) and (1.2.15).

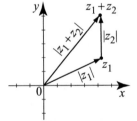

Fig. 1.12 Ineq. (1.2.16).

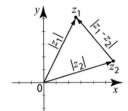

Fig. 1.13 Ineq. (1.2.20).

A geometric proof of the triangle inequality (1.2.16) is contained in Figure 1.12 where a triangle with sides $|z_1 + z_2|$, $|z_1|$, and $|z_2|$ appears. Then $|z_1 + z_2| \leq |z_1| + |z_2|$ is a consequence of the fact that the sum $|z_1| + |z_2|$ of two sides is larger than the length of the third side, which is $|z_1 + z_2|$. The triangle in Figure 1.13 with vertices at 0, z_1, and z_2 provides a geometric proof of the lower triangle inequality (1.2.20): The length $|z_1 - z_2|$ of the side of triangle is at least as big as the differences of the other two sides which are $|z_1| - |z_2|$ and $|z_2| - |z_1|$.

Since the triangle inequality (1.2.16) is fundamental in the development of complex analysis, we also offer an algebraic proof. Start by observing that

$$\overline{z_1 \, \overline{z_2}} = \overline{z_1} \, \overline{\overline{z_2}} = \overline{z_1} \, z_2.$$

For an arbitrary complex number w we have $w + \overline{w} = 2\,\mathrm{Re}\,w$, and thus we obtain

$$z_1\,\overline{z_2} + \overline{z_1}\,z_2 = z_1\,\overline{z_2} + \overline{z_1\,\overline{z_2}} = 2\,\mathrm{Re}\,(z_1\,\overline{z_2}).$$

Using this interesting identity, along with (1.2.8) and basic properties of complex conjugation, we obtain

$$
\begin{aligned}
|z_1 + z_2|^2 &= (z_1 + z_2)\overline{(z_1 + z_2)} = (z_1 + z_2)(\overline{z_1} + \overline{z_2}) \\
&= z_1\,\overline{z_1} + z_2\,\overline{z_2} + z_1\,\overline{z_2} + \overline{z_1}\,z_2 = |z_1|^2 + |z_2|^2 + z_1\,\overline{z_2} + \overline{z_1}\,z_2 \\
&= |z_1|^2 + |z_2|^2 + 2\,\mathrm{Re}\,(z_1\,\overline{z_2}) \\
&\leq |z_1|^2 + |z_2|^2 + 2|z_1\,\overline{z_2}| \qquad \text{(by (1.2.14))} \\
&= |z_1|^2 + |z_2|^2 + 2|z_1|\,|\overline{z_2}| \\
&= |z_1|^2 + |z_2|^2 + 2|z_1|\,|z_2| \qquad \text{(by (1.2.9) and } |\overline{z_2}| = |z_2|) \\
&= (|z_1| + |z_2|)^2,
\end{aligned}
$$

and (1.2.16) follows upon taking square roots on both sides. Next, notice that (1.2.17) is obtained by a repeated applications of (1.2.16), while (1.2.18) is deduced from (1.2.16) replacing z_2 by $-z_2$.

Replacing z_1 by $z_1 - z_2$ in (1.2.16), we obtain $|z_1| \leq |z_1 - z_2| + |z_2|$, and so

$$|z_1 - z_2| \geq |z_1| - |z_2|.$$

Reversing the roles of z_1 and z_2, and realizing that $|z_2 - z_1| = |z_1 - z_2|$, we also have

$$|z_1 - z_2| \geq |z_2| - |z_1|.$$

Putting these two together, we conclude $|z_1 - z_2| \geq \big||z_1| - |z_2|\big|$, which proves inequality in (1.2.20). Finally, we deduce (1.2.19) replacing z_2 by $-z_2$. ∎

The triangle inequality is used extensively in proofs to provide estimates on the sizes of complex-valued expressions. We illustrate such applications via examples.

Example 1.2.9. (Estimating the size of an absolute value) What is an upper bound for $|z^5 - 4|$ if $|z| \leq 1$?

Solution. Applying the triangle inequality, we get

$$|z^5 - 4| \leq |z^5| + 4 = |z|^5 + 4 \leq 1 + 4 = 5,$$

because $|z| \leq 1$. Hence if $|z| \leq 1$, an upper bound for $|z^5 - 4|$ is 5.

Can we find a number smaller than 5 that is also an upper bound, or is 5 the *least upper bound*? It is easy to see that for $z = -1$, we have $|z^5 - 4| = |(-1)^5 - 4| = |-1 - 4| = 5$. Thus, the upper bound 5 is best possible. You should be cautioned that, in general, the triangle inequality is considered a crude inequality, which means that it will not yield least upper bound estimates as it did in this case. See Exercise 38 for an illustration of this fact. □

Example 1.2.10. (Inequalities) Show that, for arbitrary complex numbers z and a with $|z| \neq |a|$, we have

$$\frac{1}{|a| + |z|} \leq \frac{1}{|a + z|} \leq \frac{1}{||a| - |z||}. \tag{1.2.21}$$

Solution. The triangle inequality (1.2.16) tells us that $|a + z| \leq |a| + |z|$, while (1.2.19) implies that $||a| - |z|| \leq |a + z|$. Hence the inequalities

$$||a| - |z|| \leq |a + z| \leq |a| + |z|.$$

Taking reciprocals reverses the inequalities and yields (1.2.21).   □

The last example of this section illustrates a classical trick when dealing with inequalities. It consists of adding and subtracting a number in order to transform an expression into a form that contains familiar terms.

Example 1.2.11. (Techniques with absolute values)
(a) What is an upper bound for $|z - 3|$ if $|z - i| \leq 1$?
(b) What is a lower bound for $|z - 3|$ if $|z - i| \leq 1$?

Solution. (a) We wish to estimate the size of $|z - 3|$ given some information about $|z - i|$. The trick is to add and subtract i, then use the triangle inequality as follows:

$$\begin{aligned}
|z - 3| = |z - i + i - 3| = |(z - i) + (-3 + i)| \quad &\text{(Add and subtract } i) \\
\leq |z - i| + \underbrace{|-3 + i|}_{\sqrt{10}} \leq 1 + \sqrt{10}, \quad &\text{(Triangle inequality)}
\end{aligned}$$

since $|z - i| \leq 1$. Thus an upper bound is $1 + \sqrt{10}$.
(b) In finding a lower bound, we will proceed as in (a) but use (1.2.19) instead of the triangle inequality. We have

$$\begin{aligned}
|z - 3| = |(z - i) + (-3 + i)| \quad &\text{(Add and subtract } i.) \\
\geq ||z - i| - |-3 + i|| = ||z - i| - \sqrt{10}|,
\end{aligned}$$

by (1.2.19). Now since $|z - i|$ is at most 1 and $\sqrt{10} > 1$, we see that

$$||z - i| - \sqrt{10}| = \sqrt{10} - |z - i| \geq \sqrt{10} - 1.$$

Hence, $|z - 3| \geq \sqrt{10} - 1$ if $|z - i| \leq 1$.   □

Exercises 1.2

In Exercises 1–6, plot the points z, $-z$, $\bar{z}$ and the associated vectors emanating from the origin. In each case, compute the modulus of z.

1. $1-i$ **2.** $\frac{\sqrt{2}}{2}+i\frac{\sqrt{2}}{2}$ **3.** $3i+5$ **4.** i^7 **5.** $\overline{1-i}$ **6.** $(1+i)^2$

7. Let

$$z_1 = i, \; z_2 = \frac{\sqrt{2}}{2} + i\frac{\sqrt{2}}{2}, \; z_3 = \frac{\sqrt{2}}{2} - i\frac{\sqrt{2}}{2}, \; z_4 = \frac{1}{2} + i\frac{3}{2}.$$

(a) Plot z_2, z_3, and $z_2 + z_3$ on the same complex plane. Explain in words how you constructed $z_2 + z_3$.

(b) Plot the points z_1, z_2, z_3, z_4, compute their moduli, and decide which point or points are closest to the origin.

(c) Find graphically $z_1 - z_2$, $z_2 - z_3$, and $z_3 - z_4$.

(d) Which one of the points z_2 or z_4 is closer to z_1?

8. Let $z_1 = i$ and $z_2 = 1 + 2i$.

(a) On the same complex plane, plot z_1, z_2, $z_1 + z_2$, $z_1 - z_2$, $z_2 - z_1$.

(b) How does vector $z_1 - z_2$ compare with $z_2 - z_1$?

(c) On the same complex plane, plot the vectors z_2 and iz_2. How do these vectors compare? Describe in general the vector iz in comparison with vector z.

(d) Describe in general the vector z/i in comparison with z.

In Exercises 9–14, compute the moduli.

9. $|(1+i)(1-i)(1+3i)|$ **10.** $\left|(2+3i)^8\right|$ **11.** $\left|\left(\frac{\sqrt{2}}{2}+i\frac{\sqrt{2}}{2}\right)^{27}\right|$

12. $\left|\dfrac{1+i}{(1-i)(1+3i)}\right|$ **13.** $\left|\dfrac{i}{\overline{2-i}}\right|$ **14.** $\left|\dfrac{(1+i)^5}{(-2+2i)^5}\right|$

In Exercises 15–26, describe in words and then plot the set of points satisfying the equations or inequalities.

15. $|z-4| = 3$ **16.** $|z+2+i| = 1$ **17.** $|z-i| = -1$

18. $|z+1|+|z-1| = 4$ **19.** $|z-i|+|z| = 2$ **20.** $|2z|+|2z-1| = 4$

21. $|z-1| \leq 4$ **22.** $|z-1-i| > 1$ **23.** $|z-2+i| \geq 2$

24. $1 < |z-2i| \leq 4$ **25.** $0 < |z-1-i| < 1$ **26.** $|z+i| \leq 0$

27. Derive the equation of the ellipse in Exercise 20.

28. Derive the equation of the circle in Exercise 16.

29. (Lines in the complex plane) (a) Show that the set of points z in the complex plane with $\operatorname{Re} z = a$ (a real) is the vertical line $x = a$.

(b) Show that the set of points z in the complex plane with $\operatorname{Im} z = b$ (b real) is the horizontal line $y = b$.

(c) Let $z_1 \neq z_2$ be two points in the complex plane. Show that the set of points z such that $z = z_1 + t(z_2 - z_1)$, where t is real, is the line going through z_1 and z_2. Illustrate your answer graphically by plotting the vectors z_1, $z_1 + z_2$, and $z_1 + t(z_2 - z_1)$ for several values of t.

30. Show that three distinct points z_1, z_2, and z_3 lie on the same line if and only if

$$\frac{z_1 - z_2}{z_1 - z_3} = t,$$

where t is real. (Compare with Exercise 29(c).)

31. (Parabolas) Recall from geometry that a parabola is the set of points in the plane that are equidistant from a fixed line (called the **directrix**) and a fixed point not on the line (called the **focus**).

(a) Using this description and the geometric interpretation of the absolute value, argue that the set of points z satisfying

$$|z - 1 - i| = \operatorname{Re}(z) + 1$$

is a parabola. Find its directrix and its focus and then plot it.

(b) More generally, describe the set of points z such that

$$|z - z_0| = \operatorname{Re}(z) - a,$$

where a is a real number.

32. (a) With the help of Exercise 31, describe the set of points z such that $|z - i| = \operatorname{Im}(z) + 2$.

(b) More generally, describe the set of points z such that $|z - z_0| = \operatorname{Im}(z) - a$, where a is a real number.

33. (The multiplicative inverse) Show that if $z \neq 0$, then

$$z^{-1} = \frac{\bar{z}}{|z|^2}.$$

34. (Equality in the triangle inequality) When do we have $|z_1 + z_2| = |z_1| + |z_2|$? [*Hint:* To answer this question, review the algebraic proof of the triangle inequality and note that the string of equalities in the proof was broken when we replaced $2\operatorname{Re}(z_1 \bar{z_2})$ by $2|z_1||z_2|$. Thus we have an equality in the triangle inequality if and only if $\operatorname{Re}(z_1 \bar{z_2}) = |z_1||z_2|$. Show that this is the case if and only either z_1 or z_2 is zero or $z_1 = \alpha z_2$, where α is a positive real number. Geometrically, this states that z_1 and z_2 are on the same side of the ray issued from the origin.]

35. Show that a complex number z satisfies $z^2 = |z|^2$ if and only if $\operatorname{Im} z = 0$.

36. (Parallelogram identity) Consider an arbitrary parallelogram with sides a, b, and diagonals c and d, as shown in the adjacent figure.

(a) Prove the following identity:

$$c^2 + d^2 = 2(a^2 + b^2).$$

(The identity states the well-known fact that, in a parallelogram, the sum of the squares of the diagonals is equal to the sum of the squares of the sides.) In your proof, use the law of cosines $r^2 + s^2 - 2rs\cos\alpha = a^2$, where α is the angle opposite the side a, and r and s are the other two sides.

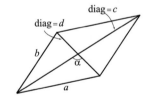

(b) Let u and v be arbitrary complex numbers. Form the parallelogram with sides u and v. Plot $u + v$ and $u - v$ as diagonals of this parallelogram.

(c) Using parts (a) and (b) and geometric considerations show that for arbitrary complex numbers u and v we have

$$|u + v|^2 + |u - v|^2 = 2\left(|u|^2 + |v|^2\right).$$

This is known as the **parallelogram identity** for complex numbers.

(d) Prove the parallelogram identity algebraically using (1.2.8).

37. (a) Notice that

$$|\cos\theta + i\sin\theta| \leq |\cos\theta| + |i\sin\theta| = |\cos\theta| + |\sin\theta| \leq 2.$$

Thus, a straightforward application of the triangle inequality yields $|\cos\theta + i\sin\theta| \le 2$.
(b) Use the definition of the absolute value to show $|\cos\theta + i\sin\theta| = 1$. Hence the estimate in (a) is not the best possible.

38. (a) Use the triangle inequality to show that $|z - 1| \le 2$ for $|z| \le 1$.
(b) Explain your result in (a) geometrically.
(c) Is the upper bound in (a) best possible?

39. Show that $|z - 4| \le 6$ if $|z - 3i| \le 1$.

40. Show that $|z - 4| \ge 4$ if $|z - 3i| \le 1$.

41. Show that $\left|\frac{1}{z-4}\right| \le \frac{1}{2}$ if $|z - 1| \le 1$. [*Hint*: Find a lower bound for $|z - 4|$.]

42. Show that $\left|\frac{1}{1-z}\right| \le \frac{2}{2\sqrt{2}-1}$ if $|z - i| \le \frac{1}{2}$.

43. Show that $\left|\frac{1}{z^2+z+1}\right| \le 4$ if $|z| \le \frac{1}{2}$.

44. Show that $\left|\frac{1}{z^2-iz+12}\right| \ge \frac{1}{18}$ if $|z - 2i| \le 1$. [*Hint*: Factor the quadratic.]

45. Project Problem: Cauchy-Schwarz inequality. Suppose that $v_1, v_2, \ldots, v_n$ and $w_1, w_2, \ldots, w_n$ are arbitrary complex numbers. The **Cauchy-Schwarz inequality** states that

$$\left|\sum_{j=1}^{n} \overline{v_j} w_j\right| \le \sqrt{\sum_{j=1}^{n} |v_j|^2} \sqrt{\sum_{j=1}^{n} |w_j|^2}. \tag{1.2.22}$$

(a) Show that in order to prove (1.2.22) it is enough to prove

$$\sum_{j=1}^{n} |v_j| |w_j| \le \sqrt{\sum_{j=1}^{n} |v_j|^2} \sqrt{\sum_{j=1}^{n} |w_j|^2}. \tag{1.2.23}$$

(b) Prove (1.2.23) in the case $\sum_{j=1}^{n} |v_j|^2 = 1$ and $\sum_{j=1}^{n} |w_j|^2 = 1$. [*Hint*: Start with the inequality $0 \le \sum_{j=1}^{n} (|v_j| - |w_j|)^2$. Expand and simplify.]
(c) Prove (1.2.23) in the general case. [*Hint*: Let $v = (v_1, v_2, \ldots, v_n)$ and $w = (w_1, w_2, \ldots, w_n)$, and think of v and w as vectors. Without loss of generality, v and w are not identically 0. Let $\|v\| = \sqrt{\sum_{j=1}^{n} |v_j|^2}$ and define $\|w\|$ similarly. Show that you can apply the result of (b) to the vectors $\frac{1}{\|v\|} v$ and $\frac{1}{\|w\|} w$.]

1.3 Polar Form

In the previous section, we represented complex numbers as points in the plane with Cartesian coordinates. In this section, we introduce an alternative way to describe complex numbers, the *polar representation* or *polar form*. This form enables us to take advantage of geometric insight as follows: We think of a complex number $z = x + iy$ as a point (x, y) in the complex plane. If $P = (x, y) \ne (0, 0)$ is a point in Cartesian coordinates, we identify it with the pair (r, θ), where r is the distance from P to the origin O, and θ is the angle between the x-axis and the ray OP. We allow θ to be negative if the point P lies in the lower half space.

Definition 1.3.1. (Polar Form of Complex Numbers) Let $z = x + iy$ be a nonzero complex number. We define a number $r > 0$ by setting

$$r = \sqrt{x^2 + y^2} > 0, \tag{1.3.1}$$

and we let θ be an angle[2] such that

$$\cos\theta = \frac{x}{\sqrt{x^2 + y^2}} = \frac{x}{r}, \quad \sin\theta = \frac{y}{\sqrt{x^2 + y^2}} = \frac{y}{r} \quad (r \neq 0). \tag{1.3.2}$$

Then we call r the **modulus** of z and θ **an argument** of z. Then we can write

$$z = r(\cos\theta + i\sin\theta), \tag{1.3.3}$$

and this is called the **polar representation** of a complex number $z \neq 0$. The argument of z is not defined when $z = 0$ or equivalently when $r = 0$.

Identities (1.3.2) are derived from trigonometrical considerations. In fact, r is the distance from P to the origin O and θ can be chosen to be the angle between the ray OP and the positive x-semiaxis. To visualize the situation consider the case $x, y > 0$. Then consider the triangle formed by the point P, the origin, and the projection of P onto the x-axis. The hypotenuse of this triangle is r while x is equal to r times the cosine of the adjacent angle, while y is equal to r times the sine of the opposite angle; in both cases the angle is θ. See Figure 1.14

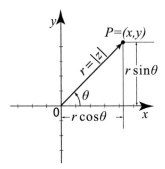

Fig. 1.14 Polar representation of the complex number $z = x + iy$.

Rewriting the identities (1.3.2) as follows

$$\operatorname{Re} z = r\cos\theta, \qquad \operatorname{Im} z = r\sin\theta, \tag{1.3.4}$$

and dividing we obtain

$$\tan\theta = \frac{\operatorname{Im} z}{\operatorname{Re} z} = \frac{y}{x} \quad (x \neq 0). \tag{1.3.5}$$

Using (1.3.1) and (1.2.8) we obtain

$$r = \sqrt{x^2 + y^2} = |z| = \sqrt{z\bar{z}}. \tag{1.3.6}$$

It is easy to see that

---

[2] such a θ exists since x/r and y/r satisfy $(x/r)^2 + (y/r)^2 = 1$.

$$|z| = 1 \Leftrightarrow r = 1 \Leftrightarrow z = \cos\theta + i\sin\theta, \text{ for some real } \theta. \qquad (1.3.7)$$

A complex number z such that $|z| = 1$ is called **unimodular**. See Exercise 30 for basic properties of unimodular numbers.

We now shift our attention to the argument, whose choice is more delicate. If θ is an angle that satisfies (1.3.2), then $\theta + 2k\pi$ ($k = 0, \pm 1, \pm 2, \ldots$) will also satisfy (1.3.2), because the cosine and sine are 2π-periodic functions. Thus, relations (1.3.2) do not determine a unique value of argument z. If we restrict the choice of θ to the interval $-\pi < \theta \leq \pi$, then there is a unique value of θ that satisfies (1.3.2).

Definition 1.3.2. The **principal value** of the argument of a complex number $z = x + iy$ is the unique number $\mathrm{Arg}\,z$ with the following properties:

$$-\pi < \mathrm{Arg}\,z \leq \pi, \quad \cos(\mathrm{Arg}\,z) = \frac{x}{|z|}, \quad \sin(\mathrm{Arg}\,z) = \frac{y}{|z|}. \qquad (1.3.8)$$

The set of all values of the argument is denoted by

$$\arg z = \{\mathrm{Arg}\,z + 2k\pi : k = 0, \pm 1, \pm 2, \ldots\}. \qquad (1.3.9)$$

Unlike $\mathrm{Arg}\,z$, which is single-valued, $\arg z$ is multi-valued or a set-valued function. For a set S and a number c we write

$$c + S = \{c + s : s \in S\}$$

and

$$cS = \{cs : s \in S\}.$$

With this notation in mind, expressions of the form $\arg z + \pi$ or $-\arg z$ make sense.

Sometimes we abuse notation and write $\arg z = \theta + 2k\pi$ or simply $\arg z = \theta$ to denote a specific value of the argument from the set $\arg z$.

Table 1 contains some special trigonometric values that will be useful in the examples and exercises.

θ	0	$\frac{\pi}{6}$	$\frac{\pi}{4}$	$\frac{\pi}{3}$	$\frac{\pi}{2}$	$\frac{2\pi}{3}$	$\frac{3\pi}{4}$	$\frac{5\pi}{6}$	π	$\frac{7\pi}{6}$	$\frac{5\pi}{4}$	$\frac{4\pi}{3}$	$\frac{3\pi}{2}$	$\frac{5\pi}{3}$	$\frac{7\pi}{4}$	$\frac{11\pi}{6}$
$\cos\theta$	1	$\frac{\sqrt{3}}{2}$	$\frac{\sqrt{2}}{2}$	$\frac{1}{2}$	0	$-\frac{1}{2}$	$-\frac{\sqrt{2}}{2}$	$-\frac{\sqrt{3}}{2}$	-1	$-\frac{\sqrt{3}}{2}$	$-\frac{\sqrt{2}}{2}$	$-\frac{1}{2}$	0	$\frac{1}{2}$	$\frac{\sqrt{2}}{2}$	$\frac{\sqrt{3}}{2}$
$\sin\theta$	0	$\frac{1}{2}$	$\frac{\sqrt{2}}{2}$	$\frac{\sqrt{3}}{2}$	1	$\frac{\sqrt{3}}{2}$	$\frac{\sqrt{2}}{2}$	$\frac{1}{2}$	0	$-\frac{1}{2}$	$-\frac{\sqrt{2}}{2}$	$-\frac{\sqrt{3}}{2}$	-1	$-\frac{\sqrt{3}}{2}$	$-\frac{\sqrt{2}}{2}$	$-\frac{1}{2}$
$\tan\theta$	0	$\frac{\sqrt{3}}{3}$	1	$\sqrt{3}$	not defined	$-\sqrt{3}$	-1	$-\frac{\sqrt{3}}{3}$	0	$\frac{\sqrt{3}}{3}$	1	$\sqrt{3}$	not defined	$-\sqrt{3}$	-1	$-\frac{\sqrt{3}}{3}$

Table 1. Special trigonometric values.

Example 1.3.3. (Polar form) Find the modulus, argument, and polar form of the complex numbers:

(a) $z_1 = 5$ (b) $z_2 = -3i$
(c) $z_3 = \sqrt{3}+i$ (d) $z_4 = 1+i$
(e) $z_5 = 1-i$ (f) $z_6 = -1-i$

These numbers are shown in Figure 1.15 plotted in the complex plane.

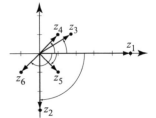

Fig. 1.15 z_1 through z_6.

Solution. (a) From (1.3.1), $r = |z_1| = \sqrt{5^2} = 5$. An argument of z_1 is clearly $\theta = 0$, hence $\{2k\pi : k = 0, \pm1, \pm2, \ldots\}$ is the set of all arguments of z_1. Its polar representation is

$$z_1 = 5 = 5(\cos 0 + i \sin 0).$$

(b) Here $r = |z_2| = |-3i| = 3$, and, from Figure 1.15, $\arg z_2 = \frac{3\pi}{2} + 2k\pi$; thus the polar representation of z_2 is

$$z_2 = -3i = 3\left(\cos\frac{3\pi}{2} + i\sin\frac{3\pi}{2}\right).$$

(c) We have $r = |z_3| = |\sqrt{3}+i| = \sqrt{3+1} = 2$. So from (1.3.2), we have

$$\cos\theta = \frac{x}{r} = \frac{\sqrt{3}}{2} \quad \text{and} \quad \sin\theta = \frac{y}{r} = \frac{1}{2}.$$

From Table 1, we find that $\theta = \frac{\pi}{6}$. Thus, $\arg z_3 = \frac{\pi}{6} + 2k\pi$, and the polar representation is

$$z_3 = \sqrt{3}+i = 2\left(\cos\frac{\pi}{6} + i\sin\frac{\pi}{6}\right).$$

(d) We have $r = |z_4| = \sqrt{1+1} = \sqrt{2}$. Factoring the modulus of $1+i$, we obtain from Table 1

$$z_4 = 1+i = \sqrt{2}\left(\frac{1}{\sqrt{2}} + i\frac{1}{\sqrt{2}}\right) = \sqrt{2}\left(\frac{\sqrt{2}}{2} + i\frac{\sqrt{2}}{2}\right) = \sqrt{2}\left(\cos\frac{\pi}{4} + i\sin\frac{\pi}{4}\right)$$

which is the polar form of z_4, with $\arg z_4 = \frac{\pi}{4}$.

(e) It is clear from Figure 1.15 that z_5 is the reflection of $z_4 = 1+i$ about the x-axis (equivalently, $z_5 = \overline{z_4}$). Hence z_5 and z_4 have the same moduli and opposite arguments. So $r = \sqrt{2}$, $\arg z_5 = -\frac{\pi}{4}$, and the polar representation is

$$z_5 = 1-i = \sqrt{2}\left(\cos\left(-\frac{\pi}{4}\right) + i\sin\left(-\frac{\pi}{4}\right)\right).$$

(f) Note that z_6 is the reflection of $z_4 = 1+i$ about the origin (equivalently, $z_6 = -z_4$). Hence z_6 and z_4 have the same moduli and their arguments are related by the identity $\arg z_6 = \arg z_4 + \pi$. So, $r = \sqrt{2}$, $\arg z_6 = \frac{\pi}{4} + \pi = \frac{5\pi}{4}$, and the polar representation is

$$z_6 = -1 - i = \sqrt{2}\left(\cos\frac{5\pi}{4} + i\sin\frac{5\pi}{4}\right).$$ □

The following properties of the argument were illustrated in Example 1.3.3(e) and (f); for a proof we refer to Exercise 26:

$$\arg\bar{z} = -\arg z, \tag{1.3.10}$$
$$\arg(-z) = \arg z + \pi. \tag{1.3.11}$$

Example 1.3.4. (Principal argument) Compute $\operatorname{Arg} z_j$, where z_j is as in the preceding example for $j = 1, \ldots, 6$.

Solution. In each case we recall z_j and $\arg z_j$ from the previous example. To get $\operatorname{Arg} z_j$ we pick the value of $\arg z$ that lies in the interval $(-\pi, \pi]$. Refer to Figure 1.15 for illustration.

(a) $z_1 = 5$, $\arg(5) = 0 + 2k\pi$. Since 0 is in the interval $(-\pi, \pi]$, we also have $\operatorname{Arg} z_1 = 0$.

(b) $z_2 = -3i$, $\arg z_2 = \frac{3\pi}{2} + 2k\pi$. To determine $\operatorname{Arg} z_2$, we must pick the unique value of $\arg z_2$ that lies in the interval $(-\pi, \pi]$. From Figure 1.15, we see that $\operatorname{Arg} z_2 = -\frac{\pi}{2}$, which is the value of $\arg z_2 = \frac{3\pi}{2} + 2k\pi$ that corresponds to $k = -1$.

(c) $z_3 = \sqrt{3} + i$, $\arg z_3 = \frac{\pi}{6} + 2k\pi$, and so $\operatorname{Arg} z_3 = \frac{\pi}{6}$, because $\frac{\pi}{6}$ is in $(-\pi, \pi]$.

(d) $z_4 = 1 + i$, $\arg z_4 = \frac{\pi}{4} + 2k\pi$, and so $\operatorname{Arg} z_4 = \frac{\pi}{4}$.

(e) $z_5 = 1 - i$, $\arg z_5 = -\frac{\pi}{4}$, and so $\operatorname{Arg} z_5 = -\frac{\pi}{4}$.

(f) $z_6 = -1 - i$, $\arg z_6 = \frac{5\pi}{4} + 2k\pi$, and so $\operatorname{Arg} z_6 = -\frac{3\pi}{4}$, which is the value of $\arg z_6$ that corresponds to $k = -1$.   □

It is important to keep in mind that while $\operatorname{Arg} z$ is a particular value of $\arg z$, the function $\operatorname{Arg} z$ does not necessarily satisfy the same properties as $\arg z$. In particular, identities (1.3.10), (1.3.11), and several others discussed in this section that hold for $\arg z$ may not hold for $\operatorname{Arg} z$. Parts (d) and (f) of Example 1.3.3 show that (1.3.11) is not true if we use $\operatorname{Arg} z$ in place of $\arg z$.

It is tempting to compute the argument θ of a complex number $z = x + iy$ by taking the inverse tangent on both sides of (1.3.5) and writing $\theta = \tan^{-1}\left(\frac{y}{x}\right)$. This formula is true only if θ is in the interval $(-\frac{\pi}{2}, \frac{\pi}{2})$, because the inverse tangent takes its values in the interval $(-\frac{\pi}{2}, \frac{\pi}{2})$ (see Figure 1.16); and so it will not yield values of θ that are outside this interval. For example, if $\theta = \frac{4\pi}{3}$, from Table. 1 we have $\tan\frac{4\pi}{3} = \sqrt{3}$. However, $\tan^{-1}\sqrt{3} = \frac{\pi}{3}$ and not $\frac{4\pi}{3}$. To overcome this problem, recall that the tangent is π-periodic; that is, for all θ,

$$\tan\theta = \tan(\theta + k\pi) \qquad (k = 0, \pm 1, \pm 2, \ldots).$$

Since $\tan\theta = \frac{y}{x}$, we conclude that

$$\theta = \tan^{-1}\left(\frac{y}{x}\right) + k\pi \quad (x \neq 0), \tag{1.3.12}$$

where the choice of k depends on z. You can check that, for $z = x + iy$ with $x \neq 0$, we have

$$x > 0 \qquad\qquad \Leftrightarrow \qquad z \text{ lies in the first or fourth quadrants;}$$

$$x < 0 \text{ and } y > 0 \qquad \Leftrightarrow \qquad z \text{ lies in the second quadrant;}$$

$$x < 0 \text{ and } y < 0 \qquad \Leftrightarrow \qquad z \text{ lies in the third quadrant,}$$

and also that for $z = x + iy$ we have

$$\text{Arg} \, z = \begin{cases} \tan^{-1} \left(\frac{y}{x} \right) & \text{if } x > 0; \\[2mm] \tan^{-1} \left(\frac{y}{x} \right) + \pi & \text{if } x < 0 \text{ and } y \geq 0; \\[2mm] \tan^{-1} \left(\frac{y}{x} \right) - \pi & \text{if } x < 0 \text{ and } y < 0. \end{cases} \qquad (1.3.13)$$

When x is zero, we cannot use (1.3.12). In this case,

$$\text{Arg} \, z = \text{Arg} \, (iy) = \begin{cases} \frac{\pi}{2} & \text{if } y > 0; \\[2mm] -\frac{\pi}{2} & \text{if } y < 0. \end{cases} \qquad (1.3.14)$$

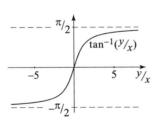

Fig. 1.16 The inverse tangent takes values in $\left(\frac{-\pi}{2}, \frac{\pi}{2} \right)$.

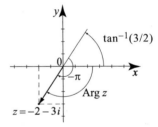

Fig. 1.17 Computing $\text{Arg} \, z$ using (1.3.13).

For example, in Figure 1.17, the point $z_1 = 2 + 3i$ lies in the first quadrant. Using a calculator, we find $\text{Arg} \, z_1 = \tan^{-1} \frac{3}{2} \approx 0.983$. Also $z_2 = -2 - 3i$ lies in the third quadrant and $\text{Arg} \, z_2 = \tan^{-1} \frac{3}{2} - \pi \approx -2.159$. (All angles are measured in radians.)

Multiplication, Inverses, and Division in Polar Form

Let $z_1 = r_1 (\cos \theta_1 + i \sin \theta_1)$ and $z_2 = r_2 (\cos \theta_2 + i \sin \theta_2)$ be two nonzero complex numbers. We compute their product directly, using $i^2 = -1$:

$$z_1 z_2 = r_1 (\cos \theta_1 + i \sin \theta_1) r_2 (\cos \theta_2 + i \sin \theta_2)$$
$$= r_1 r_2 [(\cos \theta_1 \cos \theta_2 - \sin \theta_1 \sin \theta_2) + i(\sin \theta_1 \cos \theta_2 + \cos \theta_1 \sin \theta_2)].$$

We recognize the trigonometric expressions as the sum angle formulas for the cosine and sine:

$$\cos(\theta_1 + \theta_2) = \cos\theta_1 \cos\theta_2 - \sin\theta_1 \sin\theta_2$$

and

$$\sin(\theta_1 + \theta_2) = \sin\theta_1 \cos\theta_2 + \cos\theta_1 \sin\theta_2.$$

Therefore, the **polar form of the product** is

$$z_1 z_2 = r_1 r_2 [\cos(\theta_1 + \theta_2) + i\sin(\theta_1 + \theta_2)]. \qquad (1.3.15)$$

Taking the modulus of both sides of (1.3.15) we obtain

$$|z_1 z_2| = r_1 r_2, \qquad (1.3.16)$$

which is already known from Section 1.2. Examining the argument of both sides of (1.3.15) we deduce

$$\arg(z_1 z_2) = \theta_1 + \theta_2 = \arg z_1 + \arg z_2. \qquad (1.3.17)$$

More precisely, we have $\arg(z_1 z_2) = \{\theta_1 + \theta_2 + 2k\pi : k = 0, \pm 1, \pm 2, \ldots\}$. Identities (1.3.16) and (1.3.17) tell us that when we multiply two complex numbers in polar form, we multiply their moduli and add their arguments. See Figure 1.18.

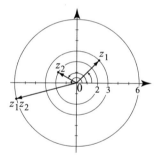

Fig. 1.18 To multiply in polar form, add the arguments and multiply the moduli.

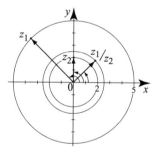

Fig. 1.19 To divide in polar form, subtract the arguments and divide the moduli.

Suppose that $z_1 \neq 0$, and $z_2 = z_1^{-1}$ in (1.3.15). Since $z_1 z_2 = z_1 z_1^{-1} = 1$, we obtain

$$1 = r_1 r_2 [\cos(\theta_1 + \theta_2) + i\sin(\theta_1 + \theta_2)].$$

Hence from (1.3.7), $r_1 r_2 = 1$ and $\theta_1 + \theta_2 = 0$, because the modulus of 1 is 1 and its argument is 0. Hence $r_2 = \frac{1}{r_1}$ and $\theta_2 = -\theta_1$. So the **polar form of the inverse** of a complex number z is

$$z^{-1} = \frac{1}{r}\big(\cos(-\theta) + i\sin(-\theta)\big) = \frac{1}{r}(\cos\theta - i\sin\theta). \qquad (1.3.18)$$

Consider now arbitrary z_1 and $z_2 \neq 0$. Since $\frac{z_1}{z_2} = z_1 z_2^{-1}$, then by (1.3.15) and (1.3.18), we obtain the **polar form of a quotient**

$$\frac{z_1}{z_2} = \frac{r_1}{r_2}(\cos(\theta_1 - \theta_2) + i\sin(\theta_1 - \theta_2)). \tag{1.3.19}$$

Thus to divide two complex numbers in polar form, we divide their moduli and subtract their arguments; see Figure 1.19.

Example 1.3.5. (Multiplication and division in polar form)
(a) Let $z_1 = 3(\cos\frac{\pi}{4} + i\sin\frac{\pi}{4})$ and $z_2 = 2(\cos\frac{5\pi}{6} + i\sin\frac{5\pi}{6})$. Calculate $z_1 z_2$.
(b) Let $z_1 = 5(\cos\frac{3\pi}{4} + i\sin\frac{3\pi}{4})$ and $z_2 = 2i$. Calculate $\frac{z_1}{z_2}$.

Solution. (a) Reading off the values $r_1 = 3$, $\theta_1 = \frac{\pi}{4}$, $r_2 = 2$, $\theta_2 = \frac{5\pi}{6}$, we use (1.3.15) to obtain

$$z_1 z_2 = 2 \cdot 3\left[\cos\left(\frac{\pi}{4} + \frac{5\pi}{6}\right) + i\sin\left(\frac{\pi}{4} + \frac{5\pi}{6}\right)\right] = 6\left[\cos\frac{13\pi}{12} + i\sin\frac{13\pi}{12}\right].$$

This is the multiplication shown in Figure 1.18.
(b) Writing z_2 in polar form as $2i = 2(\cos\frac{\pi}{2} + i\sin\frac{\pi}{2})$, we perform the division by dividing moduli and subtracting arguments:

$$\frac{z_1}{z_2} = \frac{5}{2}\left[\cos\left(\frac{3\pi}{4} - \frac{\pi}{2}\right) + i\sin\left(\frac{3\pi}{4} - \frac{\pi}{2}\right)\right]$$

$$= \frac{5}{2}\left(\cos\frac{\pi}{4} + i\sin\frac{\pi}{4}\right) = \frac{5}{2}\left(\frac{\sqrt{2}}{2} + i\frac{\sqrt{2}}{2}\right).$$

This is the division shown in Figure 1.19.   □

De Moivre's Identity

Multiplying the number $z = 1(\cos\theta + i\sin\theta)$ by itself, using (1.3.15), we obtain

$$zz = (1 \cdot 1)[\cos(\theta + \theta) + i\sin(\theta + \theta)] = \cos 2\theta + i\sin 2\theta.$$

Computing successive powers of z, a pattern emerges:

$$z = \cos\theta + i\sin\theta,$$
$$z^2 = \cos 2\theta + i\sin 2\theta,$$
$$z^3 = \cos 3\theta + i\sin 3\theta,$$
$$\vdots$$

More generally, we have the following useful identity.

Proposition 1.3.6. (De Moivre's Identity) *For a positive integer n and a real number θ we have*

$$(\cos\theta + i\sin\theta)^n = \cos n\theta + i\sin n\theta. \qquad (1.3.20)$$

Proof. We prove (1.3.20) by mathematical induction. For $n = 1$, the statement is true, since we have trivially $(\cos\theta + i\sin\theta)^1 = \cos 1 \cdot \theta + i\sin 1 \cdot \theta$. Now for the inductive step: We assume the statement is true for n, and prove that it is true for $n+1$. Let us compute

$$\begin{aligned}
(\cos\theta + i\sin\theta)^{n+1} &= (\cos\theta + i\sin\theta)^n(\cos\theta + i\sin\theta) \\
&= (\cos n\theta + i\sin n\theta)(\cos\theta + i\sin\theta) \\
&= \cos(n+1)\theta + i\sin(n+1)\theta,
\end{aligned}$$

where the second equality holds by the induction hypothesis and the third equality is a consequence of (1.3.15). Thus (1.3.20) holds for $n+1$ in place of n and thus it holds for all positive integers n by mathematical induction. ∎

We can use De Moivre's identity to calculate powers of complex numbers of arbitrary modulus $r \neq 0$. For if $z = r(\cos\theta + i\sin\theta)$, then

$$\begin{aligned}
z^n &= [r(\cos\theta + i\sin\theta)]^n \\
&= r^n(\cos\theta + i\sin\theta)^n \qquad (1.3.21) \\
&= r^n(\cos n\theta + i\sin n\theta).
\end{aligned}$$

Example 1.3.7. (Polar Form and Powers) Calculate $(2+2i)^{11}$.

Solution. We use De Moivre's identity for a quick calculation. First, write the number $2+2i$ in polar form: $2+2i = 2^{3/2}(\cos\frac{\pi}{4} + i\sin\frac{\pi}{4})$. Then, from (1.3.21) we obtain our answer in polar form:

$$(2+2i)^{11} = 2^{33/2}\left(\cos\frac{11\pi}{4} + i\sin\frac{11\pi}{4}\right).$$

Subtracting a multiple of 2π from the angle, we can simplify our answer and put it in Cartesian coordinates as follows:

$$\cos\frac{11\pi}{4} = \cos\left(\frac{11\pi}{4} - 2\pi\right) = \cos\frac{3\pi}{4}.$$

Then we have

$$(2+2i)^{11} = 2^{16}\sqrt{2}\left(\cos\frac{3\pi}{4} + i\sin\frac{3\pi}{4}\right) = 2^{16}(-1+i). \qquad \square$$

Example 1.3.8. (Double-angle identities) Use De Moivre's identity with $n = 2$ to derive the double-angle formulas for $\cos 2\theta$ and $\sin 2\theta$.

Solution. From De Moivre's identity with $n = 2$,

$$\cos 2\theta + i \sin 2\theta = (\cos \theta + i \sin \theta)^2 = \cos^2 \theta - \sin^2 \theta + i 2 \sin \theta \cos \theta.$$

Equating real and imaginary parts, we get the double-angle formulas

$$\cos 2\theta = \cos^2 \theta - \sin^2 \theta \quad \text{and} \quad \sin 2\theta = 2 \sin \theta \cos \theta. \qquad \square$$

Roots of Complex Numbers

Definition 1.3.9. Let $w \neq 0$ be a complex number and n a positive integer. A number z is called an nth **root** of w if $z^n = w$.

De Moivre's identity can in essence be worked "backwards" to find roots of complex numbers. Let $w = \rho(\cos \phi + i \sin \phi)$ and $z = r(\cos \theta + i \sin \theta)$. In view of (1.3.21), equation $z^n = w$ tells us that

$$r^n(\cos n\theta + i \sin n\theta) = \rho(\cos \phi + i \sin \phi). \qquad (1.3.22)$$

Two complex numbers are equal only if their moduli are equal, so $r^n = \rho$. Thus, take $r = \rho^{1/n}$ (meaning the real root of a real number). Also, when two complex numbers are equal, their arguments must differ by $2k\pi$, where k is an integer. So

$$n\theta = \phi + 2k\pi, \quad \text{or} \quad \theta = \frac{\phi}{n} + \frac{2k\pi}{n}.$$

If we take the values $k = 0, 1, \ldots, n-1$, we get n values of θ that yield n distinct roots of w. Any other value of k produces a root identical to one of these, since when k increases by n, the argument of z increases by 2π. Thus we have a formula for the n roots of a complex number w.

Proposition 1.3.10. *Let $w = \rho(\cos \phi + i \sin \phi) \neq 0$. The nth roots of w are the solutions of the equation $z^n = w$. These are*

$$z_{k+1} = \rho^{1/n} \left[\cos \left(\frac{\phi}{n} + \frac{2k\pi}{n} \right) + i \sin \left(\frac{\phi}{n} + \frac{2k\pi}{n} \right) \right], \qquad (1.3.23)$$

$k = 0, 1, \ldots, n-1.$

The unique number z such that $z^n = w$ and $\operatorname{Arg} z = \frac{\operatorname{Arg} w}{n}$ is called the **principal nth root** of w. The principal root is obtained from (1.3.23) by taking $\phi = \operatorname{Arg} w$ and $k = 0$.

Example 1.3.11. (Sixth roots of unity) Find and plot all numbers z such that $z^6 = 1$. What is the principal sixth root of 1?

Solution. The modulus of 1 is 1 and the argument of 1 is 0. According to (1.3.23), the six roots have $r = 1^{1/6} = 1$ and $\theta = \frac{0}{6} + \frac{k\pi}{3}$, for $k = 0, 1, 2, 3, 4$, and 5. Hence the roots are

$$z_{k+1} = \cos\frac{k\pi}{3} + i\sin\frac{k\pi}{3}, \ k = 0, 1, \ldots, 5.$$

The principal root is clearly $z_1 = 1$. We can list the roots explicitly, with the help of Table 1:

$$z_1 = 1, \quad z_2 = \tfrac{1}{2} + i\tfrac{\sqrt{3}}{2}, \quad z_3 = -\tfrac{1}{2} + i\tfrac{\sqrt{3}}{2},$$
$$z_4 = -1, z_5 = -\tfrac{1}{2} - i\tfrac{\sqrt{3}}{2}, z_6 = \tfrac{1}{2} - i\tfrac{\sqrt{3}}{2}.$$

The six roots are displayed in Figure 1.20. Since they have the same modulus, they all lie on the same circle centered at the origin. They have a common angular separation of $\pi/3$. This particular set of roots is symmetric about the x-axis because the polynomial equation $z^6 = 1$ has real coefficients, and nonreal solutions come in conjugate pairs (see Exercise 42 in Section 1.1). □

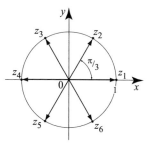

Fig. 1.20 The sixth roots of unity.

One may check directly that the sum of all sixth roots of unity in Example 1.3.11 equals 0. This interesting fact holds for all nth roots of unity. (See Exercise 63.)

Example 1.3.12. (Finding roots of complex numbers) Find and plot all numbers z such that $(z+1)^3 = 2 + 2i$.

Solution. Change variables to $w = z + 1$. We must now solve $w^3 = 2 + 2i$. The polar form of $2 + 2i$ is $2 + 2i = 2^{3/2}(\cos\frac{\pi}{4} + i\sin\frac{\pi}{4})$ and the equation becomes $w^3 = 2^{3/2}(\cos\frac{\pi}{4} + i\sin\frac{\pi}{4})$.

Appealing to (1.3.23) with $n = 3$ we find

$$w_1 = \sqrt{2}\left(\cos\frac{\pi}{12} + i\sin\frac{\pi}{12}\right),$$

$$w_2 = \sqrt{2}\left(\cos\frac{9\pi}{12} + i\sin\frac{9\pi}{12}\right),$$

$$w_3 = \sqrt{2}\left(\cos\frac{17\pi}{12} + i\sin\frac{17\pi}{12}\right).$$

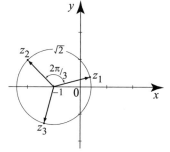

Since $z = w - 1$, we conclude that the solutions z_1, z_2, and z_3 of $(z+1)^3 = 2 + 2i$ are

Fig. 1.21 The three solutions of $(z+1)^3 = 2 + 2i$

$$z_1 = \sqrt{2}\,\cos\frac{\pi}{12} - 1 + i\sqrt{2}\sin\frac{\pi}{12}$$

$$z_2 = \sqrt{2}\cos\frac{9\pi}{12} - 1 + i\sqrt{2}\sin\frac{9\pi}{12}$$

$$z_3 = \sqrt{2}\cos\frac{17\pi}{12} - 1 + i\sqrt{2}\sin\frac{17\pi}{12}.$$

The roots z_1, z_2, z_3 are plotted in Figure 1.21. $\qquad\qquad\qquad\qquad\qquad\qquad$ □

As a final application, we revisit the quadratic equation

$$az^2 + bz + c = 0 \qquad (a \neq 0), \tag{1.3.24}$$

where a, b, c are now complex numbers. The algebraic manipulations for solving the equation with real coefficients lead to the solution

$$z = \frac{-b \pm \sqrt{b^2 - 4ac}}{2a}, \tag{1.3.25}$$

where now $\pm\sqrt{b^2 - 4ac}$ represents the two complex square roots of $b^2 - 4ac$. These square roots can be computed by appealing to (1.3.23) with $n = 2$, thus yielding the solutions of (1.3.24).

Example 1.3.13. (Quadratic equation with complex coefficients) Solve the equation $z^2 - 2iz + 3 + i = 0$.

Solution. From (1.3.25), we have

$$z = \frac{2i \pm \sqrt{(-2i)^2 - 4(3+i)}}{2} = i \pm \sqrt{-4 - i}.$$

We have

$$-4 - i = \sqrt{17}\left(-\frac{4}{\sqrt{17}} - \frac{i}{\sqrt{17}}\right) = \sqrt{17}\left(\cos\theta + i\sin\theta\right),$$

where θ is the angle in the third quadrant (we may take $\pi < \theta < \frac{3\pi}{2}$) such that

$$\cos\theta = -\frac{4}{\sqrt{17}} \quad\text{and}\quad \sin\theta = -\frac{1}{\sqrt{17}}.$$

Appealing to (1.3.23) to compute w_1 and w_2, the two square roots of $-4 - i$, we find

$$w_1 = 17^{1/4}\left(\cos\frac{\theta}{2} + i\sin\frac{\theta}{2}\right)$$

and

$$w_2 = 17^{1/4} \left(\cos \left(\frac{\theta}{2} + \pi \right) + i \sin \left(\frac{\theta}{2} + \pi \right) \right)$$

$$= 17^{1/4} \left(- \cos \frac{\theta}{2} - i \sin \frac{\theta}{2} \right) = -w_1,$$

as you would have expected w_2 to be related to w_1. Thus,

$$z_1 = i + w_1 \quad \text{and} \quad z_2 = i - w_1.$$

To compute z_1 and z_2 explicitly, we must determine the values of $\cos \frac{\theta}{2}$ and $\sin \frac{\theta}{2}$ from the values of $\cos \theta$ and $\sin \theta$. This can be done with the help of the half-angle formulas:

$$\cos^2 \frac{\theta}{2} = \frac{1 + \cos \theta}{2} \quad \text{and} \quad \sin^2 \frac{\theta}{2} = \frac{1 - \cos \theta}{2}. \tag{1.3.26}$$

Since $\pi < \theta < \frac{3\pi}{2}$, we have $\frac{\pi}{2} < \frac{\theta}{2} < \frac{3\pi}{4}$; so $\cos \frac{\theta}{2} < 0$ and $\sin \frac{\theta}{2} > 0$. Hence from (1.3.26)

$$\cos \frac{\theta}{2} = -\sqrt{\frac{1 + \cos \theta}{2}} \quad \text{and} \quad \sin \frac{\theta}{2} = \sqrt{\frac{1 - \cos \theta}{2}}.$$

Using the explicit value of $\cos \theta$, we get

$$\cos \frac{\theta}{2} = -\sqrt{\frac{\sqrt{17} - 4}{2\sqrt{17}}} \quad \text{and} \quad \sin \frac{\theta}{2} = \sqrt{\frac{\sqrt{17} + 4}{2\sqrt{17}}},$$

and hence

$$z_1 = i + \frac{1}{\sqrt{2}} \left[-\sqrt{\sqrt{17} - 4} + i\sqrt{\sqrt{17} + 4} \right],$$

$$z_2 = i - \frac{1}{\sqrt{2}} \left[-\sqrt{\sqrt{17} - 4} + i\sqrt{\sqrt{17} + 4} \right].$$

Observe that the roots z_1 and z_2 are not complex conjugates. Compare this with the result of Exercise 42 in Section 1.1. □

Exercises 1.3

In Exercises 1–4, draw the complex numbers given in polar form.

1. $3 \left(\cos \frac{7\pi}{12} + i \sin \frac{7\pi}{12} \right)$
2. $\sqrt{2} \left(\cos \frac{-\pi}{2} + i \sin \frac{-\pi}{2} \right)$

3. $\frac{1}{2} \left(\cos \frac{64\pi}{3} + i \sin \frac{64\pi}{3} \right)$
4. $3 \left(\cos \frac{-72\pi}{11} + i \sin \frac{-72\pi}{11} \right)$

In Exercises 5–12, represent the complex numbers in polar form.

5. $-3 - 3i$
6. $-\frac{\sqrt{3}}{2} + \frac{i}{2}$
7. $-1 - \sqrt{3}i$
8. $1 + i$

9. $-\frac{i}{2}$
10. $\frac{1+i}{1+\sqrt{3}i}$
11. $\frac{1+i}{1-i}$
12. $\frac{i}{10 + 10i}$

In Exercises 13–16, compute the principal arguments of the complex numbers using (1.3.13) *or* (1.3.14); *then find their general arguments. If needed, use a calculator to compute* $\tan^{-1}$.

13. $13 + 2i$ **14.** $-3 - 32i$ **15.** $-1 + \frac{1}{2}i$ **16.** $-\frac{3\pi}{2}i$

In Exercises 17–20, find the polar form of the complex number.

17. $(-\sqrt{3} + i)^3$ **18.** $(-2 - 3i)^{17}$ **19.** $\left(\dfrac{1-i}{1+i}\right)^{10}$ **20.** $\dfrac{i}{(1 + 2\sqrt{3}i)^5}$

In Exercises 21–24, find the real and imaginary parts of the complex numbers.

21. $(1 + i)^{30}$ **22.** $\left(\cos\dfrac{2\pi}{17} + i\sin\dfrac{2\pi}{17}\right)^{170}$ **23.** $\left(\dfrac{1-i}{1+i}\right)^4$ **24.** $\dfrac{-i}{(1+i)^5}$

25. For each part, find two complex numbers z_1 and z_2 that illustrate the statement.
(a) $\mathrm{Arg}(z_1 z_2) \neq \mathrm{Arg}(z_1) + \mathrm{Arg}(z_2)$.
(b) $\mathrm{Arg}\left(\frac{z_1}{z_2}\right) \neq \mathrm{Arg}(z_1) - \mathrm{Arg}(z_2)$.
(c) $\mathrm{Arg}(\overline{z_1}) \neq -\mathrm{Arg}(z_1)$.
(d) $\mathrm{Arg}(-z_1) \neq \mathrm{Arg}(z_1) + \pi$.

26. Prove properties (1.3.10) and (1.3.11).

27. Show that
$$\left(\frac{1 + i\tan\theta}{1 - i\tan\theta}\right)^n = \frac{1 + i\tan n\theta}{1 - i\tan n\theta}.$$

28. Use polar forms to show that, for $z \neq 0$, $\overline{z}$ and z^{-1} represent parallel vectors. Find the positive constant α such that $\overline{z} = \alpha z^{-1}$.

29. Suppose $z_1, z_2, \ldots, z_n$ are complex numbers with respective moduli $r_1, r_2, \ldots, r_n$ and arguments $\theta_1, \theta_2, \ldots, \theta_n$. Show that
$$z_1 z_2 \cdots z_n = r_1 r_2 \cdots r_n [\cos(\theta_1 + \theta_2 + \cdots + \theta_n) + i\sin(\theta_1 + \theta_2 + \cdots + \theta_n)].$$

30. (a) Show that if $z_1, z_2, \ldots, z_n$ are all unimodular, then so is $z_1 z_2 \cdots z_n$.
(b) Let n be a positive integer and z a nonzero complex number. Show that z is unimodular if and only if all its nth roots are unimodular.

31. (a) Show that two nonzero complex numbers z_1 and z_2 represent perpendicular vectors only if $\arg\left(\frac{z_1}{z_2}\right) = \frac{\pi}{2} + k\pi$, where k is an integer.
(b) Show that two nonzero complex numbers z_1 and z_2 represent parallel vectors only if $\arg\left(\frac{z_1}{z_2}\right) = k\pi$, where k is an integer.

32. Suppose $z = r(\cos\theta + i\sin\theta)$. What will become of its polar coordinates r and θ if we multiply it by
(a) a positive real number α?
(b) a negative real number $-\alpha$?
(c) a unimodular complex number $\cos\phi + i\sin\phi$?

In Exercises 33–40, solve the equations and plot the solutions. In each case, determine the principal root.

33. $z^2 = i$ **34.** $z^3 = i$ **35.** $z^4 = i$

36. $z^5 = -1$ **37.** $z^7 = -7$ **38.** $z^{10} = -3i$

39. $z^2 = 1 + i$ **40.** $z^3 = -1 + i$ **41.** $z^4 = -1 - i$

42. $z^5 = 2 - 3i$ **43.** $z^8 = 3 - 4i$ **44.** $z^{10} = \cos\frac{\pi}{11} + i\sin\frac{\pi}{11}$

In Exercises 45–48, solve the equations.

45. $(z+2)^3 = 3i$

46. $(z-i)^4 = 1$

47. $(z-5+i)^3 = -125$

48. $(3z-2)^4 = 11$

In Exercises 49–56, solve the equations.

49. $z^2 + z + 1 - i = 0$

50. $z^2 + 3z + 3 + i = 0$

51. $z^2 + (1+i)z + i = 0$

52. $z^2 + iz + 1 = 0$

53. $z^4 - (1+i)z^2 + i = 0$

54. $z^4 - z^2 + 1 + i = 0$

55. $z^4 - z^2 + 1 - i = 0$

56. $z^4 + z^2 + 1 + i = 0$

In Exercises 57–60, use De Moivre's identity to derive the trigonometric identities. (More generally, see Exercise 66.)

57. $\cos(3\theta) = \cos^3 \theta - 3 \cos \theta \sin^2 \theta$

58. $\sin(3\theta) = 3 \cos^2 \theta \sin \theta - \sin^3 \theta$

59. $\cos(4\theta) = \cos^4 \theta - 6 \cos^2 \theta \sin^2 \theta + \sin^4 \theta$

60. $\sin(4\theta) = 4 \cos^3 \theta \sin \theta - 4 \cos \theta \sin^3 \theta$

61. (Roots of unity) Let n be a positive integer. Solve $z^n = 1$. These n values of z are called the *n*th **roots of unity** and are denoted by $\omega_1, \ldots, \omega_n$.

62. Use the fact that

$$\cos\left(\frac{\phi}{n} + \frac{2k\pi}{n}\right) + i \sin\left(\frac{\phi}{n} + \frac{2k\pi}{n}\right) = \left[\cos\left(\frac{\phi}{n}\right) + i \sin\left(\frac{\phi}{n}\right)\right]\left[\cos\left(\frac{2k\pi}{n}\right) + i \sin\left(\frac{2k\pi}{n}\right)\right]$$

to show that the roots of the equation $z^n = w$ are $w_p^{1/n}\omega_j$ where $w_p^{1/n}$ is the principal root of w and ω_j is an *n*th root of unity, $j = 1, 2, \ldots, n$.

63. (Summing roots of unity) Let $\omega_1, \omega_2, \ldots, \omega_n$ denote the *n*th roots of unity where $n \geq 2$; that is $\omega_j^n = 1$ for $j = 1, 2, \ldots, n$. Pick and fix a root $\omega_0 \neq 1$ from the set $(\omega_j)_{j=1}^n$.
(a) Show that $\omega_0 \omega_j$ is an *n*th root of unity for $j = 1, 2, \ldots, n$. [*Hint:* Verify the equation $z^n = 1$.]
(b) Show that $\omega_0 \omega_j \neq \omega_0 \omega_k$ if $j \neq k$. Conclude that the set $(\omega_0 \omega_j)_{j=1}^n$ is the same as the set of all n roots of unity.
(c) Show that the sum of the n roots of unity is zero. That is, show that

$$\omega_1 + \omega_2 + \cdots + \omega_n = 0.$$

[*Hint:* $\omega_1 + \omega_2 + \cdots + \omega_n = \omega_0 \omega_1 + \omega_0 \omega_2 + \cdots + \omega_0 \omega_n$; why? Factor ω_0 and conclude that the sum has to be 0.]
(d) Show directly that

$$1 + \omega_0 + \omega_0^2 + \cdots + \omega_0^{n-1} = 0.$$

[*Hint:* Multiply the left side by $1 - \omega_0 \neq 0$.]

64. Project Problem: Our goal is to solve the equation

$$z^n = (z+1)^n. \tag{1.3.27}$$

This polynomial equation is actually of order $n - 1$, since upon expanding, the terms in z^n will cancel.
(a) Divide both sides of (1.3.27) by $(z+1)^n$ (evidently, $z+1$ cannot be zero) and conclude that $\frac{z}{z+1}$ must be one of the *n*th roots of unity (see Exercise 61). Hence $z = (z+1)\omega_k$, where $\omega_k = \cos\left(\frac{2k\pi}{n}\right) + i \sin\left(\frac{2k\pi}{n}\right)$, for $k = 1, 2, \ldots, n-1$. We must throw out $k = 0$ because $z = z+1$ cannot be correct.
(b) Write $z = x + iy$ and solve for x and y by equating real and imaginary parts in $z = (z+1)\omega_k$.

Obtain the answers

$$x = -\frac{1}{2} \quad \text{and} \quad y_k = \frac{\sin(2k\pi/n)}{2(1 - \cos(2k\pi/n))} = \frac{1}{2}\cot(k\pi/n).$$

(c) Apply the result of (b) to solve $(z+1)^7 = z^7$.

In the remaining problems, we present a family of polynomials, known as the **Chebyshev polynomials**. These polynomials have useful applications in numerical analysis. Our presentation uses De Moivre's identity and the binomial formula.

65. (The binomial formula) Use mathematical induction to prove that for arbitrary complex numbers a and b and for a positive integer n we have

$$(a+b)^n = a^n + \binom{n}{1}a^{n-1}b^1 + \binom{n}{2}a^{n-2}b^2 + \cdots + \binom{n}{n-1}a^1 b^{n-1} + b^n, \qquad (1.3.28)$$

where for $0 \le m \le n$, the **binomial coefficient** $\binom{n}{m}$ (read as "n choose m") is defined by

$$\binom{n}{m} = \frac{n!}{(n-m)!m!},$$

with $0! = 1$ as a convention. The binomial formula is also written as

$$(a+b)^n = \sum_{m=0}^{n} \binom{n}{m} a^{n-m}b^m.$$

[*Hint*: You should come up with two sums. Shift the index on one summation so the summand looks like the other. Pull off the a^{n+1} and b^{n+1} terms. Then use the identity

$$\binom{n}{m-1} + \binom{n}{m} = \binom{n+1}{m},$$

referred to as Pascal's identity.]

66. (a) Use De Moivre's identity and the binomial formula to show that, for $n = 1, 2, \ldots$,

$$\cos n\theta = \sum_{k=0}^{\left[\frac{n}{2}\right]} \binom{n}{2k}(\cos\theta)^{n-2k}(-1)^k(\sin\theta)^{2k} \qquad (1.3.29)$$

and

$$\sin n\theta = \sum_{k=0}^{\left[\frac{n-1}{2}\right]} \binom{n}{2k+1}(\cos\theta)^{n-2k-1}(-1)^k(\sin\theta)^{2k+1}, \qquad (1.3.30)$$

where, for a real number s, $[s]$ denotes the greatest integer not larger than s.

(b) Show that

$$\cos n\theta = \sum_{k=0}^{\left[\frac{n}{2}\right]} \binom{n}{2k}(\cos\theta)^{n-2k}(-1)^k(1 - \cos^2\theta)^k. \qquad (1.3.31)$$

(c) Derive the results of Exercises 57 and 58 from (a).

67. (Chebyshev polynomials) It is clear from (1.3.31) that $\cos n\theta$ can be expressed as a polynomial of degree n in $\cos\theta$. So, for $n = 0, 1, 2, \ldots$, we define the nth **Chebyshev polynomial**[3] T_n by the formula

---

[3] In the Latin alphabet, Chebyshev is often transliterated as Tchebyshev. Hence T_n.

$$T_n(\cos\theta) = \cos n\theta, \quad n = 0, 1, 2, \ldots. \tag{1.3.32}$$

(a) Obtain the formula

$$T_n(x) = \sum_{k=0}^{\left[\frac{n}{2}\right]} \binom{n}{2k} (-1)^k x^{n-2k} (1-x^2)^k. \tag{1.3.33}$$

(b) Use (1.3.33) to derive the following list of Chebyshev polynomials:

$$T_0(x) = 1 \qquad\qquad T_1(x) = x \qquad\qquad T_2(x) = -1 + 2x^2$$

$$T_3(x) = -3x + 4x^3 \qquad T_4(x) = 1 - 8x^2 + 8x^4 \qquad T_5(x) = 5x - 20x^3 + 16x^5$$

68. (Properties of the Chebyshev polynomials) Derive the following properties of Chebyshev polynomials. These properties are characteristic of many other families of functions that we will encounter when solving important differential equations such as the Legendre, Laguerre, and Bessel differential equations.

(a) $T_n(1) = 1$ and $T_n(-1) = (-1)^n$.

(b) $T_n(0) = \cos(\frac{n\pi}{2}) = 0$ if n is odd and $(-1)^k$ if $n = 2k$.

(c) $|T_n(x)| \le 1$ for all x in $[-1,1]$.

(d) $T_n(x)$ is a polynomial of degree n. [*Hint*: Induction on n and (1.3.33).]

(e) Derive the **recurrence relation** $T_{n+1}(x) + T_{n-1}(x) = 2xT_n(x)$. [*Hint*: Use the trigonometric identities $\cos(a \pm b) = \cos a \cos b \mp \sin a \sin b$ and (1.3.32).]

(f) Show that $T_n(x)$ has n simple zeros in $[-1,1]$ at the points $\alpha_k = \cos\left(\frac{2k-1}{2n}\pi\right)$, $k = 1, 2, \ldots, n$.

(g) Show that $T_n(x)$ assumes its absolute extrema in $[-1,1]$ at the points $\beta_k = \cos\left(\frac{k}{n}\pi\right)$, $k = 0, 1, 2, \ldots, n$, with $T_n(\beta_k) = (-1)^k$.

(h) Show that for $m \ne n$

$$\int_{-1}^1 T_m(x) T_n(x) \frac{dx}{\sqrt{1-x^2}} = 0.$$

[*Hint*: Change variables: $x = \cos\theta$.]

(i) Show that for $n = 0, 1, 2, \ldots$,

$$\int_{-1}^1 |T_n(x)|^2 \frac{dx}{\sqrt{1-x^2}} = \begin{cases} \pi & \text{if } n = 0; \\ \frac{\pi}{2} & \text{if } n \ge 1. \end{cases}$$

1.4 Complex Functions

A **complex-valued function** f of a complex variable is a relation that assigns to each complex number z in a set S a unique complex number $f(z)$. The set S is a subset of the complex numbers and is called the **domain of definition** of f. The unique complex number $f(z)$ is called the **value** of f at z and is sometimes written $w = f(z)$. For example, the function $f(z) = z^3$ assigns to each complex number z the complex number $w = z^3$. When a function is given by a formula and the domain is not specified, the domain is taken to be the largest set on which the formula makes sense. For instance, the domain of $f(z) = z^3$ is the set of all complex numbers $\mathbb{C}$, while the domain of $g(z) = (z^2 - 1)^{-1}$ is $\mathbb{C} \setminus \{-1, 1\}$.

In calculus, a real-valued function $y = g(x)$ of a real variable x was represented as a graph in the (x,y)-plane. The graph provides a pictorial representation of the function and its properties and contains vital information. To visualize a complex-valued function $z \mapsto f(z)$ unfortunately requires four dimensions: two dimensions for the variable z and two for the values $w = f(z)$. Since a four-dimensional picture is not possible, we use two planes, the z-plane and the w-plane, and view the function as a **mapping** from a subset of one plane to the other (see Figure 1.22). If we write $z = x + iy$ and $w = u + iv$, as a convention, the z-plane axes are labeled by x and y and the w-plane axes by u and v. The **image** $f[S]$ of a set S under a mapping f is the set of all points w such that $w = f(z)$ for some z in S. We illustrate the mapping process with basic examples including some familiar geometric transformations.

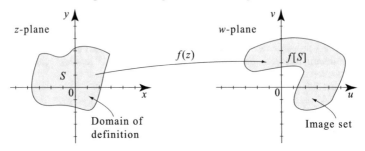

Fig. 1.22 To visualize a mapping by a complex-valued function $w = f(z)$, we use two planes: the z-plane or (x, y)-plane for the domain of definition and the w-plane or (u, v)-plane for the image.

Example 1.4.1. (Translation) Let S denote the disk $S = \{z : |z| \leq 1\}$. Find the image of S under the mapping $f(z) = z + 2 + i$.

Solution. For z in S, the number $f(z)$ is found by adding z to $2 + i$. If z is represented by (x, y), then $f(z)$ will be represented by $(x + 2, y + 1)$. Hence the function translates the point z two units to the right and one unit up. The image of S, then, is the set S translated two units to the right and one unit up (see Figure 1.23). □

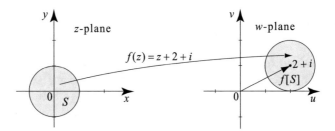

Fig. 1.23 A translation is a mapping of the form $f(z) = z + b$, where b is a complex number. In Example 1.4.1, $b = 2 + i$. Consequently, the image of the disk is a disk of the same radius, centered at $2 + i$ (the image of the original center). Hence, $f[S] = \{w : |w - 2 - i| \leq 1\}$.

The next example deals with functions of the form $f(z) = az$, where a is a nonzero complex constant. To understand these mappings, recall that when we

multiply two complex numbers, we multiply their moduli and add their arguments. Writing a in polar form as $a = r(\cos\theta + i\sin\theta)$, we see that the mapping
$z \mapsto r(\cos\theta + i\sin\theta)z$ multiplies the modulus of z by $r > 0$ (a dilation) and adds θ
to the argument of z (a rotation). Since multiplication is commutative, these operations of dilation and rotation may be applied in either order. When $r = 1$, we obtain
the mapping $z \mapsto (\cos\theta + i\sin\theta)z$, which is a **rotation** by the angle θ. Mappings of
the form $z \mapsto rz$, where $r > 0$, are called **dilations** by a factor r.

Example 1.4.2. (Dilations and rotations) Let S be the closed square of side length
2 with sides parallel to the axes centered at the point 2 (Figure 1.24).
(a) What is the image of S under the mapping $f(z) = 3z$?
(b) What is the image of S under the mapping $f(z) = 2iz$?

Solution. (a) For z in S the function $f(z) = 3z$ has the effect of tripling the modulus
and leaving the argument unchanged. Hence $f(z)$ lies on the ray extending from the
origin to z, at three times the distance from the origin to z. In particular, the image
of the square is another square whose corners are the images of the corners of the
original square.

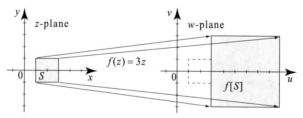

Fig. 1.24 The images of the corners of S under $f(z) = 3z$ are
$f(1+i) = 3 + 3i$, $f(1-i) = 3 - 3i$, $f(3+i) = 9 + 3i$, $f(3-i) = 9 - 3i$.

As seen from Figure 1.24, the image of S is a square of side length 6 centered at the
point 6 on the real axis.
(b) In polar form, we have $i = \cos\frac{\pi}{2} + i\sin\frac{\pi}{2}$, and so $f(z) = 2(\cos\frac{\pi}{2} + i\sin\frac{\pi}{2})z$.

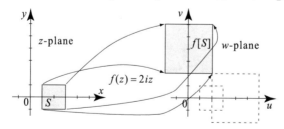

Fig. 1.25 The mapping $f(z) = 2iz$ is a composition of two
mappings: a dilation by a factor of 2, followed by a counterclockwise rotation by an angle of $\pi/2$. The angle of rotation
is the argument of $2i$.

Hence for z in S, $f(z)$ has the effect of doubling the modulus and adding $\frac{\pi}{2}$ to the argument. To determine $f[S]$, take the set S, dilate it by a factor of 2, then rotate it counterclockwise by $\pi/2$ (Figure 1.25).   □

Examples 1.4.1 and 1.4.2 deal with mappings of the type $f(z) = az + b$, where a and b are complex constants and $a \neq 0$. These are called **linear transformations** and can always be thought of in terms of a dilation, a rotation, and a translation. These transformations map regions to geometrically similar regions. It is important that $a \neq 0$ because otherwise the transformation would be a constant. The transformation $f(z) = z$ is called the **identity transformation** for obvious reasons.

The next transformation is not linear.

Example 1.4.3. (Inversion) Find the image of the following sets under the mapping $f(z) = 1/z$.
(a) $S = \{z : 0 < |z| < 1,\ 0 \le \arg z \le \pi/2\}$.
(b) $S = \{z : 2 \le |z|,\ 0 \le \arg z \le \pi\}$.

Solution. (a) We know from (1.3.18) that for $z = r(\cos\theta + i\sin\theta) \neq 0$, we have

$$\frac{1}{z} = \frac{1}{r}(\cos(-\theta) + i\sin(-\theta)).$$

According to this formula, the modulus of the number $1/z$ is the reciprocal of the modulus of z and the argument of $f(z)$ is the negative of the argument of z. Consequently, numbers inside the unit circle ($|z| \le 1$) get mapped to numbers outside the unit circle ($\frac{1}{|z|} \ge 1$), and numbers in the upper half-plane get mapped to numbers in the lower half-plane. Looking at S, as the modulus of z goes from 1 down to 0, the modulus of $f(z)$ goes from 1 up to infinity. As the argument of z goes from 0 up to $\pi/2$, the argument of $1/z$ goes from 0 down to $-\pi/2$. Hence $f[S]$ is the set of all points in the fourth quadrant, including the border axes, that lie outside the unit circle (see Figure 1.26):

$$f[S] = \{w : 1 < |w|,\ -\pi/2 \le \arg z \le 0\}.$$

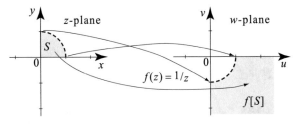

Fig. 1.26 The inversion $w = \frac{1}{z}$ has the effect of inverting the modulus and changing the sign of the argument, i.e., $|w| = \frac{1}{|z|}$ and $\operatorname{Arg} w = -\operatorname{Arg} z$.

(b) As the modulus of z increases from 2 up to infinity, the modulus of $1/z$ decreases from $1/2$ down to zero (but never equals zero). As the argument of z goes from 0 up to π, the argument of $1/z$ goes from 0 down to $-\pi$.

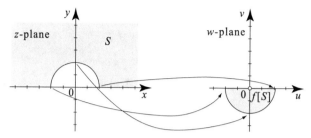

Fig. 1.27 Under the inversion $f(z) = \frac{1}{z}$, points outside the circle of radius 2, $|z| \geq 2$, get mapped to points inside the circle of radius $\frac{1}{2}$, $|w| \leq \frac{1}{2}$.

Hence $f[S]$ is the set of nonzero points in the lower half-plane, including the real axis, with $0 < |w| < 1/2$ (see Figure 1.27):

$$f[S] = \{w : \ 0 < |w| < 1/2, -\pi \leq \arg z \leq 0\}. \qquad \square$$

The function $f(z) = 1/z$ in Example 1.4.3 is a special case of a general type of mapping of the form

$$w = \frac{az+b}{cz+d} \qquad (ad \neq bc), \tag{1.4.1}$$

known as a **linear fractional transformation** or **Möbius transformation**. Here a, b, c, d are complex numbers, and you can check that when $ad = bc$, w is constant.

Real and Imaginary Parts of Functions

For a complex function f, let $u = \operatorname{Re} f$ and $v = \operatorname{Im} f$. The functions u and v are real-valued functions, and we may think of them as functions of two real variables. With a slight abuse of notation, we sometimes write $u(z) = u(x+iy) = u(x, y)$, and $v(z) = v(x+iy) = v(x, y)$. Thus

$$f(z) = f(x+iy) = u(x, y) + iv(x, y). \tag{1.4.2}$$

For example, for $f(z) = z^2 = (x+iy)^2 = x^2 - y^2 + 2ixy$, we have

$$u(x, y) = x^2 - y^2 \qquad \text{and} \qquad v(x, y) = 2xy.$$

As we illustrate in the sequel, the functions $u(x, y)$ and $v(x, y)$ may be used to determine algebraically the image of a set when the answer is not geometrically obvious.

Example 1.4.4. (Squaring) Let S be the vertical strip

$$S = \{z = x + iy : 1 \leq x \leq 2\}.$$

Find the image of S under the mapping $f(z) = z^2$.

Solution. As before, write

$$f(z) = z^2 = x^2 - y^2 + i2xy.$$

Thus the real part of $f(z)$ is $u(x, y) = x^2 - y^2$ and the imaginary part of $f(z)$ is $v(x, y) = 2xy$. Let us fix $1 \leq x_0 \leq 2$, and find the image of the vertical line $x = x_0$. A point (x_0, y) on the line maps to the point (u, v), where $u = x_0^2 - y^2$, $v = 2x_0 y$. To determine the equation of the curve that is traced by the point (u, v) as y varies from $-\infty$ to ∞, we will eliminate y and get an algebraic relation between u and v. From $v = 2x_0 y$, we obtain $y = \frac{v}{2x_0}$. Plugging into the expression for u, we obtain

$$u = x_0^2 - \frac{v^2}{4x_0^2}.$$

This gives u as a quadratic function of v. Hence the graph is a leftward-facing parabola with a vertex at $(u, v) = (x_0^2, 0)$ and v-intercepts at $(0, \pm 2x_0^2)$. As x_0 ranges from 1 up to 2, the corresponding parabolas in the w-plane sweep out a parabolic region, as illustrated in Figure 1.28.

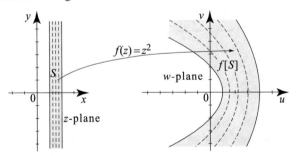

Fig. 1.28 For $1 \leq x_0 \leq 2$, the image of a line $x = x_0$ under the mapping $f(z) = z^2$ is a parabola $u = x_0^2 - \frac{v^2}{4x_0^2}$. As we vary x_0 from 1 to 2, these parabolas sweep out a parabolic region, which determines $f[S]$.

Since all points lie to the right of the parabola, where $x_0 = 1$, and to the left of the parabola, where $x_0 = 2$, we have

$$f[S] = \left\{ w = u + iv : \ 1 - \frac{v^2}{4} \leq u \leq 4 - \frac{v^2}{16} \right\}. \qquad \Box$$

Example 1.4.5. (Inversion) Let $0 < a < b$ be real numbers.

(a) Show that the mapping $f(z) = 1/z$ takes the vertical line $x = x_0 > 0$ in the z-plane into the circle $(u - \frac{1}{2x_0})^2 + v^2 = \frac{1}{(2x_0)^2}$ in the w-plane.

(b) Determine the image of the vertical strip $S = \{z = x + iy : a \le x \le b\}$ under the mapping $f(z) = 1/z$.

Solution. (a) For $z = x + iy$, in view of (1.1.3), we have $\frac{1}{z} = u(x, y) + iv(x, y)$ where

$$u(x, y) = \frac{x}{x^2 + y^2} \quad \text{and} \quad v(x, y) = \frac{-y}{x^2 + y^2}.$$

The image of a point (x, y) on the line $x = x_0$ is the point (u, v) with

$$u = \frac{x_0}{x_0^2 + y^2} \quad \text{and} \quad v = \frac{-y}{x_0^2 + y^2}.$$

It follows

$$u^2 + v^2 = \frac{x_0^2 + y^2}{(x_0^2 + y^2)^2} = \frac{1}{x_0^2 + y^2} = \frac{u}{x_0}$$

which implies $u^2 - \frac{u}{x_0} + v^2 = 0$. Adding $\frac{1}{(2x_0)^2}$ to both sides we obtain:

$$u^2 - \frac{u}{x_0} + \frac{1}{(2x_0)^2} + v^2 = \frac{1}{(2x_0)^2} \quad \text{or} \quad \left(u - \frac{1}{2x_0}\right)^2 + v^2 = \frac{1}{(2x_0)^2}$$

This is the equation of a circle in the w-plane, centered at $(\frac{1}{2x_0}, 0)$, with radius $\frac{1}{2x_0}$ (Figure 1.29). The point $(0, 0)$ is on the circle but is not part of the image of the line $x = x_0$.

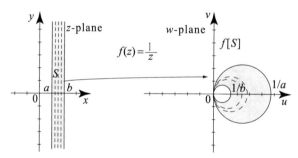

Fig. 1.29 The inversion $f(z) = 1/z$ maps the line $x = a$ onto a circle.

(b) Notice that as x_0 varies from a to b, the line $x = x_0$ sweeps the vertical strip S, and the image of the line $x = x_0$ sweeps the image of S. It is now clear from Figure 1.29 that the image of S is the annular region bounded by the outer circle with radius $\frac{1}{2a}$ centered at $(\frac{1}{2a}, 0)$ and the inner circle with radius $\frac{1}{2b}$ centered at $(\frac{1}{2b}, 0)$. $\square$

Mappings in Polar Coordinates

Some complex functions and regions are more naturally suited to polar coordinates. In this case it may be advantageous to write $z = r\cos\theta + ir\sin\theta$ and express $w = f(z)$ in polar coordinates as

$$w = \rho(\cos\phi + i\sin\phi).$$

Then we identify the polar coordinates of w as functions of the polar coordinates (r, θ) of z:

$$\rho(r, \theta) = |f(r\cos\theta + ir\sin\theta)| \quad \text{and} \quad \phi(r, \theta) = \arg\left(f(r\cos\theta + ir\sin\theta)\right).$$

The next example uses such polar coordinates to track the mapping of circular sectors.

Example 1.4.6. (Mapping sectors) Let S be the sector

$$S = \left\{z : |z| \le \frac{3}{2}, \ 0 \le \arg z \le \frac{\pi}{4}\right\}.$$

Find the image of S under the mapping $f(z) = z^3$.

Solution. If we write $z = r(\cos\theta + i\sin\theta)$, then $z^3 = r^3(\cos 3\theta + i\sin 3\theta)$. Hence the polar coordinates of $w = f(z) = \rho(\cos\phi + i\sin\phi)$ are $\rho = r^3$ and $\phi = 3\theta$. As r increases from 0 up to $\frac{3}{2}$, ρ increases from 0 up to $\frac{27}{8}$; and, as θ goes from 0 up to $\frac{\pi}{4}$, ϕ goes from 0 up to $\frac{3\pi}{4}$.

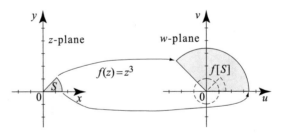

Fig. 1.30 The mapping $w = z^3$ has the effect of cubing the norm and tripling the argument: $|w| = |z|^3$; $\arg w = 3\operatorname{Arg} z$.

Hence the image of S is the set of all w with modulus less than $\frac{27}{8}$ and arguments between 0 and $\frac{3\pi}{4}$, i.e., $f[S] = \{w : |w| \le \frac{27}{8}, \ 0 \le \arg w \le \frac{3\pi}{4}\}$ (see Figure 1.30). $\square$

All the mappings considered in the examples have been **one-to-one**, in that distinct points z_1 and z_2 always map to distinct points $f(z_1)$ and $f(z_2)$. It is possible that more than one point on the z-plane will map to the same point on the w-plane. Such mappings are not one-to-one, and we are already familiar with some of them.

For example, the function $f(z) = z^2$ with domain of definition $\mathbb{C}$ maps z and $-z$ to the same point in the w-plane.

Exercises 1.4

In Exercises 1–4, compute $f(z)$ for the following values of z. Give your answer in the form $a + bi$, where a and b are real.

1. $f(z) = iz + 2 + i$, $z = 1 + i$, $-1 + i$, $-1 - i$, $1 - i$

2. $f(z) = (3 + 4i)z$, $z = 1, i, -1, -i$

3. $f(z) = z^2 + 2iz - 1$, $z = -1 - i$, $1 + i$, 0, $2i$

4. $f(z) = \dfrac{z - i}{z + i}$, $z = i, 1 + i, -2i, 3 + \dfrac{i}{2}$

5. Refer to Exercise 1. The given values of z represent the four corners of a square. Plot the square and then determine and plot its image under the mapping f.

6. Repeat Exercise 5, using the data from Exercise 2.

In Exercises 7–10, describe the linear mapping as a composition of a rotation, dilation, and translation. Be specific in your description and the order of the composition.

7. $f(z) = (-2 + 2i)z + 1 - 7i$ 8. $f(z) = -iz + 5 + 3i$

9. $f(z) = (-1 - i)z + 3 + i$ 10. $f(z) = \frac{z}{1+i} + 1 + i$

In Exercises 11–18, express $f(z)$ in the form $u(z) + iv(z)$, where u and v are the real and imaginary parts of f.

11. $f(z) = iz + 2 - i$ 12. $f(z) = z + 2 + i$ 13. $f(z) = z^2 - 2z + i$

14. $f(z) = (1 + i)z^2 - 2iz$ 15. $f(z) = \dfrac{z - 1}{z + 1}$ 16. $f(z) = i|z|$

17. $f(z) = 3 \, \mathrm{Arg}\, z$ 18. $f(z) = \mathrm{Re}\, z + \mathrm{Im}\, z$

In Exercises 19–20, find the largest subset of $\mathbb{C}$ on which the expression is a well-defined function.

19. (a) $\dfrac{i - z}{2 - i - z}$ (b) $3 + iz^2$

20. (a) $\dfrac{1}{1 + z^2}$ (b) $2 + i \, \mathrm{Arg}\, (z - 1)$

21. Find a linear transformation $f(z)$ such that $f(1) = 3 + i$ and $f(3i) = -2 + 6i$.

22. Find a linear transformation $f(z)$ such that $f(2 - i) = -3 - 3i$ and $f(2) = -2 - 2i$.

In Exercises 23–26, find the image $f[S]$ under the linear transformations. Draw a picture of S and $f[S]$, and depict arrows mapping select points.

23. $S = \{z : |z| < 1\}$, $f(z) = 4z$.

24. $S = \{z : \mathrm{Re}\, z > 0\}$, $f(z) = iz + i$.

25. $S = \{z : \mathrm{Re}\, z > 0, \, \mathrm{Im}\, z > 0\}$, $f(z) = -z + 2i$.

26. $S = \{z : |z| \le 2, \, 0 \le \mathrm{Arg}\, z \le \frac{\pi}{2}\}$, $f(z) = iz + 2$.

In Exercises 27–30, we describe a set S and its image $f[S]$ by a linear mapping $w = f(z)$. (a) Plot S and $f[S]$. (b) Find the linear mapping f.

27. S is the rectangle with corners at $(0, 0)$, $(2, 0)$, $(2, 1)$, $(0, 1)$; $f[S]$ is obtained by rotating S counterclockwise by $\pi/4$, then translating by 2 units up and 1 unit to the left.

28. S is the rectangle with corners at $(-1, 0)$, $(1, 0)$, $(1, 2)$, $(-1, 2)$; $f[S]$ is obtained by translating S by 2 units down and 1 unit to the right, then rotating clockwise by $\pi/4$.

29. S is the square with corners at $(1, 1)$, $(-1, 1)$, $(-1, -1)$, $(1, -1)$; $f[S]$ is obtained by translating S 1 unit to the right, rotating clockwise by $\frac{\pi}{2}$, and then dilating by a factor of 3.

30. S is the circle centered at the origin with radius 2; $f[S]$ is obtained by translating S by 2 units to the right, then rotating clockwise by π.

In Exercises 31–34, use the method of Example 1.4.3 to find the image $f[S]$ under the inversion $f(z) = \frac{1}{z}$. Draw a picture of S and $f[S]$ and depict arrows mapping select points.

31. $S = \{z : 0 < |z| \leq 1\}$

32. $S = \{z : |z| \geq 1\}$

33. $S = \{z : 0 < |z| \leq 3, \frac{\pi}{3} \leq \operatorname{Arg} z \leq \frac{2\pi}{3}\}$

34. $S = \{z : z \neq 0, 0 \leq \operatorname{Arg} z \leq \frac{\pi}{2}\}$

In Exercises 35–38, use the method of Example 1.4.4 to find the image $f[S]$ under the mapping $f(z) = z^2$. Draw a picture of S and $f[S]$ and depict arrows mapping select points.

35. S is the square with corners at $(0, 0)$, $(1, 0)$, $(1, 1)$, $(0, 1)$.

36. $S = \{z : 0 \leq \operatorname{Im} z \leq 1\}$

37. $S = \{z : -2 \leq \operatorname{Re} z \leq 0\}$

38. $S = \{z : \operatorname{Re} z > 0 \text{ and } \operatorname{Im} z > 0\}$

In Exercises 39–42, use the method of Example 1.4.4 to find the image $f[S]$ under the mapping $f(z) = z^2$. Draw a picture of $f[S]$ to illustrate your answer and be specific about the images of the boundary lines of S.

39.   **40.**

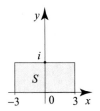

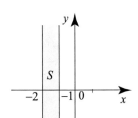

41.   **42.**

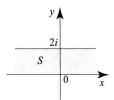

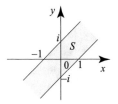

In Exercises 43–46, use polar coordinates as in Example 1.4.6 to find the image $f[S]$ under the mapping. Draw a picture of $f[S]$ to illustrate your answer and be specific about the images of the boundary lines of S.

43. $f(z) = z^3$

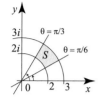

44. $f(z) = z^2$

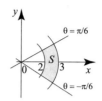

45. $f(z) = iz^2$

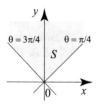

46. $f(z) = \dfrac{i}{z^2}$

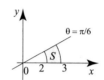

47. Consider the mapping $z \mapsto \dfrac{1}{z}$.

(a) Show that the image of the circle $|z| = a > 0$ is the circle $|w| = \dfrac{1}{a}$.

(b) Show that the image of the ray $\operatorname{Arg} z = \alpha$ $(z \neq 0)$ is the ray $\operatorname{Arg} w = -\alpha$ $(w \neq 0)$.

48. Show that for $f(z) = 1/z$, $f(z) = g(h(z)) = h(g(z))$, where $g(z) = \bar{z}$ and $h(z) = \dfrac{z}{|z|^2}$.

49. (a) Suppose that $f(z) = az + b$ and $g(z) = cz + d$ are linear transformations. Show that $f(g(z))$ is also a linear transformation.

(b) Show that every linear transformation $f(z) = az + b$ $(a \neq 0)$ can be written in the form $g_1(g_2(g_3(z)))$, where g_1 is a translation, g_2 is a rotation, and g_3 is a dilation.

50. Let $f(z) = az$ and $g(z) = z + b$. Show that $f(g(z)) = g(f(z))$ for all z if and only if $a = 1$ or $b = 0$.

51. Construct a linear transformation that rotates the entire plane by an angle ϕ about a point z_0.

52. Let $a \neq 0$. Show that for each linear transformation $f(z) = az + b$, there exists a linear transformation $g(z)$ such that $f(g(z)) = g(f(z)) = z$ for all complex z.

53. Find the image of the set $S = \{z : z \text{ is real}\}$ under the mapping $f(z) = \operatorname{Arg} z$.

54. Find the image of the set $S = \{z : |z| \leq 1\}$ under the mapping $f(z) = z + \bar{z}$.

55. Find a linear fractional transformation $f(z) = \dfrac{az+b}{cz+d}$ such that $f(0) = -5 + i$, $f(-2i) = 5 - 2i$, $f(2) = -\dfrac{8}{5} - i\dfrac{11}{5}$. [*Hint*: In solving for a, b, c, d, keep in mind that these are not uniquely determined. Once you determine that a coefficient is not zero, you may set it equal to 1.]

56. Find a linear fractional transformation $f(z) = \dfrac{az+b}{cz+d}$ such that $f(0) = -2i$, $f(9i) = -\dfrac{i}{5}$, and $f(4 - i) = \dfrac{1}{2}$.

A **fixed point** *of a function f is a complex number z_0 satisfying $f(z_0) = z_0$. In Exercises 57–60, determine the fixed points of the functions.*

57. $f(z) = \dfrac{1}{z}$

58. $f(z) = az + b$.   [*Hint*: Consider the cases: $a = 1$ and $a \neq 1$]

59. $f(z) = 2\left(z + \frac{1}{z}\right)$

60. $f(z) = \frac{-6i + (2+3i)z}{z}$

61. Define the set of **lattice points** in the plane as $L = \{z: z = m + in, \ m \text{ and } n \text{ integers}\}$. Consider the mapping $f(z) = z^2$ and the image $f[L]$ of f under L.
(a) Show that if w is in $f[L]$, $-w$, $\overline{w}$, and $-\operatorname{Re}w + i\operatorname{Im}w$ are in $f[L]$.
(b) Show that if w is in $f[L]$, w is also in L.
(c) Show that if w is in $f[L]$, $f(w)$ is in $f[L]$.

1.5 Sequences and Series of Complex Numbers

A **sequence** of complex numbers is a function whose domain of definition is the set of positive integers $\{1, 2, \ldots, n, \ldots\}$ and whose range is a subset of $\mathbb{C}$. Thus a sequence is an ordered list of complex numbers $a(1), a(2), a(3), \ldots, a(n), \ldots$. It is customary to write a_n instead of $a(n)$ and to denote the sequence by $\{a_n\}_{n=1}^{\infty}$. Sometimes we index a sequence by nonnegative integers, such as $\{a_n\}_{n=0}^{\infty}$.

Many analytical expressions involving sequences of complex numbers look identical to those for real sequences. The difference is that, in the complex case, the absolute value refers to distance in the plane, and sequences of complex numbers can be thought of as sequences of points in the plane, which converge by eventually staying inside small disks centered at the limit point (Figure 1.31).

Convergence of Sequences

We start by investigating the notions of convergence and divergence of complex sequences.

Definition 1.5.1. We say that a sequence $\{a_n\}_{n=1}^{\infty}$ **converges** to a complex number L or has **limit** L as n tends to infinity, and write

$$\lim_{n \to \infty} a_n = L,$$

if for every $\varepsilon > 0$ there is an integer N such that

$$|a_n - L| < \varepsilon \qquad \text{for all } n \geq N.$$

If the sequence $\{a_n\}_{n=1}^{\infty}$ does not converge, then we say that it **diverges**.

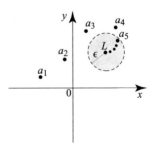

Fig. 1.31 A sequence $\{a_n\}$ converging to a complex number L.

If $L = \lim_{n \to \infty} a_n$, then we often use the notation $a_n \to L$ as $n \to \infty$ or merely $a_n \to L$. It is immediate from the definition that a sequence $a_n \to L$ if and only if the real sequence $|a_n - L| \to 0$. For sequences that converge to zero, we have

$$\lim_{n \to \infty} a_n = 0 \quad \Leftrightarrow \quad \lim_{n \to \infty} |a_n| = 0. \tag{1.5.1}$$

Proposition 1.5.2. *If a limit of a complex sequence exists, then it is unique.*

Proof. Indeed, if $a_n \to L$ and $a_n \to L'$ with $L' \neq L$, then for

$$\varepsilon = \frac{1}{2}|L - L'| > 0$$

there is N such that $|a_n - L| < \varepsilon$ and $|a_n - L'| < \varepsilon$ for all $n \geq N$. The triangle inequality gives $|L - L'| < 2\varepsilon$, and this contradicts the choice of $\varepsilon = |L - L'|/2$. Thus we must have $L = L'$. ∎

Example 1.5.3. Determine whether or not the sequences $a_n = \frac{1}{n}\cos(n\frac{\pi}{4}) + i\frac{1}{n}\sin(n\frac{\pi}{4})$ and $b_n = \cos(n\frac{\pi}{4}) + i(n\frac{\pi}{4})$ converge as $n \to \infty$, and if they do, find their limits.

Solution. The first terms of the sequence $\{b_n\}_{n=1}^{\infty}$ are $\frac{\sqrt{2}}{2} + i\frac{\sqrt{2}}{2}$, i, $-\frac{\sqrt{2}}{2} + i\frac{\sqrt{2}}{2}$, -1, $-\frac{\sqrt{2}}{2} - i\frac{\sqrt{2}}{2}$, $-i$, $\frac{\sqrt{2}}{2} - i\frac{\sqrt{2}}{2}$, 1, $\frac{\sqrt{2}}{2} + i\frac{\sqrt{2}}{2}$, ... The sequence is clearly not converging, since its terms cycle over the first eight terms indefinitely. The first few terms of the sequence $\{a_n\}_{n=1}^{\infty}$ are shown in Figure 1.32. The figure suggests that this sequence converges to 0. To prove this we note that $|b_n| = 1$ and

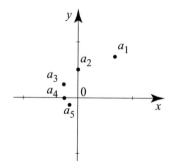

Fig. 1.32 The first five terms of the sequence $a_n = \frac{1}{n}\cos(n\frac{\pi}{4}) + i\frac{1}{n}\sin(n\frac{\pi}{4})$, and its limit 0. Note that the arguments of two successive terms differ by $\frac{\pi}{4}$.

$$|a_n| = \frac{1}{n}|b_n| = \frac{1}{n}.$$

Since $\frac{1}{n} \to 0$ as $n \to \infty$, we conclude that $|a_n| \to 0$ as $n \to \infty$, and so does a_n by (1.5.1). □

Definition 1.5.4. A sequence of complex numbers $\{a_n\}_{n=1}^{\infty}$ is said to be **bounded** if there is a positive number $M > 0$ such that $|a_n| \leq M$ for all n.

The next theorem is proved as in the real case.

Theorem 1.5.5. *Every convergent sequence of complex numbers is bounded.*

The following theorem is also analogous to one from calculus. Its proof is omitted.

Theorem 1.5.6. *Let $\{a_n\}_{n=1}^{\infty}$ and $\{b_n\}_{n=1}^{\infty}$ be sequences of complex numbers.*
(i) Suppose that $\lim_{n\to\infty} a_n = 0$ and $|b_n| \le |a_n|$ for all $n \ge n_0$. Then $\lim_{n\to\infty} b_n = 0$.
(ii) If $\lim_{n\to\infty} a_n = 0$ and $\{b_n\}_{n=1}^{\infty}$ is a bounded sequence, then $\lim_{n\to\infty} a_n b_n = 0$.

The proof of the next theorem is also left to the reader.

Theorem 1.5.7. *If $\{a_n\}_{n=1}^{\infty}$ and $\{b_n\}_{n=1}^{\infty}$ are convergent sequences of complex numbers and α and β are complex numbers, then*

$$\lim_{n\to\infty} (\alpha a_n + \beta b_n) = \alpha \lim_{n\to\infty} a_n + \beta \lim_{n\to\infty} b_n;$$

$$\lim_{n\to\infty} (a_n b_n) = \lim_{n\to\infty} a_n \lim_{n\to\infty} b_n;$$

$$\lim_{n\to\infty} \frac{a_n}{b_n} = \frac{\lim\limits_{n\to\infty} a_n}{\lim\limits_{n\to\infty} b_n} \quad if \lim_{n\to\infty} b_n \ne 0$$

$$\lim_{n\to\infty} \overline{a_n} = \overline{\lim_{n\to\infty} a_n}$$

$$\lim_{n\to\infty} |a_n| = \left| \lim_{n\to\infty} a_n \right|.$$

Theorem 1.5.8. *Suppose that $\{z_n\}_{n=1}^{\infty}$ is a sequence of complex numbers and write $z_n = x_n + iy_n$, where $x_n = \operatorname{Re} z_n$ and $y_n = \operatorname{Im} z_n$. Then for x, y real numbers we have*

$$\lim_{n\to\infty} z_n = x + iy \quad \Leftrightarrow \quad \lim_{n\to\infty} x_n = x \quad and \quad \lim_{n\to\infty} y_n = y.$$

Proof. Suppose that $z_n \to x + iy$. Then by Theorem 1.5.7 we have that $\overline{z_n} \to \overline{x + iy}$. Using again Theorem 1.5.7 we obtain $z_n + \overline{z_n} \to x + iy + \overline{x + iy} = 2x$ and $z_n - \overline{z_n} \to x + iy - \overline{(x + iy)} = 2iy$. Thus $2x_n \to 2x$ and $2iy_n \to 2iy$ which implies that $x_n \to x$ and $y_n \to y$ as $n \to \infty$. Conversely, if $x_n \to x$ and $y_n \to y$ as $n \to \infty$, then by Theorem 1.5.7 we have $iy_n \to iy$ and adding yields $x_n + iy_n \to x + iy$. ∎

Next we show how to use the preceding results along with our knowledge of real-valued sequences to compute limits of complex-valued sequences.

Example 1.5.9. (A useful limit) Show that

$$\lim_{n\to\infty} z^n = \begin{cases} 0 & \text{if } |z| < 1, \\ 1 & \text{if } z = 1. \end{cases}$$

Moreover, show that the limit does not exist for all other values of z; that is, if $|z| > 1$, or $|z| = 1$ and $z \ne 1$, then $\lim_{n\to\infty} z^n$ does not exist.

Solution. Recall that for a real number $r \ge 0$, we have

$$\lim_{n \to \infty} r^n = \begin{cases} 0 & \text{if } 0 \le r < 1, \\ 1 & \text{if } r = 1, \\ \infty & \text{if } r > 1. \end{cases}$$

Consequently, for a complex number $|z| < 1$, we have $\lim_{n \to \infty} |z|^n = 0$. Since $|z|^n = |z^n|$, we conclude from (1.5.1) that $\lim_{n \to \infty} z^n = 0$. For $|z| > 1$, we have $|z|^n \to \infty$ as $n \to \infty$. Hence, by Theorem 1.5.5 the sequence $\{z^n\}_{n=1}^{\infty}$ cannot converge because it is not bounded. Now we deal with the case $|z| = 1$. If $z = 1$, the sequence $\{z^n\}_{n=1}^{\infty}$ is the constant sequence 1, which is trivially convergent. The case $|z| = 1$, $z \ne 1$ is not as simple because the sequence, though bounded, does not converge. We will prove that if this sequence converges, then $z = 1$. First note that if $\lim_{n \to \infty} z^n = L$, then $|L| = |\lim_{n \to \infty} z^n| = \lim_{n \to \infty} |z|^n = 1$, hence $L \ne 0$. Also, if $z^n \to L$, then $z^{n+1} \to L$, as $n \to \infty$. But $z^{n+1} = z^n z$, and by taking limits on both sides of this equality we obtain $L = Lz$. Dividing by L (which is nonzero) we deduce $z = 1$, as claimed. $\square$

Example 1.5.9 can be used to show that for θ not an integer multiple of π, $\lim_{n \to \infty} \cos n\theta$ and $\lim_{n \to \infty} \sin n\theta$ do not exist (Exercise 8).

We now introduce a fundamental concept, which is crucial in establishing convergence when the limit is not known.

Definition 1.5.10. A sequence $\{a_n\}_{n=1}^{\infty}$ is said to be a **Cauchy sequence** if for every $\varepsilon > 0$ there is an integer N such that

$$|a_n - a_m| < \varepsilon \qquad \text{for all } m, n \ge N. \tag{1.5.2}$$

Thus the terms of a Cauchy sequence become arbitrarily close together. It is not hard to see that a convergent sequence is a Cauchy sequence; this is because if the terms are close to a limit, they must be close to each other.

The converse (that a Cauchy sequence converges) is also true and to prove it, we appeal to the well-known fact that a Cauchy sequence of real numbers must converge. This is a consequence of the **completeness property** of real numbers, which states that if S is a nonempty subset of the real line with an **upper bound** M ($x \le M$ for all x in S), then S has a **least upper bound** b. That is, b is an upper bound for S and if M is any other upper bound for S, then $b \le M$. The completeness property is an axiom in the construction of the real number system, and it is equivalent to the statement that all real Cauchy sequences are convergent.

Using this property of real numbers, we can prove a corresponding one for complex numbers.

Theorem 1.5.11. *A sequence of complex numbers $\{a_n\}_{n=1}^{\infty}$ converges if and only if it is a Cauchy sequence.*

Proof. Suppose that $\{a_n\}_{n=1}^{\infty}$ converges to a limit L. Given $\varepsilon > 0$, let N be such that $n \ge N$ implies $|a_n - L| < \frac{\varepsilon}{2}$. For $m, n \ge N$, we have by the triangle inequality (Figure 1.33):

$$|a_m - a_n| = |(a_m - L) + (L - a_n)| \le |a_m - L| + |L - a_n| < \frac{\varepsilon}{2} + \frac{\varepsilon}{2} = \varepsilon.$$

Hence $\{a_n\}_{n=1}^{\infty}$ is a Cauchy sequence.

Conversely, suppose that $\{a_n\}_{n=1}^{\infty}$ is a Cauchy sequence. Then for given $\varepsilon > 0$ there is an N such that $|a_n - a_m| < \varepsilon$ for all $m, n \geq N$. Write $a_n = x_n + i y_n$, where x_n, y_n are real. The inequalities $|\operatorname{Re} z| \leq |z|$ and $|\operatorname{Im} z| \leq |z|$ imply that $|x_n - x_m| \leq |a_n - a_m| < \varepsilon$ and $|y_n - y_m| \leq |a_n - a_m| < \varepsilon$, which in turn implies that $\{x_n\}$ and $\{y_n\}$ are Cauchy sequences of real numbers. By the completeness property of real numbers, $\{x_n\}_{n=1}^{\infty}$ and $\{y_n\}_{n=1}^{\infty}$ are convergent sequences.

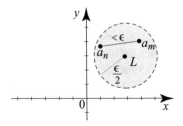

Fig. 1.33 If both $|a_m - L| < \frac{\varepsilon}{2}$ and $|a_n - L| < \frac{\varepsilon}{2}$ hold, then $|a_m - a_n| < \varepsilon$.

Hence $\{a_n\}_{n=1}^{\infty}$ is convergent by Theorem 1.5.8.   ∎

Complex Series

An **infinite complex series** is an expression of the form

$$\sum_{n=0}^{\infty} a_n, \tag{1.5.3}$$

where $\{a_n\}_{n=0}^{\infty}$ is an infinite sequence of complex numbers. The indexing set may not always start at $n = 0$; for example, we sometimes work with expressions of the form

$$\sum_{n=1}^{\infty} a_n, \ \sum_{n=2}^{\infty} a_n, \ \sum_{n=3}^{\infty} a_n, \ldots \text{ etc.}$$

The number a_n is called the **nth term** of the series. To each series $\sum_{n=0}^{\infty} a_n$ we associate a **sequence of partial sums** $\{s_n\}_{n=0}^{\infty}$, where

$$s_n = \sum_{j=0}^{n} a_j = a_0 + a_1 + \cdots + a_n. \tag{1.5.4}$$

Definition 1.5.12. We say that a series $\sum_{n=0}^{\infty} a_n$ **converges**, or is **convergent**, to a complex number s, and we write $s = \sum_{n=0}^{\infty} a_n$, if the sequence of partial sums converges to s, i.e., if $\lim_{n \to \infty} s_n = s$. Otherwise, we say that the series $\sum_{n=0}^{\infty} a_n$ **diverges** or is **divergent**.

So in order to establish the convergence or divergence of a series, we must study the behavior of the sequence of partial sums.

Example 1.5.13. Show that if $|z| < 1$ the **geometric series** $\sum_{n=0}^{\infty} z^n$ converges and

$$\sum_{n=0}^{\infty} z^n = \frac{1}{1-z}.$$

Show that the series diverges for all other values of z.

Solution. Consider the partial sum (a typical case is shown in Figure 1.34):

$$s_n = 1 + z + z^2 + \cdots + z^n.$$

We multiply by z and add 1 to obtain

$$1 + zs_n = 1 + z + z^2 + z^3 + \cdots + z^{n+1}$$

which is equal to $s_n + z^{n+1}$. If $z \neq 1$, we solve for s_n to find

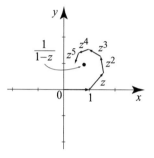

Fig. 1.34 Terms in a convergent geometric series ($|z| < 1$). To get a partial sum s_n, add the vectors $1, z, \ldots, z^n$.

$$s_n = \frac{(1 + z + z^2 + \cdots + z^n)(1-z)}{1-z} = \frac{1 - z^{n+1}}{1-z}.$$

From Example 1.5.9, the sequence $\{z^{n+1}\}_{n=0}^{\infty}$ converges to 0 if $|z| < 1$ and diverges if $|z| > 1$ or $|z| = 1$ and $z \neq 1$. This implies that $\{s_n\}_{n=0}^{\infty}$ converges to $\frac{1}{1-z}$ if $|z| < 1$ and diverges for all other values of z, which is what we wanted to show. $\square$

Geometric series may appear in disguise. Basically, whenever you see a series of the form $\sum_{n=0}^{\infty} w^n$ you should be able to use the geometric series to sum it. However, you have to be careful with the region of convergence.

Example 1.5.14. (Geometric series in disguise) Determine the largest region in which the series

$$\sum_{n=0}^{\infty} \frac{1}{(4 + 2z)^n}$$

is convergent and find its sum.

Solution. This is a geometric series of the form $\sum_{n=0}^{\infty} w^n$ where $w = \frac{1}{4+2z}$.

This series converges to $\frac{1}{1-w}$ if and only if $|w| < 1$. Expressing these results in terms of z, we find that the series converges to

$$\frac{1}{1 - \frac{1}{4+2z}} = \frac{4+2z}{3+2z}$$

if and only if

$$\left| \frac{1}{4+2z} \right| < 1 \quad \Leftrightarrow \quad 1 < |4+2z|.$$

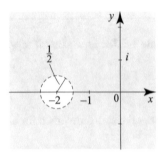

Fig. 1.35 The shaded region describes all z such that $\frac{1}{2} < |z+2|$. These are the points where the series in Example 4 converges.

To better understand the region of convergence, we write the inequality using expressions of the form $|z - z_0|$ and interpret the latter as a distance in the usual way.

We have

$$1 < |4+2z| \quad \Leftrightarrow \quad \frac{1}{2} < |2+z| \quad \Leftrightarrow \quad \frac{1}{2} < |z-(-2)|.$$

This describes the set of all z whose distance to -2 is strictly larger than $\frac{1}{2}$. Thus the series converges outside the closed disk shown in Figure 1.35, with center at -2 and radius $\frac{1}{2}$. □

Properties of Series and Tests of Convergence

Finding a closed expression for the partial sums of a series, as for geometric series, is rather rare and in many cases not possible. Therefore, we have to rely on properties of series to establish their convergence or divergence. Because a series is really a sequence of partial sums, all of the results about sequences can be restated for series. For convenience and ease of reference, we state some of these results along with several tests of convergence that are similar to ones for real series. The proofs are omitted in most cases.

Theorem 1.5.15. *If $\sum_{n=0}^{\infty} a_n$ and $\sum_{n=0}^{\infty} b_n$ are convergent series of complex numbers and α, β are complex numbers, then*

(i) $\sum_{n=0}^{\infty} (\alpha a_n + \beta b_n) = \alpha \sum_{n=0}^{\infty} a_n + \beta \sum_{n=0}^{\infty} b_n$;

(ii) $\overline{\sum_{n=0}^{\infty} a_n} = \sum_{n=0}^{\infty} \overline{a_n}$;

(iii) $\mathrm{Re} \left(\sum_{n=0}^{\infty} a_n \right) = \sum_{n=0}^{\infty} \mathrm{Re}\,(a_n)$ *and* $\mathrm{Im} \left(\sum_{n=0}^{\infty} a_n \right) = \sum_{n=0}^{\infty} \mathrm{Im}\,(a_n)$.

We can use complex series to sum real series.

Example 1.5.16. Show that $\displaystyle\sum_{n=0}^{\infty} \frac{\cos n\theta}{2^n}$ converges for all θ and find the sum.

Solution. We recognize $\cos n\theta$ as the real part of $(\cos\theta + i\sin\theta)^n$, and so the given series is the real part of the geometric series

$$\sum_{n=0}^{\infty} z^n \quad \text{where} \quad z = \frac{1}{2}(\cos\theta + i\sin\theta).$$

From Example 1.5.13, since $|z| = 1/2 < 1$ we have

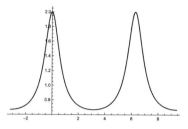

Fig. 1.36 Graph of the series $\sum_{n=0}^{\infty} \frac{\cos n\theta}{2^n}$ over $[-\pi, 3\pi]$.

$$\sum_{n=0}^{\infty} z^n = \frac{1}{1-z} = \frac{1-\bar{z}}{(1-z)(1-\bar{z})} = \frac{1 - \frac{1}{2}\cos\theta + \frac{i}{2}\sin\theta}{(1 - \frac{1}{2}\cos\theta)^2 + (\frac{1}{2}\sin\theta)^2} = \frac{4 - 2\cos\theta + 2i\sin\theta}{5 - 4\cos\theta}.$$

Taking real parts and using Theorem 1.5.15(*iii*), we obtain

$$\sum_{n=0}^{\infty} \frac{\cos n\theta}{2^n} = \text{Re}\left(\frac{4 - 2\cos\theta + 2i\sin\theta}{5 - 4\cos\theta}\right) = \frac{4 - 2\cos\theta}{5 - 4\cos\theta}.$$

The series is plotted in Figure 1.36 as a function of θ. □

Theorem 1.5.17. (The nth Term Test for Divergence) *If $\sum_{n=0}^{\infty} a_n$ is convergent, then $\displaystyle\lim_{n\to\infty} a_n = 0$. Equivalently, if $\lim_{n\to\infty} a_n \neq 0$ or $\lim_{n\to\infty} a_n$ does not exist, then $\sum_{n=0}^{\infty} a_n$ diverges.*

Proof. Let $s_n = \sum_{m=0}^{n} a_m$. If $s_n \to s$, then also $s_{n-1} \to s$, and so $s_n - s_{n-1} \to s - s = 0$. But $s_n - s_{n-1} = a_n$, and so $a_n \to 0$. ∎

Applying the nth term test, we see right away that the geometric series $\sum_{n=0}^{\infty} z^n$ is divergent if $|z| = 1$ or $|z| > 1$.

For $m \geq 1$, the expression $t_m = \sum_{n=m+1}^{\infty} a_n$ is called a **tail of the series** $\sum_{n=0}^{\infty} a_n$. For fixed m, the tail t_m is itself a series, which differs from the original series by finitely many terms. So it is obvious that a series converges if and only if all its tails converge. As $m \to \infty$, we are dropping more and more terms from the tail series; as a result, we have the following useful fact.

Proposition 1.5.18. *If $\sum_{n=0}^{\infty} a_n$ is convergent, then $\lim_{m\to\infty} \sum_{n=m+1}^{\infty} a_n = 0$. Hence if a series converges, then its tail tends to 0.*

Proof. Let $s = \sum_{n=0}^{\infty} a_n$, $t_m = \sum_{n=m+1}^{\infty} a_n$, and $s_m = \sum_{n=1}^{m} a_n$. Since s_m is a partial sum of $\sum_{n=0}^{\infty} a_n$, we have $s_m \to s$ as $m \to \infty$. For each m, we have

$$s_m + t_m = \sum_{n=0}^{\infty} a_n = s \quad \Rightarrow \quad t_m = s - s_m.$$

Let $m \to \infty$ and use $s_m - s \to 0$ to obtain that $t_m \to 0$, as desired. ∎

Definition 1.5.19. A complex series $\sum_{n=0}^{\infty} a_n$ is said to be **absolutely convergent** if the series $\sum_{n=0}^{\infty} |a_n|$ is convergent.

A well-known consequence of the completeness property of real numbers is that every bounded monotonic sequence (increasing or decreasing) converges. Since the partial sums of a series with nonnegative terms are increasing, we conclude that if these partial sums are bounded, then the series is convergent. Thus, if for a complex series we have $\sum_{n=1}^{N} |a_n| \leq M$ for all N, then the series $\sum_{n=1}^{\infty} a_n$ is absolutely convergent.

Recall that, for series with real terms, absolute convergence implies convergence. The same is true for complex series.

Theorem 1.5.20. *Absolutely convergent series are convergent, i.e., for $a_n \in \mathbb{C}$*

$$\sum_{n=0}^{\infty} |a_n| < \infty \quad \Rightarrow \quad \sum_{n=0}^{\infty} a_n \text{ converges}.$$

Proof. Let $s_n = a_0 + a_1 + \cdots + a_n$ and $v_n = |a_0| + |a_1| + \cdots + |a_n|$. By Theorem 1.5.11, it is enough to show that the sequence of partial sums $\{s_n\}_{n=0}^{\infty}$ is Cauchy. For $n > m \geq 0$, using the triangle inequality, we have

$$|s_n - s_m| = \left| \sum_{j=m+1}^{n} a_j \right| \leq \sum_{j=m+1}^{n} |a_j| = v_n - v_m.$$

Since $\sum_{n=0}^{\infty} |a_n|$ converges, the sequence $\{v_n\}_{n=0}^{\infty}$ converges and hence it is Cauchy. Thus, given $\varepsilon > 0$ we can find N so that, $v_n - v_m < \varepsilon$ for $n > m \geq N$, implying that $|s_n - s_m| < \varepsilon$ for $n > m \geq N$. Hence $\{s_n\}_{n=0}^{\infty}$ is a Cauchy sequence. ∎

For a complex series $\sum_{n=0}^{\infty} a_n$, consider the series $\sum_{n=0}^{\infty} |a_n|$ whose terms are real and nonnegative. If we can establish the convergence of the series $\sum_{n=0}^{\infty} |a_n|$ using one of the tests of convergence for series with nonnegative terms, then using Theorem 1.5.20, we can infer that the series $\sum_{n=0}^{\infty} a_n$ is convergent. Thus, all known tests of convergence for series with nonnegative terms can be used to test the (absolute) convergence of complex series. We list a few such convergence theorems.

Theorem 1.5.21. *Suppose that a_n are complex numbers, b_n are real numbers, $|a_n| \leq b_n$ for all $n \geq n_0$, and $\sum_{n=0}^{\infty} b_n$ is convergent. Then $\sum_{n=0}^{\infty} a_n$ is absolutely convergent.*

Proof. By the comparison test for real series, we have that $\sum_{n=0}^{\infty} |a_n|$ is convergent. By Theorem 1.5.20, it follows that $\sum_{n=0}^{\infty} a_n$ is convergent. ∎

Here is a simple application of the comparison test, which illustrates the passage from complex to real series in establishing the convergence of complex series.

Example 1.5.22. (Comparison test) The series $\sum_{n=0}^{\infty} \frac{2\cos(n\theta)+2i\sin(n\theta)}{n^2+3}$ is convergent by comparison to the convergent series $\sum_{n=1}^{\infty} \frac{2}{n^2}$, because

$$\left| \frac{2\cos(n\theta)+2i\sin(n\theta)}{n^2+3} \right| \leq \frac{2\left|\cos(n\theta)+i\sin(n\theta)\right|}{n^2} = \frac{2}{n^2}. \qquad \square$$

Theorem 1.5.23. (Ratio Test) *Let a_n be nonzero complex numbers and suppose that*

$$\rho = \lim_{n\to\infty} \left| \frac{a_{n+1}}{a_n} \right| \tag{1.5.5}$$

exists or is infinite. Then the complex series $\sum_{n=0}^{\infty} a_n$ converges absolutely if $\rho < 1$ and diverges if $\rho > 1$. If $\rho = 1$ the test is inconclusive.

Example 1.5.24. (Ratio test and the exponential series) The series $\sum_{n=0}^{\infty} \frac{z^n}{n!}$ converges absolutely for all z. The series is obviously convergent if $z = 0$. For $z \neq 0$,

$$\rho = \lim_{n\to\infty} \left| \frac{a_{n+1}}{a_n} \right| = \lim_{n\to\infty} \left| \frac{z^{n+1}\, n!}{z^n\,(n+1)!} \right| = \lim_{n\to\infty} \frac{|z|}{n+1} = 0.$$

Since $\rho < 1$, the series is absolutely convergent by the ratio test, hence it is convergent. $\qquad \square$

Theorem 1.5.25. (Root Test) *Let a_n be complex numbers and suppose that*

$$\rho = \lim_{n\to\infty} |a_n|^{1/n} \tag{1.5.6}$$

either exists or is infinite. Then the complex series $\sum_{n=0}^{\infty} a_n$ converges absolutely if $\rho < 1$ and diverges if $\rho > 1$. If $\rho = 1$ the test is inconclusive.

In general, the ratio test is easier to apply than the root test. But there are situations that call naturally for the root test. Here is an example.

Example 1.5.26. Test the series $\sum_{n=0}^{\infty} \frac{z^n}{(n+1)^n}$ for convergence.

Solution. The presence of the exponent n in the terms suggests using the root test. We have

$$\rho = \lim_{n\to\infty} \left| \frac{z^n}{(n+1)^n} \right|^{\frac{1}{n}} = \lim_{n\to\infty} \frac{|z|}{n+1} = 0.$$

Since $\rho < 1$, the series is absolutely convergent for all z. $\qquad \square$

Definition 1.5.27. The product $\sum_{n=0}^{\infty} c_n$ of two series $\sum_{n=0}^{\infty} a_n$ and $\sum_{n=0}^{\infty} b_n$ is defined as the series with coefficients

$$c_n = a_0 b_n + a_1 b_{n-1} + \cdots + a_{n-1} b_1 + a_n b_0 = \sum_{j=0}^{n} a_j b_{n-j}. \tag{1.5.7}$$

The series $\sum_{n=0}^{\infty} c_n$ is called the **Cauchy product** of $\sum_{n=0}^{\infty} a_n$ and $\sum_{n=0}^{\infty} b_n$.

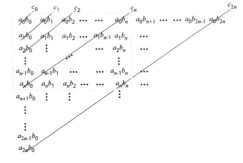

To better understand this definition, imagine we are able to cross multiply all the terms of the series $\sum_{n=0}^{\infty} a_n$ by those of $\sum_{n=0}^{\infty} b_n$. We get terms of the form $a_j b_k$, where j and k range over $0, 1, 2, \ldots$. We list the terms $a_j b_k$ in an array as shown in Figure 1.37.
The nth term of a Cauchy product:
$a_0 b_n + a_1 b_{n-1} + \cdots + a_{n-1} b_1 + a_n b_0$
$= \sum_{j=0}^{n} a_j b_{n-j} = c_n$. By summing all the c_n's, we pick up all the terms of form $a_j b_k$, but in a very special order: $(a_0 b_0) + (a_1 b_0 + a_1 b_1) + \cdots$.

Fig. 1.37 In the array $a_j b_k$, the c_n's is obtained by summing the terms along the shown slanted lines.

The term c_n in the Cauchy product is obtained by summing terms $a_j b_k$ along the diagonal $j + k = n$, as shown in Figure 1.37. If we sum over all the diagonals, as prescribed by the Cauchy product, we eventually collect all the terms $a_j b_k$. Does the Cauchy product series converge to the ordinary product of the two series (where we just multiply two complex numbers)? If the two series are absolutely convergent, the answer is yes.

Theorem 1.5.28. *Suppose that $\sum_{n=0}^{\infty} a_n$ and $\sum_{n=0}^{\infty} b_n$ are absolutely convergent complex series. Let c_n be as in (1.5.7). Then the series $\sum_{n=0}^{\infty} c_n$ is absolutely convergent and we have*

$$\sum_{n=0}^{\infty} c_n = \left(\sum_{n=0}^{\infty} a_n \right) \left(\sum_{n=0}^{\infty} b_n \right). \tag{1.5.8}$$

Proof. First we will show that the Cauchy product is absolutely convergent, then we will show that it converges to the right limit. For the first assertion, observe that

$$\sum_{k=0}^{n} |c_k| \leq \left(\sum_{k=0}^{n} |a_k| \right) \left(\sum_{k=0}^{n} |b_k| \right) \leq \left(\sum_{k=0}^{\infty} |a_k| \right) \left(\sum_{k=0}^{\infty} |b_k| \right) < \infty.$$

The first inequality follows because all the terms on the left are on or above the diagonal in a $(n+1) \times (n+1)$-array of nonnegative numbers, while the terms on the right are all the terms in the $(n+1) \times (n+1)$-array (see Figure 1.37). The second inequality follows since for series with nonnegative terms, a partial sum is smaller than the sum of all terms. Hence the partial sums of $\sum_{n=0}^{\infty} |c_n|$ are bounded, and so the series converges in view of Theorem 1.5.20. Next we show that $s_n = \sum_{k=0}^{n} c_k$

converges to $\left(\sum_{n=0}^{\infty} a_n\right)\left(\sum_{n=0}^{\infty} b_n\right)$. We have already established that s_n converges, and so, it will be enough to prove that the subsequence $\{s_{2n}\}$ converges to this limit. We have

$$\left|\sum_{k=0}^{2n} c_k - \left(\sum_{k=0}^{n} a_k\right)\left(\sum_{k=0}^{n} b_k\right)\right| \leq \sum_{k=0}^{\infty} |a_k| \sum_{k=n+1}^{\infty} |b_k| + \sum_{k=0}^{\infty} |b_k| \sum_{k=n+1}^{\infty} |a_k|. \quad (1.5.9)$$

To see this, notice that in Figure 1.37, $\sum_{k=0}^{2n} c_k$ is the sum of all terms along the slanted lines while $\left(\sum_{k=0}^{n} a_k\right)\left(\sum_{k=0}^{n} b_k\right)$ is the sum of all terms in the rectangular box. Letting $n \to \infty$ in (1.5.9) and using the fact that the tails of the absolutely convergent series $\sum_{k=n+1}^{\infty} |b_k|$ and $\sum_{k=n+1}^{\infty} |a_k|$ tend to zero (Proposition 1.5.18), we see that the right side tends to 0 as $n \to \infty$, and this implies (1.5.8). ∎

Exercises 1.5

In Exercises 1–6, determine whether or not the sequence $\{a_n\}_{n=1}^{\infty}$ converges, and find its limit if it does converge.

1. $a_n = \dfrac{i\sin(n\frac{\pi}{2})}{n}$.

2. $a_n = \dfrac{\cos(n\frac{\pi}{2}) + in}{n}$.

3. $a_n = \dfrac{1}{n+i}$.

4. $a_n = \dfrac{\cos(\ln n) - i}{\sqrt{n}}$.

5. $a_n = \dfrac{\cos n - in}{n^2}$.

6. $a_n = \dfrac{(1+2i)n^2 + 2n - 1}{3in^2 + i}$.

7. Evaluate the limit of the sequence $\dfrac{4^n + i5^n + (2-i)6^{-n}}{5^{n+1} - 2i4^n + 1}$ as $n \to \infty$.

8. Using the result of Example 1.5.9 show that, for θ not an integer multiple of π, $\lim_{n\to\infty} \cos n\theta$ and $\lim_{n\to\infty} \sin n\theta$ do not exist. [*Hint*: Use the addition formula for the cosine to show that, for $\theta \neq k\pi$, $\lim_{n\to\infty} \cos n\theta$ exists if and only if $\lim_{n\to\infty} \sin n\theta$ exists. Then use Theorem 1.5.8.] What happens when θ is an even multiple of π or an odd multiple of π?

9. (a) Show that $\lim_{n\to\infty} a_n = \lim_{n\to\infty} a_{n+1}$ for a convergent sequence $\{a_n\}$.
(b) Define $a_1 = i$ and $a_{n+1} = \frac{3}{2+a_n}$. Suppose that $\{a_n\}$ is convergent and find its limit.

10. (a) Let $a_n = n^{\frac{1}{n}} - 1$. Use the binomial expansion to show that for $n > 1$

$$0 < \frac{n(n-1)}{2} a_n^2 < (1+a_n)^n = n.$$

(b) Conclude that $\lim_{n\to\infty} a_n = 0$.
(c) Derive the useful limit: $\lim_{n\to\infty} n^{\frac{1}{n}} = 1$.

In Exercises 11–20, determine whether the series is convergent or divergent, and find its sum if it is convergent.

11. $\displaystyle\sum_{n=0}^{\infty} \frac{\cos(n\frac{\pi}{2}) + i\sin(n\frac{\pi}{2})}{3^n}$

12. $\displaystyle\sum_{n=0}^{\infty} \left(\frac{1+i}{2}\right)^n$

13. $\displaystyle\sum_{n=3}^{\infty} \frac{3-i}{(1+i)^n}$

14. $\displaystyle\sum_{n=0}^{\infty} \frac{\cos n\theta}{3^n}$

15. $\displaystyle\sum_{n=0}^{\infty} \frac{3+\sin n\theta}{10^n}$

16. $\displaystyle\sum_{n=0}^{\infty} \frac{\cos n\theta + (2i)^n}{3^n}$

17. $\displaystyle\sum_{n=2}^{\infty} \frac{1}{(n+i)((n-1)+i)}$

18. $\displaystyle\sum_{n=0}^{\infty} \frac{n^2}{(n+i)(n+200+2i)}$

19. $\displaystyle\sum_{n=2}^{\infty} \frac{e^{-n} - e^n}{2in^2}$

20. $\displaystyle\sum_{n=2}^{\infty} \frac{e^{-n} - e^n}{2ie^n}$

In Exercises 21–32, determine whether the series is convergent or divergent.

21. $\displaystyle\sum_{n=0}^{\infty}\left(\frac{1+3i}{4}\right)^n$

22. $\displaystyle\sum_{n=1}^{\infty}(-1)^n\frac{2^n+4^n}{(1+3i)^n}$

23. $\displaystyle\sum_{n=0}^{\infty}\frac{3i^n}{4+in^2}$

24. $\displaystyle\sum_{n=0}^{\infty}\left(\frac{3+10i}{4+5in}\right)^n$

25. $\displaystyle\sum_{n=1}^{\infty}\frac{(1+2in)^n}{n^n}$

26. $\displaystyle\sum_{n=1}^{\infty}\frac{\left(\frac{2+i}{2-i}\right)^n}{n^2}$

27. $\displaystyle\sum_{n=1}^{\infty}\mathrm{Re}\left[\left(\cos(\tfrac{1}{n^3})+i\sin(\tfrac{1}{n^3})\right)^n\right]$

28. $\displaystyle\sum_{n=1}^{\infty}\mathrm{Im}\left[\left(\cos(\tfrac{1}{n^3})+i\sin(\tfrac{1}{n^3})\right)^n\right]$

29. $\displaystyle\sum_{n=1}^{\infty}\frac{e^n-ie^{-n}}{e^{n^2}}$

30. $\displaystyle\sum_{n=1}^{\infty}\frac{(3+10i)n^n}{n!}$

31. $\displaystyle\sum_{n=0}^{\infty}\frac{(2+3i)^n}{n!}$

32. $\displaystyle\sum_{n=1}^{\infty}\frac{1}{3+i^n}$

In Exercises 33–40, use the geometric series to determine the largest region in which the series converges and find the value of the infinite sum.

33. $\displaystyle\sum_{n=0}^{\infty}\frac{z^n}{2^n}$

34. $\displaystyle\sum_{n=1}^{\infty}(1+z)^n$

35. $\displaystyle\sum_{n=0}^{\infty}\left(\frac{(3+i)z}{4-i}\right)^n$

36. $\displaystyle\sum_{n=0}^{\infty}\frac{(2+i)^n}{z^n}$

37. $\displaystyle\sum_{n=1}^{\infty}\frac{1}{(2-10z)^n}$

38. $\displaystyle\sum_{n=0}^{\infty}\frac{2^{n+1}}{(2+i-z)^n}$

39. $\displaystyle\sum_{n=0}^{\infty}\left\{\left(\frac{2}{z}\right)^n+\left(\frac{z}{3}\right)^n\right\}$

40. $\displaystyle\sum_{n=0}^{\infty}\left\{\frac{1}{(1-z)^n}-z^n\right\}$

41. The nth partial sum of a series is $s_n=\frac{i}{n}$. Does the series converge or diverge? If it does converge, what is its limit?

42. Show that if $\sum_{n=0}^{\infty}a_n$ is absolutely convergent, then $|\sum_{n=0}^{\infty}a_n|\le\sum_{n=0}^{\infty}|a_n|$.

43. Let $t>0$ and x be real numbers. Find the sum $\sum_{n=0}^{\infty}e^{-nt}\cos nx$. [*Hint*: Proceed as in Example 1.5.16.]

44. (a) The nth term of a series is $1-\frac{1}{n}$. Does the series converge?
(b) The nth partial sum of a series is $1+\frac{1}{n}$. Does the series converge?

45. The terms of a series are defined recursively by

$$a_1=2+i,\qquad a_{n+1}=\frac{(7+3i)n}{1+2in^2}a_n.$$

Does the series $\sum_{n=1}^{\infty}a_n$ converge or diverge?

46. The terms of a series are defined recursively by

$$a_1=i,\qquad a_{n+1}=\frac{\cos(\frac{1}{n})+i\sin(\frac{1}{n})}{\sqrt{n}}a_n.$$

Does the series $\sum_{n=1}^{\infty}a_n$ converge or diverge?

1.6 The Complex Exponential

For a real number x, we recall how to express e^x in a series as follows:

$$e^x=1+\frac{x}{1!}+\frac{x^2}{2!}+\frac{x^3}{3!}+\cdots\qquad(-\infty<x<\infty).\qquad(1.6.1)$$

Looking for an extension of the exponential function e^x ($x \in \mathbb{R}$) to the complex plane, we wonder whether we are able to substitute x with a complex number z in (1.6.1). Note that the series in (1.6.1) converges absolutely by an easy application of the ratio test (Example 1.5.24). Consequently, for a complex z we have

$$\sum_{n=0}^{\infty} \frac{|z|^n}{n!} = 1 + \frac{|z|}{1!} + \frac{|z|^2}{2!} + \frac{|z|^3}{3!} + \cdots = e^{|z|} < \infty.$$

Thus the series of complex numbers $\sum_{n=0}^{\infty} \frac{z^n}{n!}$ converges absolutely and therefore it converges by Theorem 1.5.20.

Definition 1.6.1. We define the **complex exponential function** $\exp(z)$ or e^z as the convergent series

$$e^z = \sum_{n=0}^{\infty} \frac{z^n}{n!} = 1 + \frac{z}{1!} + \frac{z^2}{2!} + \frac{z^3}{3!} + \cdots \quad \text{for all } z \in \mathbb{C}. \tag{1.6.2}$$

We discuss some fundamental properties of the complex exponential function.

Theorem 1.6.2. *Let z and w be arbitrary complex numbers. We have that*

$$e^{z+w} = e^z e^w \tag{1.6.3}$$

Moreover, e^z is never zero and

$$e^{-z} = \frac{1}{e^z} \tag{1.6.4}$$

$$e^{z-w} = \frac{e^z}{e^w} \tag{1.6.5}$$

Proof. We have $e^z = \sum_{n=0}^{\infty} \frac{z^n}{n!}$ and $e^w = \sum_{n=0}^{\infty} \frac{w^n}{n!}$, where both series converge absolutely. Applying Theorem 1.5.28 we obtain

$$\left(\sum_{k=0}^{\infty} \frac{z^k}{k!} \right) \left(\sum_{m=0}^{\infty} \frac{w^m}{m!} \right) = \sum_{n=0}^{\infty} c_n \tag{1.6.6}$$

and c_n is defined in (1.5.7) by

$$c_n = \sum_{j=0}^{n} \frac{z^j}{j!} \frac{w^{n-j}}{(n-j)!} = \frac{1}{n!} \overbrace{\sum_{j=0}^{n} \frac{n!}{j!(n-j)!} z^j w^{n-j}}^{(z+w)^n} = \frac{(z+w)^n}{n!}, \tag{1.6.7}$$

where the last equality is a consequence of the binomial identity (Exercise 65 in Section 1.3). Now the left-hand side of (1.6.6) is $e^z e^w$, but, in view of (1.6.7), the right-hand side is e^{z+w}, hence (1.6.3) holds.

Now (1.6.4) is a consequence of (1.6.3) and of the fact that $e^0 = 1$ since

$$1 = e^{z-z} = e^z e^{-z}.$$

It follows from this that e^z is never zero and that its reciprocal is e^{-z}. Finally (1.6.5) is obtained from (1.6.3) and (1.6.4) by noticing that

$$e^{z-w} = e^z e^{-w} = e^z \frac{1}{e^w} = \frac{e^z}{e^w}.$$

This concludes the proof of the assertions.   ∎

Note that the exponential function reduces to the familiar function e^x when z in (1.6.2) is a real number x. Now let us compute e^z when z is purely imaginary. The result is an important identity.

Proposition 1.6.3. *If $z = i\theta$, where θ is real, then*

$$e^{i\theta} = \cos\theta + i\sin\theta \tag{1.6.8}$$

This is known as **Euler's identity**.

Proof. We use the definition (1.6.2) with $z = i\theta$ and get

$$e^{i\theta} = 1 + i\theta + \frac{(i\theta)^2}{2!} + \frac{(i\theta)^3}{3!} + \frac{(i\theta)^4}{4!} + \frac{(i\theta)^5}{5!} + \cdots$$

$$= 1 + i\theta - \frac{\theta^2}{2!} - i\frac{\theta^3}{3!} + \frac{\theta^4}{4!} + i\frac{\theta^5}{5!} + \cdots$$

$$= \left(1 - \frac{\theta^2}{2!} + \frac{\theta^4}{4!} - \cdots\right) + i\left(\theta - \frac{\theta^3}{3!} + \frac{\theta^5}{5!} - \cdots\right)$$

$$= \cos\theta + i\sin\theta,$$

where in the last step we have recognized two familiar power series expansions from calculus:

$$\cos\theta = \sum_{n=0}^{\infty} \frac{(-1)^n \theta^{2n}}{(2n)!}$$

and

$$\sin\theta = \sum_{n=0}^{\infty} \frac{(-1)^n \theta^{2n+1}}{(2n+1)!},$$

which converge for all real θ.   ∎

Euler's identity has many important applications that we will explore in this text. Let us use it to express e^z in terms of familiar functions. Write $z = x + iy$, where x and y are real. By (1.6.3), we have

$$e^z = e^{x+iy} = e^x e^{iy},$$

and, by Euler's identity, $e^{iy} = \cos y + i\sin y$. Putting this together, we obtain the desired expression of e^z in terms of the basic functions: e^x, $\cos x$, and $\sin x$.

Corollary 1.6.4. *For* $z = x + iy$, *with* x, y *real, we have*

$$e^z = e^x(\cos y + i \sin y) = e^x \cos y + i e^x \sin y. \tag{1.6.9}$$

Taking real and imaginary parts in (1.6.9), we find

$$\text{Re}\,(e^z) = e^x \cos y \qquad \text{and} \qquad \text{Im}\,(e^z) = e^x \sin y. \tag{1.6.10}$$

Example 1.6.5. Compute e^z for the following values of z.

(a) $2 + i\pi$ (b) $3 - i\frac{\pi}{3}$ (c) $-1 + i\frac{\pi}{2}$ (d) $i\frac{5\pi}{4}$ (e) $2 + 3\pi i$

Solution. (a) Using (1.6.9) we write

$$e^{2+i\pi} = e^2(\cos \pi + i \sin \pi) = -e^2.$$

This number is purely real and negative.

(b) From (1.6.9),

$$e^{3-i\frac{\pi}{3}} = e^3 \left(\cos \frac{\pi}{3} + i \sin \left(-\frac{\pi}{3} \right) \right) = e^3 \left(\frac{1}{2} - i \frac{\sqrt{3}}{2} \right).$$

(c) From (1.6.9),
$$e^{-1+i\frac{\pi}{2}} = e^{-1} \left(\cos \frac{\pi}{2} + i \sin \frac{\pi}{2} \right) = \frac{i}{e}.$$

This number is imaginary.

(d) Here z is purely imaginary; we may use Euler's identity. From (1.6.8),

$$e^{i\frac{5\pi}{4}} = \cos \frac{5\pi}{4} + i \sin \frac{5\pi}{4} = -\frac{\sqrt{2}}{2} - i \frac{\sqrt{2}}{2}.$$

This number is unimodular.

(e) From (1.6.9),

$$e^{2+3\pi i} = e^2(\cos(3\pi) + i \sin(3\pi)) = -e^2.$$

This value is the same as the one we found in (a) for $e^{2+\pi i}$.   □

Example 1.6.5 shows that the complex exponential function can take on negative real values (part (a)) and complex values (parts (b)–(d)) as opposed to the real exponential function, e^x, which is always positive. Also, in contrast with e^x, e^z is not one-to-one, as illustrated by parts (a) and (e) of Example 1.6.5.

Looking back at (1.6.9) and using the fact that $e^x > 0$ for all x, it follows immediately that (1.6.9) is the polar form, as in (1.3.3), of e^z.

Next, we make some remarks about the modulus and the argument of the exponential function. It follows from the polar form of the exponential that for $z = x + iy$, the modulus of e^z is e^x and the argument of e^z is y. This is displayed in Figure 1.38. Consequently, e^z is never zero. Moreover, $y + 2k\pi$ is also an argument of e^z for an integer k. We summarize these facts:

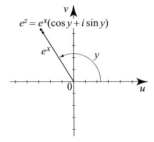

Fig. 1.38 The modulus and argument of e^z.

$$|e^z| = e^x > 0 \qquad (1.6.11)$$

and the argument of e^z is

$$\arg(e^z) = y + 2k\pi \qquad (k \text{ an integer}). \qquad (1.6.12)$$

Example 1.6.6. Compute $|e^z|$ and $\text{Arg}\,(e^z)$ for the following values of z.

(a) $1 + i$ (b) $-1 - 6i$ (c) $i\frac{\pi}{2}$ (d) $-\pi$

Solution. In all cases, we use (1.6.11) to compute $|e^z|$. We use (1.6.12) to compute $\arg(e^z)$ and then, by adding an integer multiple of 2π, we find the value of $\arg(e^z)$ that lies in the interval $(-\pi, \pi]$, which is $\text{Arg}\,(e^z)$

(a) $z = 1 + i$, $e^z = e^{1+i} = ee^i = e(\cos 1 + i \sin 1)$. Thus $|e^{1+i}| = e$, $\arg(e^{1+i}) = 1 + 2k\pi$. Since 1 is in the interval $(-\pi, \pi]$, we also have $\text{Arg}\,(e^{1+i}) = 1$.

(b) $z = -1 - 6i$, $e^z = e^{-1-6i} = e^{-1}e^{-6i} = e(\cos(-6) + i\sin(-6))$. Thus $|e^{-1-6i}| = e^{-1}$, $\arg(e^{-1-6i}) = -6 + 2k\pi$. Now -6 is not in the interval $(-\pi, \pi]$, but $-6 + \pi$ is in this interval; so $\text{Arg}\,(e^{-1-6i}) = -6 + 2\pi$.

(c) $z = i\frac{\pi}{2}$, $e^z = e^{i\frac{\pi}{2}} = \cos(\frac{\pi}{2}) + i\sin(\frac{\pi}{2})) = i$. Thus $|e^{i\frac{\pi}{2}}| = e^0 = 1$, $\arg(e^{i\frac{\pi}{2}}) = \frac{\pi}{2} + 2k\pi$; $\text{Arg}\,(e^{i\frac{\pi}{2}}) = \frac{\pi}{2}$.

(d) In this case, z is real and so e^z is real and positive: $e^z = e^{-\pi} > 0$. Thus $|e^{-\pi}| = e^{-\pi}$; $\arg(e^{-\pi}) = 0 + 2k\pi$; and so $\text{Arg}\,(e^{-\pi}) = 0$.   □

We observed following Example 1.6.5 that e^z is not one-to-one. In fact, let us call a complex-valued function $f(z)$ **periodic** with period $\tau \neq 0$ if for all z in the domain of definition of f, we have $f(z + \tau) = f(z)$. For all complex numbers z,

$$e^{z+2\pi i} = e^z e^{2\pi i} = e^z(\cos(2\pi) + i\sin(2\pi)) = e^z. \qquad (1.6.13)$$

Hence the exponential function e^z is periodic with period $2\pi i$.

Proposition 1.6.7. *We have*

$$e^z = 1 \qquad \text{if and only if} \qquad z = 2k\pi i \text{ for some integer } k. \qquad (1.6.14)$$

Also,

$$e^{z_1} = e^{z_2} \qquad \text{if and only if} \qquad z_1 = z_2 + 2k\pi i \text{ for some integer } k. \qquad (1.6.15)$$

Proof. Since $e^z = e^x \cos y + i e^x \sin y$, equating real and imaginary parts, we find

$$1 = e^z \Leftrightarrow 1 = e^x \cos y + i e^x \sin y \Leftrightarrow 1 = e^x \cos y \text{ and } e^x \sin y = 0.$$

Since $e^x > 0$, if we have $e^x \sin y = 0$, then $\sin y = 0$ which implies that $y = l\pi$ for some integer l. Hence $\cos y = \cos(l\pi) = (-1)^l$ and so $1 = e^x \cos y = e^x(-1)^l$. It follows that l must be an even integer, say $l = 2k$. Then $1 = e^x(-1)^{2k}$, hence $x = 0$. This establishes the difficult direction in (1.6.14). To prove (1.6.15), notice that $e^{z_1} = e^{z_2} \Leftrightarrow e^{z_1 - z_2} = 1$, which by (1.6.14) holds if and only if $z_1 - z_2 = 2k\pi i$ or $z_1 = z_2 + 2k\pi i$ for some integer k. ∎

Exponential and Polar Representations

Euler's identity (1.6.8) provides yet another convenient way to represent complex numbers. Indeed, if $z = r(\cos\theta + i\sin\theta)$ is a complex number in polar form, then, since $e^{i\theta} = \cos\theta + i\sin\theta$, we obtain the **exponential representation** or polar form $z = re^{i\theta}$. From this representation it is clear that

$$|z| = 1 \Leftrightarrow r = 1 \Leftrightarrow z = e^{i\theta}, \qquad \arg z = \theta.$$

Thus the complex numbers $e^{i\theta}$, θ real, are exactly the unimodular complex numbers. Because their distance to the origin always equals 1, all complex numbers of the form $e^{i\theta}$ lie on the unit circle. All other nonzero complex numbers are positive multiples of some $e^{i\theta}$. This fact is illustrated in Figure 1.39, where the ray from the origin to $z = re^{i\theta}$ intersects the unit circle at the point $e^{i\theta}$. Hence, to plot a point in exponential notation $z = re^{i\theta}$, we move a distance r along the ray extending from the origin to $e^{i\theta}$.

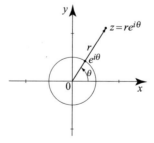

Fig. 1.39 Plotting $z = re^{i\theta}$.

Many operations on complex numbers are simplified by using the exponential notation, $z = re^{i\theta}$.

Proposition 1.6.8. (Exponential Representation) *Let $z = r(\cos\theta + i\sin\theta)$ with $r = |z| > 0$, θ real, and $\arg z = \theta + 2k\pi$. Then*

$$z = re^{i\theta} \tag{1.6.16}$$

$$\bar{z} = re^{-i\theta} \tag{1.6.17}$$

$$z^{-1} = \frac{1}{r}e^{-i\theta}. \tag{1.6.18}$$

If $z_1 = r_1 e^{i\theta_1}$ and $z_2 = r_2 e^{i\theta_2}$, then

$$z_1 z_2 = r_1 r_2 e^{i(\theta_1 + \theta_2)} \tag{1.6.19}$$

$$\frac{z_1}{z_2} = \frac{r_1}{r_2} e^{i(\theta_1 - \theta_2)} \qquad (z_2 \neq 0.) \tag{1.6.20}$$

Proof. The expression (1.6.16) is a consequence of Euler's identity (1.6.8). For (1.6.17), we have

$$\bar{z} = \overline{r(\cos\theta + i\sin\theta)} = r(\cos\theta - i\sin\theta) = r(\cos(-\theta) + i\sin(-\theta)) = re^{-i\theta}.$$

To prove (1.6.18), write $z = re^{i\theta}$ with $r \neq 0$. Then

$$z\frac{1}{r}e^{-i\theta} = re^{i\theta}\frac{1}{r}e^{-i\theta} = e^{i(\theta-\theta)} = e^0 = 1.$$

The proofs of (1.6.19) and (1.6.20) are immediate from (1.6.3) and (1.6.18). ∎

Multiplication and division of unimodular numbers are particularly easy via the complex exponential. Indeed, if $z_1 = e^{i\theta_1}$ and $z_2 = e^{i\theta_2}$, then from (1.6.19) and (1.6.20) we obtain

$$z_1 z_2 = e^{i\theta_1} e^{i\theta_2} = e^{i(\theta_1 + \theta_2)}$$

$$\frac{z_1}{z_2} = \frac{e^{i\theta_1}}{e^{i\theta_2}} = e^{i(\theta_1 - \theta_2)}.$$

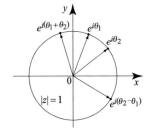

Fig. 1.40 Multiplication and division of unimodular numbers.

Thus to multiply two unimodular numbers we add their arguments, and to divide them we subtract their arguments.

Example 1.6.9. (Exponential representation) Let $z_1 = -7\sqrt{3} + 7i$ and $z_2 = 1 + i$. Find the exponential form of the expressions below. Additionally, provide your answer in Cartesian form in (d) and (e).

(a) z_1 (b) z_2 (c) $z_1 z_2$ (d) $\dfrac{1}{z_2}$ (e) $\dfrac{z_1}{z_2}$

Solution. (a) We will use (1.6.16). We have $|z_1| = \sqrt{(-7\sqrt{3})^2 + 7^2} = 14$. To compute a value of $\arg z_1$, we appeal to (1.3.13). Since z_1 is in the second quadrant, we have $\operatorname{Arg} z_1 = \tan^{-1}(\frac{7}{-7\sqrt{3}}) + \pi = \frac{5\pi}{6}$. Hence

$$z_1 = 14\left(\cos\frac{5\pi}{6} + i\sin\frac{5\pi}{6}\right) = 14e^{i\frac{5\pi}{6}}.$$

(b) Following the steps in (a), we find $|z_2| = \sqrt{1^2 + 1^2} = \sqrt{2}$ and $\operatorname{Arg} z_2 = \tan^{-1}(1) = \frac{\pi}{4}$. Hence

$$z_2 = \sqrt{2}\left(\cos\frac{\pi}{4} + i\sin\frac{\pi}{4}\right) = \sqrt{2}e^{i\frac{\pi}{4}}.$$

(c) In view of (a), (b), and (1.6.19) we obtain

$$z_1 z_2 = 14e^{i\frac{5\pi}{6}}\sqrt{2}e^{i\frac{\pi}{4}} = 14\sqrt{2}e^{i(\frac{5\pi}{6} + \frac{\pi}{4})} = 14\sqrt{2}e^{i\frac{13\pi}{12}}.$$

(d) We use (b) and (1.6.18) and get

$$\frac{1}{z_2} = z_2^{-1} = \frac{1}{\sqrt{2}}e^{-i\frac{\pi}{4}}.$$

This is the polar form of $\frac{1}{z_2}$. To obtain the Cartesian form, we use Euler's identity:

$$\frac{1}{z_2} = \frac{1}{\sqrt{2}}\left(\cos\frac{-\pi}{4} + i\sin\frac{-\pi}{4}\right) = \frac{1}{\sqrt{2}}\left(\frac{\sqrt{2}}{2} - i\frac{\sqrt{2}}{2}\right) = \frac{1}{2} - i\frac{1}{2}.$$

Checking our answer, we have

$$z_2\frac{1}{z_2} = (1+i)\left(\frac{1}{2} - i\frac{1}{2}\right) = \left(\frac{1}{2} + \frac{1}{2}\right) + i\left(\frac{1}{2} - \frac{1}{2}\right) = 1,$$

as it should be.

(e) We use (a) and (b) and (1.6.20) and get

$$\frac{z_1}{z_2} = \frac{14}{\sqrt{2}}e^{i(\frac{5\pi}{6} - \frac{\pi}{4})} = 7\sqrt{2}e^{i\frac{7\pi}{12}}.$$

To deduce the Cartesian form, we use Euler's identity:

$$\frac{z_1}{z_2} = 7\sqrt{2}\left(\cos\frac{7\pi}{12} + i\sin\frac{7\pi}{12}\right)$$
$$= 7\sqrt{2}\left(\frac{1-\sqrt{3}}{2\sqrt{2}} + i\frac{1+\sqrt{3}}{2\sqrt{2}}\right) = \frac{7}{2}\left((1-\sqrt{3}) + i(1+\sqrt{3})\right),$$

where we have used the subtraction formulas for the cosine and sine to compute the exact values of $\cos\frac{7\pi}{12}$ and $\sin\frac{7\pi}{12}$. For example,

$$\cos\frac{7\pi}{12} = \cos\left(\frac{5\pi}{6} - \frac{\pi}{4}\right) = \cos\frac{5\pi}{6}\cos\frac{\pi}{4} + \sin\frac{5\pi}{6}\sin\frac{\pi}{4}$$

$$= -\frac{\sqrt{3}}{2}\frac{\sqrt{2}}{2} + \frac{1}{2}\frac{\sqrt{2}}{2} = \frac{1-\sqrt{3}}{2\sqrt{2}}.$$

The value of $\sin\frac{7\pi}{12} = \frac{1+\sqrt{3}}{2\sqrt{2}}$ can be derived similarly.   □

Example 1.6.10. Solve the equation $e^z = 1+i$.

Solution. This problem is asking us to find the inverse image of $1+i$ by the mapping e^z. We know from (1.6.15) that we have infinitely many solutions, all differing by $2k\pi i$. Write $z = x+iy$, $e^z = e^x e^{iy}$, and use the exponential representation of $1+i$ from the previous example; then

$$e^z = 1+i \Leftrightarrow e^x e^{iy} = \sqrt{2}e^{i\frac{\pi}{4}}$$

$$\Leftrightarrow e^x = \sqrt{2} \text{ and } e^{iy} = e^{i\frac{\pi}{4}}$$

$$\Leftrightarrow x = \ln(\sqrt{2}) = \frac{1}{2}\ln 2 \text{ and } y = \frac{\pi}{4} + 2k\pi.$$

Thus the solutions to the equation $e^z = 1+i$ are

$$z = \frac{1}{2}\ln 2 + i\left(\frac{\pi}{4} + 2k\pi\right) = \left(\frac{1}{2}\ln 2 + \frac{\pi}{4}i\right) + 2k\pi i,$$

where k is an integer. As expected, any two solutions differ by $2k\pi i$, $k \in \mathbb{Z}$.   □

The Exponential as a Mapping

Equation (1.6.12) has as a consequence that the argument of e^z is equal to the imaginary part of z. Because of this property, we expect the exponential function to map line segments to circular arcs and rectangular regions to circular regions. We illustrate these ideas with an example.

Example 1.6.11. (An exponential mapping) Consider the rectangular area

$$S = \{z = x+iy : -1 \le x \le 1, \ 0 \le y \le \pi\}.$$

Find the image of S under the mapping $f(z) = e^z$.

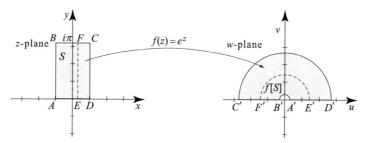

Fig. 1.41 We denote the image of a point P in the xy-plane by the point P' in the uv-plane. The mapping $w = e^z$ takes the vertical line segment EF to a semicircle in the uv-plane with u-intercepts at E' and F'.

Solution. Fix x_0 in the interval $[-1, 1]$ and consider EF, the vertical line segment $x = x_0$ inside S. Points on the segment EF are of the form $z = x_0 + iy$, where $0 \leq y \leq \pi$. For such z,

$$f(z) = e^z = e^{x_0} e^{iy} = e^{x_0} (\cos y + i \sin y).$$

The point $w = e^{x_0} e^{iy}$ has modulus e^{x_0} and argument y. In particular, w lies on the circle of radius e^{x_0} with center at 0. As y varies from 0 to π, the point w traces the upper semicircle. Thus e^z maps the line segment EF to the upper semicircle with center at 0 and radius e^{x_0}.

Now, as we vary x_0 from -1 to 1, e^{x_0} varies from e^{-1} to e. As a consequence, the corresponding semicircles increase in radius and fill the semi-annular area between the semicircle of radius e^{-1} with center at 0 and the semicircle of radius e with center at 0 (Figure 1.41). $\qquad \square$

Exercises 1.6

Write the expressions in the form $a + bi$, where a and b are real numbers.

1. (a) $e^{i\pi}$ (b) $e^{2i\pi}$ (c) $3e^{-1+200i\pi}$ (d) $\dfrac{e^{\ln(3)+201\,i\frac{\pi}{2}}}{3}$

2. (a) $e^{i\frac{3\pi}{4}}$ (b) $e^{2-i\frac{\pi}{4}}$ (c) $e^{-1-i\frac{\pi}{6}}$ (d) $-2e^{i+\pi}$

3. (a) $3e^{3+i\frac{\pi}{2}}$ (b) $e^{-701\,i\frac{\pi}{4}}$ (c) $-ie^{-i\frac{\pi}{3}}$ (d) $e^{e^{i\frac{\pi}{4}}}$

4. (a) $\dfrac{e^{-1-i}}{2i}$ (b) $e^{-\ln(7)+i\frac{\pi}{4}}$ (c) $(1-i)e^{-i\frac{\pi}{2}}$ (d) $e^{2-e^{i\frac{\pi}{3}}}$

Write the expressions in the form $e^{i\alpha}$, where α is a real number. Take θ to be real in all occurrences.

5. (a) $\cos\theta - i\sin\theta$ (b) $\sin\theta + i\cos\theta$

    (c) $\dfrac{1}{\cos\theta + i\sin\theta}$ (d) $\dfrac{\cos\theta - i\sin\theta}{\cos(3\theta) + i\sin(3\theta)}$

6. (a) $\sin 2\theta - i\cos 2\theta$ (b) $(\cos\theta - i\sin\theta)^8$

    (c) $\dfrac{1}{2}\cos\theta + i\dfrac{\sqrt{3}}{2}\sin\theta$ (d) $\dfrac{1}{\cos\theta - i\sin\theta}$

7. (a) $\dfrac{\sqrt{2}}{2}\cos\theta - i\dfrac{\sqrt{2}}{2}\sin\theta$ (b) $(\cos\theta + i\sin\theta)(\cos 2\theta - i\sin 2\theta)$

(c) $-i$ (d) $\dfrac{\cos\theta + i\sin\theta}{\cos\theta - i\sin\theta}$

8. (a) 1   (b) $(\sin\theta - i\cos\theta)^{11}$

(c) $\dfrac{1}{\left(\frac{\sqrt{2}}{2}\cos\theta - i\frac{\sqrt{2}}{2}\sin\theta\right)^3}$ (d) $\left(\dfrac{\sqrt{3}}{2}\cos\theta - i\dfrac{1}{2}\sin\theta\right)^7$

Evaluate the following expressions and write your answer in the form $a + bi$, where a and b are real numbers. Take $z_1 = 1 + i$, $z_2 = 1 - i$, $z_3 = 2 + 5i$.

9. (a) e^{z_1} (b) $3ie^{z_2}$ (c) $e^{z_1}e^{z_2}$ (d) $\dfrac{e^{z_1}}{e^{z_2}}$

10. (a) $e^{z_1}e^{z_2}e^{z_3}$ (b) $\dfrac{1}{e^{z_1}}$ (c) $(e^{z_1}e^{z_2})^{10}$ (d) $\dfrac{e^{z_1} + e^{z_2}}{e^{z_3}}$

11. (a) $\operatorname{Arg}(e^{z_1})$ (b) $\operatorname{Arg}(e^{z_3})$ (c) $|e^{z_1}e^{z_2}|$ (d) $\left|\dfrac{e^{z_1}}{e^{z_2}}\right|$

12. (a) $\operatorname{Re}(e^{z_1})$ (b) $\operatorname{Im}(e^{z_1})$ (c) $\operatorname{Re}(e^{z_1}e^{z_2})$ (d) $\operatorname{Im}(e^{z_1}e^{z_2}e^{z_3})$

Write the complex numbers in the exponential form $re^{i\theta}$.

13. (a) $-3 - 3i$ (b) $-\dfrac{\sqrt{3}}{2} + \dfrac{i}{2}$ (c) $-1 - \sqrt{3}i$ (d) $-3e^{2i}$

14. (a) $-\dfrac{i}{2}$ (b) $\dfrac{1 + i}{1 + \sqrt{3}i}$ (c) $\dfrac{1 + i}{1 - i}$ (d) $\dfrac{i}{10 + 10i}$

15. Let $z = x + iy$ with x, y real numbers. Find the real and imaginary parts of the functions in terms of x, y.

(a) e^{3z} (b) e^{z^2} (c) $e^{\bar{z}}$ (d) e^{iz}

16. Let z be a complex number. Show that

(a) $(e^z)^n = e^{nz}, n = 0, \pm 1, \pm 2, \ldots$ (b) $\overline{e^z} = e^{\bar{z}}$ (c) $e^{z + i\pi} = -e^z$

In Exercises 17–22, show that the shaded area S in the z-plane is mapped to the shaded area in the w-plane by the corresponding mappings.

17.

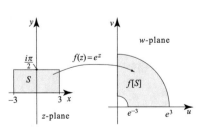

18.

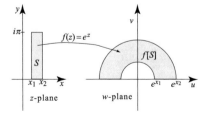

19.

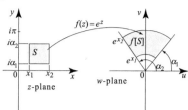

20.

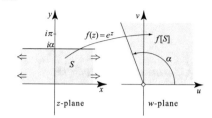

21.

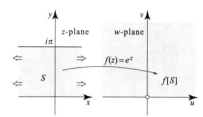

22.

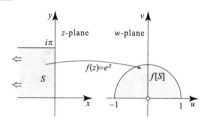

Consider the complex numbers in exponential form $z_1 = 2e^{i\frac{\pi}{6}}$, $z_2 = \frac{1}{2}e^{i\frac{\pi}{3}}$, *and* $z_3 = 2e^{-i\frac{\pi}{6}}$. *Plot the indicated complex numbers.*

23. (a) z_1, z_2, z_3 (b) $z_1 z_2$ **24.** (a) $\dfrac{z_1}{z_2}$ (b) $\dfrac{z_1}{z_3}$

Solve the equations.

25. (a) $e^z = 2 - 2i$ (b) $e^{2z} = i$ **26.** (a) $e^{-z+1} = -3 - 4i$ (b) $(1-i)e^z = 1+i$

27. Show that for all complex numbers z, we have $|e^z| \leq e^{|z|}$. When does equality hold?

28. In the text, we pointed out a few similarities and differences between e^z and e^x. In what follows we consider some additional ones. For each part, either prove your answer or provide an example to show that the statement is false.
(a) The function e^x is one-to-one. Is e^z one-to-one?
(b) The function e^x is increasing (if $x_1 < x_2$ then $e^{x_1} < e^{x_2}$). If $|z_1| < |z_2|$ do we have $|e^{z_1}| < |e^{z_2}|$?
(c) The function e^x never vanishes. Can e^z vanish?
(d) The function e^x is always positive. Is e^z always real and positive?
(e) The modulus or absolute value of e^x is e^x. What is the absolute value of e^z?
(f) We have $e^x = 1 \Leftrightarrow x = 0$. Do we have $e^z = 1 \Leftrightarrow z = 0$?

29. Show that $|e^z| \leq 1$ if and only if $\operatorname{Re} z \leq 0$. When does equality hold?

30. Let $w \neq 0$. Show that the equation $e^z = w$ has infinitely many solutions. [*Hint*: Proceed as in Example 1.6.10.]

1.7 Trigonometric and Hyperbolic Functions

In this section we extend the definitions of the trigonometric and hyperbolic functions from real numbers to complex numbers. As in the previous section, we could define these functions using convergent series expressions, but it also convenient to actually define them in terms of the complex exponential function.

Trigonometric Functions

To relate the trigonometric functions to the exponential function, recall Euler's identity

$$e^{i\theta} = \cos\theta + i\sin\theta$$

where θ is a real number. Replacing θ by $-\theta$ we obtain

$$e^{-i\theta} = \cos\theta - i\sin\theta.$$

Adding these two identities and dividing by 2, we deduce

$$\cos\theta = \frac{e^{i\theta} + e^{-i\theta}}{2}. \tag{1.7.1}$$

Similarly, subtracting and dividing by $2i$, we get

$$\sin\theta = \frac{e^{i\theta} - e^{-i\theta}}{2i}. \tag{1.7.2}$$

Since the right side of each identity involves the exponential function, which we have already extended to complex numbers, we may use (1.7.1) and (1.7.2) to define the complex cosine and sine functions for all complex numbers z.

Definition 1.7.1. For a complex number z, we set

$$\cos z = \frac{e^{iz} + e^{-iz}}{2} \tag{1.7.3}$$

and

$$\sin z = \frac{e^{iz} - e^{-iz}}{2i} \tag{1.7.4}$$

These are new functions, even though they are named after familiar functions; they share some similar properties with the usual cosine and sine functions but also differ in many respects.

Example 1.7.2. Compute (a) $\cos(2 + i\pi)$ and (b) $\sin(i\frac{5\pi}{4})$.

Solution. We use the definitions. For (a), we have from (1.7.3),

$$\cos(2+i\pi) = \frac{1}{2}\left(e^{i(2+i\pi)} + e^{-i(2+i\pi)}\right)$$

$$= \frac{1}{2}\left(e^{-\pi+2i} + e^{\pi-2i}\right) = \frac{1}{2}\left(e^{-\pi}e^{2i} + e^{\pi}e^{-2i}\right)$$

$$= \frac{1}{2}\left(e^{-\pi}[\cos(2) + i\sin(2)] + e^{\pi}[\cos(2) - i\sin(2)]\right)$$

$$= \cos(2)\frac{e^{\pi} + e^{-\pi}}{2} - i\sin(2)\frac{e^{\pi} - e^{-\pi}}{2}$$

$$= \cos(2)\cosh\pi - i\sin(2)\sinh\pi,$$

where $\cosh t = \frac{1}{2}(e^t + e^{-t})$ and $\sinh t = \frac{1}{2}(e^t - e^{-t})$ are the real hyperbolic functions. For (b), we use (1.7.4) and proceed in a similar way:

$$\sin\left(i\frac{5\pi}{4}\right) = \frac{1}{2i}\left(e^{i\left(i\frac{5\pi}{4}\right)} - e^{-i\left(i\frac{5\pi}{4}\right)}\right)$$

$$= \frac{-i}{2}\left(e^{-\frac{5\pi}{4}} - e^{\frac{5\pi}{4}}\right) = i\sinh\left(\frac{5\pi}{4}\right).$$

The appearance of the hyperbolic functions in the expressions of the real and imaginary parts of the complex cosine and sine functions was not a coincidence. In fact, general formulas of this nature are derived later in this section. □

A function $f(z)$ is said to be **even** if $f(z) = f(-z)$ and **odd** if $f(-z) = -f(z)$, for all z in the complex plane. We can show from their definitions that the cosine is even while the sine is odd; also, both of them are 2π-periodic. In fact, the complex trigonometric functions satisfy many identities that we are familiar with for real trigonometric functions.

Proposition 1.7.3. (Properties of Trigonometric Functions) *The following identities are valid for a complex number z:*

$$\cos(-z) = \cos z \qquad \sin(-z) = -\sin z \qquad (1.7.5)$$

$$\cos(z+2\pi) = \cos z \qquad \sin(z+2\pi) = \sin z \qquad (1.7.6)$$

$$\sin(z+\tfrac{\pi}{2}) = \cos z \qquad \cos(z+\tfrac{\pi}{2}) = -\sin z \qquad (1.7.7)$$

$$e^{iz} = \cos z + i\sin z \qquad (1.7.8)$$

$$\cos^2 z + \sin^2 z = 1. \qquad (1.7.9)$$

Proof. For the first identity in (1.7.5) we appeal to (1.7.3):

$$\cos(-z) = \frac{e^{i(-z)} + e^{-i(-z)}}{2} = \frac{e^{-iz} + e^{iz}}{2} = \frac{e^{iz} + e^{-iz}}{2} = \cos z.$$

In proving the first identity in (1.7.6), we use the fact that $e^{\pm 2\pi i} = 1$:

$$\cos(z+2\pi) = \frac{e^{i(z+2\pi)} + e^{-i(z+2\pi)}}{2} = \frac{e^{iz}e^{2\pi i} + e^{-iz}e^{-2\pi i}}{2} = \frac{e^{iz} + e^{-iz}}{2} = \cos z.$$

For the first identity in (1.7.7), we have the following calculation

$$\sin\left(z+\frac{\pi}{2}\right) = \frac{e^{i(z+\pi/2)} - e^{-i(z+\pi/2)}}{2i} = \frac{ie^{iz} - (-i)e^{-iz}}{2i} = \cos z.$$

We recognize (1.7.8) as Euler's identity (1.6.8), in which the real argument θ is replaced by a complex argument z. The second identities in (1.7.5), (1.7.6), and (1.7.7) are proved similarly. Next, to prove (1.7.8), we multiply (1.7.4) by i and add the resulting identity to (1.7.3).

Identity (1.7.9) is the analog of the famous Pythagorean identity relating the real cosine and sine functions. Using (1.7.8), we have

$$1 = e^{iz}e^{-iz} = \overbrace{(\cos z + i\sin z)}^{e^{iz}}\overbrace{(\cos z - i\sin z)}^{e^{-iz}} = \cos^2 z + \sin^2 z,$$

proving (1.7.9).   ∎

The familiar angle-addition and half-angle formulas also apply to the complex cosine and sine.

Proposition 1.7.4. (Trigonometric Identities) *Let* z, z_1, z_2 *be complex numbers. Then we have*

$$\cos(z_1 + z_2) = \cos z_1 \cos z_2 - \sin z_1 \sin z_2 ; \tag{1.7.10}$$
$$\sin(z_1 + z_2) = \sin z_1 \cos z_2 + \cos z_1 \sin z_2 ; \tag{1.7.11}$$
$$\cos^2 z = \frac{1 + \cos(2z)}{2} ; \tag{1.7.12}$$
$$\sin^2 z = \frac{1 - \cos(2z)}{2} . \tag{1.7.13}$$

Proof. Expanding the right side of (1.7.10), we find

$$\frac{(e^{iz_1} + e^{-iz_1})(e^{iz_2} + e^{-iz_2})}{2^2} - \frac{(e^{iz_1} - e^{-iz_1})(e^{iz_2} - e^{-iz_2})}{(2i)^2}.$$

Expanding the numerators and adding the fractions, all terms in $e^{i(z_1-z_2)}$ and $e^{i(z_2-z_1)}$ cancel and we are left with

$$\frac{2e^{i(z_1+z_2)} + 2e^{-i(z_1+z_2)}}{4},$$

which is the same as $\cos(z_1 + z_2)$. The proof of (1.7.11) is similar. Now, setting $z_1 = z_2 = z$ in (1.7.10) yields

$$\cos 2z = \cos^2 z - \sin^2 z.$$

Replacing $\sin^2 z$ by $1 - \cos^2 z$, we obtain (1.7.12). Replacing $\cos^2 z$ by $1 - \sin^2 z$, we deduce (1.7.13).   ∎

Up to this point the properties of the complex trigonometric functions have been similar to those of the real trigonometric functions. But there are some differences. Taking $z = iy$ in (1.7.3), where y is a real number, we obtain

$$\cos(iy) = \frac{e^{i(iy)} + e^{-i(iy)}}{2} = \frac{e^y + e^{-y}}{2} = \cosh y. \qquad (1.7.14)$$

Similarly, inserting $z = iy$ in (1.7.4), we obtain

$$\sin(iy) = \frac{e^{i(iy)} - e^{-i(iy)}}{2i} = i\frac{e^y - e^{-y}}{2} = i\sinh y. \qquad (1.7.15)$$

It follows from these identities that the complex cosine and sine function are unbounded functions. This should be contrasted with the fact that $|\cos x| \le 1$ and $|\sin x| \le 1$ for all x (real).

Example 1.7.5. Show that $\cos z$ and $\sin z$ are not bounded functions.

Solution. Take $z = iy$, where y is real. Then, from (1.7.14), $|\cos z| = |\cos iy| = \cosh y$. Since $\cosh y \to \infty$ as $y \to \pm\infty$, we see that $|\cos z|$ cannot be bounded by a finite number; hence $\cos z$ is not a bounded function. Similarly, using (1.7.15), we have $|\sin iy| = |i\sinh y| = |\sinh y|$, and since $|\sinh y| \to \infty$ as $y \to \pm\infty$, it follows that $\sin z$ is not a bounded function. □

The connection between the complex trigonometric functions and the real hyperbolic functions becomes more explicit once we compute the real and imaginary parts of $\cos z$ and $\sin z$.

Proposition 1.7.6. *Let $z = x + iy$ be a complex number (x, y real). Then we have*

$$\cos z = \cos x \cosh y - i\sin x \sinh y \qquad (1.7.16)$$
$$\sin z = \sin x \cosh y + i\cos x \sinh y \qquad (1.7.17)$$
$$|\cos z| = \sqrt{\cos^2 x + \sinh^2 y} \qquad (1.7.18)$$
$$|\sin z| = \sqrt{\sin^2 x + \sinh^2 y}. \qquad (1.7.19)$$

Proof. To prove (1.7.16), we appeal to (1.7.10) and (1.7.14)–(1.7.15) and write

$$\cos z = \cos(x + iy)$$
$$= \cos x \cos(iy) - \sin x \sin(iy)$$
$$= \cos x \cosh y - i\sin x \sinh y.$$

The proof of (1.7.17) is similar. To prove (1.7.18), we use (1.7.16) and (1.2.1). We also use the identity $\cosh^2 y - \sinh^2 y = 1$ for real hyperbolic functions:

$$\begin{aligned}
|\cos z|^2 &= \cos^2 x \cosh^2 y + \sin^2 x \sinh^2 y \\
&= \cos^2 x (1 + \sinh^2 y) + \sin^2 x \sinh^2 y \\
&= \cos^2 x + \sinh^2 y (\cos^2 x + \sin^2 x) \\
&= \cos^2 x + \sinh^2 y.
\end{aligned}$$

The proof of (1.7.19) is similar and is left as an exercise. ■

Example 1.7.7. (Zeros of the sine and cosine functions)
(a) Show that $\sin z = 0 \Leftrightarrow z = k\pi$, for some integer k.
(b) Show $\cos z = 0 \Leftrightarrow z = \frac{\pi}{2} + k\pi$, for some integer k.
Thus $\cos z$ and $\sin z$ have the same zeros as their real counterparts, $\cos x$ and $\sin x$.

Solution. (a) Suppose that $\sin z = 0$; then $|\sin z| = 0$, and so by (1.7.19), we have $\sin x = 0$ and $\sinh y = 0$. The real function $\sinh y$ equals zero $\Leftrightarrow y = 0$, and the real function $\sin x$ equals zero $\Leftrightarrow x = k\pi$ for some integer k. Hence (a) holds.
(b) Knowing the zeros of the sine, we find the zeros of the cosine using (1.7.7). We have $\cos z = 0$ if and only if $\sin\left(z + \frac{\pi}{2}\right) = 0$. This happens exactly when $z + \frac{\pi}{2} = k\pi$ for some integer k or equivalently when $z = -\frac{\pi}{2} + k\pi$ for some integer k. Replacing k by $k + 1$, we deduce (b). □

Next we study certain images under the mapping $w = \sin z$.

Example 1.7.8. (The mapping $w = \sin z$) Find the image under the mapping $f(z) = \sin z$ of the semi-infinite strip

$$S = \left\{ z = x + iy : \ -\frac{\pi}{2} \leq x \leq \frac{\pi}{2}, \ y \geq 0 \right\}.$$

Solution. As in previous examples of mappings, we first find the image under f of a simple curve in the domain of definition, often a line segment or line. Then we sweep the domain of definition with this curve and keep track of the area swept by the image. Fix $0 \leq y_0 < \infty$ and consider the horizontal line segment defined by: $y = y_0$, $-\frac{\pi}{2} \leq x \leq \frac{\pi}{2}$. Let $u + iv$ denote the image of a point $z = x + iy_0$ on this line segment. Using (1.7.17), we get $u + iv = \sin(x + iy_0) = \sin x \cosh y_0 + i \cos x \sinh y_0$. Hence $u = \sin x \cosh y_0$ and $v = \cos x \sinh y_0$. If $y_0 = 0$, we see that $v = 0$ and $u = \sin x$, which shows that the image of the interval $-\frac{\pi}{2} \leq x \leq \frac{\pi}{2}$ under the mapping $\sin z$ is the interval $-1 \leq u \leq 1$. The case $y_0 > 0$ is more interesting. In this case, we have

$$\frac{u}{\cosh y_0} = \sin x \quad \text{and} \quad \frac{v}{\sinh y_0} = \cos x. \tag{1.7.20}$$

Note that $v \geq 0$ because $\cos x \geq 0$ for $-\frac{\pi}{2} \leq x \leq \frac{\pi}{2}$. Squaring both equations in (1.7.20) then adding them, we get

$$\left(\frac{u}{\cosh y_0}\right)^2 + \left(\frac{v}{\sinh y_0}\right)^2 = \sin^2 x + \cos^2 x = 1.$$

Hence as x varies in the interval $-\frac{\pi}{2} \le x \le \frac{\pi}{2}$, the point (u, v) traces the upper semi-ellipse

$$\left(\frac{u}{\cosh y_0}\right)^2 + \left(\frac{v}{\sinh y_0}\right)^2 = 1, \qquad v \ge 0.$$

The u-intercepts of the ellipse are at $u = \pm\cosh y_0$ and the v-intercept is at $v = \sinh y_0$. As $y_0 \to \infty$, $\cosh y_0$ and $\sinh y_0$ tend to ∞. And as $y_0 \to 0$, $\sinh y_0 \to 0$ and $\cosh y_0 \to 1$. So, as y_0 varies in the interval $0 < y_0 < \infty$, the upper semi-ellipses fill the upper half w-plane $v \ge 0$, including the u-axis (Figure 1.42).

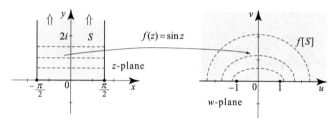

Fig. 1.42 The mapping $w = \sin z$ takes the horizontal line segment $y = y_0 > 0$, $-\frac{\pi}{2} \le x \le \frac{\pi}{2}$ onto the semi-ellipse $\left(\frac{u}{\cosh y_0}\right)^2 + \left(\frac{v}{\sinh y_0}\right)^2 = 1$, for $v \ge 0$.

One can verify (Exercise 23) that the boundary of S is mapped to the boundary of $f[S]$, namely the u-axis. ☐

The other trigonometric functions are defined for complex variables in terms of the cosine and sine in accordance with the real definitions.

Definition 1.7.9. (Other Trigonometric Functions) For a complex number z define

$$\tan z = \frac{\sin z}{\cos z} = -i\frac{e^{iz} - e^{-iz}}{e^{iz} + e^{-iz}} \qquad (\cos z \ne 0), \tag{1.7.21}$$

$$\cot z = \frac{\cos z}{\sin z} = i\frac{e^{iz} + e^{-iz}}{e^{iz} - e^{-iz}} \qquad (\sin z \ne 0), \tag{1.7.22}$$

$$\sec z = \frac{1}{\cos z} = \frac{2}{e^{iz} + e^{-iz}} \qquad (\cos z \ne 0), \tag{1.7.23}$$

$$\csc z = \frac{1}{\sin z} = \frac{2i}{e^{iz} - e^{-iz}} \qquad (\sin z \ne 0). \tag{1.7.24}$$

Like the complex cosine and sine functions, these functions share several properties with their real counterparts. The following is one illustration.

Example 1.7.10. (tan z is π-**periodic**) Show that $\tan z_1 = \tan z_2$ if and only if $z_1 = z_2 + k\pi$, where k is an integer.

Solution. Note that $\tan z$ is not defined for $z = \frac{\pi}{2} + k\pi$. For $z_1, z_2 \neq \frac{\pi}{2} + k\pi$, we have

$$
\begin{aligned}
\tan z_1 = \tan z_2 &\Leftrightarrow \frac{\sin z_1}{\cos z_1} = \frac{\sin z_2}{\cos z_2} \\
&\Leftrightarrow \sin z_1 \cos z_2 - \cos z_1 \sin z_2 = 0 \\
&\Leftrightarrow \sin(z_1 - z_2) = 0 \quad \text{(use (1.7.11) with z_2 replaced by $(-z_2)$)} \\
&\Leftrightarrow z_1 - z_2 = k\pi \quad \Leftrightarrow \quad z_1 = z_2 + k\pi,
\end{aligned}
$$

where the step before last follows from Example 1.7.7(a). □

Hyperbolic Functions

The real hyperbolic functions have complex extensions as well.

Definition 1.7.11. We define the hyperbolic cosine and the hyperbolic sine of a complex number z as follows:

$$
\cosh z = \frac{e^z + e^{-z}}{2} \quad \text{and} \quad \sinh z = \frac{e^z - e^{-z}}{2}. \tag{1.7.25}
$$

We also define the hyperbolic tangent, hyperbolic secant, hyperbolic cosecant, and hyperbolic cotangent of z in terms of $\cosh z$ and $\sinh z$ as follows:

$$
\tanh z = \frac{\sinh z}{\cosh z} \quad (\cosh z \neq 0), \tag{1.7.26}
$$

$$
\operatorname{sech} z = \frac{1}{\cosh z} \quad (\cosh z \neq 0), \tag{1.7.27}
$$

$$
\operatorname{csch} z = \frac{1}{\sinh z} \quad (\sinh z \neq 0), \tag{1.7.28}
$$

$$
\coth z = \frac{\cosh z}{\sinh z} \quad (\sinh z \neq 0). \tag{1.7.29}
$$

The hyperbolic functions satisfy interesting identities that relate to the trigonometric functions.

Proposition 1.7.12. *For an arbitrary complex number $z = x + iy$ (x, y real) we have*

$$
\cosh(iz) = \cos z \qquad \cos(iz) = \cosh z \tag{1.7.30}
$$

$$\sinh(iz) = i\sin z \qquad \sin(iz) = i\sinh z \tag{1.7.31}$$

$$\cosh^2 z - \sinh^2 z = 1 \tag{1.7.32}$$

$$\cosh z = \cosh x \cos y + i\sinh x \sin y \tag{1.7.33}$$

$$\sinh z = \sinh x \cos y + i\cosh x \sin y \tag{1.7.34}$$

$$\tanh(iz) = i\tan z \qquad \tan(iz) = i\tanh z \tag{1.7.35}$$

$$\coth(iz) = -i\cot z \qquad \cot(iz) = -i\coth z. \tag{1.7.36}$$

These identities can be proved from the definitions (1.7.25). See Exercises 35–51.

Exercises 1.7

Evaluate $\cos z$ *and* $\sin z$ *for the following values of* z, *using the definitions* (1.7.3) *and* (1.7.4). *Then verify that your answers satisfy* (1.7.16) *and* (1.7.17).

1. (a) i (b) $\dfrac{\pi}{2}$ (c) $\pi + i$ (d) $\dfrac{\pi}{2} + 2\pi i$

2. (a) $\dfrac{\pi}{4} + i\dfrac{\pi}{4}$ (b) $-i\dfrac{\pi}{4}$ (c) $-\dfrac{\pi}{2} + i\dfrac{\pi}{3}$ (d) π

In Exercises 3 and 4, use (1.7.16) *and* (1.7.17) *to establish the stated fact.*

3. For all complex z show that (a) $\overline{\cos z} = \cos \overline{z}$ and (b) $\overline{\sin z} = \sin \overline{z}$.

4. For all complex z show that (a) $\cos(z + 2\pi) = \cos z$ and (b) $\sin(z + 2\pi) = \sin z$. (In other words, the complex cosine and sine functions are 2π-periodic.)

In Exercises 5–8, for the following values of z, *(a) evaluate* $\cos z$, $\sin z$, *and* $\tan z$, *using* (1.7.16) *and* (1.7.17). *(b) Compute* $|\cos z|$ *and* $|\sin z|$.

5. $1 + i$ **6.** $1 - i$ **7.** $\dfrac{3\pi}{2} + i$ **8.** $\dfrac{\pi}{6} - i$

In Exercises 9–14, compute the real and imaginary parts of the functions.

9. $\sin(2z)$ **10.** $\cos(z^2)$ **11.** $z\sin z$

12. $z\cos z$ **13.** $\tan z$ **14.** $\sec z$

In Exercises 15–20, show that the shaded area S in the z-plane is mapped to the shaded area in the w-plane under the complex mappings indicated in the figures.

15.

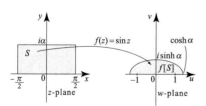

16.

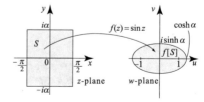

17.

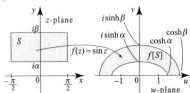

18.

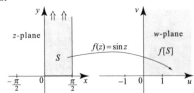

19.

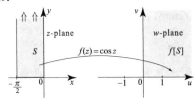

20.

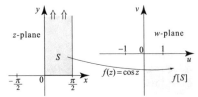

21. Establish identities (1.7.17) and (1.7.19).

22. Let S be the horizontal strip $\{z = x + iy;\ x \geq 0\ -\frac{\pi}{2} \leq y \leq \frac{\pi}{2}\}$. Find the image of S by the mapping $f(z) = \sinh z$. [*Hint*: Express $\sinh z$ in terms of $\sin z$.]

23. We study properties of the mapping $z \mapsto \sin z$.
(a) Show that the half-line $x = \frac{\pi}{2}$, $y \geq 0$, is mapped to the half-line $u \geq 1$, $v = 0$.
(b) Show that the half-line $x = \frac{-\pi}{2}$, $y \geq 0$, is mapped to the half-line $u \leq -1$, $v = 0$.
(c) Conclude that the boundary of the set S in Example 4 is mapped to the boundary of the set $f[S]$.
(d) Recall from your calculus course that an ellipse of the form $\frac{x^2}{a^2} + \frac{y^2}{b^2} = 1$ with $0 < b < a$ has its foci at $x = \pm\sqrt{a^2 - b^2}$. Show that all the ellipses in Example 1.7.8 have the same foci located on the u-axis at $u = \pm 1$.

24. (Zeros of hyperbolic functions) Let $z \in \mathbb{C}$. Show that

$$\sinh z = 0 \quad \Leftrightarrow \quad z = ik\pi,\ k \text{ an integer};$$

and

$$\cosh z = 0 \quad \Leftrightarrow \quad z = i\left(k + \frac{1}{2}\right)\pi,\ k \text{ an integer}.$$

[*Hint*: z is a zero of $\sin z \Leftrightarrow iz$ is a zero of the hyperbolic sine (why?). Reason in the same way for the cosine.]

25. (Linearization) Let m and n be nonnegative integers such that $m + n = p$. We discuss how to express the product $\cos^m \theta \sin^n \theta$, as a linear combination of terms involving $\cos(j\theta)$ and $\sin(k\theta)$, where $1 \leq j,\ k \leq p$. For example, the identity $\cos^3 \theta = \frac{1}{4}\left(\cos 3\theta + 3\cos\theta\right)$ is called the linearization of $\cos^3 \theta$. Derive this identity by raising both sides of (1.7.3) to the third power and simplifying.

26. Linearize $\sin^4 \theta$.

In Exercises 27–52, establish the stated identities. Here z_1, z_2, z are complex numbers and $z = x + iy$ with x, y real. Working with hyperbolic functions, you may want to use the corresponding one for trigonometric functions and (1.7.30) and (1.7.31).

27. (a) $\sin(-z) = -\sin z$ (b) $\sin(z + 2\pi) = \sin z$

28. (a) $\cos(z + \pi) = -\cos z$ (b) $\sin(z + \pi) = -\sin z$

29. $\sin(z_1 + z_2) = \sin z_1 \cos z_2 + \cos z_1 \sin z_2$

30. $\cos 2z = \cos^2 z - \sin^2 z = 2\cos^2 z - 1 = 1 - 2\sin^2 z$

31. $\sin 2z = 2\sin z \cos z$

32. $2\cos z_1 \cos z_2 = \cos(z_1 - z_2) + \cos(z_1 + z_2)$

33. $2\sin z_1 \sin z_2 = \cos(z_1 - z_2) - \cos(z_1 + z_2)$

34. $2\sin z_1 \cos z_2 = \sin(z_1 + z_2) + \sin(z_1 - z_2)$

35. $\cosh(-z) = \cosh z$ and $\sinh(-z) = -\sinh z$

36. $\cosh(z + 2\pi i) = \cosh z$ and $\sinh(z + 2\pi i) = \sinh z$

37. $\cosh(z + \pi i) = -\cosh z$ and $\sinh(z + \pi i) = -\sinh z$

38. $\sinh(z + \frac{i\pi}{2}) = i\cosh z$ and $\cosh(z + \frac{i\pi}{2}) = i\sinh z$

39. $e^z = \cosh z + \sinh z$ **40.** $\cosh^2 z - \sinh^2 z = 1$

41. $\cosh 2z = \cosh^2 z + \sinh^2 z = 2\cosh^2 z - 1 = 1 + 2\sinh^2 z$

42. $\sinh 2z = 2\sinh z \cosh z$

43. $\cosh^2 z = \dfrac{1 + \cosh 2z}{2}$ **44.** $\sinh^2 z = \dfrac{-1 + \cosh 2z}{2}$

45. $\cosh(z_1 + z_2) = \cosh z_1 \cosh z_2 + \sinh z_1 \sinh z_2$

46. $\sinh(z_1 + z_2) = \sinh z_1 \cosh z_2 + \cosh z_1 \sinh z_2$

47. $2\cosh z_1 \cosh z_2 = \cosh(z_1 + z_2) + \cosh(z_1 - z_2)$

48. $2\sinh z_1 \sinh z_2 = \cosh(z_1 + z_2) - \cosh(z_1 - z_2)$

49. $2\sinh z_1 \cosh z_2 = \sinh(z_1 + z_2) + \sinh(z_1 - z_2)$

50. $\cosh z = \cosh x \cos y + i \sinh x \sin y$ and $\sinh z = \sinh x \cos y + i \cosh x \sin y$

51. $|\cosh z| = \sqrt{\sinh^2 x + \cos^2 y}$ **52.** $|\sinh z| = \sqrt{\sinh^2 x + \sin^2 y}$

53. Project Problem: The Dirichlet kernel. (a) For $z \neq 1$ and $n = 0, 1, 2, \ldots$, show that

$$1 + z + z^2 + \cdots + z^n = \frac{1 - z^{n+1}}{1 - z}.$$

(b) Take $z = e^{i\theta}$, where θ is a real number $\neq 2k\pi$ (k an integer), and obtain

$$1 + e^{i\theta} + e^{2i\theta} + \cdots + e^{in\theta} = \frac{i}{2} \frac{(1 - e^{i(n+1)\theta})e^{-i\frac{\theta}{2}}}{\sin \frac{\theta}{2}}.$$

[*Hint:* After substituting $z = e^{i\theta}$, multiply and divide by $e^{-i\frac{\theta}{2}}$; then use (1.7.2).]

(c) Taking the real and imaginary parts of the identity in (b), obtain

$$\frac{1}{2} + \cos\theta + \cos 2\theta + \cdots + \cos n\theta = \frac{\sin[(n + \frac{1}{2})\theta]}{2\sin \frac{\theta}{2}}$$

and

$$\sin\theta + \sin 2\theta + \cdots + \sin n\theta = \frac{\cos \frac{\theta}{2} - \cos[(n + \frac{1}{2})\theta]}{2\sin \frac{\theta}{2}}.$$

The sum $D_n(\theta) = 1 + 2\cos\theta + 2\cos 2\theta + \cdots + 2\cos n\theta$ is called the **Dirichlet kernel** and plays an important role in the theory of Fourier series.

1.8 Logarithms and Powers

In this section we define complex logarithms and complex powers of complex numbers. This enables us to compute expressions like $\text{Log}\,i$ and i^i.

The logarithm was defined in elementary algebra as the inverse of the exponential function. We follow this idea to define the complex logarithm, $\log z$ for $z \neq 0$. However, we expect to encounter some difficulties here because the exponential function e^z is not one-to-one. To define the complex function $w = \log z$, set

$$w = \log z \quad \Leftrightarrow \quad e^w = z. \tag{1.8.1}$$

To determine w in terms of z, write $w = u + iv$ and $z = re^{i\theta}$, with $|z| = r > 0$ and $\theta = \arg z$. Then (1.8.1) becomes

$$e^{u+iv} = e^u e^{iv} = z = re^{i\theta},$$

and hence

$$e^u = r \quad \text{and} \quad e^{iv} = e^{i\theta}. \tag{1.8.2}$$

The first equation gives $u = \ln r$, where here $\ln r$ denotes the usual natural logarithm of the positive number r. The second equation in (1.8.2) tells us that v and θ differ by an integer multiple of 2π, because the complex exponential is $2\pi i$ periodic. So $v = \theta + 2k\pi$, where k is an integer, or simply $v = \arg z$. Putting this together, we obtain the formula for the **complex logarithm**:

$$\log z = \ln|z| + i \arg z \quad (z \neq 0). \tag{1.8.3}$$

Unlike the real logarithm, this formula defines a **multi-valued** function, because $\arg z$ takes multiple values. The complex logarithm is not a function in the standard sense, since functions assign a unique value to each variable.

Example 1.8.1. (Computing logarithms) Evaluate the following logarithms:
(a) $\log i$ (b) $\log(1+i)$ (c) $\log(-2)$.

Solution. (a) The polar form of i is $i = e^{i\frac{\pi}{2}}$. So $|i| = 1$ and $\arg i = \frac{\pi}{2} + 2k\pi$, where k is an integer. Hence, by (1.8.3),

$$\log i = \overbrace{\ln(1)}^{0} + i\left(\frac{\pi}{2} + 2k\pi\right) = i\left(\frac{\pi}{2} + 2k\pi\right).$$

As expected, $\log i$ takes on an infinite number of values, all of which happen to be purely imaginary. Any two values of $\log i$ differ by an integer multiple of $2\pi i$.
(b) We will apply (1.8.3) after putting $1 + i$ in polar form. As you can check, $1 + i = \sqrt{2}e^{i\frac{\pi}{4}}$, $|1 + i| = \sqrt{2}$, and $\arg(1 + i) = \frac{\pi}{4} + 2k\pi$. Thus, from (1.8.3),

$$\log(1+i) = \ln(\sqrt{2}) + i\frac{\pi}{4} = \frac{1}{2}\ln 2 + i\left(\frac{\pi}{4} + 2k\pi\right).$$

(c) We have

$$|-2| = 2, \ \arg(-2) = \pi + 2k\pi \Rightarrow \log(-2) = \ln 2 + i(\pi + 2k\pi) = \ln 2 + (2k+1)\pi i.$$

More explicitly, $\log(-2)$ consists of the following complex values:

$$\ln 2 + \pi i, \ \ln 2 - \pi i, \ \ln 2 + 3\pi i, \ \ln 2 - 3\pi i, \ \ln 2 \pm 5\pi i, \ldots .$$

Note that all values of $\log(-2)$ have identical real parts and their imaginary parts differ by integer multiples of $2\pi i$. These observations are true in general. □

It is clear from Example 1.8.1 and from the definition of $\log z$ that to make $\log z$ single-valued, and hence turn it into a function, it is enough to define a single-valued version of $\arg z$. For example, we can use the principal value of the argument, $\operatorname{Arg} z$ (see Definition 1.3.2) which satisfies $-\pi < \operatorname{Arg} z \le \pi$.

Definition 1.8.2. The **principal value** or **principal branch** of the complex logarithm is defined by

$$\operatorname{Log} z = \ln|z| + i \operatorname{Arg} z \qquad (z \ne 0). \tag{1.8.4}$$

Thus $\operatorname{Log} z$ is the (particular) value of $\log z$ whose imaginary part is in the interval $(-\pi, \pi]$. Because $\arg z = \operatorname{Arg} z + 2k\pi$, where k is an integer, we see from (1.8.3) and (1.8.4) that all the values of $\log z$ differ from the principal value by $2k\pi i$. Thus

$$\log z = \operatorname{Log} z + 2k\pi i \qquad (z \ne 0). \tag{1.8.5}$$

Example 1.8.3. (Principal values of the logarithm) Compute the expressions
(a) $\operatorname{Log} i$ (b) $\operatorname{Log}(1+i)$ (c) $\operatorname{Log}(-i)$ (d) $\operatorname{Log} 5$ (e) $\operatorname{Log}(e^{6\pi i})$.
Solution. If we know $\log z$, to find $\operatorname{Log} z$, it suffices to choose the value of $\log z$ with imaginary part that lies in the interval $(-\pi, \pi]$. If we do not know $\log z$, we compute $\operatorname{Log} z$ using (1.8.4). Appealing to Example 1.8.1, we have for (a) $\operatorname{Log} i = i\frac{\pi}{2}$; and for (b) $\operatorname{Log}(1+i) = \operatorname{Log}(1+i) = \frac{1}{2}\ln 2 + i\frac{\pi}{4}$. For (c), we use (1.8.4):

$$|-i| = 1, \ \operatorname{Arg}(-i) = -\frac{\pi}{2} \Rightarrow \operatorname{Log}(-i) = \ln(1) - i\frac{\pi}{2} = -i\frac{\pi}{2}.$$

For (d), we use (1.8.4):

$$|5| = 5, \ \operatorname{Arg} 5 = 0 \Rightarrow \operatorname{Log} 5 = \ln 5.$$

For (e), we use (1.8.4) and note that $e^{6\pi i}$ is a unimodular number. In fact $e^{6\pi i} = 1$. So

$$|e^{6\pi i}| = 1, \ \operatorname{Arg}(e^{6\pi i}) = 0 \Rightarrow \operatorname{Log}(e^{6\pi i}) = \ln 1 = 0. \qquad □$$

A few observations are in order to highlight some similarities and differences between the natural logarithm and the complex logarithm.

- If x is a positive real number, then $\text{Log}\,x = \ln x$.
- If x is a negative real number, then $\text{Log}\,x = \ln|x| + i\pi$.
- For all z, the identity $e^{\text{Log}\,z} = z$ is true. But, as illustrated by Example 1.8.3(e), $\text{Log}\,(e^z)$ is not always equal to z. In fact, $\text{Log}\,(e^z) = z \Leftrightarrow -\pi < \text{Im}\,z \leq \pi$.
- Many algebraic properties of $\ln x$ no longer hold for $\text{Log}\,z$. For example, the identity $\ln(x_1 x_2) = \ln x_1 + \ln x_2$, which holds for all positive real numbers x_1 and x_2, does not hold for $\text{Log}\,z$. Consider the following:

$$\text{Log}\,((-1)(-1)) = \text{Log}\,(1) = 0 \neq \text{Log}\,(-1) + \text{Log}\,(-1),$$

since $\text{Log}\,(-1) = i\pi$.

Branches of the argument and the logarithm

As we may imagine, we could have specified a different range of values of $\arg z$ in defining a logarithmic function in terms of (1.8.3). In fact, for every real number α we can specify that $\alpha < \arg z \leq \alpha + 2\pi$. This selection assigns a single value to $\arg z$, denoted by $\arg_\alpha z$, that lies in the interval $(\alpha, \alpha + 2\pi]$.

Definition 1.8.4. Let α be a fixed real number. For $z \neq 0$, we call the unique value of $\arg z$ that falls in the interval $(\alpha, \alpha + 2\pi]$ *the α-th branch of* $\arg z$ and we denote it by $\arg_\alpha z$. Precisely, we define the α-th **branch** of $\log z$ by the identity

$$\log_\alpha z = \ln|z| + i\arg_\alpha z, \qquad \text{where } \alpha < \arg_\alpha z \leq \alpha + 2\pi. \tag{1.8.6}$$

The ray through the origin along which a branch of the logarithm is discontinuous is called a **branch cut**.

When $\alpha = -\pi$, this definition leads to the principal value of the logarithm; that is, $\log_{-\pi} z = \text{Log}\,z$.

Since two values of $\arg z$ differ by an integer multiple of 2π, it follows that, for a complex number $z \neq 0$ and real numbers α and β, there is an integer k (depending on z, α, and β), such

$$n\log_\alpha z = \log_\beta z + 2k\pi i.$$

Example 1.8.5. (Different branches of the logarithm) Evaluate
(a) $\log_0 i$ (b) $\log_{\frac{\pi}{2}} i$ (c) $\log_{\frac{\pi}{2}}(-2)$

Solution. If we know $\log z$, to find $\log_\alpha z$, it suffices to choose the value of $\log z$ with imaginary part that lies in the interval $(\alpha, \alpha + 2\pi]$. If we do not know $\log z$, we compute $\log_\alpha z$ using (1.8.6).
(a) We have $\alpha = 0$ and so the imaginary part of $\log_0 z$, $\arg_0 z$, must be in the interval $(0, 2\pi]$. From Example 1.8.1, $\log i = i\frac{\pi}{2} + 2k\pi i$; and so $\log_0 i = i\frac{\pi}{2}$. Note that $\text{Log}\,i = \log_0 i$.

(b) We have $\alpha = \frac{\pi}{2}$, so $\arg_{\frac{\pi}{2}} z$ belongs to the interval $(\frac{\pi}{2}, \frac{\pi}{2} + 2\pi] = (\frac{\pi}{2}, \frac{5\pi}{2}]$. Hence

$$\arg_{\frac{\pi}{2}} i = \frac{\pi}{2} + 2\pi = \frac{5\pi}{2}$$

(and not $\frac{\pi}{2}$). Consequently, $\log_{\frac{\pi}{2}} i = i\frac{5\pi}{2}$. Note that $\operatorname{Log} i = \log_{\frac{\pi}{2}} i + 2\pi$. (c) Reasoning as in (b), we find that $\arg_{\frac{\pi}{2}}(-2) = \pi$, because this value of $\arg(-2)$ does lie in the interval $(\frac{\pi}{2}, \frac{5\pi}{2}]$. So

$$\log_{\frac{\pi}{2}}(-2) = \ln 2 + i \arg_{\frac{\pi}{2}}(-2) = \ln 2 + i\pi$$

Note that $\operatorname{Log}(-2) = \log_{\frac{\pi}{2}}(-2)$.   □

The Logarithm as a Map

In studying mapping properties of the logarithm, we recall that the exponential function maps rectangular regions to circular regions. Here we expect the logarithm to do the opposite, namely to map circular regions to rectangular regions.

Example 1.8.6. (The map $\operatorname{Log} z$) Let $0 < a < b$ and $0 \le \alpha_1 < \alpha_2 \le \pi$. Find the image of the circular region $S = \{z : a \le |z| \le b, \ \alpha_1 \le \operatorname{Arg} z \le \alpha_2\}$ under the mapping $f(z) = \operatorname{Log} z$.

Solution. The region S is bounded by the rays at angle $0 \le \alpha_1$ and $\alpha_2 \le \pi$, and the circular arcs with radii a and b, as shown in Figure 1.43.

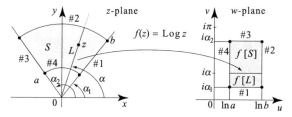

Fig. 1.43 $\operatorname{Log} z$ is a one-to-one mapping of the circular region onto a rectangle. The boundary of the circular region is mapped onto the boundary of the rectangle as follows: side #j in the domain is mapped to side #j in the range.

Consider a line segment L on the ray at angle α, where $\alpha_1 \le \alpha \le \alpha_2$. For z on this ray, we have $\operatorname{Arg} z = \alpha$ and so $\operatorname{Log} z = \ln |z| + i \operatorname{Arg} z = \ln |z| + i\alpha$. As $|z|$ varies from a to b, $\ln |z|$ varies from $\ln a$ to $\ln b$, and thus $\operatorname{Log} z$ describes the horizontal line segment $u + i\alpha$, where $\ln a \le u \le \ln b$. By letting α vary from α_1 to α_2, L sweeps S and the image of L sweeps the rectangular area with vertices $(\ln a, \alpha_1)$, $(\ln b, \alpha_1)$, $(\ln b, \alpha_2)$ and $(\ln a, \alpha_2)$, as shown in Figure 1.43.   □

Complex Powers

In analogy with calculus, for a complex number $z \neq 0$, we define the **complex power**

$$z^a = e^{a\log z} \qquad (z \neq 0), \tag{1.8.7}$$

where $\log z$ is the complex logarithm (1.8.3). Since $\log z$ is multi-valued, it follows from (1.8.7) that z^a is in general multi-valued. By specifying a branch of the logarithm, we obtain a single-valued   branch of the complex power function from (1.8.7). In particular, if we choose the principal logarithm (1.8.4), we obtain the **principal value** of z^a:

$$z^a = e^{a \operatorname{Log} z} \qquad (z \neq 0). \tag{1.8.8}$$

Example 1.8.7. (Evaluating complex powers) Compute $(-i)^{1+i}$ using (a) the principal branch of the logarithm; (b) the branch of the logarithm with a branch cut at angle $\alpha = 0$.

Solution. (a) Using (1.8.8), we find

$$(-i)^{1+i} = e^{(1+i)\operatorname{Log}(-i)} = e^{(1+i)(-\frac{i\pi}{2})} = e^{\frac{\pi}{2}} e^{-\frac{i\pi}{2}} = -ie^{\frac{\pi}{2}}.$$

(b) Using the logarithm with a branch cut at angle 0 in (1.8.7), we have

$$(-i)^{1+i} = e^{(1+i)\log_0(-i)} = e^{(1+i)\frac{3i\pi}{2}} = e^{-\frac{3\pi}{2}} e^{3\frac{i\pi}{2}} = -ie^{-\frac{3\pi}{2}},$$

which is a different value from the one we found in (a). $\square$

For $z \neq 0$, is the function z^a, defined by (1.8.7), always multi-valued? To answer this question, let us use the formula (1.8.5) for $\log z$ and write

$$z^a = e^{a\log z} = e^{a(\operatorname{Log} z + 2k\pi i)} = e^{a\operatorname{Log} z} e^{2ka\pi i},$$

where k is an integer. To determine the number of distinct values of z^a, we must determine the number of distinct values taken by $e^{2ka\pi i}$ as k varies over the integers. We distinguish three cases.

Case (i): a is an integer. Then $2ka\pi i$ is an integer multiple of $2\pi i$, and hence $e^{2ka\pi i} = 1$ for all integers k. The expression z^a has only one value. This result is in concordance with our notion of z^n, z^{-1}, etc., as being single-valued functions.

Case (ii): a is a (real) rational number. Write $a = \frac{p}{q}$, where p and q are integers and have no common factors. The quantity $e^{2ka\pi i} = e^{2\pi i \frac{pk}{q}}$ will have q distinct values, for $k = 0, 1, \ldots, q-1$ (see Exercise 53). Thus, for each value of $k = 0, 1, \ldots, q-1$, we obtain a distinct power function

$$z^{\frac{p}{q}} = e^{\frac{p}{q}\operatorname{Log} z} e^{2\pi i \frac{pk}{q}} \tag{1.8.9}$$

called a **branch** of $z^{\frac{p}{q}}$. The branch for $k = 0$ is called the **principal branch** of $z^{\frac{p}{q}}$.
This result is in accordance with our notion of nth roots $z^{1/n}$; there are n of them.
Note also that case (i) is just case (ii) with $q = 1$.

Case (iii): a is a complex number not of either of the preceding two types. This is
the case when a is an irrational real number or a complex number with a nonzero
imaginary part. Then the quantities $e^{2ka\pi i}$ are distinct for all integers k (Exercise 54),
and z^a has an infinite number of values. As in case (ii), each value of k determines a
branch of z^a, except that here we have infinitely many distinct branches.

Note that our definition of a complex power is inconsistent with our definition of
the complex exponential e^z. According to (1.8.7), we can take e and raise it to the
power a, resulting in

$$e^a = e^{a \log e} = e^{a \ln e} e^{a 2 k \pi i},$$

which is, in general, multi-valued. We must distinguish this concept of "raising e to
the power a" from our previous, single-valued definition of the "exponential function." As a convention, e^z will always refer to the exponential function, unless otherwise stated.

The last example of this section deals with inverse trigonometric functions. These
may be computed using complex powers and logarithms, and so they are multi-
valued in general. For example, the inverse sine of z is an arbitrary complex number
$w = \sin^{-1} z$ with $\sin w = z$. The solutions of the last equation form an infinite set,
and so $\sin^{-1} z$ is infinite-valued.

Example 1.8.8. (Computing the inverse sine function) Show that

$$\sin^{-1} z = -i \log \left(iz + (1 - z^2)^{\frac{1}{2}} \right), \tag{1.8.10}$$

where the complex logarithm is defined in (1.8.3). (For every $z \neq 0$, the square
root takes two values, and the complex logarithm is infinite-valued. As a result the
inverse sine is infinite-valued.)

Solution. We want to solve $\sin w = z$. Recalling the definition of $\sin w$ from (1.7.4)
we have

$$z = \frac{e^{iw} - e^{-iw}}{2i},$$

or, after multiplying both sides by e^{iw} and simplifying,

$$\left(e^{iw} \right)^2 - 2iz e^{iw} - 1 = 0.$$

This is a quadratic equation in e^{iw}, which is solved solve via the quadratic formula
(1.3.25):

$$e^{iw} = iz + (1 - z^2)^{\frac{1}{2}},$$

where the square root has two values in general. The desired identity (1.8.10) follows now upon taking logarithms.   □

Formulas for the inverses of other trigonometric and hyperbolic functions can be derived in a similar fashion. See Exercises 48–52.

Exercises 1.8

In Exercises 1–4, evaluate $\log z$ for the following values of z.

1. (a) $2i$ (b) $-3-3i$ (c) $5e^{i\frac{\pi}{7}}$ (d) -3

2. (a) 1 (b) $2e^{i\frac{2\pi}{3}}$ (c) $4+4i$ (d) $-2e^{2+i\frac{2\pi}{11}}$

3. (a) $-\frac{1}{2}$ (b) $(1-i)^7$ (c) $1+i\sqrt{3}$ (d) $e^{-1-i\frac{\pi}{7}}$

4. (a) $\dfrac{1}{(1+i)^4}$ (b) $(1-i)^{11}$ (c) $(1+i\sqrt{3})^8$ (d) $\dfrac{e^{-i\frac{\pi}{7}}}{2e^{-i\frac{\pi}{3}}}$

5. Compute $\operatorname{Log} z$, where z is as in Exercise 1(a)–(d).

6. Compute $\operatorname{Log} z$, where z is as in Exercise 4(a)–(d).

7. Compute $\log_0 z$, where z is as in Exercise 3(a)–(d).

8. Compute $\log_\pi z$, where z is as in Exercise 2(a)–(d).

In Exercises 9–12, evaluate the logarithms.

9. $\log_6 1$ **10.** $\log_{\sqrt{3}}(1+i)$ **11.** $\log_5(-5i)$ **12.** $\log_{2\pi} i$

In Exercises 13–18, solve the equations.

13. $e^z = 3$ **14.** $e^{-z} = 1+i$ **15.** $e^{z+3} = i$

16. $e^{2z}+3e^z+2=0$ **17.** $e^{2z}+5=0$ **18.** $e^z = \frac{1+i}{1-i}$

19. (a) Compute $\operatorname{Log}(e^{i\pi})$, $\operatorname{Log}(e^{3i\pi})$, and $\operatorname{Log}(e^{5i\pi})$.
(b) Show that $\operatorname{Log}(e^z) = z$ if and only if $-\pi < \operatorname{Im} z \le \pi$.

20. Show that $\log z = i \arg z$ if and only if z is unimodular.

21. Compute $\operatorname{Log}(-1)$, $\operatorname{Log} i$, and $\operatorname{Log}(-i)$ and conclude that, unlike the identity for the usual real logarithm, in general, $\operatorname{Log}(z_1 z_2)$ is not equal to $\operatorname{Log} z_1 + \operatorname{Log} z_2$.

22. Show that $\operatorname{Log} z$ is one-to-one for all $z \ne 0$. Generalize this statement to $\log_\alpha$, where α is an arbitrary real number.

23. For which real α is $\log_\alpha 1 = 0$?   **24.** For which real α is $\log_\alpha(1+i) = \frac{\pi}{4}$?

25. Solve $\operatorname{Log} z + \operatorname{Log}(2z) = \frac{3\pi}{2}$ and verify your answers.

26. Solve the equation $\operatorname{Log}(iz) - i\operatorname{Log} z = \frac{3\pi}{2}$. Verify your answers.

27. (a) Let α be a positive real number. Show that $\operatorname{Log}(\alpha z) = \ln \alpha + \operatorname{Log} z$ for all $z \ne 0$.
(b) Show that the identity in (a) fails if α is not real and positive.

28. Consider the equality $\operatorname{Log} z = \log_0 z + 2k\pi i$. What values of k are needed as z takes on all nonzero complex numbers?

In Exercises 29–32, evaluate the principal value of the powers.

29. 5^i **30.** $(1+i)^{3+i}$ **31.** i^i **32.** $\left(\frac{1+i}{1-i}\right)^i$

In Exercises 33–36, state how many values the given power takes and find them.

33. $(3i)^4$ **34.** $(1+i\sqrt{3})^{\frac{2}{7}}$ **35.** $(-i)^i$ **36.** $(-e)^{\frac{i}{2}}$

37. Give and example to show that, in general, $\operatorname{Log} \bar{z} \ne \overline{\operatorname{Log} z}$.

38. (a) Let α be a real number. Show that the exponential function maps the strip

$$S_\alpha = \{z = x+iy: \ \alpha < y \le \alpha + 2\pi\} \tag{1.8.11}$$

onto $\mathbb{C} \setminus \{0\}$ and is one-to-one on this region. The strip S_α is called a **fundamental region** for e^z.
(b) Conclude that the branch of the logarithm $\log_\alpha z$ maps the punctured plane $\mathbb{C} \setminus \{0\}$ back onto the fundamental region S_α (see the figure below).

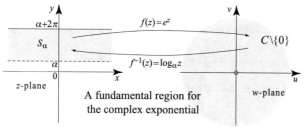

A fundamental region for
the complex exponential

In Exercises 39–42, refer to the figure above and use the result of Exercise 38, if necessary. For α real, the fundamental region S_α of e^z is the infinite horizontal strip $S_\alpha = \{z : \alpha < \operatorname{Im} z \le \alpha + 2\pi\}$. The upper boundary line, $\operatorname{Im} z = \alpha + 2\pi$, is included in S_α, but the lower boundary line, $\operatorname{Im} z = \alpha$, is not. (If we were to include the line $\operatorname{Im} z = \alpha$, then e^z would cease to be one-to-one in the region.)

39. Take $\alpha = -\pi$ and determine the image of $\mathbb{C} \setminus \{0\}$ under the mapping $\operatorname{Log} z$.

40. Take $\alpha = 0$ and determine the image of the punctured plane $\mathbb{C} \setminus \{0\}$ under the mapping $\log_0 z$.

41. Take $\alpha = 3\pi$ and determine the image of $\mathbb{C} \setminus \{0\}$ under the mapping $\log_{3\pi} z$.

42. What is the image of $\mathbb{C} \setminus \{0\}$ under the mapping $\log_{n\pi} z$, where n is an integer.

In Exercises 43–46, determine the image of the region under the mapping $\operatorname{Log} z$. Be specific about the image of the boundary.

43.

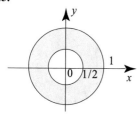

44.

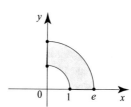

45.

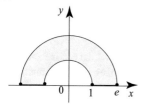

46.

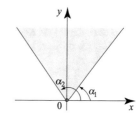

47. Find all complex solutions of the equation $\cos z = \sin z$.

48. (a) Use (1.7.3) to derive the formula $\cos^{-1} z = -i \log \left(z + (z^2 - 1)^{\frac{1}{2}} \right)$.
(b) Derive an expression for $\cos^{-1} z$ using (1.7.7) and (1.8.10).

49. (a) Verify from (1.7.21) that for a complex number w with $e^{iw} + e^{-iw} \ne 0$ we have

$$1 + i \tan w = \frac{2e^{iw}}{e^{iw} + e^{-iw}}.$$

(b) Likewise for w as in part (a) verify that

$$1 - i\tan w = \frac{2e^{-iw}}{e^{iw} + e^{-iw}}.$$

(c) Divide the quantities in parts (a) and (b) to conclude the formula

$$\tan^{-1} z = \frac{i}{2} \log \frac{1 - iz}{1 + iz} \quad (z \neq \pm i). \qquad (1.8.12)$$

50. Verify that for a complex number z we have

$$\sinh^{-1} z = \log\left(z + (z^2 + 1)^{\frac{1}{2}}\right). \qquad (1.8.13)$$

51. Verify that for a complex number z we have

$$\cosh^{-1} z = \log\left(z + (z^2 - 1)^{\frac{1}{2}}\right). \qquad (1.8.14)$$

52. Verify that for a complex number $z \neq \pm 1$ we have

$$\tanh^{-1} z = \frac{1}{2} \log \frac{1 + z}{1 - z}. \qquad (1.8.15)$$

[*Hint*: Consider $1 + \tanh w$ and $1 - \tanh w$.]

53. Project Problem: Rational powers. We prove that for rational $a = \frac{p}{q}$, p and q having no common factors, the expression z^a has exactly q distinct values.
(a) Show that all values for z^a are of the form $e^{(p/q)\operatorname{Log} z} e^{2k p \pi i / q}$.
(b) Define $E_n = e^{2n p \pi i / q}$, for all integers n. Argue that $E_n = E_{n+q}$ for all n and hence there can be at most q distinct values for z^a. Without loss of generality we need only consider $0 \leq n \leq q - 1$.
(c) Suppose that $E_j = E_l$ for some $0 \leq j < l \leq q - 1$. Use (1.6.15) to conclude that $\frac{p(l-j)}{q} = k$ for some integer k.
(d) Argue that $\frac{p(l-j)}{q}$ cannot be an integer; for $\frac{p(l-j)}{q}$ to be an integer, all the prime factors of q must be canceled by terms in the numerator. However, p has none of them, and $l - j$ cannot have all of them since $l - j < q$. Conclude that all E_n are distinct for $0 \leq n \leq q - 1$, and hence $z^{\frac{p}{q}}$ has q distinct values.

54. Project Problem: Nonreal and irrational powers. In this problem we prove that if $\operatorname{Im} a \neq 0$ or $\operatorname{Re} a$ is irrational, z^a has an infinite number of values.
(a) Write $a = \alpha + i\beta$ and show that all values for z^a are of the form $e^{a \operatorname{Log} z} e^{-\beta 2 k \pi} e^{\alpha 2 k \pi i}$.
(b) Define $E_n = e^{\beta 2 n \pi} e^{\alpha 2 n \pi i}$ for all integers n. Argue that if $\beta \neq 0$, then $|E_n|$ are all distinct, and hence all values of E_n are distinct. Thus z^a has an infinite number of values.
(c) Otherwise, if $\beta = 0$ and α is irrational, we have $E_n = e^{\alpha 2 n \pi i}$. Suppose $E_j = E_l$ for some $j < l$. Use (1.6.15) to conclude that $\alpha(l - j) = k$ for some integer k. However, this is impossible because α is irrational. Hence all E_n are distinct and z^a has an infinite number of values.

55. Show that formula (1.8.12) for the inverse tangent holds for real z using the geometric interpretation of the tangent function in Figure 1.44.

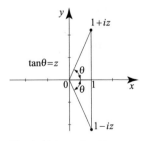

Fig. 1.44 Exercise 55.

Chapter 2
Analytic Functions

> *I get up at 4 o'clock each morning and I am busy from then on... Today I drew the plans for forges that I am to have built in granite. I am also constructing two lighthouses, one on each of the two piers that are located at the entrance of the harbor. I do not get tired of working; on the contrary, it invigorates me and I am in perfect health...*
>
> -Augustin-Louis Cauchy (1789–1857)

Complex analysis is concerned with the study of analytic functions. These are functions that have a complex derivative in an open planar set. The notion of analyticity is fundamental in complex analysis; it lays the foundational cornerstone and sets the stage for the development of the subject. Most of the theory of analytic functions is due to Augustin-Louis Cauchy (1789–1857) and was prepared for a course that he taught at the Institut de France in 1814 and later at the Ecole Polytechnique. Cauchy single-handedly defined the derivative and integral of complex functions and developed one of the most fruitful theories of mathematics. In the process of developing his theory, he defined for the first time the notion of limit for functions and gave rigorous definitions of continuity and differentiability for real-valued functions, as are known nowadays in calculus. He also developed a solid groundwork for the theory of definite integrals and series and he established the theoretical aspects of complex analysis. In doing so, he paid great attention to rigorous mathematical proof, a trait that characterizes pure mathematics today.

So who was Cauchy? Augustin-Louis Cauchy was born on August 21, 1789, in Paris. He received his early education from his father, Louis-François Cauchy, a master of classical studies. Cauchy entered the Ecole Polytechnique in Paris in 1805 and continued his education as a civil engineer at the Ecole des Ponts et Chaussées. He began his career as a military engineer, working in Napoleon's administration from 1810 to 1813. His mathematical talents were soon discovered by leading mathematicians, among them was Joseph Louis Lagrange, who persuaded Cauchy to leave his career as an engineer and devote himself to mathematics. Cauchy's mathematical output was phenomenal. He is considered to be one of the greatest mathematicians. His contributions cover many areas of pure and applied mathematics, including the theory of heat, the theory of light, the mathematical theory of elasticity, and fluid dynamics. Cauchy's contributions to modern calculus are so fundamental that he "has come to be regarded as the creator of calculus in the modern sense," according to *The History of Mathematics, An Introduction*, 3rd edition, by David M. Burton (McGraw-Hill, 1997).

© Springer International Publishing AG, part of Springer Nature 2018
N. H. Asmar and L. Grafakos, *Complex Analysis with Applications*,
Undergraduate Texts in Mathematics, https://doi.org/10.1007/978-3-319-94063-2_2

2.1 Regions of the Complex Plane

In the previous chapter, we defined some elementary functions of a complex variable. An important part of a function is its domain of definition. In calculus, functions are usually defined over intervals. For functions of a complex variable, intervals are replaced by subsets of the complex plane. For this reason, in order to develop the theory of functions of a complex variable, it is necessary to understand basic properties of subsets of the complex plane.

Open Sets

One very useful definition is that of a neighborhood of a point. It is the analog of an open interval in one dimension.

Definition 2.1.1. (Neighborhoods) Let $r > 0$ be a positive real number and z_0 a point in the plane. The r-**neighborhood** of z_0 is the set of all complex numbers z satisfying $|z - z_0| < r$. We denote[1] this set by $B_r(z_0)$.

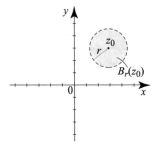
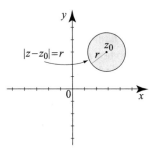

Fig. 2.1 An r-neighborhood $B_r(z_0)$ does not include the points on the circle $|z - z_0| = r$.

Fig. 2.2 A closed disk includes its boundary, the circle $|z - z_0| = r$.

Interpreting the absolute value as a distance, we visualize $B_r(z_0)$ as the disk centered at z_0 with radius $r > 0$. The fact that the inequality $|z - z_0| < r$ is strict expresses the property that the points on the circle $|z - z_0| = r$ are not part of this disk. In Figure 2.1, the dashed circle $|z - z_0| = r$ indicates exactly the noninclusion of the circle $|z - z_0| = r$ in the r-neighborhood $B_r(z_0)$ of z_0.

A neighborhood of z_0 from which we have deleted the center z_0 is called a **deleted neighborhood** or **punctured neighborhood** of z_0 and is denoted by $B'_r(z_0)$. Thus

$$B'_r(z_0) = \{z : 0 < |z - z_0| < r\}.$$

---

[1] The letter B is used for an open disk because its higher-dimensional analog is an "open ball."

Definition 2.1.2. Let S be a subset of $\mathbb{C}$. A point z_0 in S is called an **interior point** of S if we can find a neighborhood of z_0 that is wholly contained in S. A point z in the complex plane is called a **boundary point** of S if every neighborhood of z contains at least one point in S and at least one point not in S. The set of all boundary points of S is called the **boundary** of S.

Definition 2.1.2 implies that every point in S is either an interior point or a boundary point. If a point is an interior point of S, then it cannot be a boundary point of S. Also, while an interior point of S is necessarily a point in S, a boundary point of S need not be in S. The geometric concepts in Definition 2.1.2 are intuitively clear; however, dealing with them often requires delicate handling of the absolute value.

Example 2.1.3. (Interior and boundary points of r-neighborhoods $B_r(z_0)$)
(a) Show that every point z of $B_r(z_0)$ is an interior point.
(b) Show that the boundary of $B_r(z_0)$ is the circle $|z - z_0| = r$.

Solution. (a) Pick a point z in $B_r(z_0)$. As one can see in Figure 2.3, we can find a disk centered at z, which lies entirely in $B_r(z_0)$. Hence z is an interior point of $B_r(z_0)$. To give an analytic proof of this geometric argument, let $\delta = |z - z_0|$. By the definition of $B_r(z_0)$, we have $0 \leq \delta < r$. Let $\delta' = r - \delta$. For w in the neighborhood $B_{\delta'}(z)$, we have $|w - z| < \delta'$ and so, by the triangle inequality,

$$|w - z_0| \leq |w - z| + |z - z_0| < \delta' + \delta = r.$$

Hence w belongs to $B_r(z_0)$, and since this is true of every w in $B_{\delta'}(z)$, we conclude that $B_{\delta'}(z)$ is contained in $B_r(z_0)$.

(b) Pick a point z_1 on the circle $|z - z_0| = r$. It is clear from Figure 2.3 that every disk centered at z_1 contains (infinitely many) points in $B_r(z_0)$ and (infinitely many) points not in $B_r(z_0)$. Hence every point on the circle $|z - z_0|$ is a boundary point of $B_r(z_0)$. We now show that no other points are boundary points. Since points inside the circle are interior points, they cannot be boundary points. Also, points outside the circle can be enclosed in open disks that do not intersect $B_r(z_0)$. Hence such points are not boundary points either.

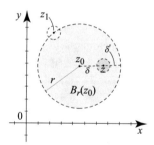

Fig. 2.3 The point z is an interior point of $B_r(z_0)$, while z_1 is a boundary point.

In this example none of the boundary points of $B_r(z_0)$ belong to $B_r(z_0)$. ☐

Definition 2.1.4. (Open Sets) A subset S of the complex numbers is called **open** if every point in S is an interior point of S.

Thus S is open if around each point z in S you can find a neighborhood $B_r(z)$ that is entirely contained in S. The radius of $B_r(z)$ depends on z and may be chosen

as small as we wish. Also, if we can find one value of r, say r_0, such that $B_{r_0}(z)$
is contained in S, then we can find infinitely many values of r such that $B_r(z)$ is
contained in S: Just take $0 < r < r_0$. Here are some useful examples to keep in mind.

- The empty set, denoted as usual by $\emptyset$, is open. Because there are no points in $\emptyset$,
  the definition of open sets is vacuously satisfied.
- The set of all complex numbers $\mathbb{C}$ is open.
- An r-neighborhood, $B_r(z_0)$, is open. We just verified in Example 2.1.3(a) that
  every point in $B_r(z_0)$ is an interior point.
- The set of all z such that $|z - z_0| > r$ is open. This set is called a **neighborhood
  of** ∞.

An r-neighborhood, $B_r(z_0)$, is more commonly called an **open disk** of radius r,
centered at z_0.

One can show that a set is open if and only if it contains none of its boundary
points (Exercise 18). Sets that contain all of their boundary points are called **closed**.
The complex plane $\mathbb{C}$ and the empty set $\emptyset$ are closed since they trivially contain
their empty sets of boundary points. The disk $\{z : |z - z_0| \leq r\}$ is closed because it
contains all its boundary points consisting of the circle $|z - z_0| = r$ (Figure 2.2). We
refer to such a disk as the **closed disk** of radius r, centered at z_0. The smallest closed
set that contains a set A is called the **closure** of A and is denoted by $\overline{A}$. For instance
the closure of the open disk $B_r(z_0)$ is the closed disk $\overline{B_r(z_0)} = \{z : |z - z_0| \leq r\}$.
The punctured open disk $B'_r(z_0)$ also has the same closure. A point z_0 is called an
accumulation point of a set A if $B'_r(z) \cap A \neq \emptyset$ for every $r > 0$. For instance, every
boundary point of an open disk is an accumulation point of it.

Some sets are neither open nor closed.
For example, the set

$$S = \{z : |z - z_0| \leq r; \ \text{Im } z > 0\}$$

contains the boundary points on the upper semicircle, but it does not contain its
boundary points that lie on the x-axis.
Hence, this set is neither open nor closed.
See Figure 2.4.

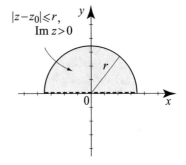

Fig. 2.4 S is neither open nor closed.

Next, we introduce some set notation for convenience. If a point z is in a set S,
we say that z is an **element** of S and write $z \in S$. If z does not belong to S, we will
write $z \notin S$. Let A and B be two sets of complex numbers. The **union** of A and B,
denoted $A \cup B$, is the set

$$A \cup B = \{z : z \in A \text{ or } z \in B\}.$$

The **intersection** of A and B, denoted $A \cap B$, is the set

$$A \cap B = \{z : z \in A \text{ and } z \in B\}.$$

Two sets A and B are **disjoint** if $A \cap B = \emptyset$. The **set difference** between A and B is the set

$$A \setminus B = \{z : z \in A \text{ and } z \notin B\}.$$

We say that A is a **subset** of B or that B contains A and we write $A \subset B$ or $B \supset A$ if every element of A is also an element of B: $z \in A \Rightarrow z \in B$.

Connected Sets

A basic result from calculus states that if the derivative of a differentiable function is constant on open interval, then the function is constant on that interval. This result is not true if the domain of definition of the function is not connected. For example, consider the function

$$h(t) = \begin{cases} 1 & \text{if } 0 < t < 1, \\ -1 & \text{if } 2 < t < 3, \end{cases}$$

whose domain of definition is $(0,1) \cup (2,3)$. We have $h'(t) = 0$ for all t in the union $(0,1) \cup (2,3)$, but clearly h is not constant. This example shows that the connectedness of the domain is required in order to have consistent behavior and be able to derive important conclusions for functions. For subsets of the plane, one way to define connectedness is as follows.

Definition 2.1.5. A **polygonal line** is a finite union of closed line segments L_j, $j = 1, \ldots, m$, such that the end of each L_j coincides with the beginning of L_{j+1} ($j = 1, \ldots, m-1$). A subset Ω of the complex plane is called **polygonally connected** if any two points in Ω can be joined by a polygonal line entirely contained in it.

We denote by $[z_0, z_1]$ the line segment

$$\{(1-t)z_0 + tz_1 : t \in [0,1]\}$$

that joins two fixed points $z_0, z_1 \in \mathbb{C}$. A polygonal line $\cup_{j=1}^{m}[z_{j-1}, z_j]$ can be thought of as a map L from the interval $[0, m]$ to the complex plane defined as follows:

$$L(t) = (j-t)z_{j-1} + (t-j+1)z_j$$

for $t \in [j-1, j]$, $j = 1, \ldots, m$. Then

$$L(j) = z_j,$$

i.e., $L(t)$ passes through the point z_j at "time" $t = j$. See Figure 2.5.

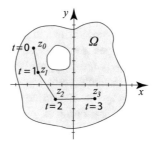

Fig. 2.5 A polygonally connected subset and a polygonal line ($m = 3$).

Definition 2.1.6. A nonempty open and polygonally connected subset of the complex plane is called a **region**.[2]

Polygonally connected open sets are exactly the connected sets known from topology. This characterization is proved now.

Proposition 2.1.7. (Characterization of Regions) *A nonempty open subset Ω of the complex plane is polygonally connected if and only if it cannot be written as the disjoint union of two nonempty open subsets. Consequently, if Ω is a region and $\Omega = A \cup B$, where A and B are open and disjoint, then either $A = \emptyset$ or $B = \emptyset$.*

Proof. It is best to prove the equivalence of the negations of the claimed assertions. Suppose we can write $\Omega = A \cup B$, where A, B are disjoint nonempty open sets. If $L(t)$ is a polygonal line that joins a point $z_0 \in A$ to a point $z_m \in B$, note that the set $\{t \in [0,m] : L(t) \in A\}$ is nonempty as it contains 0 and is bounded above by m. We define

$$t_* = \sup\{t \in [0,m] : L(t) \in A\}.$$

Now $L(t_*)$ should lie in A or B. If $L(t_*)$ lies in B, then by the openness of B, $A \cap B \neq \emptyset$. If $L(t_*)$ lies in A, then by the openness of A, there is another point $t' > t_*$ with $L(t') \in A$, contradicting the definition of t_*. Hence we obtain that the polygonal line L is not contained in $\Omega = A \cup B$ and consequently, Ω is not polygonally connected.

Conversely, suppose that Ω is not polygonally connected. Then there exist $z_0, z_1 \in \Omega$ that are not connected via a polygonal line. We show that there exist open disjoint nonempty sets A and B such that $\Omega = A \cup B$. We define

$$A = \{z \in \Omega : z \text{ is connected to } z_0 \text{ via a polygonal line}\}$$
$$B = \{z \in \Omega : z \text{ is not connected to } z_0 \text{ via a polygonal line}\}.$$

Since $z_0 \in A$, $z_1 \in B$, these sets are nonempty. Also A and B are open, since if z is (or is not) connected to z_0 via a polygonal line, then so does an arbitrary point w that lies in an open disk contained in Ω centered at z. ∎

In topology, connected sets in the plane are exactly the ones that cannot be written as a union of two disjoint nonempty open sets. Proposition 2.1.7 indicates that open sets are connected if and only if they are polygonally connected. This characterization provides an intuitive way to determine which open subsets of the complex plane are regions.

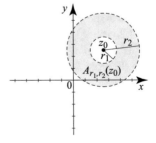

Fig. 2.6 The open annulus $A_{r_1,r_2}(z_0)$ is a region.

---

[2] some authors prefer the term **domain** for regions

Here are some useful examples of regions:

- An open disk $B_r(z_0)$ is a region.
- A punctured disk centered at z_0, $B'_r(z_0) = \{z : 0 < |z - z_0| < r\}$ is a region.
- An open **annulus** centered at z_0, $A_{r_1, r_2}(z_0) = \{z : r_1 < |z - z_0| < r_2\}$ is a region (Figure 2.6).
- The open upper half-plane $\{z : \operatorname{Im} z > 0\}$ is a region.
- The complex plane is a region.

Here are sets that are not regions.

- A closed disk is not a region because it is not open.
- The union of two disjoint open disks, for example, $B_1(0) \cup B_{\frac{1}{2}}(2i)$, is not a region because it is not connected.
- An interval (a, b) is not a region because it is not an open subset of the complex plane.

Stereographic Projection

Suppose that a sphere of radius one, called a **Riemann sphere**, is positioned on the complex plane with its equator coinciding with the unit circle (Figure 2.7). Let N be the north pole of the sphere and let z be a point in the complex plane. The line from N to z intersects the sphere at one other point $z^\star$. Conversely, if $z^\star$ is a point of the sphere other than the north pole, then the line from N to $z^\star$ will intersect the plane at a single point z. It is not difficult to see that the mapping P that takes $z^\star$ to z is one-to-one from the sphere minus the north pole onto the complex plane.

This mapping, known as the **stereographic projection**, was introduced by the German mathematician Bernhard Riemann (1826–1866). It enables us to represent points in the complex plane by points on the sphere, and vice versa. This also suggests that we introduce the **point at infinity**, $z = \infty$, as the image of the north pole by the stereographic projection. Thus $P(N) = \infty$. The complex plane together with this point at infinity is called the **extended complex plane** and written $\mathbb{C} \cup \{\infty\}$. It is in one-to-one correspondence with the whole sphere. Thinking of the set $\mathbb{C} \cup \{\infty\}$ as the sphere allows the incorporation of ∞ in the complex number system.

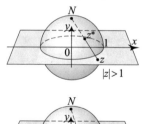

Fig. 2.7 Stereographic projection and the Riemann sphere. For $|z| > 1$, the point $z^\star$ is in the northern hemisphere. For $|z| < 1$, the point $z^\star$ is in the southern hemisphere. For $|z| = 1$, the points z and $z^\star$ coincide.

For example, for R large, the open set $|z| > R$ consisting of all complex numbers z that are exterior to the circle $|z| = R$, is mapped by the stereographic projection onto points that are near the north pole. For this reason, the set $|z| > R$ is called a neighborhood of ∞. See Exercise 22 for additional properties.

Exercises 2.1

In Exercises 1–4, identify the interior points and boundary points of the sets.

1. $\{z : |z| \leq 1\}$
2. $\{z : 0 < |z| \leq 1\}$
3. $\{z = x + iy : 0 < x < 1, y = 0\}$
4. $\{z : 1 < |z - i| \leq 2\}$

In Exercises 5–12, draw the sets of points. Are the sets open? closed? connected? regions? Justify your answers.

5. $\{z : \operatorname{Re} z > 0\}$
6. $\{z : \operatorname{Im} z \leq 1\}$
7. $B_1(i) \cup B_1(0)$
8. $\{z : z \neq 0, |\operatorname{Arg} z| < \frac{\pi}{4}\}$
9. $\{z : z \neq 0, |\operatorname{Arg} z| < \frac{\pi}{4}\} \cup \{0\}$
10. $\{z : |z + 5 + i| < 1\}$
11. $\{z : |\operatorname{Re}(z + 3 + i)| > 1\}$
12. $\{z : |z - 3i| > 1\}$

In Exercises 13–16, construct an example to illustrate the statements.

13. The union of two connected sets need not be connected.

14. A set with an infinite number of points need not have interior points.

15. If A is a subset of B, then the boundary of A need not be contained in the boundary of B.

16. The boundary of a region could be empty.

17. Prove that a set S is open if and only if its **complement**, $\mathbb{C} \setminus S$, is closed.

18. Show that a set S is open if and only if it contains none of its boundary points.

19. Suppose that $A_1, A_2, \ldots$ are open sets. Show that their union

$$\bigcup_{n=1}^{\infty} A_n = \{z : z \in A_n \text{ for some } n\}$$

is also open.

20. (a) Suppose that $A_1, A_2, \ldots$ are open sets. Show that a finite intersection

$$\bigcap_{n=1}^{N} A_n = \{z : z \in A_n \text{ for all } 1 \leq n \leq N\}$$

is also open. [*Hint*: Pick a neighborhood that is contained in all the A_n's.]
(b) Show that the infinite intersection

$$\bigcap_{n=1}^{\infty} A_n = \{z : z \in A_n \text{ for all } n\}$$

may not be open. [*Hint*: Consider $A_n = \{z : |z| < \frac{1}{n}\}$.]

21. Suppose that A and B are two regions with nonempty intersection. Show that $A \cup B$ is also a region.

22. Project Problem: Stereographic projection. Answer parts (a)–(e) by using geometric reasoning with the help of Figure 2.7.
(a) Consider a circle C on the sphere that is parallel to the complex plane (these are called parallels of latitude). What is its image under P?
(b) Which points on the sphere are mapped to the set of all z in the plane such that $|z| > R$. Can

you now justify the terminology "neighborhood of infinity"?

(c) What is the image under P of a great circle passing through the poles?

(d) What is the image under P of a circle passing through the north pole but not the south pole?

(e) Argue geometrically that z^* approaches N if and only if $P(z^*) \to \infty$.

Answers in (a)–(e) can also be justified with the help of the formulas derived in parts (f)–(h).

(f) Let $z^* = (x_1, x_2, x_3)$ and $P(z^*) = x + iy = (x, y)$. Show that the equation of the line through z^* and z is

$$\frac{x_1 - 0}{x - 0} = \frac{x_2 - 0}{y - 0} = \frac{x_3 - 1}{0 - 1}.$$

(g) Use (f) and the equation of the Riemann sphere $x_1^2 + x_2^2 + x_3^2 = 1$ to derive

$$x_1 = \frac{2x}{x^2 + y^2 + 1}, \qquad x_2 = \frac{2y}{x^2 + y^2 + 1}, \qquad x_3 = 1 - \frac{2}{x^2 + y^2 + 1}.$$

(h) Conversely, solve for x and y in (f) and obtain

$$x = \frac{x_1}{1 - x_3}, \qquad y = \frac{x_2}{1 - x_3}.$$

2.2 Limits and Continuity

Before studying differentiation it will be important to understand limits of complex functions. Using limits we also define and study continuous functions.

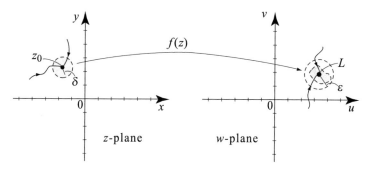

Fig. 2.8 To say that $f(z) \to L$ as $z \to z_0$ is a strong assertion; it means that no matter how z approaches z_0, the distance $|f(z) - L|$ tends to 0.

Definition 2.2.1. (Limits of Complex-valued Functions) Let f be a complex-valued function defined on subset S of the complex plane and let z_0 be an accumulation point of S. We say that a complex number L is a **limit** of $f(z)$ as z approaches z_0, if for an arbitrary $\varepsilon > 0$ there exists a $\delta > 0$ such that

$$z \in S \quad \text{and} \quad 0 < |z - z_0| < \delta \quad \Longrightarrow \quad |f(z) - L| < \varepsilon. \tag{2.2.1}$$

If such a complex number L exists, then we say that $\lim_{z \to z_0} f(z)$ exists and is equal to L, and in this case we write $\lim_{z \to z_0} f(z) = L$ or we say that $f(z) \to L$ as $z \to z_0$.

Geometrically, interpreting the absolute value $|f(z) - L|$ as the distance between $f(z)$ and L, we see from (2.2.1) that the function $f(z)$ has limit L as $z \to z_0$ if and only if the distance from $f(z)$ to L tends to zero as z tends to z_0. See Figure 2.8. Thus, $\lim_{z \to z_0} f(z) = L$ if and only if

$$\lim_{z \to z_0} |f(z) - L| = 0. \tag{2.2.2}$$

Note that in (2.2.1) the function f need not be defined at z_0.

Proposition 2.2.2. (Uniqueness of Limits) *If f is defined on a subset S of $\mathbb{C}$, z_0 is an accumulation point of S, and $\lim_{z \to z_0} f(z)$ exists, then it is a unique.*

Proof. Suppose that $\lim_{z \to z_0} f(z) = L$ and $\lim_{z \to z_0} f(z) = L'$, where L, L' are different complex numbers. Then for $\varepsilon = |L - L'|/2 > 0$ there is a $\delta > 0$ such that

$$0 < |z - z_0| < \delta \implies |f(z) - L| < \frac{1}{2}|L - L'|$$

$$0 < |z - z_0| < \delta \implies |f(z) - L'| < \frac{1}{2}|L - L'|.$$

Adding, we obtain that for $0 < |z - z_0| < \delta$ we must have

$$|L - L'| \le |f(z) - L| + |f(z) - L'| < \frac{1}{2}|L - L'| + \frac{1}{2}|L - L'| = |L - L'|,$$

which is a contradiction since $|L - L'| < |L - L'|$ is impossible. ∎

Example 2.2.3. Prove that:

(a) $\lim_{z \to z_0} z = z_0$ (b) $\lim_{z \to z_0} c = c$, where c is a constant.

Solution. (a) Given $\varepsilon > 0$, we want to find a $\delta > 0$ so that

$$0 < |z - z_0| < \delta \quad \Rightarrow \quad |f(z) - L| < \varepsilon.$$

Identifying $f(z) = z$ and $L = z_0$, this becomes

$$0 < |z - z_0| < \delta \quad \Rightarrow \quad |z - z_0| < \varepsilon.$$

Clearly, the choice $\delta = \varepsilon$ will do.

(b) The inequality $|f(z) - L| = |c - c| < \varepsilon$ holds for any choice of $\delta > 0$, and this shows that $\lim_{z \to z_0} c = c$. □

Definition 2.2.4. A function g is called **bounded** on a set S if there is a positive real number M such that $|g(z)| \le M$ for all z in S.

For example, the function $3z + 2 + i$ is bounded on the disk $S = \{z : |z| < 5\}$ since

$$|3z+2+i| \leq 3|z|+|2+i| \leq (3)(5)+\sqrt{5} = 15+\sqrt{5},$$

by the triangle inequality and the fact that $|z| < 5$.

Theorem 2.2.5. (Squeeze Theorem) *Let f, g be defined on a subset S of $\mathbb{C}$ and let z_0 be an accumulation point of S.*
(i) Suppose that $f(z) \to 0$ as $z \to z_0$ and $|g(z)| \leq |f(z)|$ in a deleted neighborhood of z_0. Then $g(z) \to 0$ as $z \to z_0$.
(ii) Suppose that $f(z) \to 0$ as $z \to z_0$ and $g(z)$ is bounded in a deleted neighborhood of z_0. Then $f(z)g(z) \to 0$ as $z \to z_0$.

Proof. (i) Since $f(z) \to 0$, given $\varepsilon > 0$, there is a δ such that $|f(z)| < \varepsilon$ whenever $0 < |z - z_0| < \delta$. But since $|g(z)| \leq |f(z)|$, we also have $|g(z) - 0| < \varepsilon$ whenever $0 < |z - z_0| < \delta$, which implies that $g(z) \to 0$ as $z \to z_0$.
(ii) Since g is bounded in a deleted neighborhood of z_0, we can find $M > 0$ and $r > 0$ such that $|g(z)| \leq M$ for $0 < |z - z_0| < r$. For $0 < |z - z_0| < r$, we have

$$0 \leq |f(z)g(z)| \leq M|f(z)|.$$

Since $M|f(z)| \to 0$, it follows from (i) that $f(z)g(z) \to 0$. ∎

Example 2.2.6. Compute $\lim\limits_{z \to 0}(\operatorname{Im} z)e^{i/|z|}$.

Solution. For $z = x + iy$, x, y real, let $f(z) = y$ and $g(z) = e^{i/|z|}$. As $z \to 0$, $f(z) \to 0$. Also, for $z \neq 0$, since $1/|z|$ is a purely real number, we have $|e^{i/|z|}| = 1$, by (1.6.11). Thus, applying Theorem 2.2.5(ii), we conclude that $\lim\limits_{z \to 0} ye^{i/|z|} = 0$. □

Computing more complicated limits by recourse to the (ε, δ)-definition (2.2.1) is not always easy. To simplify this task, we will use properties of limits.

Theorem 2.2.7. (Operations with Limits) *Let f, g be functions defined on a subset S of the complex plane and let z_0 be an accumulation point of S. Suppose that both $\lim\limits_{z \to z_0} f(z)$ and $\lim\limits_{z \to z_0} g(z)$ exist and that c_1, c_2 are complex constants. Then*

$$\lim_{z \to z_0}[c_1 f(z) + c_2 g(z)] = c_1 \lim_{z \to z_0} f(z) + c_2 \lim_{z \to z_0} g(z). \qquad (2.2.3)$$

Moreover, f, g are bounded in some neighborhood of z_0 and we have

$$\lim_{z \to z_0}[f(z)g(z)] = \lim_{z \to z_0} f(z) \lim_{z \to z_0} g(z), \qquad (2.2.4)$$

$$\lim_{z \to z_0}\left[\frac{f(z)}{g(z)}\right] = \frac{\lim\limits_{z \to z_0} f(z)}{\lim\limits_{z \to z_0} g(z)}, \quad provided \lim_{z \to z_0} g(z) \neq 0. \qquad (2.2.5)$$

Proof. The proofs of (2.2.3), (2.2.4), and (2.2.5) are similar to the proofs of the corresponding results from calculus and are left as exercises. ∎

Example 2.2.8. Suppose that $\lim\limits_{z \to i} f(z) = 2 + i$ and $\lim\limits_{z \to i} g(z) = 3 - i$. Find

$$L = \lim_{z \to i}\left[(f(z))^2 + \frac{(3+i)g(z)}{z}\right]. \tag{2.2.6}$$

Solution. Since the limits of $f(z)$, $g(z)$, and z all exist as $z \to i$, and the denominator in the expression (2.2.6) tends to $i \neq 0$, we conclude that

$$L = \left(\lim_{z \to i} f(z)\right)^2 + (3+i)\frac{\lim\limits_{z \to i} g(z)}{\lim\limits_{z \to i} z} = (2+i)^2 + (3+i)\overbrace{\frac{3-i}{i}}^{-i(10)}$$

$$= 3 + 4i - 10i = 3 - 6i. \qquad\qquad \square$$

It is often advantageous to study the limit of a complex-valued function by studying properties of the limits of the real and imaginary parts of it. This is possible because of the following simple characterization.

Theorem 2.2.9. *Let* u, v *be real-valued functions defined on a subset* S *of* $\mathbb{C}$ *and let* z_0 *be an accumulation point of* S *and* $L = a + ib$ *be a complex number. Then we have that the function* $f = u + iv$ *satisfies*

$$\lim_{z \to z_0} f(z) = L \iff \lim_{z \to z_0} u(z) = a \quad and \quad \lim_{z \to z_0} v(z) = b. \tag{2.2.7}$$

Proof. We have

$$\lim_{z \to z_0} f(z) = L \iff \lim_{z \to z_0} |f(z) - L| = 0 \iff \lim_{z \to z_0} |f(z) - L|^2 = 0.$$

But

$$|f(z) - L|^2 = |\overbrace{u(z) - a}^{\mathrm{Re}\,(f(z) - L)}|^2 + |\overbrace{v(z) - b}^{\mathrm{Im}\,(f(z) - L)}|^2.$$

The right side of this equality is the sum of two nonnegative terms; it tends to zero if and only if both terms tend to zero. Hence

$$\lim_{z \to z_0} |f(z) - L|^2 = 0 \iff \lim_{z \to z_0} |u(z) - a|^2 = 0 \quad \text{and} \quad \lim_{z \to z_0} |v(z) - b)|^2 = 0$$

$$\iff \lim_{z \to z_0} |u(z) - a| = 0 \quad \text{and} \quad \lim_{z \to z_0} |v(z) - b)| = 0$$

$$\iff \lim_{z \to z_0} u(z) = a \quad \text{and} \quad \lim_{z \to z_0} v(z) = b$$

which proves the theorem. $\blacksquare$

We have avoided thus far dealing with limits that involve ∞. What do we mean by statements such as $\lim_{z \to z_0} f(z) = \infty$ or $\lim_{z \to \infty} f(z) = L$ or even $\lim_{z \to \infty} f(z) = \infty$? We answer these questions by introducing the following definitions.

Definition 2.2.10. (Limits Involving Infinity)
(i) If f is defined in a deleted neighborbood of z_0 we write $\lim_{z \to z_0} f(z) = \infty$ to mean that for every $M > 0$ there is a $\delta > 0$ such that $0 < |z - z_0| < \delta \Rightarrow |f(z)| > M$.
(ii) If f is defined in the complement of a ball centered at the origin we write $\lim_{z \to \infty} f(z) = L$ to mean that for every $\varepsilon > 0$ there is an $R > 0$ such that $|z| > R \Rightarrow |f(z) - L| < \varepsilon$.
(iii) If f is defined in the complement of a ball centered at the origin we write $\lim_{z \to \infty} f(z) = \infty$ to mean that for every $M > 0$ there is an $R > 0$ such that $|z| > R \Rightarrow |f(z)| > M$.

Looking at these definitions, we see that $z \to \infty$ means that the real quantity $|z| \to \infty$, and similarly $f(z) \to \infty$ means that $|f(z)| \to \infty$. Hence

$$\lim_{z \to z_0} f(z) = \infty \quad \Leftrightarrow \quad \lim_{z \to z_0} |f(z)| = \infty; \tag{2.2.8}$$

$$\lim_{z \to \infty} f(z) = L \quad \Leftrightarrow \quad \lim_{|z| \to \infty} |f(z) - L| = 0; \tag{2.2.9}$$

$$\lim_{z \to \infty} f(z) = \infty \quad \Leftrightarrow \quad \lim_{|z| \to \infty} |f(z)| = \infty. \tag{2.2.10}$$

Limits at infinity can also be reduced to limits at $z_0 = 0$ by means of the inversion $1/z$. The idea is that taking the limit as $z \to \infty$ of $f(z)$ is the same procedure as taking the limit as $z \to 0$ of $f\left(\frac{1}{z}\right)$. It is straightforward to verify that

$$\lim_{z \to \infty} f(z) = L \quad \Leftrightarrow \quad \lim_{z \to 0} f\left(\frac{1}{z}\right) = L; \tag{2.2.11}$$

and

$$\lim_{z \to \infty} f(z) = \infty \quad \Leftrightarrow \quad \lim_{z \to 0} f\left(\frac{1}{z}\right) = \infty. \tag{2.2.12}$$

These equivalent statements are sometimes useful. For example, appealing to (2.2.12), we have

$$\lim_{z \to \infty} \frac{1}{z} = \lim_{z \to 0} \frac{1}{1/z} = \lim_{z \to 0} z = 0. \tag{2.2.13}$$

Similarly, for a constant c and a positive integer n, we have

$$\lim_{z \to \infty} \frac{c}{z^n} = \lim_{z \to 0} \frac{c}{1/z^n} = \lim_{z \to 0} cz^n = 0.$$

Example 2.2.11. Evaluate:
(a) $\displaystyle \lim_{z \to \infty} \frac{z - 1}{z + i}$ and (b) $\displaystyle \lim_{z \to \infty} \frac{2z + 3i}{z^2 + z + 1}$.

Solution. (a) Since we are concerned with the behavior of the function for $|z|$ large, it is safe to divide both numerator and denominator of $\frac{z-1}{z+i}$ by z, and we conclude

$$\lim_{z \to \infty} \frac{z-1}{z+i} = \lim_{z \to \infty} \frac{1 - \frac{1}{z}}{1 + \frac{i}{z}}$$

$$= \frac{1 - \lim_{z \to \infty} \frac{1}{z}}{1 + i \lim_{z \to \infty} \frac{1}{z}} \qquad \text{[by (2.2.5) and (2.2.3)]}$$

$$= 1 \qquad \text{[by (2.2.13)].}$$

(b) Dividing both numerator and denominator by z^2 we write

$$\lim_{z \to \infty} \frac{2z + 3i}{z^2 + z + 1} = \lim_{z \to \infty} \frac{\frac{2}{z} + \frac{3i}{z^2}}{1 + \frac{1}{z} + \frac{1}{z^2}} = \frac{0 + 0}{1 + 0 + 0} = 0. \qquad \square$$

   While we have successfully used skills from calculus to compute complex-valued limits, real-variable intuition may not always apply. For example, the limit $\lim_{z \to \infty} e^{-z}$ is not 0; in fact, this limit does not exist (Exercise 21).

Continuous Functions

Often, the limit of a function as the variable approaches a point equals with the value of the function at this point. This property is called continuity.

Definition 2.2.12. Let f be defined on an a subset S of $\mathbb{C}$ and let z_0 be a point in S. We say that f is **continuous at** z_0 if for every $\varepsilon > 0$ there is a $\delta > 0$ such that

$$z \in S, \quad |z - z_0| < \delta \implies |f(z) - f(z_0)| < \varepsilon. \qquad (2.2.14)$$

The function f is called **continuous on** S if it is continuous at every point in S.

   If z_0 is not an accumulation point of S there is a $\delta > 0$ with $B'(z_0, \delta) \cap S = \emptyset$; then $z \in S$ and $|z - z_0| < \delta \implies z = z_0$, thus f is continuous at z_0, as (2.2.14) is satisfied for any $\varepsilon > 0$. If $z_0 \in S$ happens to be an accumulation point of S, then (2.2.1) [with $L = f(z_0)$] is equivalent to (2.2.14), since obviously (2.2.14) holds when $z = z_0$; in this case f is continuous at z_0 if and only if $\lim_{z \to z_0} f(z) = f(z_0)$.
   Since continuity is defined in terms of limits, many properties of limits extend to continuous functions.

Theorem 2.2.13. *Let f, g be complex-valued functions defined on a subset S of $\mathbb{C}$ and let z_0 be a point in S. Suppose that f, g are continuous at z_0. Let c_1, c_2 be complex constants. Then the following assertions are valid:*

(i) Re f *and* Im f *are continuous at* z_0.
(ii) $c_1 f + c_2 g$, fg *are continuous at* z_0.
(iii) f/g *is continuous at* z_0, *provided* $g(z_0) \neq 0$.
(iv) *If* h *is defined on a set containing* $f(z_0)$ *and is continuous at* $f(z_0)$, *then the composition* $h \circ f$ *is continuous at* z_0.

Proof. Assertion *(i)* is a consequence of Theorem 2.2.9. The proofs of *(ii)* and *(iii)* are immediate consequences of Theorem 2.2.7. To prove *(iv)*, in view of the continuity of h at $f(z_0)$, given $\varepsilon > 0$ we find a $\delta > 0$ such that

$$w \in f[S] \quad \text{and} \quad |w - f(z_0)| < \delta \quad \Longrightarrow \quad |h(w) - h(f(z_0))| < \varepsilon.$$

For the positive δ we found, by the continuity of f at z_0, there is a $\eta > 0$ such that

$$z \in S \quad \text{and} \quad |z - z_0| < \eta \quad \Longrightarrow \quad |f(z) - f(z_0)| < \delta.$$

Putting these implications together and using that $z \in S$ implies $f(z) \in f[S]$, we obtain:

$$z \in S \quad \text{and} \quad |z - z_0| < \eta \quad \Longrightarrow \quad |h(f(z)) - h(f(z_0))| < \varepsilon,$$

thus proving *(iv)*.   ∎

Example 2.2.14. (Polynomial and rational functions)
(a) Let $a_j \in \mathbb{C}$. Show that a polynomial $p(z) = a_n z^n + a_{n-1} z^{n-1} + \cdots + a_0$ is continuous at all points in the plane.
(b) A **rational** function is a function of the form

$$r(z) = \frac{p(z)}{q(z)},$$

where p and q are polynomials. This is defined for all $z \in \mathbb{C}$ for which $q(z) \neq 0$. Show that a rational function is continuous at all points where $q(z) \neq 0$.

Solution. (a) Since the function $f(z) = z$ is continuous, we use the fact that the product of two continuous functions is continuous to conclude that $z^2, z^3, \ldots, z^n$ are continuous. Then, by repeated applications of the fact that a linear combination of continuous functions is continuous, we conclude that $a_n z^n + a_{n-1} z^{n-1} + \cdots + a_0$ is a continuous function.
(b) By part (a), the polynomials p and q are continuous in the entire plane. Hence $r(z) = \frac{p(z)}{q(z)}$ is continuous on $\mathbb{C}$ except at those points where $q(z) = 0$.   □

Example 2.2.15. (Limits and continuity of rational functions) Compute the limits and determine whether the rational functions are continuous at the given points.
(a) $\displaystyle \lim_{z \to 2i} \frac{2z^2 - i}{z + 2}$ (b) $\displaystyle \lim_{z \to i} \frac{z - i}{z^2 + 1}$

Solution. (a) Since $z + 2 \neq 0$ when $z = 2i$, the function in (a) is continuous at $z = 2i$ by Example 2.2.14(b). The limit as $z \to 2i$ is then computed by evaluation:

$$\lim_{z \to 2i} \frac{2z^2 - i}{z + 2} = f(2i) = \frac{2(2i)^2 - i}{2i + 2} = \frac{-8 - i}{2 + 2i} = \frac{1}{4}(-9 + 7i).$$

(b) The denominator $z^2 + 1$ vanishes at $z = i$ and so the rational function in (b) is not continuous at $z = i$. Does the limit exist at $z = i$? We have

$$f(z) = \frac{z - i}{z^2 + 1} = \frac{z - i}{(z - i)(z + i)}.$$

For $z \neq i$, we cancel the factor $z - i$ and obtain

$$\lim_{z \to i} \frac{z - i}{z^2 + 1} = \lim_{z \to i} \frac{1}{z + i} = \frac{1}{2i} = \frac{-i}{2}. \qquad \square$$

In Example 2.2.15(b), the $\lim_{z \to i} f(z)$ exists at the point of discontinuity $z = i$, and for this reason the discontinuity of f at this point can be removed by redefining $f(i) = -i/2$. Such a point of discontinuity is called a **removable discontinuity**.

If the discontinuity at a point cannot be removed, then it is called a **nonremovable discontinuity**. The examples below concern nonremovable discontinuities. We use the uniqueness of the limit to show that a limit fails to exist. The method works as follows: If one can show that $f(z) \to L$ as z approaches z_0 on curve C but $f(z) \to L'$ as z approaches z_0 on curve C', and $L \neq L'$, then, by the uniqueness of limits, we conclude that $\lim_{z \to z_0} f(z)$ does not exist. See Figure 2.9.

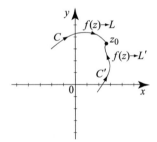

Fig. 2.9 $\lim\limits_{z \to z_0} f(z)$ cannot exist if $L \neq L'$.

Example 2.2.16. (A nonremovable discontinuity) Show that $\frac{\operatorname{Re} z}{z}$ has a nonremovable discontinuity at $z = 0$.

Solution. Write $z = x + iy$ where x and y are not both equal to 0. We follow the strategy outlined before the example and show that $\lim_{z \to 0} f(z)$ does not exist by letting z approach 0 in two different ways: Once along the x-axis (that is, $y = 0$ and $z = x$) and once along the y-axis (that is, $x = 0$ and $z = iy$). For $z = x$, we have

$$\lim_{z \to 0} \frac{\operatorname{Re} z}{z} = \lim_{x \to 0} \frac{x}{x} = \lim_{x \to 0} 1 = 1.$$

For $z = iy$ with y real, we have $\text{Re}\, z = 0$, and so

$$\lim_{z \to 0} \frac{\text{Re}\, z}{z} = \lim_{iy \to 0} \frac{0}{iy} = 0.$$

Since we have obtained different limits as we approached 0 in different ways, we conclude that the function $\frac{\text{Re}\, z}{z}$ has no limit as $z \to 0$.   □

The next example involves a function with infinitely many nonremovable discontinuities.

Example 2.2.17. (The nonremovable discontinuities of $\text{Arg}\, z$**)** The principal branch of the argument $\text{Arg}\, z$ takes the value of argument z that is in the interval $-\pi < \text{Arg}\, z \leq \pi$. It is not defined at $z = 0$ and hence $\text{Arg}\, z$ is not continuous at $z = 0$. We show that $z = 0$ is not a removable discontinuity of $\text{Arg}\, z$ by showing that $\lim_{z \to 0} \text{Arg}\, z$ does not exist.

Indeed, if $z = x > 0$, then $\text{Arg}\, z = 0$ and so $\lim_{z=x \downarrow 0} \text{Arg}\, z = 0$, where the down-arrow denotes the limit from the right, also denoted as $\lim_{z=x \to 0^+} \text{Arg}\, z$. However, if $z = x < 0$, then $\text{Arg}\, z = \pi$ and so $\lim_{z=x \uparrow 0} \text{Arg}\, z = \pi$, where the up-arrow denotes the limit from the left, also denoted as $\lim_{z=x \to 0^-} \text{Arg}\, z$. By the uniqueness of limits, we conclude that $\lim_{z \to 0} \text{Arg}\, z$ does not exist. Also, for a point on the negative x-axis, $z_0 = x_0 < 0$, we have $\text{Arg}\, z_0 = \pi$. If z approaches z_0 from the second quadrant, say along a curve C as in Figure 2.10, we have $\lim_{z \to z_0} \text{Arg}\, z = \pi = \text{Arg}\, z_0$. But if z approaches z_0 from the third quadrant, say along curve C' as shown in Figure 2.10, we have $\lim_{z \to z_0} \text{Arg}\, z = -\pi$.

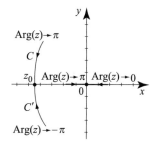

Fig. 2.10 $\text{Arg}\, z$ has nonremovable discontinuities at $z = 0$ and at all negative real z.

Hence $\text{Arg}\, z$ is not continuous at z_0 and the discontinuity is not removable, because $\lim_{z \to z_0} \text{Arg}\, z$ does not exist for such z_0. It is not hard to show, using geometric considerations, that for $z \neq 0$ and z not on the negative x-axis, $\text{Arg}\, z$ is continuous. Since the set of points of continuity of $\text{Arg}\, z$ is the complex plane $\mathbb{C}$ minus the interval $(-\infty, 0]$ on the real line, the principal branch of the argument is continuous on $\mathbb{C} \setminus (\infty, 0]$.   □

Many important functions of several variables are made up of products, quotients, and linear combinations of functions of a single variable. For example, the function $u(x, y) = e^x \cos y$ is the product of two functions of a single variable each; namely, e^x and $\cos y$. The exponential function $e^z = e^x(\cos y + i \sin y)$ is a linear combination of two products of functions of a single variable. In establishing the continuity of such functions, the following simple observations are very useful.

Proposition 2.2.18. *Suppose that $\phi(x)$ is a continuous function of a single variable x defined on an interval (a, b). Then the function $f(x, y) = \phi(x)$ is continuous at (x_0, y_0) whenever x_0 is in (a, b). Similarly, $g(x, y) = \phi(y)$ is continuous at (x_0, y_0) whenever y_0 is in (a, b).*

Proof. If $(x, y) \to (x_0, y_0)$, then $x \to x_0$ and so $\phi(x) \to \phi(x_0)$ and, consequently, $f(x, y) = \phi(x) \to \phi(x_0) = f(x_0, y_0)$. Thus f is continuous at (x_0, y_0) as claimed. The second part of the proposition follows similarly.   ∎

Combined with Theorem 2.2.13, Proposition 2.2.18 becomes a handy tool. Here are some interesting applications.

Example 2.2.19. (Exponential and trigonometric functions) Show that the following are continuous functions of z.
(a) e^z (b) $\cos z$

Solution. (a) We know from calculus that the functions e^x, $\cos x$, and $\sin x$ are continuous for all x. By Proposition 2.2.18, the functions $f_1(x, y) = e^x$, $f_2(x, y) = \cos y$, and $f_3(x, y) = \sin y$ are continuous for all (x, y). Appealing to Theorem 2.2.13, we see that $e^x \cos y + i e^x \sin y = e^z$ is continuous for all (x, y).
(b) The function e^{iz} is continuous because it is the composition of two continuous functions; namely, iz and e^z. Similarly, e^{-iz} is continuous, and hence the linear combination $\frac{e^{iz} + e^{-iz}}{2} = \cos z$ is also continuous.   □

Example 2.2.20. Show that the absolute value $|z|$ is continuous.

Solution. We show that $\lim_{z \to z_0} |z| = |z_0|$ for arbitrary z_0. By the lower estimate (1.2.19) we have $\big||z| - |z_0|\big| \leq |z - z_0|$. As $z \to z_0$, $|z - z_0| \to 0$, and so $\big||z| - |z_0|\big| \to 0$ by the squeeze theorem.   □

Continuity of the Logarithms

Understanding the behavior of the logarithm is crucial to certain applications. In view of our knowledge of $\operatorname{Arg} z$ and $|z|$, the following result is not a surprise.

Theorem 2.2.21. (Continuity of the Logarithm) *The principal branch of the logarithm*

$$\operatorname{Log}(z) = \ln|z| + i\operatorname{Arg}(z), \quad -\pi < \operatorname{Arg} z \leq \pi \quad (z \neq 0), \tag{2.2.15}$$

is continuous for all z in $\mathbb{C} \setminus (-\infty, 0]$. For z in $(-\infty, 0]$, the discontinuities of $\operatorname{Log} z$ are not removable.

Proof. We showed in Example 2.2.20 that $|z|$ is a continuous function of z. Composing the continuous function $\ln x$, for $x > 0$, with the real-valued function $|z|$, it follows from Theorem 2.2.13(iv) that $\ln|z|$ is continuous for all $z \neq 0$. We showed in Example 2.2.17 that $\text{Arg}\,z$ is continuous except for nonremovable discontinuities at $z = 0$ and z on the negative x-axis. Appealing to Theorem 2.2.13 parts (i) and (ii), we see that a discontinuity at $z = z_0$ of a function $f(z) = u(z) + iv(z)$ is removable if and only if z_0 is a removable discontinuity of both u and v (Exercise 42). Thus, in view of the nonremovable discontinuities of $\text{Arg}\,z$, it follows that $z = y$, $y \leq 0$, are also nonremovable discontinuities of $\text{Log}\,z$. Thus the set of points of continuity of $\text{Log}\,z$ is $\mathbb{C} \setminus (-\infty, 0]$.   ∎

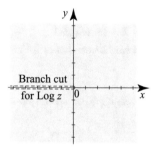

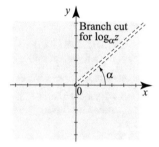

Fig. 2.11 $\text{Log}\,z$ has nonremovable discontinuities at $z = 0$ and at all negative real z. For all other z, $\text{Log}\,z$ is continuous.

Fig. 2.12 The branch cut of $\log_\alpha z$ is the ray at angle α. The branch cut is the set of nonremovable discontinuities of $\log_\alpha z$.

A discussion similar to the preceding one shows that a branch of the logarithm, $\log_\alpha z$, is continuous at all z except for nonremovable discontinuities at $z = 0$ and z on the ray at angle α. The set of nonremovable discontinuities of $\log_\alpha z$ is called a **branch cut**. For example, the branch cut of $\text{Log}\,z$ is $(-\infty, 0]$ (Figure 2.11), and the branch cut of $\log_\alpha z$ is the ray at angle α (Figure 2.12).

Exercises 2.2

In Exercises 1–12, evaluate the limits justifying each step in terms of properties of limits from this section.

1. $\displaystyle\lim_{z \to i} 3z^2 + 2z - 1$

2. $\displaystyle\lim_{z \to 2+i} z + \frac{1}{z}$

3. $\displaystyle\lim_{z \to 0} \frac{z}{\cos z}$

4. $\displaystyle\lim_{z \to 2} \frac{z^4 - 16}{z - 2}$

5. $\displaystyle\lim_{z \to i} \frac{1}{z - i} - \frac{1}{z^2 + 1}$

6. $\displaystyle\lim_{z \to 0} z\,\text{Arg}\,z$

7. $\displaystyle\lim_{z \to 0} z e^{i\,\text{Re}\,z}$

8. $\displaystyle\lim_{z \to i} \text{Re}\,(z) \sin z$

9. $\displaystyle\lim_{z \to -3} (\text{Arg}\,z)^2$

10. $\displaystyle\lim_{z \to 1} (z + 1)\,\text{Im}\,(iz)$

11. $\displaystyle\lim_{z \to 0} \sin \bar{z}$

12. $\displaystyle\lim_{z \to 0} z e^{i/|z|^2}$

In Exercises 13–18, evaluate the limits involving ∞. Justify your steps.

13. $\displaystyle\lim_{z \to \infty} \frac{z + 1}{3iz + 2}$

14. $\displaystyle\lim_{z \to \infty} \frac{z^2 + i}{z^3 + 3z^2 + z + 1}$

15. $\displaystyle\lim_{z \to \infty} \left(\frac{z^3 + i}{z^3 - i} \right)^2$

16. $\lim\limits_{z\to i}\dfrac{1}{z^2+1}$　　　　　**17.** $\lim\limits_{z\to 1}\dfrac{-1}{(z-1)^2}$　　　　　**18.** $\lim\limits_{z\to\infty}\dfrac{\mathrm{Log}\,z}{z}$

In Exercises 19–26, *show that the limits do not exist by approaching* z_0 *from different directions. If the limit involves* $z\to\infty$, *try some of the following directions: the positive x-axis, the negative x-axis, the positive y-axis* ($z=iy$, $y>0$), *or the negative y-axis* ($z=iy$, $y<0$).

19. $\lim\limits_{z\to-3}\mathrm{Arg}\,z$　　　　　**20.** $\lim\limits_{z\to-1}\mathrm{Log}\,z$　　　　　**21.** $\lim\limits_{z\to\infty}e^{-z}$

22. $\lim\limits_{z\to 0}\dfrac{\bar{z}}{z}$　　　　　**23.** $\lim\limits_{z\to 0}e^{1/z}$　　　　　**24.** $\lim\limits_{z\to 0}\dfrac{\mathrm{Re}\,z}{|z|^2}$

25. $\lim\limits_{z\to 0}\dfrac{z}{|z|}$　　　　　**26.** $\lim\limits_{z\to 0}\dfrac{\mathrm{Im}\,z}{z}$

27. Show that if a function has a limit as z tends to z_0, then it is bounded in a neighborhood of z_0. Then use this fact and the triangle inequality to derive (2.2.4).

28. Derive property (2.2.5). [*Hint*: You may assume first that $f=1$ and then use (2.2.4).]

29. Prove (2.2.11).　　　　　　　　　　　**30.** Prove (2.2.12).

31. Assume that the function f does not vanish on a deleted neighborhood of $z_0\in\mathbb{C}$. Show that $\lim_{z\to z_0}f(z)=0$ if and only if $\lim_{z\to z_0}\frac{1}{f(z)}=\infty$.

32. Use the result of Exercise 31 to evaluate $\lim_{z\to 0}\frac{\cos z}{z}$.

In Exercises 33–40, *determine the set of points where the functions are continuous. For a point of discontinuity, determine whether it is removable or not.*

33. $\dfrac{z-i}{z+1+3i}$　　　　　**34.** $\dfrac{2z+1}{z^2+3z+2}$　　　　　**35.** $\bar{z}$

36. $\mathrm{Log}\,(z+1)$　　　　　**37.** $\sin z$　　　　　**38.** $(\mathrm{Arg}\,z)^2$

39. $z(\mathrm{Arg}\,z)^2$　　　　　**40.** $\dfrac{z}{|z|}$

41. (Pre-image of sets) Let f be a complex-valued function defined on a subset S of $\mathbb{C}$. If A is a subset of $\mathbb{C}$, the **pre-image** or **inverse image** of A under f is the set $f^{-1}[A]=\{z\in S:\ f(z)\in A\}$. (a) Show that f is continuous if and only if $f^{-1}[A]$ is open whenever A is open. (b) Show that f is continuous if and only if $f^{-1}[A]$ is closed whenever A is closed.

42. Show that a discontinuity at $z=z_0$ of a function $f(z)=u(z)+iv(z)$ is removable if and only if z_0 is a removable discontinuity of both u and v.

2.3 Analytic Functions

The derivative of a real-variable function $h(x)$ at a point x_0 is defined as $h'(x_0)=\lim_{x\to x_0}\frac{h(x)-h(x_0)}{x-x_0}$, when the limit exists. The definition of the derivative of a complex function is a natural extension of the real one.

Definition 2.3.1. (Complex Derivative) Let f be defined on an open subset U of $\mathbb{C}$ and let $z_0\in U$. We say that f has a complex derivative at the point z_0 if the limit

$$\lim_{z\to z_0}\frac{f(z)-f(z_0)}{z-z_0}\qquad(2.3.1)$$

exists. If this is the case, then the number in (2.3.1) is called the **complex derivative** of f at z_0 and is denoted by $f'(z_0)$.

We say that f is **analytic** on U if it has a complex derivative at every point in U. We also say that f **is analytic at a point** w **in** U if it is analytic on some open neighborhood of w contained in U.

While this extension looks similar to the definition of a derivative of functions defined on the real line, we are asking a lot more in the complex case. In the real case, x can only approach x_0 from either the right or the left. In the complex case, z can approach z_0 from any number of directions. For the derivative to exist, we are requiring that the limit exists no matter how we approach z_0 in (2.3.1).

Definition 2.3.2. An analytic function defined on the complex plane $\mathbb{C}$ is said to be **entire**.

Example 2.3.3. Show that the functions below are entire and find their derivatives.
(a) $f(z) = 2 + 4i$.   (b) $g(z) = (3 - i)z$.   (c) $h(z) = z^2$.

Solution. We use the difference quotient as in (2.3.1) to calculate the derivatives.
(a) Fix z_0 in the plane. Since $f(z) = 2 + 4i$ is constant, $f(z) = f(z_0)$ for all z. Hence $\frac{f(z) - f(z_0)}{z - z_0} = 0$. Taking the limit as $z \to z_0$, we get $f'(z_0) = 0$. Thus $f'(z) = 0$ for all z, and hence f is analytic on $\mathbb{C}$, or entire.
(b) Fix z_0 in the plane. We have

$$g'(z_0) = \lim_{z \to z_0} \frac{g(z) - g(z_0)}{z - z_0} = \lim_{z \to z_0} \frac{(3 - i)z - (3 - i)z_0}{z - z_0} = 3 - i.$$

Thus $g'(z) = 3 - i$ for all z, and so f is entire.
(c) Fix z_0 in the plane. We have

$$h'(z_0) = \lim_{z \to z_0} \frac{z^2 - z_0^2}{z - z_0} = \lim_{z \to z_0} \frac{(z - z_0)(z + z_0)}{z - z_0} = \lim_{z \to z_0} z + z_0 = 2z_0.$$

Thus $h'(z) = 2z$ for all z and so h is entire.   □

More generally, if c is a constant, we have $f(z) = c \Rightarrow f'(z) = 0$. We also have

$$g(z) = cz \Rightarrow g'(z) = c. \tag{2.3.2}$$

The well-known result from calculus that differentiable functions are continuous has an analog in the complex case.

Theorem 2.3.4. *An analytic function defined on an open subset of the complex plane is continuous.*

Proof. Let z_0 be a point in an open set U and let f be an analytic function on U. We must show that $\lim_{z \to z_0} f(z) = f(z_0)$. Using the fact that the limit of a product is the product of the limits [Property (2.2.4)], we write

$$\lim_{z \to z_0} \left(f(z) - f(z_0) \right) = \lim_{z \to z_0} \frac{f(z) - f(z_0)}{z - z_0} (z - z_0)$$

$$= \lim_{z \to z_0} \frac{f(z) - f(z_0)}{z - z_0} \lim_{z \to z_0} (z - z_0)$$

$$= f'(z_0) \cdot 0 = 0.$$

Thus f is continuous at z_0. ∎

Theorem 2.3.4 yields that a discontinuous function at z_0 does not have a complex derivative at z_0. The converse of Theorem 2.3.4 is not true: the function $\bar{z}$ is continuous on $\mathbb{C}$ but does not have a complex derivative at any point (Example 2.3.9).

Since complex derivatives are modeled after real ones, it should not come as a surprise that many of the properties of real derivatives hold for analytic functions.

Theorem 2.3.5. (Properties of Analytic Functions) *Suppose that f and g are analytic functions on an open subset U of the complex plane and let c_1, c_2 be complex constants.*
(i) Then $c_1 f + c_2 g$ and fg are analytic on U and for all $z \in U$

$$(c_1 f + c_2 g)'(z) = c_1 f'(z) + c_2 g'(z). \tag{2.3.3}$$

(ii) The function fg is analytic on U and for all $z \in U$

$$(fg)'(z) = f'(z)g(z) + f(z)g'(z). \tag{2.3.4}$$

(iii) The function $\dfrac{f}{g}$ is analytic on $W = U \setminus \{w \in U : g(w) = 0\}$ and for all $z \in W$

$$\left(\frac{f}{g} \right)'(z) = \frac{f'(z)g(z) - f(z)g'(z)}{(g(z))^2}. \tag{2.3.5}$$

Proof. We leave the proof of (2.3.3) as an exercise and we prove (2.3.4). Using the definition of the derivative and the continuity of g (Theorem 2.3.4), we have

$$(fg)'(z_0) = \lim_{z \to z_0} \frac{f(z)g(z) - f(z_0)g(z_0)}{z - z_0}$$

$$= \lim_{z \to z_0} \frac{f(z)g(z) - f(z_0)g(z) + f(z_0)g(z) - f(z_0)g(z_0)}{z - z_0}$$

$$= \lim_{z \to z_0} g(z) \frac{f(z) - f(z_0)}{z - z_0} + \lim_{z \to z_0} f(z_0) \frac{g(z) - g(z_0)}{z - z_0}$$

$$= g(z_0)f'(z_0) + f(z_0)g'(z_0).$$

Next we prove (2.3.5) when $f = 1$. We notice that since $g(z_0)$ is not zero at z_0, then for $\varepsilon = |g(z_0)|/2 > 0$ there is a $\delta > 0$ such that $|z - z_0| < \delta$ we have $|g(z) - g(z_0)| < |g(z_0)|/2$, which implies $|g(z_0)|/2 < |g(z)|$, hence $g(z) \neq 0$. Then for $|z - z_0| < \delta$ we write

$$(1/g)'(z_0) = \lim_{z \to z_0} \frac{1/g(z) - 1/g(z_0)}{z - z_0}$$

$$= \lim_{z \to z_0} \frac{-1}{g(z)g(z_0)} \frac{g(z) - g(z_0)}{z - z_0}$$

$$= -\lim_{z \to z_0} \frac{1}{g(z)g(z_0)} \lim_{z \to z_0} \frac{g(z) - g(z_0)}{z - z_0}$$

$$= -\frac{1}{g(z_0)^2} g'(z_0),$$

where we used the continuity of g at z_0 and the fact that $g(z) \neq 0$ for $|z - z_0| < \delta$. To prove (2.3.5) for a general f we combine (2.3.4) and (2.3.5) for $f = 1$. ∎

Example 2.3.6. Show that z^n $(n = 1, 2, \ldots)$ is entire and

$$\frac{d}{dz} z^n = n z^{n-1}. \tag{2.3.6}$$

Solution. We give a proof by induction. The case $n = 1$ was already stated in (2.3.2). Suppose as an induction hypothesis that (2.3.6) holds for n; we will prove that it holds for $n + 1$. Setting $h(z) = z^{n+1}$, we write $h(z) = z^n z$. Applying the product rule for differentiation (2.3.4) and the induction hypothesis, we get

$$h'(z) = (z^n)'z + z^n z' = n z^{n-1} z + z^n = (n+1)z^n,$$

as desired. Hence z^n is entire and (2.3.6) is valid. □

Example 2.3.7. (Analyticity of a rational function) Find the complex derivative of

$$h(z) = \frac{(z+1)(z+i)^2}{z+1-3i}$$

and determine where h is analytic.

Solution. The formal manipulations are exactly as if we were working with a real function and treating the complex numbers as real constants. We use the quotient and product rules for differentiation, (2.3.5) and (2.3.4), and get

$$h'(z) = \frac{((z+i)^2 + (z+1)2(z+i))(z+1-3i) - (z+1)(z+i)^2}{(z+1-3i)^2}$$

$$= \frac{2z^3 + (4-7i)z^2 + (14-2i)z + (6+5i)}{(z+1-3i)^2}.$$

This function is analytic on $\mathbb{C} \setminus \{-1 + 3i\}$ since at $z = -1 + 3i$ the denominator of h vanishes. □

Using linear combinations of powers of z and appealing to Example 2.3.6 and the linearity of the derivative (2.3.3), we conclude that polynomials are entire functions. Appealing to the quotient rule, as in Example 2.3.7, we see that a rational function is analytic at all z where $g(z) \neq 0$. The proof of the next proposition is omitted.

Proposition 2.3.8. *Let n be a nonnegative integer. A polynomial of degree n, $p(z) = a_n z^n + a_{n-1} z^{n-1} + \cdots + a_1 z + a_0$, is entire. Its derivative is*

$$p'(z) = n a_n z^{n-1} + (n-1) a_{n-1} z^{n-2} + \cdots + a_1. \tag{2.3.7}$$

A rational function $f(z) = p(z)/q(z)$, where $p(z)$ and $q(z)$ are polynomials, is analytic at all points z where $q(z) \neq 0$. Its derivative is

$$f'(z) = \frac{p'(z)q(z) - p(z)q'(z)}{q(z)^2}. \tag{2.3.8}$$

Not every complex-valued function is analytic. For instance the functions $\operatorname{Re} z$, $\operatorname{Im} z$, $\bar{z}$, and $|z|$ are not analytic.

Example 2.3.9. (Functions that are nowhere analytic) Show that the functions
(a) $\bar{z}$ and   (b) $\operatorname{Re} z$
do not have complex derivatives at any point in $\mathbb{C}$.

Solution. We establish these claims by showing that the limit (2.3.1) does not exist for any complex number z_0. We achieve this approaching z_0 from two different directions and showing that the limits obtained are not equal.
(a) Fix a point $z_0 = x_0 + i y_0$ in the plane. Our goal is to show that the limit

$$\lim_{z \to z_0} \frac{\bar{z} - \overline{z_0}}{z - z_0} \tag{2.3.9}$$

does not exist. We will approach z_0 from the two directions as indicated in Figure 2.13. For z on C, we have

$$z = z_0 + t; \quad z - z_0 = t; \quad \bar{z} - \overline{z_0} = \overline{z - z_0} = \bar{t} = t,$$

because t is real. Thus,

$$\lim_{\substack{z \to z_0 \\ z \text{ on } C}} \frac{\bar{z} - \overline{z_0}}{z - z_0} = \lim_{t \to 0} \frac{t}{t} = 1.$$

For z on C', we have

$$z = z_0 + it; \quad z - z_0 = it; \quad \bar{z} - \overline{z_0} = \overline{it} = -it$$

since it is purely imaginary. Thus,

$$\lim_{\substack{z \to z_0 \\ z \text{ on } C'}} \frac{\bar{z} - \overline{z_0}}{z - z_0} = \lim_{t \to 0} \frac{-it}{it} = -1.$$

Since the limit along C is not equal to the limit along C', we conclude that the limit in (2.3.9) does not exist. Hence the function $\bar{z}$ is not analytic at z_0. Since z_0 is arbitrary, it follows that $\bar{z}$ is nowhere analytic.

(b) We follow the approach in (a) and use the same directions along C and C'. For z on C, $\operatorname{Re} z - \operatorname{Re} z_0 = x_0 + t - x_0 = t$, and, for z on C', $\operatorname{Re} z - \operatorname{Re} z_0 = x_0 - x_0 = 0$.

Thus we write

$$\lim_{\substack{z \to z_0 \\ z \text{ on } C}} \frac{\operatorname{Re} z - \operatorname{Re} z_0}{z - z_0} = \lim_{t \to 0} \frac{t}{t} = 1$$

and

$$\lim_{\substack{z \to z_0 \\ z \text{ on } C'}} \frac{\operatorname{Re} z - \operatorname{Re} z_0}{z - z_0} = \lim_{t \to 0} \frac{0}{it} = 0.$$

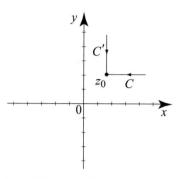

So the derivative of $\operatorname{Re} z$ does not exist at z_0. Since z_0 is arbitrary, we conclude that $\operatorname{Re} z$ is nowhere analytic.

Fig. 2.13 For $z \in C$ we have $z - z_0 = t$ while for $z \in C'$, $z - z_0 = it$.

There is also a quick proof of (b) based on the result of (a) and the identity $\bar{z} = 2 \operatorname{Re} z - z$. In fact, if $\operatorname{Re} z$ has a derivative at z_0, then by the properties of the derivative it would follow that $\bar{z}$ has a derivative at z_0, which contradicts (a). □

Suppose that $f(z)$ has a complex derivative at a point z_0 and let

$$\varepsilon(z) = \frac{f(z) - f(z_0)}{z - z_0} - f'(z_0). \tag{2.3.10}$$

Then $\varepsilon(z) \to 0$ as $z \to z_0$, because the difference quotient in (2.3.10) tends to $f'(z_0)$. Solving for $f(z)$ in (2.3.10) we obtain

$$f(z) = \overbrace{f(z_0) + f'(z_0)(z - z_0)}^{\text{linear function of } z} + \varepsilon(z)(z - z_0). \tag{2.3.11}$$

This expression shows that, near a point where f is analytic, $f(z)$ is approximately a linear function. The converse is also true.

Proposition 2.3.10. *Let U be an open subset of $\mathbb{C}$. A function f on U has a complex derivative at a point $z_0 \in U$ if and only if there is a complex number A and a function $\varepsilon(z)$ such that*

$$f(z) = f(z_0) + A(z - z_0) + \varepsilon(z)(z - z_0), \tag{2.3.12}$$

and $\varepsilon(z) \to 0$ as $z \to z_0$. If this is the case, then $A = f'(z_0)$.

Proof. We have already one direction. For the other direction, suppose that $f(z)$ can be written as in (2.3.12). Then, for $z \neq z_0$,

$$\frac{f(z) - f(z_0)}{z - z_0} = A + \varepsilon(z). \tag{2.3.13}$$

Taking the limit as $z \to z_0$ and using the fact that $\varepsilon(z) \to 0$, we conclude that $f'(z_0)$ exists and equals A. ∎

So far we have been successful in differentiating polynomials and rational functions. To go beyond these examples we need more tools, such as composition of functions. The formalism of Proposition 2.3.10 greatly simplifies the proofs related to compositions of analytic functions.

Theorem 2.3.11. (Chain Rule) *Suppose that g is analytic on an open set U and that f is analytic and an open set containing $g[U]$. Then $f \circ g$ is an analytic function on U. Moreover, for z_0 in U the chain rule identity holds*

$$(f \circ g)'(z_0) = f'(g(z_0))g'(z_0). \tag{2.3.14}$$

Proof. Suppose g is analytic at z_0 and f is analytic at $g(z_0)$. We want to show that

$$(f \circ g)'(z_0) = f'(g(z_0))g'(z_0). \tag{2.3.15}$$

Since g is analytic at z_0, appealing to Proposition 2.3.10, we write

$$\frac{g(z) - g(z_0)}{z - z_0} = g'(z_0) + \varepsilon(z), \qquad \varepsilon(z) \to 0 \text{ as } z \to z_0. \tag{2.3.16}$$

Also, f is analytic at $g(z_0)$, by Proposition 2.3.10 again we write

$$f(w) - f(g(z_0)) = f'(g(z_0))(w - g(z_0)) + \eta(w)(w - g(z_0)),$$

where $\eta(w) \to 0$ as $w \to g(z_0)$.

Replacing w by $g(z)$, dividing by $z - z_0$, and using (2.3.16), we obtain

$$\frac{f(g(z)) - f(g(z_0))}{z - z_0} = f'(g(z_0))(g'(z_0) + \varepsilon(z)) + \eta(g(z))(g'(z_0) + \varepsilon(z)). \tag{2.3.17}$$

As $z \to z_0$, $\varepsilon(z) \to 0$, $g(z) \to g(z_0)$ by continuity, and so $\eta(g(z)) \to 0$. Using this in (2.3.17), we conclude that

$$\lim_{z \to z_0} \frac{f(g(z)) - f(g(z_0))}{z - z_0} = f'(g(z_0))g'(z_0),$$

as asserted by the chain rule. ∎

The following inside-out chain rule (illustrated in Figure 2.14) is useful when dealing with inverse functions such as logarithms and powers.

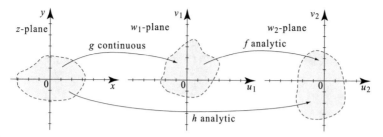

Fig. 2.14 In the reverse chain rule we suppose that g is continuous and that $h = f \circ g$ is analytic and we conclude that g is analytic.

Theorem 2.3.12. (Reverse Chain Rule) *Suppose that g is a continuous function on a region Ω and f is an analytic function on a region U that contains $g[\Omega]$. Suppose that $h = f \circ g$ is analytic on Ω and that $f'(g(z)) \neq 0$ for all z in Ω. Then g is analytic on Ω and*

$$g'(z) = \frac{h'(z)}{f'(g(z))}, \qquad z \in \Omega. \tag{2.3.18}$$

Proof. Let z_0 be in Ω. We know that $h(z) = f(g(z))$ analytic at $z = z_0$, f is analytic at $g(z_0)$ with $f'(g(z_0)) \neq 0$, and g is continuous at z_0. We want to show that

$$g'(z_0) = \frac{h'(z_0)}{f'(g(z_0))}. \tag{2.3.19}$$

Applying Proposition 2.3.10 to $h(z) = f(g(z))$, we write

$$f(g(z)) = f(g(z_0)) + h'(z_0)(z - z_0) + \varepsilon(z)(z - z_0), \quad \varepsilon(z) \to 0 \text{ as } z \to z_0. \tag{2.3.20}$$

Applying Proposition 2.3.10 to f at $g(z_0)$, we have

$$f(g(z)) = f(g(z_0)) + f'(g(z_0))(g(z) - g(z_0)) + \eta(g(z))(g(z) - g(z_0)), \tag{2.3.21}$$

where $\eta(g(z)) \to 0$ as $g(z) \to g(z_0)$ or, equivalently, as $z \to z_0$ by continuity of g at z_0. Subtract (2.3.21) from (2.3.20) and rearrange the terms to get

$$\frac{g(z) - g(z_0)}{z - z_0} = \frac{h'(z_0) + \varepsilon(z)}{f'(g(z_0)) + \eta(g(z))}. \tag{2.3.22}$$

As $z \to z_0$, $\varepsilon(z) \to 0$ and $\eta(g(z)) \to 0$, implying (2.3.19). Notice that the denominator in (2.3.22) does not vanish for z sufficiently close to z_0. ∎

Example 2.3.13. (Analyticity of nth roots) Show that the principal branch of the nth root, $g(z) = z^{1/n}$ $(n = 0, 1, 2, \ldots)$, is analytic in the region Ω (Figure 2.15) described by $\mathbb{C} \setminus \{z \in \mathbb{C} : \operatorname{Re} z \leq 0, \ \operatorname{Im} z = 0\}$. Also show that for z in Ω, we have

$$g'(z) = \frac{1}{n} z^{(1-n)/n}.$$

Solution. From (1.8.9), we have

$$g(z) = e^{\frac{1}{n} \operatorname{Log} z},$$

where $\operatorname{Log} z$ is the principal branch of the logarithm. We showed in the previous section that e^z is continuous for all z, and since $\operatorname{Log} z$ is continuous in Ω, it follows that $g(z)$ is continuous in Ω, being the composition of two continuous functions. Taking $f(z) = z^n$, $h(z) = f(g(z)) = (z^{1/n})^n = z$, we see clearly that f and h are analytic, and thus the hypotheses of Theorem 2.3.12 are satisfied. Consequently, $g(z)$ is analytic on Ω and

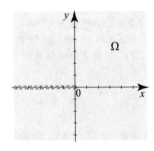

Fig. 2.15 Branch cut of $\operatorname{Log} z$.

$$g'(z) = \frac{h'(z)}{f'(g(z))} = \frac{1}{n(z^{1/n})^{n-1}} = \frac{1}{n} z^{(1-n)/n}. \qquad \square$$

Exercises 2.3

In Exercises 1–12, determine the largest set on which the functions are analytic and compute their complex derivatives. In Exercises 9–12, use the principal branch of the power.

1. $3(z-1)^2 + 2(z-1)$

2. $z^3 + \dfrac{z}{1+i}$

3. $\operatorname{Im} z$

4. $\left(\dfrac{z-2+i}{z-1+i}\right)^2$

5. $\dfrac{1}{z^3+1}$

6. $8\bar{z} + i$

7. $\dfrac{1}{z^2 - (1-2i)z - 3 - i}$

8. $\dfrac{1}{z^2 + (1+2i)z + 3 - i}$

9. $z^{2/3}$

10. $(z-1)^{\frac{1}{2}}$

11. $(z-3+i)^{1/10}$

12. $\dfrac{1}{(z+1)^{1/2}}$

In Exercises 13–16, evaluate the limit by identifying it with a complex derivative at a point. In Exercises 15–16, use the principal branch of the power.

13. $\displaystyle\lim_{z \to 1} \frac{z^{100} - 1}{z - 1}$

14. $\displaystyle\lim_{z \to i} \frac{z^{99} + i}{z - i}$

15. $\displaystyle\lim_{z \to 0} \frac{1}{z\sqrt{1+z}} - \frac{1}{z}$

16. $\displaystyle\lim_{z \to 1} \frac{z^{1/3} - 1}{z - 1}$

17. Determine the set on which the function

$$h(z) = \begin{cases} z & \text{if } |z| \le 1 \\ z^2 & \text{if } |z| > 1 \end{cases}$$

is analytic and compute its complex derivative. Justify your answer.

18. Let $g(z) = |z|^2$. Show that the complex derivative $g'(z_0)$ exists if and only if $z_0 = 0$. [*Hint:* Proceed as in Example 2.3.9.]

19. Show that the function $f(z) = |z|$ does not have a complex derivative at any point. [*Hint:* Compute the limit in (2.3.1) by letting $z = \alpha z_0$ with α real satisfying either $\alpha \downarrow 0$ or $\alpha \uparrow 0$.]

20. For this exercise, refer to identity (1.8.9).
(a) Show that the three branches of $z^{1/3}$ are

$$b_1(z) = e^{\frac{1}{3} \text{Log} z}; \qquad b_2(z) = e^{\frac{1}{3} \text{Log} z} e^{\frac{2\pi i}{3}}; \qquad b_3(z) = e^{\frac{1}{3} \text{Log} z} e^{\frac{4\pi i}{3}}.$$

(b) Use Theorem 2.3.12 to show that $b_j'(z) = \frac{b_j(z)}{3z}$ $\quad (j = 1, 2, 3)$.

21. Refer to identity (1.8.9). Show that for an integer p and a positive integer q we have

$$\frac{d}{dz} z^{p/q} = \frac{p}{qz} z^{p/q},$$

where we are using the same branch of the power on both sides. [*Hint:* Apply Theorem 2.3.12 with $g(z) = z^{p/q}$ and $f(z) = z^q$.]

22. Derivative of z^n. In the text it was shown that for positive integers n, $\frac{d}{dz} z^n = nz^{n-1}$. Construct a different proof starting with the definition (2.3.1) and using the identity

$$z^n - z_0^n = (z - z_0)(z^{n-1} + z^{n-2} z_0 + \cdots + z_0^{n-1}).$$

23. (Derivative of z^n, n negative) Show that the formula in Example 2.3.6 holds for negative n where $z \ne 0$. Conclude that z^n is analytic on $\mathbb{C} \setminus \{0\}$, when n is a negative integer.

24. (L'Hospital's rule) Prove the following version of L'Hospital's rule. If f and g are analytic at z_0 and $f(z_0) = g(z_0) = 0$, but $g'(z_0) \ne 0$, then

$$\lim_{z \to z_0} \frac{f(z)}{g(z)} = \frac{f'(z_0)}{g'(z_0)}.$$

$$\left[\textit{Hint: } \frac{f(z)}{g(z)} = \frac{f(z) - f(z_0)}{z - z_0} \frac{1}{\frac{g(z) - g(z_0)}{z - z_0}}. \right]$$

25. Find the following limits using L'Hospital's rule.
(a) $\displaystyle\lim_{z \to i} \frac{(z^2 + 1)^7}{z^6 + 1}$ $\qquad$ (b) $\displaystyle\lim_{z \to i} \frac{z^3 + (1 - 3i)z^2 + (i - 3)z + 2 + i}{z - i}$

2.4 Differentiation of Functions of Two Real Variables

We begin by reviewing a geometric interpretation of the derivative of a real-valued function of one variable, $\phi(x)$.

When we say that $\phi'(x_0)$ exists, we mean that the limit $\lim_{x \to x_0} \frac{\phi(x)-\phi(x_0)}{x-x_0}$ exists and equals a finite number $\phi'(x_0)$. If we set

$$r(x) = \phi'(x_0) - \frac{\phi(x)-\phi(x_0)}{x-x_0},$$

then $\lim_{x \to x_0} r(x) = 0$. Solving for $\phi(x)$, we obtain

$$\phi(x) = \phi(x_0) + \phi'(x_0)(x-x_0) + r(x)(x-x_0).$$

Let

$$\varepsilon(x) = r(x) \frac{x-x_0}{|x-x_0|},$$

then, since $\frac{x-x_0}{|x-x_0|} = \pm 1$ and $r(x) \to 0$ as $x \to x_0$, it follows that $\varepsilon(x) \to 0$ as $x \to x_0$, and we have

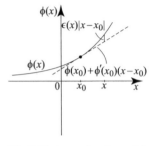

Fig. 2.16 Approximating a differentiable function by its tangent line.

$$\phi(x) = \overbrace{\phi(x_0) + \phi'(x_0)(x-x_0)}^{\text{tangent line at } x_0} + \varepsilon(x)|x-x_0|. \qquad (2.4.1)$$

This expresses the well-known geometric fact that, near a point $x = x_0$ where the function $\phi(x)$ is differentiable with derivative $\phi'(x_0)$, the tangent line approximates the graph of the function with an error that tends to 0 faster than $|x - x_0|$ (Figure 2.16).

With (2.4.1) in mind, we introduce the notion of differentiability for real-valued functions of two real variables.

Definition 2.4.1. (Differentiability of Functions of two Real Variables) A real-valued function $u(x,y)$ defined on an open subset U of $\mathbb{R}^2$ is called **differentiable** at $(x_0, y_0) \in U$ if it can be written in the form

$$u(x,y) = u(x_0, y_0) + A(x-x_0) + B(y-y_0) + \varepsilon(x,y)|(x-x_0, y-y_0)|, \qquad (2.4.2)$$

where A and B are (real) constants and $\varepsilon(x,y)$ is a function with the property $\varepsilon(x,y) \to 0$ as $(x,y) \to (x_0, y_0)$.

Definition 2.4.2. Suppose that $u(x, y)$ is a real-valued function of (x, y) defined on a nonempty open set Ω. If we fix y, we can think of u as a function of x alone and differentiate it with respect to x. This is called the **partial derivative** of u with respect to x and is denoted by $\frac{\partial u}{\partial x}$ or by u_x. Thus

$$\frac{\partial u}{\partial x} = \lim_{h \to 0} \frac{u(x+h,y) - u(x,y)}{h}. \qquad (2.4.3)$$

Similarly, fixing x and thinking of $u(x, y)$ as a function of y, differentiation with respect to y yields

$$\frac{\partial u}{\partial y} = \lim_{h \to 0} \frac{u(x,y+h) - u(x,y)}{h}. \qquad (2.4.4)$$

The limit in (2.4.3) involves the values of u at the point $(x+h, y)$. This point belongs to Ω if h is sufficiently small, because Ω is open. It is in this sense that we interpret expressions involving limits.

Theorem 2.4.3. *Suppose u is differentiable at (x_0, y_0), so that (2.4.2) holds. Then*
(i) u is continuous at (x_0, y_0); and
(ii) u_x, u_y exist at (x_0, y_0) and $u_x(x_0, y_0) = A$, $u_y(x_0, y_0) = B$.

Proof. Taking limits on both sides of (2.4.2) as $(x, y) \to (x_0, y_0)$, we obtain

$$u(x, y) = u(x_0, y_0) + \overbrace{A(x - x_0)}^{\to 0} + \overbrace{B(y - y_0)}^{\to 0} + \overbrace{\varepsilon(x, y)|(x - x_0, y - y_0)|}^{\to 0}$$

and it follows that $u(x, y) \to u(x_0, y_0)$. Hence, u is continuous at (x_0, y_0), and (i) is proved. For (ii), we only prove that $u_x(x_0, y_0) = A$, the second part being similar. To compute $u_x(x_0, y_0)$, we fix $y = y_0$ and take the derivative of $u(x, y_0)$ with respect to x. From (2.4.2) we have

$$u_x(x_0, y_0) = \lim_{x \to x_0} \frac{u(x, y_0) - u(x_0, y_0)}{x - x_0} = A + \lim_{x \to x_0} \varepsilon(x, y_0) \frac{|x - x_0|}{x - x_0} = A,$$

since the the second limit is zero; this follows from the fact that $\left| \varepsilon(x, y_0) \frac{|x - x_0|}{x - x_0} \right| = |\varepsilon(x, y_0)| \to 0$ as $(x, y) \to (x_0, y_0)$. ∎

The converse of part (ii) of Theorem 2.4.3 is not true. A function of two variables may have partial derivatives and yet fail to be differentiable at a point. In fact, the function may not even be continuous at that point. As an illustration, consider the function of two variables

$$u(x, y) = \begin{cases} \dfrac{xy}{x^2 + y^2} & \text{when } (x, y) \neq (0, 0), \\ 0 & \text{when } (x, y) = (0, 0). \end{cases} \tag{2.4.5}$$

In Exercise 1, you are asked to verify that

$$u_x(0, 0) = u_y(0, 0) = 0,$$

and that u is not continuous at $(0, 0)$. Hence by Theorem 2.4.3(i), the function u is not differentiable at the point $(0, 0)$. The graph of the function u is shown in the Figure 2.17.

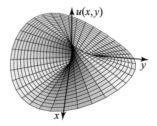

Fig. 2.17 The function $u(x, y)$ in (2.4.5).

To obtain differentiability at a point, more than the existence of the partial derivatives is needed. We have the following interesting result.

Theorem 2.4.4. (Sufficient Conditions for Differentiability) *Let u be a real-valued function defined on a neighborhood of a point (x_0, y_0) in $\mathbb{R}^2$. If*

(i) $u_x(x_0, y_0)$ *and* $u_y(x_0, y_0)$ *exist; and*

(ii) *either* $u_x(x, y)$ *or* $u_y(x, y)$ *is continuous at* (x_0, y_0),

then u is differentiable at (x_0, y_0), and (2.4.2) holds with

$$A = u_x(x_0, y_0) \qquad \text{and} \qquad B = u_y(x_0, y_0).$$

Proof. By reversing the roles of x and y, it is enough to prove either one of the cases in (ii). Let us take the case where $u_y(x, y)$ is continuous at (x_0, y_0). We have

$$u(x, y) - u(x_0, y_0) = [u(x, y_0) - u(x_0, y_0)] + [u(x, y) - u(x, y_0)]. \qquad (2.4.6)$$

For fixed y_0, we think of $u(x, y_0)$ as a function of x alone. Since this function (of one variable) has a derivative $u_x(x, y_0)$, it is differentiable and we can write

$$u(x, y_0) - u(x_0, y_0) = u_x(x_0, y_0)(x - x_0) + \varepsilon_1(x)(x - x_0), \qquad (2.4.7)$$

where $\varepsilon_1(x) \to 0$ as $x \to x_0$.

Now for fixed x, we think of $u(x, y)$ as a function of y alone, whose derivative is $u_y(x, y)$. Applying the mean value theorem from single-variable calculus, we obtain that for some number y_1 strictly between y_0 and y (as shown in Figure 2.18) we have

$$u(x, y) - u(x, y_0) = u_y(x, y_1)(y - y_0). \quad (2.4.8)$$

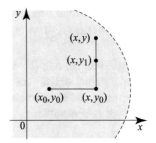

Fig. 2.18 Applying the mean value theorem.

Substituting (2.4.7) and (2.4.8) into (2.4.6), we get

$$u(x, y) - u(x_0, y_0) = [u_x(x_0, y_0) + \varepsilon_1(x)](x - x_0) + u_y(x, y_1)(y - y_0), \qquad (2.4.9)$$

where y_1 lies between y and y_0. As $(x, y) \to (x_0, y_0)$, y tends to y_0 and hence $y_1 \to y_0$, and so $u_y(x, y_1) \to u_y(x_0, y_0)$, by the continuity of u_y at (x_0, y_0). That is, $\lim_{(x,y)\to(x_0,y_0)} u_y(x, y_1) = u_y(x_0, y_0)$; equivalently,

$$u_y(x, y_1) = u_y(x_0, y_0) + \varepsilon_2(x, y), \quad \varepsilon_2(x, y) \to 0 \text{ as } (x, y) \to (x_0, y_0).$$

Inserting this in (2.4.9) and rearranging, we obtain

$$\begin{aligned} u(x, y) - u(x_0, y_0) &= [u_x(x_0, y_0) + \varepsilon_1(x)](x - x_0) + [u_y(x_0, y_0) + \varepsilon_2(x, y)](y - y_0) \\ &= u_x(x_0, y_0)(x - x_0) + u_y(x_0, y_0)(y - y_0) + \varepsilon(x, y)|(x - x_0, y - y_0)|, \end{aligned}$$

where

$$\varepsilon(x,y) = \varepsilon_1(x)\frac{x-x_0}{|(x-x_0,y-y_0)|} + \varepsilon_2(x,y)\frac{y-y_0}{|(x-x_0,y-y_0)|}. \tag{2.4.10}$$

So the required conclusion follows with $A = u_x(x_0, y_0)$ and $B = u_y(x_0, y_0)$ as long as we have that $\varepsilon(x,y) \to 0$ as $(x,y) \to (x_0, y_0)$. But this is certainly the case as the fractions in (2.4.10) are bounded by 1 and both $\varepsilon_1(x)$ and $\varepsilon_2(x,y)$ tend to zero as $(x,y) \to (x_0, y_0)$. ∎

Chain Rule and Mean Value Theorems

We give simple proofs of the chain rule and the mean value theorem in two dimensions.

Theorem 2.4.5. (Chain Rule) *Let u be a differentiable function on an open subset W of $\mathbb{R}^2$ and let x, y be differentiable functions of t defined on an open interval I on the real line such that $(x(t), y(t))$ lies in W for all $t \in I$. Then $U(t) = u(x(t), y(t))$ is differentiable for all $t \in I$ and we have*

$$\frac{dU}{dt} = \frac{\partial u}{\partial x}\frac{dx}{dt} + \frac{\partial u}{\partial y}\frac{dy}{dt}. \tag{2.4.11}$$

Proof. Fix t_0 in the interval I. For the point $(x(t_0), y(t_0)) = (x_0, y_0)$ in W, by Definition 2.4.1 and Theorem 2.4.4(ii), we have

$$u(x,y) - u(x_0, y_0) = u_x(x_0, y_0)(x-x_0) + u_y(x_0, y_0)(y-y_0) + \varepsilon(x,y)|(x-x_0, y-y_0)|,$$

and by taking $x = x(t)$ and $y = y(t)$ for t near t_0 we write

$$\frac{U(t) - U(t_0)}{t - t_0} = u_x(x_0, y_0)\frac{x(t) - x(t_0)}{t - t_0} + u_y(x_0, y_0)\frac{y(t) - y(t_0)}{t - t_0}$$
$$+ \varepsilon(x(t), y(t))\frac{|(x(t) - x_0, y(t) - y_0)|}{t - t_0}. \tag{2.4.12}$$

As $t \to t_0$ we have

$$\frac{x(t) - x(t_0)}{t - t_0} \to \frac{dx}{dt}(t_0) \quad \text{and} \quad \frac{y(t) - y(t_0)}{t - t_0} \to \frac{dy}{dt}(t_0);$$

hence (2.4.11) will be a consequence of the preceding identity once we prove that the term in (2.4.12) tends to zero as $t \to t_0$. But as $t \to t_0$, $(x(t), y(t)) \to (x_0, y_0)$ and hence $\varepsilon(x(t), y(t)) \to 0$. So it suffices to show that $\frac{|(x(t) - x_0, y(t) - y_0)|}{t - t_0}$ is bounded in a neighborhood of t_0. We have

$$\frac{|(x(t) - x_0, y - y_0)|}{|t - t_0|} = \left| \left(\frac{x(t) - x(t_0)}{t - t_0}, \frac{y(t) - y(t_0)}{t - t_0} \right) \right| \to \left| \left(\frac{dx}{dt}(t_0), \frac{dy}{dt}(t_0) \right) \right|,$$

and since this function has a limit, it is bounded in a neighborhood of t_0. ∎

There is also a version of the chain rule in the situation where x and y are differentiable functions of two variables, s and t. In that case, we set $U(s,t) = u(x(s,t), y(s,t))$, and then

$$\frac{\partial U}{\partial s} = \frac{\partial u}{\partial x}\frac{\partial x}{\partial s} + \frac{\partial u}{\partial y}\frac{\partial y}{\partial s}, \tag{2.4.13}$$

$$\frac{\partial U}{\partial t} = \frac{\partial u}{\partial x}\frac{\partial x}{\partial t} + \frac{\partial u}{\partial y}\frac{\partial y}{\partial t}. \tag{2.4.14}$$

The first formula follows by applying (2.4.11) to $U(s,t)$ while keeping t fixed, and the second follows by applying (2.4.11) while keeping s fixed.

Next, we discuss the mean value theorem in two dimensions.

Theorem 2.4.6. (Mean Value Theorem in Two Dimensions) *Let u be a differentiable function on an open subset W of $\mathbb{R}^2$. Suppose that the line segment $[z_1, z_2]$ joining $z_1 = (x_1, y_1)$ to $z_2 = (x_2, y_2)$ lies entirely in W. Then there exists a point $z = (x_0, y_0)$ on $[z_1, z_2]$ such that*

$$u(x_2, y_2) - u(x_1, y_1) = u_x(z)(x_2 - x_1) + u_y(z)(y_2 - y_1). \tag{2.4.15}$$

Proof. Parametrize the line segment $[z_1, z_2]$ by

$$x(t) = x_1 + t(x_2 - x_1), \quad y(t) = y_1 + t(y_2 - y_1), \qquad 0 \le t \le 1.$$

We have $\frac{dx}{dt} = x_2 - x_1$ and $\frac{dy}{dt} = y_2 - y_1$.
Form the function $U(t) = u(x(t), y(t))$ for $0 \le t \le 1$. We have $U(0) = u(x_1, y_1)$, $U(1) = u(x_2, y_2)$, and, by Theorem 2.4.5,

$$\frac{dU}{dt} = \frac{\partial u}{\partial x}\frac{dx}{dt} + \frac{\partial u}{\partial y}\frac{dy}{dt}$$

$$= \frac{\partial u}{\partial x}(x_2 - x_1) + \frac{\partial u}{\partial y}(y_2 - y_1).$$

By the mean value theorem in one variable applied to $U(t)$, there is a t_0 in $(0, 1)$ such that

$$U(1) - U(0) = \frac{dU}{dt}(t_0)(1 - 0).$$

Hence

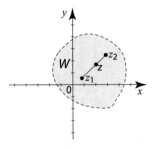

Fig. 2.19 Mean value theorem in two dimensions.

$$u(x_2, y_2) - u(x_1, y_1) = \frac{\partial u}{\partial x}(x(t_0), y(t_0))(x_2 - x_1) + \frac{\partial u}{\partial y}(x(t_0), y(t_0))(y_2 - y_1),$$

and so (2.4.15) follows with $z = (x(t_0), y(t_0))$. ∎

To motivate the next result, let us first recall the following one-dimensional result: If $f'(x) = 0$ for all x in (a, b), then $f(x)$ is a constant for all x in (a, b). To prove this, we fix a point x_0 in (a, b). Let x be in (a, b), and say $a < x < x_0 < b$. The mean value theorem asserts that there is a point x_1 in (x, x_0) such that

$$f(x_0) - f(x) = f'(x_1)(x_0 - x).$$

Since f' is identically zero in (x, x_0), we conclude that $f(x_0) - f(x) = 0$ or $f(x) = f(x_0)$. The case $x < x_0 < b$ is treated similarly, and we obtain that $f(x) = f(x_0)$ for all x in (a, b). In other words, f is constant in (a, b).

Theorem 2.4.7. *Suppose that u is a real-valued function defined over a region Ω such that $u_x(x, y) = 0$ and $u_y(x, y) = 0$ for all (x, y) in Ω. Then u is constant on Ω.*

Proof. According to the two-dimensional mean value result (Theorem 2.4.6), for any two points $A = (x_1, y_1)$ and $B = (x_2, y_2)$ in Ω such that the line segment AB is also in Ω, there exists a point $C = (x_3, y_3)$ on the line segment AB such that

$$u(x_2, y_2) - u(x_1, y_1) = u_x(x_3, y_3)(x_2 - x_1) + u_y(x_3, y_3)(y_2 - y_1). \qquad (2.4.16)$$

Fix a point (x_0, y_0) in Ω. Given a point (x, y) in Ω, connect (x_0, y_0) to (x, y) by a finite number of line segments joined end to end and wholly contained in Ω. (Here we have used the fact that Ω is a region.) Let (x_j, y_j), $j = 0, 1, \ldots, n$ denote the endpoints of the consecutive line segments, starting with (x_0, y_0) and ending with $(x_n, y_n) = (x, y)$ (see Figure 2.20). Applying (2.4.16) to each line segment and using the fact that the partial derivatives are zero, we conclude that $u(x_{j-1}, y_{j-1}) = u(x_j, y_j)$, and hence that $u(x_0, y_0) = u(x, y)$. ∎

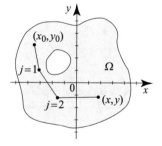

Fig. 2.20 Joining two points in a connected region by a polygonal line.

Exercises 2.4

1. Show that the partial derivatives u_x and u_y of the function u in (2.4.5) exist for all (x, y) but u is not continuous at $(0, 0)$.

2. The function $\phi(x) = x^2$ is differentiable at $x_0 = 1$. Find the function ε such that

$$\phi(x) = \phi(1) + \phi'(1)(x-1) + \varepsilon(x)|x-1|,$$

and verify directly that $\varepsilon(x) \to 0$ as $x \to 1$.

3. Using Definition 2.4.1 show that a linear function $u(x, y) = Ax + By + C$ is differentiable.

4. If $u(x,y) = yx^2 + x + y$, find the derivative of the function $U(t) = u(\ln t, t^4)$ defined for $t \in (0, \infty)$.

5. Using (2.4.1) and (2.4.2), show that if $f : \mathbb{R} \to \mathbb{R}$ is differentiable at x_0 and $g : \mathbb{R} \to \mathbb{R}$ is differentiable at y_0, then the function u defined by $u(x, y) = f(x)g(y)$ is differentiable at $(x_0, y_0) \in \mathbb{R}^2$.

6. Show that if u, v are differentiable functions on an open subset W of $\mathbb{R}^2$ and c_1, c_2 are constants, then $c_1 u + c_2 v$ and uv are also differentiable on W.

7. Recast the function $u(x, y)$ in (2.4.5) in polar coordinates by setting $x = r\cos\theta$, $y = r\sin\theta$. Show that $u(x, y) = \frac{1}{2}\sin(2\theta)$, and use this formulation to describe the behavior of the function.

8. Consider the function $g(x,y) = (x^2 + y^2)\sin\left(\frac{1}{\sqrt{x^2+y^2}}\right)$ if $(x,y) \neq 0$ and $g(0,0) = 0$. Prove that g is differentiable at $(0,0)$ even though $\frac{\partial g}{\partial x}$ and $\frac{\partial g}{\partial y}$ are both discontinuous at $(0,0)$.

9. Project Problem: Is it true that if $u_y(x, y) = 0$ for all (x, y) in a region Ω, then $u(x, y) = \phi(x)$; that is, u depends only on x? The answer is no in general, as the following counterexample shows.

(a) For (x, y) in the region Ω shown in the adjacent figure, consider the function

$$u(x, y) = \begin{cases} 0 & \text{if } x > 0, \\ \operatorname{sgn} y & \text{if } x \leq 0, \end{cases}$$

where the **signum function** is defined by $\operatorname{sgn} y = -1, 0, 1$, according as $y < 0$, $y = 0$, or $y > 0$, respectively. Show that $u_y(x, y) = 0$ for all (x, y) in Ω but that u is not a function of x alone.

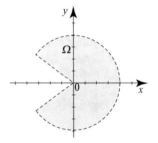

(b) Note that in the previous example u_x does not exist for $x = 0$. We now construct a function over the same region Ω for which the partials exist, $u_y = 0$, and u is not a function of x alone. Show that these properties hold for

$$u(x, y) = \begin{cases} 0 & \text{if } x \geq 0, \\ e^{-1/x^2}\operatorname{sgn} y & \text{if } x < 0. \end{cases}$$

(c) Come up with a general condition on Ω that guarantees that whenever $u_y = 0$ on Ω then u depends only on x. [*Hint*: Use the mean value theorem applied to vertical line segments in Ω.]

2.5 The Cauchy-Riemann Equations

An analytic function f, defined on an open subset U of $\mathbb{R}^2$, can be written as $u + iv$, where u, v are real-valued functions. One may wonder if the property of f being analytic is related to the property of u, v being differentiable in the sense of Definition 2.4.1. It turns out that there is such a strong relationship and, additionally, there is a special relationship between the partial derivatives of u and v.

Before we explore this relationship, we discuss consequences of analyticity. Suppose that $f(z) = f(x + iy) = u(x, y) + iv(x, y)$ is analytic in an open set U and let

$z_0 = x_0 + i y_0 = (x_0, y_0)$ be a point in U. The derivative at z_0 exists and is equal to

$$f'(z_0) = \lim_{z \to z_0} \frac{f(z) - f(z_0)}{z - z_0} = \lim_{\Delta z \to 0} \frac{f(z_0 + \Delta z) - f(z_0)}{\Delta z}. \qquad (2.5.1)$$

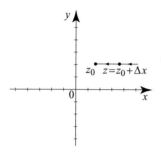

Fig. 2.21 For z approaching z_0 in the direction of the x-axis, $\Delta z = \Delta x$.

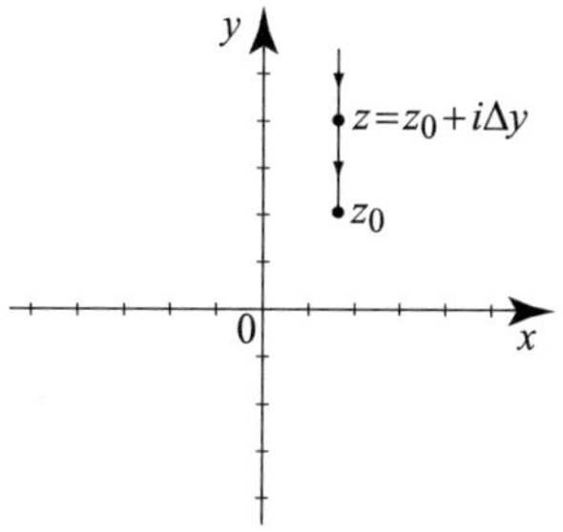

Fig. 2.22 For z approaching z_0 in the direction of the y-axis, $\Delta z = i \Delta y$.

Suppose that z approaches z_0 along the direction of the x-axis, as in Figure 2.21. Then $z = z_0 + \Delta x = (x_0 + \Delta x, y_0)$, $\Delta z = z - z_0 = \Delta x$, and (2.5.1) becomes

$$f'(z_0) = \lim_{\Delta x \to 0} \frac{f(x_0 + \Delta x + i y_0) - f(x_0 + i y_0)}{\Delta x} \qquad (2.5.2)$$

$$= \lim_{\Delta x \to 0} \left(\frac{u(x_0 + \Delta x, y_0) - u(x_0, y_0)}{\Delta x} + i \frac{v(x_0 + \Delta x, y_0) - v(x_0, y_0)}{\Delta x} \right)$$

$$= \lim_{\Delta x \to 0} \frac{u(x_0 + \Delta x, y_0) - u(x_0, y_0)}{\Delta x} + i \lim_{\Delta x \to 0} \frac{v(x_0 + \Delta x, y_0) - v(x_0, y_0)}{\Delta x},$$

where the last step is justified by Theorem 2.2.9 which asserts that the limit of a complex-valued function exists if and only if the limits of its real and imaginary parts exist. Recognizing the last two limits as the partial derivatives with respect to x of u and v [see (2.4.3) and (2.4.4)], we obtain

$$f'(z_0) = \frac{\partial u}{\partial x}(x_0, y_0) + i \frac{\partial v}{\partial x}(x_0, y_0), \qquad (2.5.3)$$

which is an expression of the derivative of f in terms of the partial derivatives with respect to x of u and v. We now repeat the preceding steps, going back to (2.5.1) and taking the limit as z approaches z_0 from the direction of the y-axis, as in Figure 2.22. Then $z = z_0 + i \Delta y = (x_0, y_0 + \Delta y)$, $\Delta z = z - z_0 = i \Delta y$. Proceeding as in (2.5.2) and noting that $i \Delta y \to 0$ if and only if $\Delta y \to 0$, we obtain

$$f'(z_0) = \lim_{\Delta y \to 0} \frac{f(x_0 + i(y_0 + \Delta y)) - f(x_0 + i y_0)}{i \Delta y} \qquad (2.5.4)$$

$$= \lim_{\Delta y \to 0} \left(\frac{u(x_0, y_0 + \Delta y) - u(x_0, y_0)}{i \Delta y} + i \frac{v(x_0, y_0 + \Delta y) - v(x_0, y_0)}{i \Delta y} \right)$$

$$= \lim_{\Delta y \to 0} \frac{v(x_0, y_0 + \Delta y) - v(x_0, y_0)}{\Delta y} - i \lim_{\Delta y \to 0} \frac{u(x_0, y_0 + \Delta y) - u(x_0, y_0)}{\Delta y},$$

where in the last step we have used $1/i = -i$ and rearranged the terms. Recognizing the partial derivatives of v and u with respect to y we obtain

$$f'(z_0) = \frac{\partial v}{\partial y}(x_0, y_0) - i\frac{\partial u}{\partial y}(x_0, y_0), \tag{2.5.5}$$

which is this time an expression of the derivative of f in terms of the partial derivatives with respect to y of u and v. Equating real and imaginary parts in (2.5.3) and (2.5.5), we deduce the following equations

$$\frac{\partial u}{\partial x} = \frac{\partial v}{\partial y} \quad \text{and} \quad \frac{\partial u}{\partial y} = -\frac{\partial v}{\partial x}. \tag{2.5.6}$$

These are called the **Cauchy-Riemann equations**. They first appeared in 1821 in the early work of Cauchy on integrals of complex-valued functions. Their connection to the existence of the complex derivative appeared in 1851 in the doctoral dissertation of the German mathematician Bernhard Riemann (1826–1866).

Theorem 2.5.1. (Cauchy-Riemann Equations) *Let U be an open subset of $\mathbb{R}^2$ and let u, v be real-valued functions defined on U. Then the complex-valued function $f(x+iy) = u(x,y) + iv(x,y)$ is analytic on U if and only if u, v are differentiable functions on U and satisfy*

$$u_x = v_y \quad \text{and} \quad u_y = -v_x \tag{2.5.7}$$

for all points in U. If this is the case, then for all $(x,y) \in U$ we have

$$f'(x+iy) = u_x(x, y) + iv_x(x, y) \quad \text{or} \quad f'(x+iy) = v_y(x, y) - iu_y(x, y). \tag{2.5.8}$$

Proof. Let us use the notation $z = x+iy$ for a general point in U and $z_0 = x_0 + iy_0$ for a fixed point in U, where x, y, x_0, y_0 are real numbers. In view of Proposition 2.3.10, if the function f is analytic, we have

$$f(z) - f(z_0) = f'(z_0)(z - z_0) + \varepsilon(z)(z - z_0) \tag{2.5.9}$$

where $\varepsilon(z)$ ends to zero as $z \to z_0$. Setting $f'(z_0) = A + iB$, where A, B are real, and splitting up real and imaginary parts in (2.5.9) we obtain

$$u(x,y) - u(x_0, y_0) = A(x - x_0) - B(y - y_0) + \varepsilon_1(x,y)|(x - x_0, y - y_0)|$$
$$v(x,y) - v(x_0, y_0) = B(x - x_0) + A(y - y_0) + \varepsilon_2(x,y)|(x - x_0, y - y_0)|,$$

where

$$\varepsilon_1(x,y) = \text{Re}\left[\varepsilon(x+iy)\frac{x - x_0 + i(y - y_0)}{|(x - x_0, y - y_0)|}\right]$$
$$\varepsilon_2(x,y) = \text{Im}\left[\varepsilon(x+iy)\frac{x - x_0 + i(y - y_0)}{|(x - x_0, y - y_0)|}\right].$$

Note that
$$|\varepsilon_1(x,y)| \le |\varepsilon(x,y)|, \qquad |\varepsilon_2(x,y)| \le |\varepsilon(x,y)|$$

and since $\varepsilon(x,y)$ tends to zero as $(x,y) \to (x_0,y_0)$, both $\varepsilon_1(x,y)$ and $\varepsilon_2(x,y)$ tend to zero as $(x,y) \to (x_0,y_0)$. Hence both u and v are differentiable at (x_0,y_0). Consequently, the partial derivatives of u, v exist (Theorem 2.4.3) and we have

$$u_x(x_0,y_0) = A \qquad u_y(x_0,y_0) = -B$$
$$v_x(x_0,y_0) = B \qquad v_y(x_0,y_0) = A.$$

The Cauchy-Riemann equations (2.5.7) follow from this.

To prove the converse direction we assume that u and v are differentiable on U and satisfy (2.5.7). By the definition of differentiability (Definition 2.4.1) we write

$$u(x,y) - u(x_0,y_0) = u_x(x_0,y_0)(x-x_0) + u_y(x_0,y_0)(y-y_0) + \varepsilon_1(x,y)|(x-x_0,y-y_0)|$$
$$v(x,y) - v(x_0,y_0) = v_x(x_0,y_0)(x-x_0) + v_y(x_0,y_0)(y-y_0) + \varepsilon_2(x,y)|(x-x_0,y-y_0)|$$

where the $\varepsilon_1(x,y)$, $\varepsilon_2(x,y)$ are functions that tend to zero as $(x,y) \to (x_0,y_0)$. Adding the displayed expressions (after multiplying the second one by i), we obtain

$$
\begin{aligned}
&f(x+iy) - f(x_0+iy_0) \\
&= (u_x(x_0,y_0) + iv_x(x_0,y_0))(x-x_0) + (u_y(x_0,y_0) + iv_y(x_0,y_0))(y-y_0) \\
&\qquad + \varepsilon_1(x,y)|(x-x_0,y-y_0)| + i\varepsilon_2(x,y)|(x-x_0,y-y_0)| \\
&= (u_x(x_0,y_0) + iv_x(x_0,y_0))(x-x_0) + (-v_x(x_0,y_0) + iu_x(x_0,y_0))(y-y_0) \\
&\qquad + \varepsilon_1(x,y)|(x-x_0,y-y_0)| + i\varepsilon_2(x,y)|(x-x_0,y-y_0)| \\
&= (u_x(x_0,y_0) + iv_x(x_0,y_0))(x-x_0 + i(y-y_0)) + E(x+iy)(x-x_0 + i(y-y_0))
\end{aligned}
$$

where in the second equality we used assumption (2.5.7) and we set

$$E(x+iy) = \varepsilon_1(x,y)\frac{|(x-x_0,y-y_0)|}{x-x_0+i(y-y_0)} + i\varepsilon_2(x,y)\frac{|(x-x_0,y-y_0)|}{x-x_0+i(y-y_0)}.$$

Notice that
$$|E(x+iy)| \le |\varepsilon_1(x,y)| + |\varepsilon_2(x,y)|,$$

which tends to 0 as $(x,y) \to (x_0,y_0)$. We have now shown that

$$f(z) - f(z_0) = (u_x(z_0) + iv_x(z_0))(z-z_0) + E(z)(z-z_0)$$

which implies that f is analytic and that $f'(z_0) = u_x(z_0) + iv_x(z_0)$. This proves one identity in (2.5.8), while the other one is a consequence of this one and (2.5.7). ∎

Corollary 2.5.2. *If u, v are real-valued functions defined on an open subset U of $\mathbb{R}^2$ which have continuous partial derivatives that satisfy $u_x = v_y$, $u_y = -v_x$, then the complex-valued function $f(x+iy) = u(x,y) + iv(x,y)$ is analytic on U.*

Proof. Apply Theorem 2.4.4. ∎

Example 2.5.3. Show that e^z is entire and

$$\frac{d}{dz}e^z = e^z.$$

Solution. We use Theorem 2.5.1. From (1.6.9) we have $e^z = e^x \cos y + i e^x \sin y$ if $z = x + i y$, with x, y real. Thus, $u(x, y) = e^x \cos y$ and $v(x, y) = e^x \sin y$. Differentiating u with respect to x and y, we find

$$\frac{\partial u}{\partial x} = e^x \cos y, \qquad \frac{\partial u}{\partial y} = -e^x \sin y.$$

Differentiating v with respect to x and y, we find

$$\frac{\partial v}{\partial x} = e^x \sin y, \qquad \frac{\partial v}{\partial y} = e^x \cos y.$$

Comparing these derivatives, we see that $\frac{\partial u}{\partial x} = \frac{\partial v}{\partial y}$ and $\frac{\partial u}{\partial y} = -\frac{\partial v}{\partial x}$. Hence the Cauchy-Riemann equations are satisfied at all points. Moreover, the functions $\frac{\partial u}{\partial x}, \frac{\partial v}{\partial x}, \frac{\partial u}{\partial x}, \frac{\partial v}{\partial y}$ are continuous. By Corollary 2.5.2, e^z is analytic at all points, or entire. To compute the derivative, we apply either formula from (2.5.8), say the first, and use the formulas that we derived for the partial derivatives of u and v. We obtain

$$\frac{d}{dz}e^z = \frac{\partial u}{\partial x} + i\frac{\partial v}{\partial x} = e^x \cos y + i e^x \sin y = e^z. \qquad \square$$

Combining the result of Example 2.5.3 with the chain rule, we deduce that $e^{f(z)}$ is analytic wherever $f(z)$ is analytic and

$$\frac{d}{dz}e^{f(z)} = f'(z)e^{f(z)}. \tag{2.5.10}$$

Example 2.5.4. Show that $\sin z$ is entire and

$$\frac{d}{dz}\sin z = \cos z.$$

Solution. We show this assertion in two ways. In view of (1.7.4) we appeal to (2.5.10) and conclude that $\sin z$ is entire and

$$\frac{d}{dz}\sin z = \frac{d}{dz}\left(\frac{e^{iz} - e^{-iz}}{2i}\right) = \frac{ie^{iz} - (-i)e^{-iz}}{2i} = \frac{e^{iz} + e^{-iz}}{2} = \cos z,$$

by (1.7.3). The second method is a bit longer and makes use of the Cauchy-Riemann equations. From (1.7.17) with $z = x + iy$, x, y real, we have

$$\sin z = \sin x \cosh y + i \cos x \sinh y.$$

Hence

$$u(x, y) = \sin x \cosh y \Rightarrow u_x(x, y) = \cos x \cosh y, \quad u_y(x, y) = \sin x \sinh y;$$
$$v(x, y) = \cos x \sinh y \Rightarrow v_x(x, y) = -\sin x \sinh y, \quad v_y(x, y) = \cos x \cosh y.$$

Comparing partial derivatives, we see that the Cauchy-Riemann equations are satisfied at all points. Moreover, the partial derivatives are continuous. Using Theorem 2.5.1 we obtain that $\sin z$ is entire and

$$\frac{d}{dz} \sin z = u_x(x, y) + i v_x(x, y) = \cos x \cosh y - i \sin x \sinh y = \cos z,$$

in view of (1.7.16). □

As we did in Example 2.5.4, we can verify the analyticity and compute the derivatives of $\cos z$, $\tan z$, and all other trigonometric and hyperbolic functions. Among the elementary functions, it remains to consider the logarithm and complex powers.

Example 2.5.5. (Log z **is analytic except on the branch cut**) Show that the principal branch of the logarithm, $\operatorname{Log} z$, is analytic on $\mathbb{C} \setminus (-\infty, 0]$, and that

$$\frac{d}{dz} \operatorname{Log} z = \frac{1}{z}. \tag{2.5.11}$$

Thus the familiar formula from calculus still holds.

Solution. In the notation of Theorem 2.3.12, set $f(z) = e^z$, $g(z) = \operatorname{Log} z$, and $h(z) = z$. Since h is analytic, f is analytic with $f'(z) = e^z \neq 0$, and g is continuous everywhere except on the branch cut, we conclude that g is analytic there with

$$\frac{d}{dz} \operatorname{Log} z = \frac{h'(z)}{f'(g(z))} = \frac{1}{e^{\operatorname{Log} z}} = \frac{1}{z}.$$

The logarithm cannot be analytic on the branch cut, because it is not continuous there (see Theorem 2.3.4). □

Using the method of Example 2.5.5, we can show that a branch of the logarithm, $\log_\alpha z$, is analytic everywhere except at its branch cut (the ray at angle α), and

$$\frac{d}{dz} \log_\alpha z = \frac{1}{z}. \tag{2.5.12}$$

Next we consider the principal branch of a power,

$$z^a = e^{a \operatorname{Log} z} \qquad \text{(where } a \neq 0 \text{ is a complex number).}$$

Since $\operatorname{Log} z$ is analytic except at its branch cut, and e^z is entire, it follows that z^a is analytic, except at the branch cut of $\operatorname{Log} z$. To compute its derivative, we use the chain rule and the derivatives of e^z and $\operatorname{Log} z$ and deduce

$$\frac{d}{dz} z^a = a z^{a-1}, \tag{2.5.13}$$

with principal branches of the power on both sides.

We saw in Example 2.3.9 that $\bar{z}$ fails to be analytic. This fact should now be obvious from the Cauchy-Riemann equations. If we write $\bar{z} = x - iy$, then $u(x,y) = x$, $v(x,y) = -y$ and thus $u_x = 1$, $v_y = -1$; this shows that the Cauchy-Riemann equations do not hold at any point. We now use the Cauchy-Riemann equations to show the failure of analyticity in less obvious situations.

Example 2.5.6. (Failure of analyticity) Show that the function of two real variables $f(x+iy) = x^2 + i(2y+x)$ is nowhere analytic.

Solution. We have $u(x,y) = x^2$ and $v(x,y) = 2y + x$. Since $u_y = 0$ and $v_x = 1$, the Cauchy-Riemann equations are not satisfied at any point and hence the function cannot have a complex derivative at any point. $\qquad\Box$

Theorem 2.5.7. *Suppose that $f = u + iv$ is analytic on a region Ω and satisfies $f'(z) = 0$ for all z in Ω. Then f is constant in Ω.*

Proof. By Theorem 2.5.1, we have

$$f'(z) = u_x(x, y) + i v_x(x, y) \text{ and } f'(z) = v_y(x, y) - i u_y(x, y), \quad z \text{ in } \Omega.$$

Since $f'(z) = 0$ it follows that $u_x = u_y = 0$ and $v_x = v_y = 0$ in Ω. Appealing to Theorem 2.4.7, we conclude that u is constant and v is constant, and hence $f = u + iv$ is constant. $\qquad\blacksquare$

We note that the connectedness of Ω in the preceding theorem is essential. For example, the function f defined on the shaded set S in Figure 2.23 via

$$f(z) = \begin{cases} 1 & \text{if } |z| < 2, \\ 0 & \text{if } |z| > 3 \end{cases}$$

has zero derivative but is not constant.

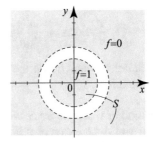

Fig. 2.23 A nonconstant function with zero derivative.

Corollary 2.5.8. *Suppose that f and g are analytic in a region Ω. If either $\operatorname{Re} f = \operatorname{Re} g$ on Ω or $\operatorname{Im} f = \operatorname{Im} g$ on Ω, then $f = g + c$ on Ω, where c is a constant.*

Proof. We only consider the case $\operatorname{Re} f = \operatorname{Re} g$ on Ω since the case $\operatorname{Im} f = \operatorname{Im} g$ is almost identical and is left to Exercise 32. Let $h = f - g = u + iv$, with u, v real-valued. We want to show that $h = c$ on Ω. Since h is analytic, it is enough by Theorem 2.4.7 to show that $h' = 0$ on Ω. We have $u = \operatorname{Re} h = \operatorname{Re} f - \operatorname{Re} g = 0$ on Ω, and so $u_x = u_y = 0$ on Ω. By the Cauchy-Riemann equations, $v_x = -u_y = 0$. Consequently, by (2.5.8), $h' = u_x + iv_x = 0$ on Ω. ∎

Exercises 2.5

In Exercises 1–14, use Corollary 2.5.2 to determine the set on which the functions are analytic and compute their complex derivative using either equation in (2.5.8).

1. z
2. z^2
3. e^{z^2}

4. $2x + 3iy$
5. $e^{\bar{z}}$
6. $\dfrac{y - ix}{x^2 + y^2}$

7. $\dfrac{1}{z+1}$
8. $z^3 - 2z$
9. ze^z

10. $\cos z$
11. $\sin(2z)$
12. $\cosh z$

13. $|z|^2$
14. $\dfrac{x^4 + i2xy(x^2 + y^2) - y^4 + x - iy}{x^2 + y^2}$

In Exercises 15–26, use properties of the derivative to compute the complex derivatives of the functions and determine the largest set on which they are analytic. In Exercises 23–26, use the principal branch of the power.

15. ze^{z^2}
16. $(1 + e^z)^5$
17. $\sin z \cos z$

18. $\operatorname{Log}(z+1)$
19. $\dfrac{\operatorname{Log}(3z-1)}{z^2 + 1}$
20. $\sinh(3z + i)$

21. $\cosh(z^2 + 3i)$
22. $\log_{\frac{\pi}{2}}(z+1)$
23. z^i

24. $(z+1)^{1/2}$
25. $\dfrac{1}{(z-i)^{1/2}}$
26. z^z

Solve Exercises 27 and 28 by identifying the limit as a complex derivative; Solve Exercises 29 and 30 using L'Hospital's rule (Exercise 24, Section 2.3).

27. $\displaystyle\lim_{z \to 0} \dfrac{\sin z}{z}$
28. $\displaystyle\lim_{z \to 0} \dfrac{e^z - 1}{z}$

29. $\displaystyle\lim_{z \to 0} \dfrac{\operatorname{Log}(z+1)}{z}$
30. $\displaystyle\lim_{z \to i} \dfrac{1 + iz}{z(z - i)}$

31. Define the **principal branch** of the inverse tangent by taking the principal branch of the logarithm as in (1.8.12):

$$\tan^{-1} z = \frac{i}{2} \operatorname{Log}\left(\frac{1 - iz}{1 + iz}\right).$$

Compute the derivative of $\tan^{-1} z$.

32. Complete the proof of Corollary 2.5.8 by treating the case $\operatorname{Im} f = \operatorname{Im} g$ on Ω.

33. Suppose that $f = u + iv$ is analytic in a region Ω. Show that that if either $\operatorname{Re} f$ or $\operatorname{Im} f$ are constant on Ω, then f must be constant on Ω.

34. Suppose that $f = u + iv$ is analytic in a region Ω. Show that
(a) $f' = u_x - iu_y$ and $f' = v_y + iv_x$;
(b) $|f'|^2 = u_x^2 + u_y^2 = v_x^2 + v_y^2$.

35. Suppose that f and $\bar{f}$ are analytic in a region Ω. Show that f must be constant in Ω. [*Hint:* Consider $f + \bar{f}$ and use Exercise 33.]

36. Suppose that $f = u + iv$ is analytic on a region Ω and $|f| = c$ is a constant on Ω. Show that f must be constant in Ω as follows:

(a) Case 1: If $c = 0$ then f is identically 0 in Ω.

(b) Case 2: If $c \neq 0$ show that $\overline{f} = c^2/f$ is analytic on Ω and conclude from Exercise 35 that f is constant in Ω.

37. Suppose that f is analytic in a region Ω and $f[\Omega]$ is a subset of a line. Show that f must be constant in Ω. [*Hint*: Rotate the line to make it horizontal or vertical and apply Exercise 33.]

38. Suppose that $f = u + iv$ is analytic in a region Ω and $\operatorname{Re} f = \operatorname{Im} f$. Show that f must be constant in Ω. [*Hint*: Use Exercise 37 or prove it directly from the Cauchy-Riemann equations.]

39. We define the partial derivatives of a complex-valued function $f = u + iv$ as $f_x = u_x + iv_x$ and $f_y = u_y + iv_y$. Show that the Cauchy-Riemann equations are equivalent to $f_x + if_y = 0$.

40. Project Problem: Cauchy-Riemann equations in polar form. In this problem we express the Cauchy-Riemann equations in polar coordinates. Recall the relationships between Cartesian and polar coordinates: $x = r\cos\theta$ and $y = r\sin\theta$. For convenience, we denote by $\frac{\partial u}{\partial r}, \frac{\partial u}{\partial \theta}$ the derivatives of $u(r\cos\theta, r\sin\theta) = u(x, y)$ with respect to r and θ, respectively.

(a) The multivariable chain rule from calculus [see (2.4.13)] states that

$$\frac{\partial u}{\partial r} = \frac{\partial u}{\partial x}\frac{\partial x}{\partial r} + \frac{\partial u}{\partial y}\frac{\partial y}{\partial r}, \frac{\partial u}{\partial \theta} = \frac{\partial u}{\partial x}\frac{\partial x}{\partial \theta} + \frac{\partial u}{\partial y}\frac{\partial y}{\partial \theta},$$
$$\frac{\partial v}{\partial r} = \frac{\partial v}{\partial x}\frac{\partial x}{\partial r} + \frac{\partial v}{\partial y}\frac{\partial y}{\partial r}, \frac{\partial v}{\partial \theta} = \frac{\partial v}{\partial x}\frac{\partial x}{\partial \theta} + \frac{\partial v}{\partial y}\frac{\partial y}{\partial \theta}$$

on some open set on which these derivatives exist. Show that

$$\frac{\partial u}{\partial r} = \cos\theta \frac{\partial u}{\partial x} + \sin\theta \frac{\partial u}{\partial y}, \frac{\partial u}{\partial \theta} = -r\sin\theta \frac{\partial u}{\partial x} + r\cos\theta \frac{\partial u}{\partial y},$$
$$\frac{\partial v}{\partial r} = \cos\theta \frac{\partial v}{\partial x} + \sin\theta \frac{\partial v}{\partial y}, \frac{\partial v}{\partial \theta} = -r\sin\theta \frac{\partial v}{\partial x} + r\cos\theta \frac{\partial v}{\partial y}.$$

(b) Derive the **polar form of the Cauchy-Riemann equations**:

$$\frac{\partial u}{\partial r} = \frac{1}{r}\frac{\partial v}{\partial \theta} \quad \text{and} \quad \frac{\partial v}{\partial r} = -\frac{1}{r}\frac{\partial u}{\partial \theta}. \tag{2.5.14}$$

Thus we can state Theorem 2.5.1 in polar form as follows: The function $f = u + iv$, u, v real-valued, is analytic on an open subset U of $\mathbb{C}$ if and only if u, v are differentiable on U and satisfy (2.5.14).

(c) Show that if $f = u + iv$ is analytic on U, then for $re^{i\theta} \in U$ we have

$$f'(re^{i\theta}) = e^{-i\theta}\left(\frac{\partial u}{\partial r} + i\frac{\partial v}{\partial r}\right). \tag{2.5.15}$$

In Exercises 41 and 42 use the polar form (2.5.14) *of the Cauchy-Riemann equations to verify the analyticity and evaluate the derivatives of the functions.*

41. $z^n = r^n(\cos(n\theta) + i\sin(n\theta))$ $(n = \pm 1, \pm 2, \ldots)$.

42. $\operatorname{Log} z = \ln|z| + i\operatorname{Arg} z$, z is not a negative real number nor zero.

Chapter 3
Complex Integration

Nature laughs at the difficulties of integration.

-Pierre-Simon de Laplace (1749–1827)

In this chapter we study integrals of complex-variable functions over paths. Paths are piecewise continuously differentiable maps from closed intervals to the complex plane. An important result proved in this chapter is an analog of the fundamental theorem of calculus for continuous functions with complex antiderivatives. This analog says that the integral of the derivative of an analytic function is equal to the difference of the values of the function at the endpoints. We also investigate an important question concerning the dependence of an integral on the path of integration.

Cauchy's theorem states that the integral over a simple closed path of an analytic function defined on an open set that contains the path and its interior must be zero. The first version (Section 3.4) is sufficient for the development of the course; this is based on Green's theorem from advanced calculus and makes the assumption that derivatives of analytic functions are continuous. In the following three sections, we prove versions of Cauchy's theorem without the continuity assumption on the derivatives. These versions involve theoretical notions, such as deformation of paths and simple connectedness, and offer geometric intuition and many applications. These sections may be omitted without interrupting the flow of the course.

In Section 3.8 we derive Cauchy's generalized integral formula, which facilitates the computation of several integrals and provides the basis for many applications. We illustrate the power of this formula by providing a simple proof of the fundamental theorem of algebra and by deriving various striking properties of analytic functions, including the mean value property and the maximum modulus principle.

3.1 Paths (Contours) in the Complex Plane

In this text a **curve** is defined as the graph of a continuous function $y = f(x)$. This curve could be written in parametric form by expressing x and y as functions of a third variable t. For example, the semicircle $y = \sqrt{1 - x^2}$, $-1 \leq x \leq 1$, in Figure 3.1 could be parametrized by the equations

$$x = x(t) = \cos t, \qquad y = y(t) = \sin t, \quad 0 \leq t \leq \pi.$$

© Springer International Publishing AG, part of Springer Nature 2018
N. H. Asmar and L. Grafakos, *Complex Analysis with Applications*,
Undergraduate Texts in Mathematics, https://doi.org/10.1007/978-3-319-94063-2_3

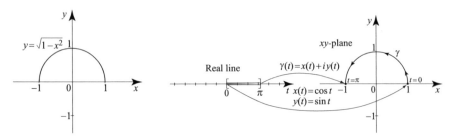

Fig. 3.1 Semicircle.   **Fig. 3.2** A parametric interval mapping to the semicircle.

Under this parametrization, the semicircle in Figure 3.1 can be thought of as a map from $[0, \pi]$ to the complex plane, as in Figure 3.2.

Definition 3.1.1. A **parametric form** of a curve is a representation of the curve by a pair of equations $x = x(t)$ and $y = y(t)$, where t ranges over a set of real numbers, usually a closed interval $[a, b]$. Each value of t determines a point $\gamma(t) = (x(t), y(t))$, which traces the curve as t moves from a to b; see Figure 3.3.

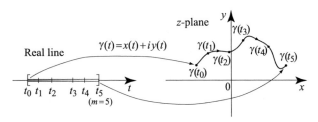

Fig. 3.3 $\gamma(t)$ traces the curve as t moves from t_0 to t_5.

Let $\gamma(t)$, $a \leq t \leq b$, be a parametrization of a curve. As t varies from a to b the point $\gamma(t)$ traces the curve in a specific direction, starting with $\gamma(a)$, the **initial point** of γ, and ending at $\gamma(b)$, the **terminal point** of γ. This direction is usually denoted by an arrow on the curve. A curve is **closed** if $\gamma(a) = \gamma(b)$. For circles and circular arcs, if the arrow points in the counterclockwise direction, we say that the curve has **positive orientation**. Curves traversed in the clockwise direction have **negative orientation** (see Figure 3.4).

It is a bit of a challenge to provide an exact definition of the notion of orientation for general closed curves and we omit this, relying on the intuition provided by our understanding of the movement of the hands of a clock in concrete examples.

A curve may have more than one parametrization. For example, the interval $[0, 1]$ can be parametrized as $\gamma_1(t) = t$, $0 \leq t \leq 1$ or $\gamma_2(t) = t^2$, $0 \leq t \leq 1$. Both γ_1 and γ_2 represent the same curve. In our analysis, we always choose and work with a specific parametrization of the curve. For that reason, it will be convenient to refer to a curve by its parametrization $\gamma(t)$ or simply γ, even though it may have more than one parametrization.

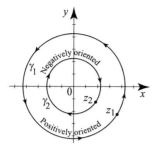

Fig. 3.4 Negative and positive orientation.

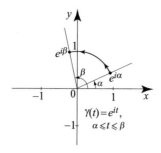

Fig. 3.5 A positively oriented arc with initial point $e^{i\alpha}$ and terminal point $e^{i\beta}$.

Since $z = x + iy$, it makes sense to adopt the notation $z(t) = x(t) + iy(t)$. In particular, we can write the parametric form of a curve γ using complex notation as

$$\gamma(t) = x(t) + iy(t), \quad a \le t \le b, \tag{3.1.1}$$

and think of the curve as the graph of a complex-valued function of a real variable t. The following examples illustrate the use of the complex notation.

Example 3.1.2. (Parametric forms of arcs, circles, and line segments)
(a) The arc in Figure 3.5 is conveniently parametrized by $\gamma(t) = e^{it} = \cos t + i \sin t$, $\alpha \le t \le \beta$. Its initial point is $e^{i\alpha}$, and its terminal point is $e^{i\beta}$.
(b) The positively oriented circle in Figure 3.6 is parametrized by

$$\gamma(t) = e^{it} = \cos t + i \sin t, \quad 0 \le t \le 2\pi.$$

The circle is a closed curve with initial point $\gamma(0) = 1$ and terminal point $\gamma(2\pi) = 1$.

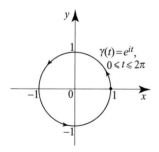

Fig. 3.6 A positively oriented circle.

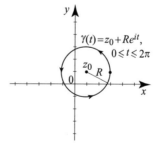

Fig. 3.7 Dilating and then translating a circle.

(c) The circle in Figure 3.7 is centered at z_0 with radius $R > 0$. We obtain its parametrization by dilating and then translating the equation in (b). This gives

$$\gamma(t) = z_0 + Re^{it} = z_0 + R(\cos t + i \sin t), \quad 0 \le t \le 2\pi.$$

For example, the equation

$$\gamma(t) = 1 + i + 2e^{it}, \quad 0 \le t \le 2\pi,$$

represents a circle centered at the point $(1, 1)$, with radius 2.

(d) Let z_1 and z_2 be arbitrary complex numbers. A **directed line segment** $[z_1, z_2]$ is the path γ over $[0, 1]$ defined by

$$\gamma(t) = (1 - t)z_1 + tz_2, \qquad 0 \le t \le 1. \tag{3.1.2}$$

Its initial point is $\gamma(0) = z_1$ and its terminal point is $\gamma(1) = z_2$.   □

As the next example illustrates, there is definitely an advantage in using the complex notation, especially when the parametric representation of a curve involves trigonometric functions.

Example 3.1.3. A **hypotrochoid** is a curve with parametric equations

$$x(t) = a\cos t + b\cos\left(\frac{at}{2}\right), \qquad y(t) = a\sin t - b\sin\left(\frac{at}{2}\right),$$

where a and b are real numbers. Express the equations in complex notation.

Solution. In complex form, we express $\gamma(t)$ as

$$x(t) + iy(t)$$
$$= \left(a\cos t + b\cos\left(\frac{at}{2}\right)\right) + i\left(a\sin t - b\sin\left(\frac{at}{2}\right)\right)$$
$$= a(\cos t + i\sin t) + b\left(\cos\left(\frac{at}{2}\right) - i\sin\left(\frac{at}{2}\right)\right)$$
$$= ae^{it} + be^{-i\frac{at}{2}}.$$

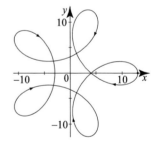

Figure 3.8 shows the hypotrochoid with constants $a = 8$ and $b = 5$ plotted over $0 \le t \le 2\pi$. Over this interval, the curve is closed.   □

Fig. 3.8 A hypotrochoid.

Definition 3.1.4. If γ is a curve parametrized by the interval $[a, b]$, the **reverse** of γ is the curve γ^* parametrized by the same interval $[a, b]$ and defined by

$$\gamma^*(t) = \gamma(b + a - t), \qquad t \in [a, b]. \tag{3.1.3}$$

The reverse of γ traces the same set of points as γ but in the opposite direction, starting with $\gamma^*(a) = \gamma(b)$ and ending with $\gamma^*(b) = \gamma(a)$.

Example 3.1.5. Find the reverse of the directed line segment $[z_1, z_2]$.

Solution. Using the parametrization for $[z_2, z_1]$ in Example 3.1.2(d) we find

$$\gamma^*(t) = \gamma(1-t) = tz_1 + (1-t)z_2, \qquad 0 \le t \le 1. \qquad \square$$

The curves in the preceding examples are all continuous, as can be seen from their graphs, but these curves also have differentiability properties. In order to study their differentiability, we investigate the derivative of a complex-valued function of a real variable. This is unrelated to the derivative of a complex-valued function of a complex variable.

Complex-Valued Functions of a Real Variable

Let f be a complex-valued function defined in an open interval (a,b). We define the derivative of f in the usual way by

$$f'(t) = \frac{d}{dt}f(t) = \lim_{h \to 0} \frac{f(t+h) - f(t)}{h}. \tag{3.1.4}$$

In view of Theorem 2.2.9, it follows that $f'(t)$ exists if and only if $(\operatorname{Re} f)'(t)$ and $(\operatorname{Im} f)'(t)$ both exist, and in this case

$$f'(t) = (\operatorname{Re} f)'(t) + i(\operatorname{Im} f)'(t). \tag{3.1.5}$$

The derivative of a complex-valued function of a real variable satisfies many properties similar to those of the usual derivative of a real-valued function.

Proposition 3.1.6. *Let f, g be complex-valued differentiable functions defined on an open interval (a,b) and let α, β be complex numbers. Then for $t \in (a,b)$ we have*

$$(\alpha f + \beta g)'(t) = \alpha f'(t) + \beta g'(t), \tag{3.1.6}$$

$$(fg)'(t) = f'(t)g(t) + g'(t)f(t), \tag{3.1.7}$$

$$\left(\frac{f}{g}\right)'(t) = \frac{f'(t)g(t) - g'(t)f(t)}{g(t)^2} \qquad \text{whenever} \quad g(t) \ne 0. \tag{3.1.8}$$

Moreover, the chain rule holds: If h is a differentiable map from (c,d) to (a,b), then

$$(f \circ h)'(t) = f'(h(t))h'(t), \qquad t \in (a,b). \tag{3.1.9}$$

These rules can be verified by splitting the real and imaginary parts; we leave their verification as an exercise.

Example 3.1.7. Suppose that $z = a + ib$ is a complex number. Show that

$$\frac{d}{dt}e^{zt} = ze^{zt}. \tag{3.1.10}$$

Solution. We have $e^{zt} = e^{at}\left(\cos(bt) + i\sin(bt)\right)$, and so using (3.1.5) we obtain

$$
\begin{aligned}
\frac{d}{dt}e^{zt} &= \frac{d}{dt}\left(e^{at}\cos(bt)\right) + i\frac{d}{dt}\left(e^{at}\sin(bt)\right) \\
&= \left(ae^{at}\cos(bt) - be^{at}\sin(bt)\right) + i\left(ae^{at}\sin(bt) + be^{at}\cos(bt)\right) \\
&= (a+ib)\left(e^{at}\cos(bt) + ie^{at}\sin(bt)\right) \\
&= ze^{zt}.
\end{aligned}
$$

$\square$

In (3.1.9) we considered the chain rule for compositions of the form

$$\mathbb{R} \to \mathbb{R} \to \mathbb{C}.$$

There is also a chain rule for compositions of the form

$$\mathbb{R} \to \mathbb{C} \to \mathbb{C},$$

where the second function is analytic. We examine this situation now.

Theorem 3.1.8. *Suppose that x,y are differentiable functions defined on an interval (a,b). Suppose that the function $\gamma = x + iy$ takes values in a region U in the complex plane and that F is an analytic function on U. Then the function $G = F \circ \gamma$ is differentiable on (a,b) and its derivative is*

$$G'(t) = F'(\gamma(t))\gamma'(t) \qquad\qquad t \in (a,b). \qquad\qquad (3.1.11)$$

Proof. Identifying $F(\gamma(t))$ with $F(x(t), y(t))$, we obtain the differentiability of $G(t)$ from Theorem 2.4.5. It remains to compute the derivative of $G(t)$. Suppose $F = u + iv$, where u, v are real-valued. Using Theorem 2.4.5 we write

$$
\begin{aligned}
\frac{dG}{dt} &= \frac{d}{dt}\left(u(x(t),y(t)) + iv(x(t),y(t))\right) \\
&= u_x(x(t),y(t))\frac{dx}{dt} + u_y(x(t),y(t))\frac{dy}{dt} + iv_x(x(t),y(t))\frac{dx}{dt} + iv_y(x(t),y(t))\frac{dy}{dt}
\end{aligned}
$$

Using the Cauchy-Riemann equations we replace u_y by $-v_x$ and v_y by u_x:

$$
\begin{aligned}
\frac{dG}{dt} &= \frac{d}{dt}\left(u(x(t),y(t)) + iv(x(t),y(t))\right) \\
&= u_x(x(t),y(t))\frac{dx}{dt} - v_x(x(t),y(t))\frac{dy}{dt} + iv_x(x(t),y(t))\frac{dx}{dt} + iu_x(x(t),y(t))\frac{dy}{dt} \\
&= u_x(x(t),y(t))\left(\frac{dx}{dt} + i\frac{dy}{dt}\right) + v_x(x(t),y(t))\left(-\frac{dy}{dt} + i\frac{dx}{dt}\right) \\
&= u_x(x(t),y(t))\left(\frac{dx}{dt} + i\frac{dy}{dt}\right) + iv_x(x(t),y(t))\left(\frac{dx}{dt} + i\frac{dy}{dt}\right) \\
&= \left(u_x(x(t),y(t)) + iv_x(x(t),y(t))\right)\left(\frac{dx}{dt} + i\frac{dy}{dt}\right).
\end{aligned}
$$

This is exactly the expression on the right in (3.1.11), in view of identity (2.5.3). ∎

Not all properties of the derivative of a real-valued function hold for complex-valued functions. Most notably, the mean value property fails for complex-valued functions. For real-variable functions, the mean value property states that if f is continuous on $[a, b]$ and differentiable on (a, b), then $f(b) - f(a) = f'(c)(b - a)$ for some c in (a, b).

This property does not hold for complex-valued functions. To see this, consider $f(x) = e^{ix}$ for x in $[0, 2\pi]$. Then f is continuous on $[0, 2\pi]$ and has a derivative $f'(x) = i e^{ix}$ on $(0, 2\pi)$. Also,

$$f(2\pi) - f(0) = 1 - 1 = 0,$$

but f' never vanishes since

$$|f'(x)| = |i e^{ix}| = |e^{ix}| = 1.$$

Hence there is no number c in $(0, 2\pi)$ such that $f(2\pi) - f(0) = (2\pi - 0)f'(c)$, and so the mean value property does not hold for complex-valued functions (Figure 3.9).

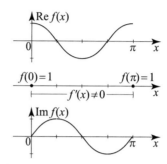

Fig. 3.9 The mean value property fails for complex-valued functions of a real variable.

Paths (Contours)

We introduce the notion of a path, which is fundamental in the theory of complex integration. Recall that a curve already contains intrinsically the notion of continuity. We now attach to it the fundamental analytical property of differentiability.

A continuous map f from $[a, b]$ to $\mathbb{C}$ is called **continuously differentiable** if it is also differentiable on $[a, b]$ and its derivative f' is continuous on $[a, b]$. The derivative f' is defined at a and b as the one-sided limits $f'(a) = \lim_{t \downarrow a} \frac{f(t) - f(a)}{t - a}$ and $f'(b) = \lim_{t \uparrow b} \frac{f(t) - f(b)}{t - b}$. We extend this notion by allowing γ to have one-sided derivatives at finitely many points in (a, b).

Definition 3.1.9. A continuous complex-valued function f defined on a closed interval $[a, b]$ is called **piecewise continuously differentiable** if there exist points $a_1 < a_2 < \cdots < a_{m-1}$ in (a, b) such that f' is continuously differentiable on each interval $[a_j, a_{j+1}]$ for $j = 0, 1, \ldots, m - 1$, where $a_0 = a$ and $a_m = b$. In other words:
(i) $f'(t)$ exists for all t in (a_j, a_{j+1}) and at the endpoints a_j, a_{j+1} as one-sided limit.
(ii) f' is continuous on each interval $[a_{j-1}, a_j]$ for $j = 1, \ldots, m$.

Note that if f' is defined on $[a, b]$, it may have jump discontinuities at some a_j.

Definition 3.1.10. A **path** or a **contour** is a curve γ defined a closed interval $[a, b]$ which is continuously differentiable or piecewise continuously differentiable. The path γ is **closed** if $\gamma(a) = \gamma(b)$.

Definition 3.1.11. Given points $a_0 < a_1 < \cdots < a_m$ and paths γ_j on $[a_{j-1}, a_j]$, $j = 1, \ldots, m$, such that $\gamma_j(a_j) = \gamma_{j+1}(a_j)$ for all $j = 1, \ldots, m-1$, the combined path

$$\Gamma = [\gamma_1, \ldots, \gamma_m]$$

is piecewise defined on $[a_0, a_m]$ by $\Gamma(t) = \gamma_j(t)$ for $t \in [a_{j-1}, a_j]$, $j = 1, \ldots, m$.

Thus, according to Definitions 3.1.10 and 3.1.11, a path or a contour γ is a finite sequence of continuously differentiable curves, $\gamma_1, \gamma_2, \ldots, \gamma_m$, joined at the endpoints, i.e., $\gamma = [\gamma_1, \ldots, \gamma_m]$. The path γ is closed if the initial point of γ_1 coincides the terminal point of γ_m, i.e., $\gamma_1(a_0) = \gamma_m(a_m)$.

Example 3.1.12. The path $\gamma = [\gamma_1, \gamma_2, \gamma_3]$ in Figure 3.10 consists of the curves: The line segment $\gamma_1 = [-2, -1]$; the semicircle γ_2; and the line segment $\gamma_3 = [1, 2]$. We can parametrize γ by the interval $[-2, 2]$ as follows:

$$\gamma(t) = \begin{cases} t & \text{if } -2 \leq t \leq -1, \\ e^{i\frac{\pi}{2}(1-t)} & \text{if } -1 \leq t \leq 1, \\ t & \text{if } 1 \leq t \leq 2. \end{cases}$$

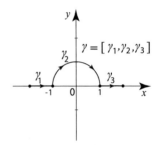

The choice of the interval $[-2, 2]$ as the domain of definition was just for convenience. Other closed intervals can be used to parametrize γ.   □

Fig. 3.10 The path of Example 3.1.12.

Example 3.1.13. (Polygonal paths) A **polygonal path** $\gamma = [z_1, z_2, \ldots, z_n]$ is the union of the line segments $[z_1, z_2]$, $[z_2, z_3]$, $\ldots$, $[z_{n-1}, z_n]$. This is a piecewise linear path with initial point z_1 and terminal point z_n and may have self intersections.

A polygonal path is called **simple** if it does not have self intersections, except possibly at the endpoints, that is, z_1 and z_n may coincide. The polygonal path is called **closed** if $z_1 = z_n$. As an illustration, let $z_1 = 0$, $z_2 = 1 + i$, and $z_3 = -1 + i$; then $\gamma = [z_1, z_2, z_3, z_1]$ is a simple closed polygonal path. To find the equation of γ, we start by finding the equations of the paths γ_1, γ_2, and γ_3, shown in Figure 3.11. From Example 3.1.2(d), we have

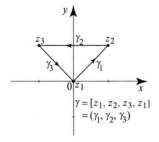

Fig. 3.11 The closed polygonal path $[z_1, z_2, z_3, z_1]$.

$$\gamma_1(t) = (1+i)t, \qquad\qquad\qquad\qquad\qquad 0 \le t \le 1;$$
$$\gamma_2(t) = (1-t)(1+i) + t(-1+i) = (1+i) - 2t, \qquad 0 \le t \le 1;$$
$$\gamma_3(t) = (1-t)(-1+i), \qquad\qquad\qquad\qquad 0 \le t \le 1.$$

We can now use these equations to parametrize γ over a closed interval, say $[0,1]$.
See Exercise 29. □

The following example shows two interesting cases.

Example 3.1.14. (Degenerate and doubly traced paths)
(a) Describe the path of the curve $\gamma(t) = z_0$, $a \le t \le b$.
(b) Describe the path of $\gamma_2(t) = Re^{it}$ $(R > 0)$, $0 \le t \le 4\pi$, and discuss how it is
different from that of the curve $\gamma_1(t) = Re^{it}$, $0 \le t \le 2\pi$.

Solution. (a) As t ranges through the interval $[a, b]$, the value of $\gamma(t)$ remains fixed
at the point z_0. Clearly γ is continuous and $\gamma' = 0$ is also continuous, so γ is a path,
which has degenerated to a single point (Figure 3.12).

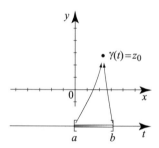

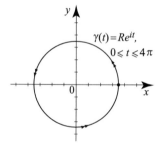

Fig. 3.12 A path that degenerates to a point.

Fig. 3.13 A doubly traced circle.

(b) Points on the path γ_2 are on the circle of radius R, centered at the origin. As
t ranges from 0 to 4π, $\gamma_2(t)$ goes around the circle twice. This path is shown in
Figure 3.13. One cannot automatically tell from the picture that the path is traced
twice and so we use double arrows to precisely indicate this fact, i.e., that the path
γ_2 is traced twice in the indicated counterclockwise direction. The path γ_1 traces
around the circle only once. The two paths have the same graph but γ_2 has double
the length of γ_1. □

As a convention, whenever we refer to a closed path, we mean the path that is
traversed only once, unless stated otherwise.

Example 3.1.15. (A curve that is not a path) Let $f(t) = t^2 \sin\frac{1}{t}$ for $t \ne 0$ and
$f(0) = 0$, and define a curve $\gamma(t) = t + if(t)$ for $-\pi \le t \le \pi$.

The graph of γ is simply the graph of f over the interval $[-\pi, \pi]$. For $t \neq 0$, we have

$$f'(t) = 2t \sin \frac{1}{t} - \cos \frac{1}{t},$$

and for $t = 0$

$$f'(0) = \lim_{t \to 0} \frac{t^2 \sin \frac{1}{t}}{t} = 0$$

by the squeeze theorem. So f is continuous and differentiable on the closed intervals $[-\pi, 0]$ and $[0, \pi]$ (with one-sided limits at the endpoints). But f' is not continuous on $[-\pi, 0]$ or on $[0, \pi]$; thus, γ is not a path. The graphs of f and f' are shown in Figure 3.14. □

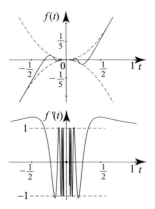

Fig. 3.14 f is a differentiable function but its derivative is not continuous at a point. Thus the graph of f is not a piecewise continuous curve, i.e., a path.

Exercises 3.1

In Exercises 1–8 parametrize the curves over suitable intervals $[a, b]$.

1. The line segment with initial point $z_1 = 1 + i$ and terminal point $z_2 = -1 - 2i$.

2. The line segment through the origin as initial point and terminal point $z = e^{i\frac{\pi}{3}}$.

3. The counterclockwise circle with center at $3i$ and radius 1.

4. The clockwise circle with center at $-2 - i$ and radius 3.

5. The positively oriented arc on the unit circle such that $-\frac{\pi}{4} \leq \operatorname{Arg} z \leq \frac{\pi}{4}$.

6. The negatively oriented arc on the unit circle such that $-\frac{\pi}{4} \leq \operatorname{Arg} z \leq \frac{\pi}{4}$.

7. The closed polygonal path $[z_1, z_2, z_3, z_1]$ where $z_1 = 0$, $z_2 = i$, and $z_3 = -1$.

8. The polygonal path $[z_1, z_2, z_3, z_4]$ where $z_1 = 1$, $z_2 = 2$, $z_3 = i$, and $z_4 = 2i$.

In Exercises 9–11, describe the parametrizations of the paths shown in the figures.

9.

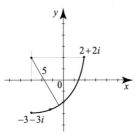

10.

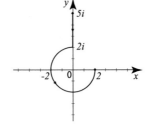

11.

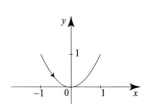

Fig. 3.15 Circular arc of radius 5 centered at $-3 + 2i$.

Fig. 3.16 Vertical line segment followed by circular arc of radius 2 centered at 0.

Fig. 3.17 Arc of the parabola $y = x^2$ for $-1 \leq x \leq 1$.

12. Find a parametrization for the reverse of the path in Exercise 9.

13. Find a parametrization for the reverse of the path in Exercise 10.

14. Find a parametrization for the reverse of the path in Exercise 11.

In Exercises 15–18, describe the circle providing its center, radius, and orientation; then plot it.

15. $\gamma(t) = 1 + i + 2e^{it}$, $0 \le t \le 2\pi$

16. $\gamma(t) = -1 + 2e^{2it}$, $0 \le t \le \pi$

17. $\gamma(t) = -i + \dfrac{1}{2}e^{-it}$, $0 \le t \le 2\pi$

18. $\gamma(t) = e^{-4it}$, $-\dfrac{\pi}{2} \le t \le 0$

In Exercises 19–24, find the derivative of the functions.

19. $f(t) = te^{-it}$

20. $f(t) = e^{2it^2}$

21. $f(t) = (2+i)\cos(3it)$

22. $f(t) = \dfrac{2+i+t}{-i-2t}$

23. $f(t) = \left(\dfrac{t+i}{t-i}\right)^2$

24. $f(t) = \mathrm{Log}\,(it)$

In Exercises 25–28, a curve is described in parametric equations. (a) Find the equation of the curve in complex form. (b) Plot the curve for specific values of a and b of your choice.

25. A hypocycloid $(a > b)$

$$x(t) = (a-b)\cos t + b\cos\left(\frac{a-b}{b}t\right)$$

$$y(t) = (a-b)\sin t - b\sin\left(\frac{a-b}{b}t\right)$$

26. An epicycloid

$$x(t) = (a+b)\cos t - b\cos\left(\frac{a+b}{b}t\right)$$

$$y(t) = (a+b)\sin t - b\sin\left(\frac{a+b}{b}t\right)$$

27. An epitrochoid

$$x(t) = a\cos t - b\cos\left(\frac{at}{2}\right)$$

$$y(t) = a\sin t - b\sin\left(\frac{at}{2}\right)$$

28. A trochoid

$$x(t) = at - b\sin t$$

$$y(t) = a - b\cos t$$

29. Verify that a parametrization of the polygonal path in Example 3.1.13 is

$$\gamma(t) = \begin{cases} 3t(1+i) & \text{if } 0 \le t \le \frac{1}{3} \\ 3+i-6t & \text{if } \frac{1}{3} \le t \le \frac{2}{3} \\ (-1+i)(3-3t) & \text{if } \frac{2}{3} \le t \le 1. \end{cases}$$

3.2 Complex Integration

We begin by extending the definition of piecewise continuous functions to complex-valued functions.

Definition 3.2.1. A complex-valued function f defined on a closed interval $[a,b]$ is said to be **piecewise continuous** if there exist points $a_1 < a_2 < \cdots < a_{m-1}$ in (a,b) such that:

(i) f is continuous at every point in $[a,b] \setminus \{a_1, a_2, \ldots, a_{m-1}\}$.

(ii) The limits $\lim_{t \uparrow a_j} f(t)$ and $\lim_{t \downarrow a_j} f(t)$ exist and are finite for every $j = 1, \ldots, m-1$.

We now extend the notion of Riemann integral to complex-valued functions. We assume from the theory of real variables that piecewise continuous real-valued functions are Riemann integrable. Since piecewise continuous complex-valued functions have piecewise continuous real and imaginary parts, they are also Riemann integrable.

Definition 3.2.2. Let f be a piecewise continuous function defined on an interval $[a,b]$ and taking values in the complex plane. We define the **Riemann integral** of f as follows:

$$\int_a^b f(t)\,dt = \int_a^b \operatorname{Re} f(t)\,dt + i \int_a^b \operatorname{Im} f(t)\,dt \tag{3.2.1}$$

As noted, if f is piecewise continuous, then so are $\operatorname{Re} f$ and $\operatorname{Im} f$, thus the integrals of these functions are well defined.

Note that as a consequence of this definition we have

$$\operatorname{Re} \int_a^b f(t)\,dt = \int_a^b \operatorname{Re} f(t)\,dt\,, \qquad \operatorname{Im} \int_a^b f(t)\,dt = \int_a^b \operatorname{Im} f(t)\,dt\,. \tag{3.2.2}$$

The statements in the following proposition are known for real-valued functions and are extended to complex-valued functions.

Proposition 3.2.3. (Properties of the Riemann Integral of Complex-valued Functions) *Let $a < b$ be real numbers. Let f and g be piecewise continuous complex-valued functions on $[a, b]$ and let β be a complex number.*
(i) Then we have

$$\int_a^b (f(t) \pm g(t))\,dt = \int_a^b f(t)\,dt \pm \int_a^b g(t)\,dt. \tag{3.2.3}$$

$$\int_a^b \beta f(t)\,dt = \beta \int_a^b f(t)\,dt. \tag{3.2.4}$$

(ii) If c lies in (a,b), then we have

$$\int_a^b f(t)\,dt = \int_a^c f(t)\,dt + \int_c^b f(t)\,dt. \tag{3.2.5}$$

(iii) We also have

$$\left| \int_a^b f(t)\,dt \right| \le \int_a^b |f(t)|\,dt. \tag{3.2.6}$$

(iv) Moreover, if f and g are differentiable on (a,b) and continuous on $[a,b]$, then

$$\int_a^b f(t)g'(t)\,dt = f(b)g(b) - f(a)g(a) - \int_a^b f'(t)g(t)\,dt\,, \tag{3.2.7}$$

that is, integration by parts holds as for real-valued functions.

Proof. Property (3.2.3) is immediate and is left to the reader. To prove (3.2.4), we write $f = f_1 + if_2$, where $f_1 = \operatorname{Re} f$ and $f_2 = \operatorname{Im} f$ and $\beta = \beta_1 + i\beta_2$, with β_1, β_2 real. Then $\beta f = \beta_1 f_1 - \beta_2 f_2 + i(\beta_1 f_2 + \beta_2 f_1)$. In view of Definition 3.2.2 we have

$$\int_a^b \beta f(t)\,dt = \int_a^b \big(\beta_1 f_1(t) - \beta_2 f_2(t)\big)\,dt + i\int_a^b \big(\beta_1 f_2(t) + \beta_2 f_1(t)\big)\,dt,$$

and in view of the linearity of the Riemann integral for real-valued functions, the preceding expression is equal to

$$\beta_1 \int_a^b f_1(t)\,dt - \beta_2 \int_a^b f_2(t)\,dt + i\left(\beta_1 \int_a^b f_2(t)\,dt + \beta_2 \int_a^b f_1(t)\,dt\right).$$

The preceding expression is equal to

$$(\beta_1 + i\beta_2)\left(\int_a^b f_1(t)\,dt + i\int_a^b f_2(t)\,dt\right) = \beta \int_a^b f(t)\,dt,$$

and this proves (3.2.4), hence (i). Property (3.2.5) is also reduced to the analogous properties of real and imaginary parts and is left to the reader. Next, in proving (3.2.6), we may assume that $\int_a^b f(t)\,dt$ is not zero. Then we set

$$e^{i\theta} = \frac{\int_a^b f(t)\,dt}{\left|\int_a^b f(t)\,dt\right|} \implies \left|\int_a^b f(t)\,dt\right| = e^{-i\theta}\int_a^b f(t)\,dt$$

for some real number θ. Using property (3.2.4), we write

$$\left|\int_a^b f(t)\,dt\right| = e^{-i\theta}\int_a^b f(t)\,dt = \int_a^b e^{-i\theta}f(t)\,dt = \mathrm{Re}\int_a^b e^{-i\theta}f(t)\,dt$$

where in the last equality we used that the expression $\int_a^b e^{-i\theta}f(t)\,dt$ is equal to a modulus, and hence, it is real. Using (3.2.2) we write

$$\left|\int_a^b f(t)\,dt\right| = \int_a^b \mathrm{Re}\left(e^{-i\theta}f(t)\right)\,dt \le \int_a^b |e^{-i\theta}f(t)|\,dt = \int_a^b |f(t)|\,dt, \quad (3.2.8)$$

thus proving (3.2.6). The only inequality in (3.2.8) uses $\mathrm{Re}\left(e^{-i\theta}f(t)\right) \le |e^{-i\theta}f(t)|$ and the monotonicity of the Riemann integral for real-valued functions.

Finally, we turn our attention to (iv). We write $f = f_1 + if_2$ and $g = g_1 + ig_2$, where f_1, f_2, g_1, g_2 are real-valued. Then the left side of (3.2.7) is equal to

$$\int_a^b (f_1 + if_2)(g_1' + ig_2')\,dt = \int_a^b (f_1 g_1' - f_2 g_2')\,dt + i\left(\int_a^b (f_2 g_1' + f_1 g_2')\,dt\right). \quad (3.2.9)$$

Integrating by parts, we write (3.2.9) as $A + iB$, where

$$A = \big[f_1(b)g_1(b) - f_1(a)g_1(a)\big] - \big[f_2(b)g_2(b) - f_2(a)g_2(a)\big] - \int_a^b (f_1' g_1 - f_2' g_2)\,dt$$

$$B = \big[f_2(b)g_1(b) - f_2(a)g_1(a)\big] - \big[f_1(b)g_2(b) - f_1(a)g_2(a)\big] - \int_a^b (f_2' g_1 - f_1' g_2)\,dt.$$

Observing that $A + iB$ is equal to the right side of (3.2.7), we complete the proof. ∎

Example 3.2.4. (A complex-valued piecewise continuous function) Evaluate the integral $\int_0^2 f(t)\,dt$, where

$$f(t) = \begin{cases} (1+i)t & \text{if } 0 \le t \le 1, \\ it^2 & \text{if } 1 < t \le 2. \end{cases}$$

Solution. We use property (3.2.5) and then property (3.2.4) to write

$$\int_0^2 f(t)\,dt = \int_0^1 f(t)\,dt + \int_1^2 f(t)\,dt = (1+i)\int_0^1 t\,dt + i\int_1^2 t^2\,dt$$

$$= \frac{1+i}{2} + \frac{i}{3}(8-1) = \frac{1}{2} + i\frac{17}{6}. \qquad \square$$

Antiderivatives of Complex-Valued Functions

Definition 3.2.5. If f is a piecewise continuous complex-valued function on $[a,b]$, we say that F is an **antiderivative** of f if $F' = f$ at all the points of continuity of f on (a,b). Hence if $f = u + iv$ and $F = U + iV$, with u,v,U,V real-valued, then the equalities $U'(x) = u(x)$ and $V'(x) = v(x)$ hold for all but finitely many x in (a,b). We say that F is a **continuous antiderivative** of f, if F is an antiderivative of f and it is a continuous function.

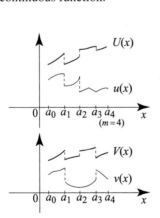

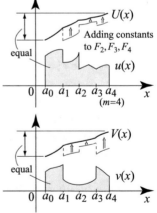

Fig. 3.18 An antiderivative of a piecewise continuous function may not be continuous.

Fig. 3.19 Selecting a continuous antiderivative.

Using the previous notation, we write $[a, b]$ as the finite union of adjacent closed subintervals $[a_0, a_1], [a_1, a_2], \ldots, [a_{m-1}, a_m]$, with $a_0 = a$ and $a_m = b$, and such that f is continuous on each subinterval. The functions U and V are continuous on each subinterval and this makes F piecewise continuous on $[a, b]$. In general, F may not be continuous at the points a_j (see Figure 3.18); but, as we show momentarily, a continuous antiderivative can always be found.

Let f_j denote the restriction of f to $[a_{j-1}, a_j]$, and let F_j denote an antiderivative of f_j over $[a_{j-1}, a_j]$. Each F_j is computed up to an arbitrary complex constant, which can be determined in such a way to make F continuous.

Start by setting the arbitrary constant in F_1 equal 0. Then determine the constant in F_2 so that $\lim_{x \uparrow a_1} F_1(x) = \lim_{x \downarrow a_1} F_2(x)$. (We use the up-arrow to denote a limit from the left and the down-arrow a limit from the right.) This determines F_2 and makes the antiderivative of f continuous on $[a, a_2]$. Continue in this fashion: Once we have found F_j, determine the constant in F_{j+1} so that $\lim_{x \uparrow a_j} F_j(x) = \lim_{x \downarrow a_j} F_{j+1}(x)$. By construction, the resulting function F will be continuous on $[a, b]$ (see Figure 3.19). The following example illustrates the method.

Example 3.2.6. (Finding a continuous antiderivative) Find a continuous antiderivative of the function in Example 3.2.4,

$$f(t) = \begin{cases} (1+i)t & \text{if} \quad 0 \leq t \leq 1, \\ it^2 & \text{if} \quad 1 < t \leq 2. \end{cases}$$

Solution. Integrating each continuous part of f, we obtain

$$F(t) = \begin{cases} \frac{1+i}{2}t^2 & \text{if} \quad 0 \leq t \leq 1, \\ \frac{i}{3}t^3 + C & \text{if} \quad 1 < t \leq 2, \end{cases}$$

where C is an arbitrary constant. Note how in the first part of F we already set the arbitrary constant equal 0. To determine C, we evaluate F at 1 using both formulas from the intervals $(0, 1)$ and $(1, 2)$ and get

$$\frac{1+i}{2} = \frac{i}{3} + C \quad \Rightarrow \quad C = \frac{1}{2} + \frac{i}{6}.$$

Hence

$$F(t) = \begin{cases} \frac{1+i}{2}t^2 & \text{if} \quad 0 \leq t \leq 1, \\ \frac{i}{3}t^3 + \frac{1}{2} + \frac{i}{6} & \text{if} \quad 1 < t \leq 2, \end{cases}$$

is a continuous antiderivative of f on $[0, 2]$. $\quad\square$

The following is an extension of the fundamental theorem of calculus to piecewise continuous complex functions.

Theorem 3.2.7. (Fundamental Theorem of Calculus) *Suppose that f is a piecewise continuous complex-valued function on the interval $[a, b]$ and let F be a continuous antiderivative of f in $[a, b]$. Then*

$$\int_a^b f(x)\,dx = F(b) - F(a). \tag{3.2.10}$$

Proof. Let $f = u + iv$ where $u = \operatorname{Re} f$ and $v = \operatorname{Im} f$. Suppose first that f is continuous on $[a,b]$. Let $F = U + iV$ be an antiderivative of f. Using (3.2.1) and the fundamental theorem of calculus, we see that

$$
\begin{aligned}
\int_a^b f(x)\,dx &= \int_a^b u(x)\,dx + i \int_a^b v(x)\,dx \\
&= \left(U(b) - U(a) \right) + i \left(V(b) - V(a) \right) \\
&= \left(U(b) + iV(b) \right) - \left(U(a) + iV(a) \right) = F(b) - F(a),
\end{aligned}
$$

and hence (3.2.10) holds.

Let $a_1 < a_2 < \cdots < a_{m-1}$ be the points of discontinuity of a piecewise continuous function f in the open interval (a,b). We use (3.2.5) and the previous case to write

$$
\begin{aligned}
\int_a^b f(x)\,dx &= \sum_{j=1}^m \int_{a_{j-1}}^{a_j} f(x)\,dx = \sum_{j=1}^m \left(F(a_j) - F(a_{j-1}) \right) \\
&= F(a_m) - F(a_0) = F(b) - F(a),
\end{aligned}
$$

which proves (3.2.10).   ∎

As an application of Theorem 3.2.7, we evaluate the integral in Example 3.2.4 using the continuous antiderivative that we found in Example 3.2.6. In the notation of Example 3.2.6, we have

$$
\int_0^2 f(t)\,dt = F(2) - F(0) = \left(\frac{i}{3} 2^3 + \frac{1}{2} + \frac{i}{6} \right) - 0 = \frac{1}{2} + \frac{17}{6} i,
$$

which agrees with the result of Example 3.2.4.

It is easy to show that two continuous antiderivatives of f differ by a complex constant on $[a,b]$. Motivated by Theorem 3.2.7, we write

$$
\int f(t)\,dt = F(t) + C \tag{3.2.11}
$$

to denote *any continuous* antiderivative of f; here C is an arbitrary complex constant. For example, if $z \neq 0$ is a complex number, then

$$
\int e^{zt}\,dt = \frac{1}{z} e^{zt} + C \qquad (z \neq 0), \tag{3.2.12}
$$

as can be checked by verifying that the derivative of the right side is equal to the integrand on the left side. This simple integral of a complex-valued function has an interesting application to the evaluation of certain integrals from calculus.

Example 3.2.8. Let a and b be arbitrary nonzero real numbers. Compute

$$I_1 = \int e^{at} \cos(bt)\, dt \qquad \text{and} \qquad I_2 = \int e^{at} \sin(bt)\, dt.$$

Solution. The idea is to compute $I = I_1 + iI_2$ and then obtain I_1 and I_2 by taking real and imaginary parts of I. We have

$$I = I_1 + iI_2 = \int e^{at} \cos(bt)\, dt + i \int e^{at} \sin(bt)\, dt$$

$$= \int e^{at} \left(\cos(bt) + i \sin(bt) \right) dt = \int e^{(a+ib)t}\, dt$$

$$= \frac{1}{a+ib} e^{(a+ib)t} + C,$$

by (3.2.12). Now we rewrite $\frac{1}{a+ib} e^{(a+ib)t}$ in terms of its real and imaginary parts. Multiplying and dividing by the conjugate $a - ib$ of the denominator we have

$$\frac{1}{a+ib} e^{(a+ib)t} = \frac{e^{at}}{a^2 + b^2}(a - ib)\left(\cos(bt) + i \sin(bt) \right)$$

$$= \frac{e^{at}}{a^2 + b^2}\left(a\cos(bt) + b\sin(bt) \right) + i\frac{e^{at}}{a^2 + b^2}\left(a\sin(bt) - b\cos(bt) \right).$$

Therefore,

$$\int e^{at} \cos(bt)\, dt = \frac{e^{at}}{a^2 + b^2}\left(a\cos(bt) + b\sin(bt) \right) + C_1 \qquad (3.2.13)$$

and

$$\int e^{at} \sin(bt)\, dt = \frac{e^{at}}{a^2 + b^2}\left(a\sin(bt) - b\cos(bt) \right) + C_2, \qquad (3.2.14)$$

where C_1 and C_2 are arbitrary real constants. $\qquad\qquad\qquad\qquad\qquad\qquad\square$

We now have all the tools that we need to construct the integral along a path in the complex plane.

Path or Contour Integrals

We discuss how to define the integral of continuous complex-valued functions over paths.

Definition 3.2.9. (Path Integral or Contour Integral) Suppose that γ is a path over a closed interval $[a, b]$ and that f is a continuous complex-valued function defined on the graph of γ. The **path** or **contour integral** of f on γ is defined as:

$$\int_\gamma f(z)\, dz = \int_a^b f(\gamma(t))\gamma'(t)\, dt. \qquad (3.2.15)$$

We now give examples of path integrals, starting with one of great importance.

Example 3.2.10. (Path integrals of $(z - z_0)^{-n}$, n an arbitrary integer) Let $C_R(z_0)$
be the positively oriented circle with center at z_0 and radius $R > 0$.
(a) Show that

$$\int_{C_R(z_0)} \frac{1}{z - z_0} \, dz = 2\pi i. \tag{3.2.16}$$

(b) Let $n \neq 1$ be an integer; show that

$$\int_{C_R(z_0)} \frac{1}{(z - z_0)^n} \, dz = 0. \tag{3.2.17}$$

Solution. (a) Parametrize $C_R(z_0)$
by $\gamma(t) = z_0 + Re^{it}$, $0 \leq t \leq 2\pi$.
Then $\gamma'(t) = iRe^{it}$ and, for $z =$
$z_0 + Re^{it}$ on γ, we have $z - z_0 =$
Re^{it}. Thus

$$\int_{C_R(z_0)} \frac{1}{z - z_0} \, dz$$

$$= \int_0^{2\pi} \frac{1}{Re^{it}} iRe^{it} \, dt$$

$$= i \int_0^{2\pi} dt = 2\pi i.$$

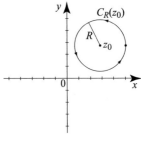

Fig. 3.20 The circle $C_R(z_0)$.

(b) We use (3.2.15) and some of the information we derived in (a) and get

$$\int_{C_R(z_0)} \frac{1}{(z - z_0)^n} \, dz = \int_0^{2\pi} \left(Re^{it}\right)^{-n} iRe^{it} \, dt = R^{-n+1} \int_0^{2\pi} ie^{i(-n+1)t} \, dt$$

$$= \frac{R^{-n+1}}{-n+1} e^{i(-n+1)t} \Big|_0^{2\pi} = \frac{R^{-n+1}}{-n+1} \left(e^{2\pi(-n+1)i} - e^0\right) = 0,$$

since $e^{2\pi(-n+1)i} = e^0 = 1$ and $n \neq 1$. ☐

In particular, if $C_1(0)$ denotes the positively oriented unit circle with center at the
origin, then (3.2.16) and (3.2.17) imply that

$$\int_{C_1(0)} \frac{1}{z} \, dz = 2\pi i \quad \text{and} \quad \int_{C_1(0)} z \, dz = 0. \tag{3.2.18}$$

Compare (3.2.18) with the following integrals involving $\bar{z}$.

Example 3.2.11. (Integrals involving $\bar{z}$) Show that

$$\int_{C_1(0)} \frac{1}{\bar{z}} dz = 0 \qquad \text{and} \qquad \int_{C_1(0)} \bar{z}\, dz = 2\pi i. \tag{3.2.19}$$

Solution. To compute the integrals we could use the definition of the path integral (3.2.15) and proceed as we did in Example 3.2.10. Another way is to use (3.2.18) and the following observations. On $C_1(0)$, $z = e^{it}$, $\bar{z} = \overline{e^{it}} = e^{-it} = \frac{1}{z}$. Thus, by (3.2.18), we have

$$\int_{C_1(0)} \frac{1}{\bar{z}} dz = \int_{C_1(0)} \frac{1}{1/z} dz = \int_{C_1(0)} z\, dz = 0,$$

and

$$\int_{C_1(0)} \bar{z}\, dz = \int_{C_1(0)} \frac{1}{z} dz = 2\pi i. \qquad \square$$

The path integral has many properties similar to the Riemann integral. We list some of them.

Proposition 3.2.12. (Properties of the Path Integral) *Suppose γ is a path on $[a,b]$, f and g are continuous functions on γ, and α and β are complex numbers. Then we have*
(i)

$$\int_\gamma (\alpha f(z) + \beta g(z))\, dz = \alpha \int_\gamma f(z)\, dz + \beta \int_\gamma g(z)\, dz. \tag{3.2.20}$$

(ii) *Let γ^* denote the reverse of γ. Then*

$$\int_{\gamma^*} f(z)\, dz = -\int_\gamma f(z)\, dz. \tag{3.2.21}$$

(iii) *For $k = 1,\ldots,m$, let γ_k be a path defined on $[k-1,k]$ such that $\gamma_k(k) = \gamma_{k+1}(k)$, $k \leq m-1$. If f is a continuous function on the combined path $\Gamma = [\gamma_1, \gamma_2, \ldots, \gamma_m]$, then*

$$\int_\Gamma f(z)\, dz = \sum_{k=1}^m \int_{\gamma_k} f(z)\, dz. \tag{3.2.22}$$

Proof. (i) This property follows by expressing the path integrals as Riemann integrals via (3.2.15) and then applying (3.2.3) and (3.2.4).
(ii) Recall the parametrization of the reverse of γ from (3.1.3): $\gamma^*(t) = \gamma(b+a-t)$, where t runs over the same interval $[a, b]$ that parametrizes γ. Then

$$(\gamma^*)'(t) = -\gamma'(b+a-t).$$

Changing variables $\tau = b+a-t$, $dt = -d\tau$ and using (3.2.15), we obtain

$$\int_{\gamma^*} f(z)\, dz = \int_a^b f(\gamma^*(t))\,(\gamma^*)'(t)\, dt = \int_a^b f(\gamma(a+b-t))\left(-\gamma'(b+a-t)\right) dt$$

$$= \int_b^a f(\gamma(\tau))\gamma'(\tau)\, d\tau = -\int_a^b f(\gamma(\tau))\gamma'(\tau)\, d\tau = -\int_\gamma f(z)\, dz.$$

(iii) The path Γ is defined piecewise on the interval $[0,m]$ by

$$\Gamma(t) = \gamma_k(t) \text{ for } k-1 \le t \le k. \tag{3.2.23}$$

Then we have

$$\int_\Gamma f(z)\,dz = \int_0^m f(\Gamma(t))\Gamma'(t)\,dt = \sum_{k=1}^m \int_{k-1}^k f(\Gamma(t))\Gamma'(t)\,dt$$

$$= \sum_{k=1}^m \int_0^1 f(\gamma_k(\tau))\gamma_k'(\tau)\,d\tau$$

$$= \sum_{k=1}^m \int_{\gamma_k} f(z)\,dz.$$

Thus we obtain (3.2.22).   ∎

Here is an illustration that uses the linearity of the path integral (3.2.20) and the familiar relations

$$x = \frac{z+\bar{z}}{2} \quad \text{and} \quad y = \frac{z-\bar{z}}{2i}.$$

Example 3.2.13. Let $C_1(0)$ denote the positively oriented unit circle traced once. Compute

$$\int_{C_1(0)} x\,dz = \int_{C_1(0)} \operatorname{Re} z\,dz.$$

Solution. Using (3.2.18) and (3.2.19), we have

$$\int_{C_1(0)} x\,dz = \int_{C_1(0)} \frac{z+\bar{z}}{2}\,dz = \frac{1}{2}\underbrace{\int_{C_1(0)} z\,dz}_{0} + \frac{1}{2}\underbrace{\int_{C_1(0)} \bar{z}\,dz}_{2\pi i} = \pi i. \qquad \square$$

The integrals in previous examples involved smooth curves. In the following example, we compute integrals over polygonal paths, which are piecewise smooth.

Example 3.2.14. (Integrals over polygonal paths) Let $z_1 = -1$, $z_2 = 1$, and $z_3 = i$ (see Figure 3.21). Compute

(a) $\displaystyle\int_{[z_1,z_2]} \bar{z}\,dz$ (b) $\displaystyle\int_{[z_2,z_3]} \bar{z}\,dz$ (c) $\displaystyle\int_{[z_3,z_1]} \bar{z}\,dz$

(d) $\displaystyle\int_{[z_1,z_2,z_3]} \bar{z}\,dz$ (e) $\displaystyle\int_{[z_1,z_2,z_3,z_1]} \bar{z}\,dz$ (f) $\displaystyle\int_{[z_1,z_3]} \bar{z}\,dz.$

Solution. Let γ_1 be a parametrization of
$[z_1, z_2]$, γ_2 be a parametrization of $[z_2, z_3]$,
and γ_3 be a parametrization of $[z_3, z_1]$.
There are several possible ways to compute γ_1, γ_2 and γ_3. To be consistent with
earlier notation we use (3.1.2). Accordingly, for $0 \le t \le 1$ we have

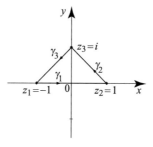

$$\gamma_1(t) = -(1-t) + t = -1 + 2t$$
$$\gamma_2(t) = 1 - t + it$$
$$\gamma_3(t) = (1-t)i - t = -t + i(1-t),$$

Fig. 3.21 The points $z_1 = -1$, $z_2 = 1, z_3 = i$.

and these imply that

$$\gamma_1'(t) = 2, \qquad \gamma_2'(t) = -1 + i, \qquad \gamma_3'(t) = -1 - i.$$

(a) Appealing to (3.2.15) we write

$$\int_{[z_1, z_2]} \overline{z} \, dz = \int_0^1 \overline{(-1 + 2t)} \, (2) \, dt = 2 \int_0^1 (-1 + 2t) \, dt = 2(-t + t^2) \Big|_0^1 = 0.$$

(b) We have

$$\int_{[z_2, z_3]} \overline{z} \, dz = \int_0^1 ((1-t) - it)(-1 + i) \, dt = (-1 + i) \int_0^1 (1 - (1+i)t) \, dt$$
$$= (-1 + i)\left(t - \frac{1+i}{2} t^2\right) \Big|_0^1 = i.$$

(c) We have

$$\int_{[z_3, z_1]} \overline{z} \, dz = \int_0^1 (-t - i(1-t))(-1 - i) \, dt = (-1 - i) \int_0^1 (-i + (-1 + i)t) \, dt$$
$$= (-1 - i)\left(-it + \frac{-1 + i}{2} t^2\right) \Big|_0^1 = i.$$

The integrals in (d)–(f) follow from (a)–(c) and properties of the path integral.
(d), (e) We use (3.2.22) to write

$$\int_{[z_1, z_2, z_3]} \overline{z} \, dz = \int_{[z_1, z_2]} \overline{z} \, dz + \int_{[z_2, z_3]} \overline{z} \, dz = 0 + i = i; \qquad (3.2.24)$$

$$\int_{[z_1, z_2, z_3, z_1]} \overline{z} \, dz = \int_{[z_1, z_2, z_3]} \overline{z} \, dz + \int_{[z_3, z_1]} \overline{z} \, dz = i + i = 2i. \qquad (3.2.25)$$

(f) We use (3.2.21) to obtain

$$\int_{[z_1,z_3]} \overline{z}\,dz = -\int_{[z_3,z_1]} \overline{z}\,dz = -i. \tag{3.2.26}$$

Parts (d) and (f) show that the integral of $\overline{z}$ over a path that joins z_1 to z_3 depends on the path. $\square$

Under what conditions is the path integral independent of the path? We address this important question in the next section. Another question that comes to mind as we work with path integrals concerns the parametrization of the path. Since a path may be parametrized in many different ways, is the integral independent of the choice of the parametrization?

Example 3.2.15. The positively oriented unit circle $C_1(0)$ can be parametrized by $\gamma_1(s) = e^{is}$, $0 \leq s \leq 2\pi$ or $\gamma_2(t) = e^{2\pi it}$, $0 \leq t \leq 1$. If f is a continuous function on $C_1(0)$, show that $\int_{\gamma_1} f(z)\,dz = \int_{\gamma_2} f(z)\,dz$.

Solution. Using the definition of path integrals and a simple change of variables $s = 2\pi t$, $ds = 2\pi\,dt$, we have

$$\int_{\gamma_2} f(z)\,dz = \int_0^1 f(e^{2\pi it})2\pi i e^{2\pi it}\,dt = \int_0^{2\pi} f(e^{is})ie^{is}\,ds = \int_{\gamma_1} f(z)\,dz. \qquad \square$$

The same proof works in general, but we have to explain what we mean by two parametrizations being the same. To do so, we need to recall the notion of a continuously differentiable map. A map from $[a,b]$ to $\mathbb{R}$ is called continuously differentiable if it is continuous on $[a,b]$ and has a continuous derivative on (a,b).

Definition 3.2.16. We say that the paths $\gamma_1(t)$, $a \leq t \leq b$, and $\gamma_2(s)$, $c \leq s \leq d$, have **equivalent parametrizations** if there is a strictly increasing continuously differentiable function ϕ from $[c,d]$ onto $[a,b]$ such that $\phi(c) = a$ and $\phi(d) = b$ and $\gamma_2(s) = \gamma_1 \circ \phi(s)$ for all s in $[c,d]$, or equivalently $\gamma_1(t) = \gamma_2 \circ \phi^{-1}(t)$ for all $t \in [a,b]$, where ϕ^{-1} is the unique inverse of ϕ.

Notice that the range of paths with equivalent parametrizations is the same. Moreover, the orientation of the trajectories of both paths is the same. Using this terminology, the two parametrizations of Example 3.2.15 are equivalent.

The next result confirms a property that we would expect from a path integral.

Proposition 3.2.17. (Independence of Parametrization) *Suppose that the paths $\gamma_1(t)$, $a \leq t \leq b$, and $\gamma_2(s)$, $c \leq s \leq d$, have equivalent parametrizations and let f be a continuous function on this path. Then*

$$\int_{\gamma_1} f(z)\,dz = \int_{\gamma_2} f(z)\,dz. \tag{3.2.27}$$

Proof. We have $\gamma_2 = \gamma_1 \circ \phi$, where ϕ is an increasing continuously differentiable function from $[c,d]$ onto $[a,b]$. Assume first that γ_1 is differentiable at every point on (a,b). Then so is $\gamma_2 = \gamma_1 \circ \phi$ on (c,d). Applying the definition of path integrals, and using $\gamma_2(s) = \gamma_1(\phi(s))$ and $\gamma_2'(s) = \gamma_1'(\phi(s))\phi'(s)$, we change variables $t = \phi(s)$ or $s = \phi^{-1}(t)$, with $dt = \phi'(s)\,ds$, to obtain

$$\int_c^d f(\gamma_2(s))\gamma_2'(s)\,ds = \int_c^d f(\gamma_1(\phi(s)))\gamma_1'(\phi(s))\phi'(s)\,ds = \int_a^b f(\gamma_1(t))\gamma_1'(t)\,dt,$$

or equivalently, (3.2.27) holds in this case.

It remains to consider the case where γ_1 is not differentiable at some points $a_1 < \cdots < a_{m-1}$ in (a, b), as in Definition 3.1.9. Set $a_0 = a$ and $a_m = b$. Then γ_1 is differentiable at every point on (a_j, a_{j+1}) for $j = 0, \ldots, m-1$. Noting that $c = \phi^{-1}(a_0)$ and $d = \phi^{-1}(a_m)$, the preceding case gives

$$\int_{\phi^{-1}(a_j)}^{\phi^{-1}(a_{j+1})} f(\gamma_2(s))\gamma_2'(s)\,ds = \int_{a_j}^{a_{j+1}} f(\gamma_1(t))\gamma_1'(t)\,dt,$$

and summing over j in $\{0, 1, \ldots, m-1\}$, we deduce

$$\int_{\phi^{-1}(a)}^{\phi^{-1}(b)} f(\gamma_2(s))\gamma_2'(s)\,ds = \int_a^b f(\gamma_1(t))\gamma_1'(t)\,dt,$$

which is just a restatement of (3.2.27). ∎

Arc Length and Bounds for Integrals

Write a smooth path $\gamma: [a, b] \to \mathbb{C}$ as $\gamma = x + iy$, where x, y are real-valued functions. The **length** of γ, denoted $\ell(\gamma)$, can be approximated by adding the length of line segments joining consecutive points on the graph of γ as in Figure 3.22. The sum of the lengths of the line segments is

$$\sum_{k=1}^m |\gamma(t_k) - \gamma(t_{k-1})| = \sum_{k=1}^m \sqrt{|x(t_k) - x(t_{k-1})|^2 + |y(t_k) - y(t_{k-1})|^2}, \qquad (3.2.28)$$

where $a = t_0 < t_1 < t_2 < \cdots < t_m = b$ is a partition of $[a, b]$.

Thus the length of γ is the limit (when it exists) of the sums on the right side of (3.2.28) as the partition of $[a, b]$ gets finer and finer. To find this limit, we use the mean value theorem and write

$$|x(t_k) - x(t_{k-1})|^2 = |x'(\alpha_k)(t_k - t_{k-1})|^2$$

and

$$|y(t_k) - y(t_{k-1})|^2 = |y'(\beta_k)(t_k - t_{k-1})|^2,$$

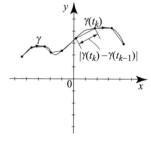

Fig. 3.22 Approximating the length of a path by adding the length of line segments.

where α_k and β_k are in $[t_{k-1}, t_k]$. Then (3.2.28) becomes

$$\sum_{k=1}^{m} \sqrt{|x'(\alpha_k)|^2 + |y'(\beta_k)|^2}\,(t_k - t_{k-1}).$$

Recognizing this sum as a Riemann sum and taking limits as the partition gets finer, we recover the formula for arc length from calculus:

$$\ell(\gamma) = \int_a^b \sqrt{|x'(t)|^2 + |y'(t)|^2}\,dt = \int_a^b |\gamma'(t)|\,dt, \qquad (3.2.29)$$

where the second equality follows from the complex notation $\gamma'(t) = x'(t) + iy'(t)$ and so $\sqrt{|x'(t)|^2 + |y'(t)|^2} = |\gamma'(t)|$.

For a piecewise smooth path γ, we repeat the preceding analysis for each smooth piece γ_j of γ and then add the lengths $\ell(\gamma_j)$'s. Definition 3.1.9 guarantees that each γ_j has a continuous derivative on the subinterval $[a_j, a_{j+1}]$ on which it is defined; thus, $\ell(\gamma_j)$ is finite, hence so is $\ell(\gamma)$. This process yields formula (3.2.29) for the arc length of a piecewise smooth path γ as well, where the integrand in this case is piecewise continuous. The element of arc length is usually denoted by ds. Thus,

$$ds = \sqrt{|x'(t)|^2 + |y'(t)|^2}\,dt. \qquad (3.2.30)$$

Example 3.2.18. (Arc length of cycloid) Let $a > 0$. Find the length of the arch of the cycloid $\gamma(t) = a(t - \sin t) + ia(1 - \cos t)$, where t ranges over the interval $[0, 2\pi]$.

The curve, illustrated in Figure 3.23, is formed by the trace of a fixed point on a moving circle that completes a full rotation.

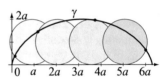

Fig. 3.23 First arch of the cycloid.

Solution. We have

$$x(t) = a(t - \sin t) \quad \Rightarrow \quad x'(t) = a(1 - \cos t);$$
$$y(t) = a(1 - \cos t) \quad \Rightarrow \quad y'(t) = a \sin t.$$

Hence

$$ds = \sqrt{x'(t)^2 + y'(t)^2}\,dt = \sqrt{a^2\left((1 - \cos t)^2 + \sin^2 t\right)}\,dt$$
$$= a\sqrt{2(1 - \cos t)}\,dt = 2a \sin\left(\frac{t}{2}\right)\,dt.$$

Applying (3.2.29), we obtain the length of the arch

$$\ell(\gamma) = \int_0^{2\pi} 2a \sin\frac{t}{2}\, dt = \left[-4a\cos\frac{t}{2} \right]_0^{2\pi} = 8a.$$
□

Recall the following inequality for the Riemann integral: If f is a continuous real-valued function of $[a, b]$ with $|f(x)| \le M$ on $[a, b]$, then

$$\left| \int_a^b f(x)\, dx \right| \le M(b-a).$$

We have a similar inequality for path integrals, called the *ML–inequality*.

Theorem 3.2.19. (Bounds for Path Integrals) *Suppose that γ is a path with length $\ell(\gamma)$, and f is a continuous function on γ such that $|f(z)| \le M$ for all z on γ. Then*

$$\left| \int_\gamma f(z)\, dz \right| \le M\ell(\gamma). \qquad (3.2.31)$$

Proof. Suppose first that the path $\gamma : [a, b] \to \mathbb{C}$ is smooth. Then we write

$$\left| \int_\gamma f(z)\, dz \right| = \left| \int_a^b f(\gamma(t))\gamma'(t)\, dt \right| \le \int_a^b |f(\gamma(t))|\,|\gamma'(t)|\, dt \le M \int_a^b |\gamma'(t)|\, dt,$$

using property (3.2.6) and the fact that $|f(z)| \le M$ for all z on γ.

If γ is piecewise smooth, we can write $\gamma = [\gamma_1, \gamma_2, \ldots, \gamma_n]$, where each γ_j is smooth. Then

$$\left| \int_\gamma f(z)\, dz \right| = \left| \sum_{j=1}^n \int_{\gamma_j} f(z)\, dz \right| \le \sum_{j=1}^n \left| \int_{\gamma_j} f(z)\, dz \right| \le M \sum_{j=1}^n \ell(\gamma_j) = M\ell(\gamma),$$

where in the last equality we used the fact that the length of γ is equal to the sum of the lengths of its parts. ∎

Example 3.2.20. (Bounding a path integral) Find an upper bound for

$$\left| \int_{C_1(0)} e^{\frac{1}{z}}\, dz \right|,$$

where $C_1(0)$ is the positively oriented unit circle.

Solution. Since the length of $C_1(0)$ is 2π, it follows from (3.2.31) that

$$\left| \int_{C_1(0)} e^{\frac{1}{z}}\, dz \right| \le 2\pi M,$$

where M is an upper bound of $|e^{\frac{1}{z}}|$ for z on the unit circle. To find M we proceed as follows. For z on $C_1(0)$, write $z = e^{it}$ where $0 \le t \le 2\pi$. Then

$$\frac{1}{z} = e^{-it} = \cos t - i\sin t,$$

and so, for $z = e^{it}$,

$$\left|e^{\frac{1}{z}}\right| = \left|e^{\cos t - i\sin t}\right| = \left|e^{\cos t}\right|\left|e^{-i\sin t}\right| = e^{\cos t} \le e^1 = e.$$

(In setting $\left|e^{-i\sin t}\right| = 1$ we have used $|e^{is}| = 1$ for real s.) Thus

$$\left|\int_{C_1(0)} e^{\frac{1}{z}}\, dz\right| \le 2\pi e.$$

Using techniques of integration from Chapter 5, we can evaluate the integral and obtain $\int_{C_1(0)} e^{\frac{1}{z}}\, dz = 2\pi i$. Thus the bound that we obtained by applying (3.2.31) is correct but not best possible, since $|2\pi i| = 2\pi$ is the best upper bound.   □

The path described by $\gamma(t) = t$ for t in $[a, b]$ is the closed interval $[a, b]$. For such a path, we have $\gamma'(t)\, dt = dt$, and if f is a function defined on γ, the path integral (3.2.15) becomes

$$\int_\gamma f(z)\, dz = \int_a^b f(t)\, dt,$$

showing that a Riemann integral of a piecewise continuous complex-valued function over $[a, b]$ is a special case of a path integral, where the path is the line segment $[a, b]$. So results about path integrals apply, in particular, to Riemann integrals.

Exercises 3.2

In Exercises 1–8, evaluate the integrals. In Exercise 6 use the principal branch of x^i.

1. $\displaystyle\int_0^{2\pi} e^{3ix}\, dx$ **2.** $\displaystyle\int_{-1}^1 (2i + 3 + ix)^2\, dx$ **3.** $\displaystyle\int_{-1}^0 \sin(ix)\, dx$

4. $\displaystyle\int_1^2 \text{Log}(ix)\, dx$ **5.** $\displaystyle\int_{-1}^1 \frac{x + i}{x - i}\, dx$ **6.** $\displaystyle\int_1^2 x^i\, dx$

7. $\displaystyle\int_{-1}^1 f(x)\, dx$,   where   **8.** $\displaystyle\int_{-1}^1 f(x)\, dx$,   where

$$f(x) = \begin{cases} (3 + 2i)x & \text{if } -1 \le x \le 0, \\ ix^2 & \text{if } 0 < x \le 1. \end{cases} \qquad f(x) = \begin{cases} e^{i\pi x} & \text{if } -1 \le x \le 0, \\ x & \text{if } 0 < x \le 1. \end{cases}$$

9. Find a continuous antiderivative of the function f in Exercise 7 and then compute $\int_{-1}^1 f(x)\, dx$ using Theorem 3.2.7.

10. Let z_1, z_2, z_3 be as in Example 3.2.14. Use properties of path integrals to evaluate the integrals:

(a) $\displaystyle\int_{[z_2, z_1]} \overline{z}\, dz$ (b) $\displaystyle\int_{[z_1, z_3, z_2]} \overline{z}\, dz$ (c) $\displaystyle\int_{[z_2, z_3, z_1, z_2]} \overline{z}\, dz$

11. Let n be an integer and let $C_R(z_0)$ denote the positively oriented circle with center at z_0 and radius $R > 0$. Show that

$$\int_{C_R(z_0)} \overline{z - z_0}^{\,n}\, dz = \begin{cases} 0 & \text{if } n \ne 1, \\ 2\pi i R^2 & \text{if } n = 1. \end{cases}$$

12. Prove Proposition 3.2.3 using properties of the Riemann integral of real-valued functions.

In Exercises 13–30, evaluate the path integrals. The positively oriented circle with center at z_0 and radius $R > 0$ is denoted $C_R(z_0)$. The polygonal path through the points $z_1, z_2, \ldots, z_n$ is denoted by $[z_1, z_2, \ldots, z_n]$.

13. $\displaystyle\int_{C_1(0)} (2z + i)\, dz.$

14. $\displaystyle\int_{C_1(i)} z^2\, dz.$

15. $\displaystyle\int_{[z_1, z_2]} \bar{z}\, dz,$ where $z_1 = 0$, $z_2 = 1 + i$.

16. $\displaystyle\int_{[z_1, z_2, z_3]} (\bar{z} - 2z)\, dz,$ where $z_1 = 0$, $z_2 = 1$, $z_3 = 1 + i$.

17. $\displaystyle\int_\gamma (z + 2\bar{z})\, dz,$ where $\gamma(t) = e^{it} + 2e^{-it}$, $0 \le t \le 2\pi$.

18. $\displaystyle\int_\gamma \frac{1}{1 + z}\, dz,$ where $\gamma(t) = (1 + i)t$, $0 \le t \le 1$.

19. $\displaystyle\int_{C_1(2+i)} \left((z - 2 - i)^3 + (z - 2 - i)^2 + \frac{i}{z - 2 - i} - \frac{3}{(z - 2 - i)^2} \right) dz.$ [*Hint*: Example 3.2.10.]

20. $\displaystyle\int_{[z_1, z_2, z_3, z_4, z_1]} \left(\mathrm{Re}\, z - 2(\mathrm{Im}\, z)^2 \right) dz,$ where $z_1 = 0$, $z_2 = 1$, $z_3 = 1 + i$, $z_4 = i$.

21. $\displaystyle\int_\gamma z\, dz,$ where γ is the contour in Exercise 20.

22. $\displaystyle\int_\gamma z\, dz,$ where γ is the semicircle e^{it}, $0 \le t \le \pi$.

23. $\displaystyle\int_\gamma e^z\, dz,$ where γ is the contour in Exercise 20.

24. $\displaystyle\int_\gamma \sin z\, dz,$ where γ is the contour in Exercise 20.

25. $\displaystyle\int_\gamma z\, dz,$ where γ is the hypotrochoid of Example 3.1.3, with $a = 8$ and $b = 5$.

26. $\displaystyle\int_\gamma \bar{z}\, dz,$ where γ is the hypotrochoid of Example 3.1.3.

27. $\displaystyle\int_\gamma \sqrt{z}\, dz,$ where $\gamma(t) = e^{it}$, $0 \le t \le \frac{\pi}{2}$, and $\sqrt{z}$ denotes the principal value of the square root.

28. $\displaystyle\int_\gamma \mathrm{Log}\, z\, dz,$ where $\gamma(t) = e^{it}$, $-\frac{3\pi}{4} \le t \le \frac{3\pi}{4}$.

29. $\displaystyle\int_{[z_1, z_2]} (x^2 + y^2)\, dz,$ where $z_1 = 2 + i$, $z_2 = -1 - i$. [*Hint*: $x^2 + y^2 = z\bar{z}$.]

30. $\displaystyle\int_{C_1(0)} (x^2 + y^2)\, dz.$

In Exercises 31–34, find the arc length of the curves.

31. $\gamma(t) = 2t + \dfrac{2i}{3} t^{3/2}$, $1 \le t \le 2$.

32. $\gamma(t) = \dfrac{1}{5} t^5 + \dfrac{i}{4} t^4$, $0 \le t \le 1$.

33. $\gamma(t) = e^{it}$, $0 \le t \le \dfrac{\pi}{6}$.

34. $\gamma(t) = (e^t - t) + 4ie^{\frac{t}{2}}$, $0 \le t \le 1$.

In Exercises 35–40, derive the estimates for the integrals.

35. $\left| \int_{\gamma} e^{z} dz \right| \leq \frac{\pi}{4}$, where $\gamma(t) = e^{it}$, $\frac{\pi}{2} \leq t \leq \frac{3\pi}{4}$.

36. $\left| \int_{C_1(0)} \operatorname{Log} z \, dz \right| \leq 2\pi^2$.

37. $\left| \int_{C_2(0)} \frac{1}{z-1} dz \right| \leq 4\pi$.

38. $\left| \int_{[z_1, z_2, z_3, z_1]} z^{-5} dz \right| \leq 5^{\frac{5}{2}} (2 + 2\sqrt{5})$, where $z_1 = -1 - i$, $z_2 = 1 - i$, $z_3 = i$.

39. $\left| \int_{\gamma} \frac{1}{z+1} dz \right| \leq \frac{2\pi}{3}$, where $\gamma(t) = 3 + e^{it}$, $0 \leq t \leq 2\pi$.

40. $\left| \int_{C_1(0)} e^{z^2 + 1} dz \right| \leq 2\pi e^2$.

41. Orthogonality of the 2π-periodic trigonometric and exponential systems. Let m and n be arbitrary integers.
(a) Show that

$$\int_{-\pi}^{\pi} e^{imx} e^{-inx} dx = \begin{cases} 0 & \text{if } m \neq n, \\ 2\pi & \text{if } m = n. \end{cases}$$

This identity states that the functions e^{imx} ($m = 0, \pm 1, \pm 2, \ldots$) are **orthogonal** on $[-\pi, \pi]$.
(b) Now suppose m and n are nonnegative integers. With the help of the identity in (a), show that

$$\int_{-\pi}^{\pi} \cos(mx) \cos(nx) \, dx = \int_{-\pi}^{\pi} \sin(mx) \sin(nx) \, dx = 0 \quad \text{if } m \neq n;$$

$$\int_{-\pi}^{\pi} \cos(mx) \sin(nx) \, dx = 0 \qquad\qquad\qquad \text{for all } m \text{ and } n;$$

$$\int_{-\pi}^{\pi} \cos^2(mx) \, dx = \int_{-\pi}^{\pi} \sin^2(mx) \, dx = \pi \qquad\quad \text{for all } m \neq 0.$$

These identities state that the functions $1, \cos x, \cos(2x), \cos(3x), \ldots, \sin x, \sin(2x), \ldots$ are **orthogonal** on the interval $[-\pi, \pi]$.

42. Orthogonality of the $2p$-periodic trigonometric and exponential systems. Let $p > 0$ be a real number and let m and n be arbitrary integers.
(a) Show that the functions $e^{i\frac{m\pi}{p}x}$ ($m = 0, \pm 1, \pm 2, \ldots$) are $2p$-periodic. The set of these functions is called the $2p$-periodic exponential system.
(b) Show that

$$\int_{-p}^{p} e^{i\frac{m\pi}{p}x} e^{-i\frac{n\pi}{p}x} dx = \begin{cases} 0 & \text{if } m \neq n, \\ 2p & \text{if } m = n. \end{cases}$$

Thus the functions in the $2p$-periodic exponential system are **orthogonal** on the interval $[-p, p]$.
(c) Now suppose m and n are nonnegative integers. With the help of the identity in (a), or by Exercise 41(b), show that

$$\int_{-p}^{p} \cos\left(\frac{m\pi}{p}x\right) \cos\left(\frac{n\pi}{p}x\right) dx = 0 \qquad\quad \text{if } m \neq n;$$

$$\int_{-p}^{p} \cos\left(\frac{m\pi}{p}x\right) \sin\left(\frac{n\pi}{p}x\right) dx = 0 \qquad\quad \text{for all } m \text{ and } n;$$

$$\int_{-p}^{p} \sin\left(\frac{m\pi}{p}x\right) \sin\left(\frac{n\pi}{p}x\right) dx = 0 \qquad\quad \text{if } m \neq n;$$

$$\int_{-p}^{p} \cos^2\left(\frac{m\pi}{p}x\right) dx = \int_{-p}^{p} \sin^2\left(\frac{m\pi}{p}x\right) dx = p \quad \text{for all } m \neq 0.$$

These identities state that the $2p$-periodic functions

$$1, \cos\left(\frac{\pi}{p}x\right), \cos\left(\frac{2\pi}{p}x\right), \ldots, \sin\left(\frac{\pi}{p}x\right), \sin\left(\frac{2\pi}{p}x\right), \ldots$$

are **orthogonal** on the interval $[-p, p]$.

43. Let a and b be nonzero real numbers. Derive the formulas

$$\int \cos(ax)\cosh(bx)\,dx = \frac{1}{a^2+b^2}\left(a\cosh(bx)\sin(ax) + b\cos(ax)\sinh(bx)\right) + C;$$

and

$$\int \sin(ax)\sinh(bx)\,dx = \frac{1}{a^2+b^2}\left(b\cosh(bx)\sin(ax) - a\cos(ax)\sinh(bx)\right) + C.$$

[*Hint*: Consider $\int \cos\left((a+ib)x\right)dx$.]

44. Let a and b be nonzero real numbers. Derive the formulas

$$\int \sin(ax)\cosh(bx)\,dx = \frac{1}{a^2+b^2}\left(-a\cos(ax)\cosh(bx) + b\sin(ax)\sinh(bx)\right) + C;$$

and

$$\int \cos(ax)\sinh(bx)\,dx = \frac{1}{a^2+b^2}\left(b\cos(ax)\cosh(bx) + a\sin(ax)\sinh(bx)\right) + C.$$

[*Hint*: Consider $\int \sin\left((a+ib)x\right)dx$.]

3.3 Independence of Paths

In calculus, the problem of computing a definite integral over an interval was often reduced to finding an antiderivative and evaluating it at the endpoints. This is achieved by the fundamental theorem of calculus which says that if g is a continuous function on an interval $[a, b]$ and G is an antiderivative of g on (a, b), then

$$\int_a^b g(t)\,dt = G(b) - G(a).$$

In this section, we investigate an analogous way to compute path integrals.

Definition 3.3.1. Let f be a continuous function on a region Ω. We say that F is an **antiderivative** of f in Ω if F is an analytic function on Ω with the property $F'(z) = f(z)$ for all z in Ω.

Since f is continuous, F is analytic and, in particular, it is continuous. Moreover, any two antiderivatives of f differ by a constant, as a consequence of Theorem 2.5.7.

To find an antiderivative of a continuous function f we try, if possible, to guess our answer based on formulas from calculus. For example, if $f(z) = ze^z$, we guess the antiderivative $F(z) = ze^z - e^z$. Then we can verify this guess by differentiation. As the following examples illustrate, part of checking the answer is to make sure that the antiderivative is analytic on the domain of definition.

Example 3.3.2. (Antiderivatives)

(a) An antiderivative of $f(z) = z^3 + 7z - 2$ is $F(z) = \frac{z^4}{4} + \frac{7}{2}z^2 - 2z$, and the identity $F'(z) = f(z)$ holds for all z in $\mathbb{C}$.

(b) The function $f(z) = \text{Log}\, z$ is continuous for z in $\mathbb{C} \setminus (-\infty, 0]$. An antiderivative is $F(z) = z\, \text{Log}\, z - z$, and the identity $F'(z) = f(z)$ holds for all z in $\mathbb{C} \setminus (-\infty, 0]$.

(c) The function $f(z) = \frac{1}{z}$ is continuous in $\mathbb{C} \setminus \{0\}$. We may guess an antiderivative $F(z) = \text{Log}\, z$. But the equality $F'(z) = f(z)$ holds for all z in $\mathbb{C} \setminus (-\infty, 0]$. So $\text{Log}\, z$ is an antiderivative of $\frac{1}{z}$ in $\mathbb{C} \setminus (-\infty, 0]$ only, even though f is continuous in $\mathbb{C} \setminus \{0\}$.

(d) Let Ω_α denote the region $\mathbb{C}$ minus the ray at angle α, and let $\log_\alpha z$ denote the branch of the logarithm with a branch cut at angle α. In view of the discussion after

Example 2.5.5, $\log_\alpha z$ is analytic in Ω_α and

$$\frac{d}{dz} \log_\alpha z = \frac{1}{z},$$

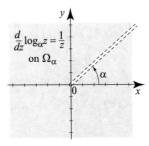

for all z in Ω_α. Thus $F(z) = \log_\alpha z$ is an antiderivative of $\frac{1}{z}$ in Ω_α. In particular, if we choose α in such a way that the ray at angle α is not the negative x-axis, then $F(z) = \log_\alpha z$ becomes an antiderivative of $\frac{1}{z}$ in a region that contains the negative x-axis. See Figure 3.24.

Fig. 3.24 $\log_\alpha z$ and its branch cut at angle α.

Thus antiderivatives of $\frac{1}{z}$ can be found on all regions Ω_α. □

We now show that we can use antiderivatives to evaluate path integrals much like we used antiderivatives in calculus to evaluate definite integrals. We achieve this goal via an important theorem; before stating it we need the following lemma.

Lemma 3.3.3. *Suppose that f is a continuous function in a region Ω that contains an open disk $B_r(z)$ for some $z \in \Omega$ and some $r > 0$. Then for $|w| < r$ the line segment $[z, z+w]$ is contained in Ω and we have*

$$\lim_{w \to 0} \frac{1}{w} \int_{[z, z+w]} f(\zeta)\, d\zeta = f(z). \tag{3.3.1}$$

Proof. Parametrize $[z, z+w]$ by $\gamma(t) = (1-t)z + t(z+w) = z + tw$, where $0 \le t \le 1$. Then $\gamma'(t) = w\, dt$ and so

$$\frac{1}{w} \int_{[z, z+w]} f(\zeta)\, d\zeta = \int_0^1 f(z+tw)\, dt. \tag{3.3.2}$$

Given $\varepsilon > 0$, by the continuity of f, there is a $\delta > 0$ (with $\delta < r$) such that

$$|w| < \delta \implies |f(z+w) - f(z)| < \varepsilon.$$

If $|w| < \delta$, we have $|tw| < \delta$ for all $t \in [0,1]$ and thus $|f(z+tw) - f(z)| < \varepsilon$. From the *ML*-inequality (3.2.31), we obtain for $|w| < \delta$

$$\left| \int_0^1 f(z+tw)\, dt - f(z) \right| = \left| \int_0^1 [f(z+tw) - f(z)]\, dt \right| \leq \int_0^1 |f(z+tw) - f(z)|\, dt \leq \varepsilon$$

and (3.3.1) follows using (3.3.2) by the (ε, δ) definition of the limit. ∎

We are now able to prove the main result of this section.

Theorem 3.3.4. (Independence of Path) *Let f be a continuous complex-valued function on a region Ω. Then the following assertions are equivalent:*
(a) There is an analytic function F on Ω such that $f(z) = F'(z)$ for all z in Ω.
(b) For arbitrary points z_1, z_2 and any path γ in Ω that joins z_1 to z_2, the integral

$$I = \int_\gamma f(z)\, dz$$

is independent of the path γ.
(c) The integral of f over all closed paths is zero.
Moreover, if (a) holds,[1] then for any path γ in Ω that joins z_1 and z_2 we have

$$\int_\gamma f(z)\, dz = F(z_2) - F(z_1). \tag{3.3.3}$$

Proof. If F is an antiderivative of f in Ω, then the complex-valued function $t \mapsto F(\gamma(t))$ is differentiable at the points t in (a,b) where $\gamma'(t)$ exists and we have

$$\frac{d}{dt} F(\gamma(t)) = F'(\gamma(t))\gamma'(t) = f(\gamma(t))\gamma'(t) \tag{3.3.4}$$

in view of Theorem 3.1.8. Now $t \mapsto f(\gamma(t))\gamma'(t)$ is piecewise continuous, because f is continuous and γ' is piecewise continuous. Also, since $F \circ \gamma$ is continuous, (3.3.4) tells us that $F \circ \gamma$ is a continuous antiderivative of $(f \circ \gamma)\gamma'$, in the sense of Theorem 3.2.7. Using this theorem, we deduce

$$\int_\gamma f(z)\, dz = \int_a^b f(\gamma(t))\gamma'(t)\, dt = F(\gamma(b)) - F(\gamma(a)) = F(z_2) - F(z_1),$$

completing the proof that (a) implies (b) and simultaneously deriving (3.3.3).

We now show that (b) implies (a). We only need to show that if I is independent of path, then f has an antiderivative F. Fix z_0 in Ω. For z in Ω, define

$$F(z) = \int_{\gamma(z_0, z)} f(\zeta)\, d\zeta, \tag{3.3.5}$$

---

[1] *and thus if (b) or (c) hold*

where $\gamma(z_0, z)$ is a path joining z_0 to z. Such a path exists since Ω is polygonally connected and the function in (3.3.5) is well defined in view of the property that the integral is independent of path.

Since Ω is open and $z \in \Omega$, there is an open disk $B_r(z)$ centered at z which is entirely contained in Ω. For $z + w$ in $B_r(z)$, the line segment $[z, z + w]$ lies in Ω.

Since the integral of f is independent of path, we do not choose an arbitrary path that joins z_0 to $z + w$ but we choose the path

$$\gamma(z_0, z + w) = \big[\gamma(z_0, z), [z, z + w]\big],$$

fomed by the path $\gamma(z_0, z)$ followed by the line segment $[z, z + w]$, as shown in Figure 3.25.

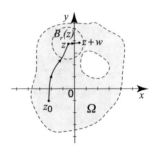

Fig. 3.25 Picture of the proof.

Then, in view of Proposition 3.2.12 (iii), we have

$$\int_{\gamma(z_0, z+w)} f(\zeta) \, d\zeta = \int_{\gamma(z_0, z)} f(\zeta) \, d\zeta + \int_{[z, z+w]} f(\zeta) \, d\zeta$$

and so

$$F(z+w) - F(z) = \int_{\gamma(z_0, z+w)} f(\zeta) \, d\zeta - \int_{\gamma(z_0, z)} f(\zeta) \, d\zeta = \int_{[z, z+w]} f(\zeta) \, d\zeta .$$

Then,

$$\frac{F(z+w) - F(z)}{w} = \frac{1}{w} \int_{[z, z+w]} f(\zeta) \, d\zeta . \qquad (3.3.6)$$

Taking the limit as $w \to 0$ and appealing to Lemma 3.3.3, we obtain that the function F has a complex derivative at z and $F'(z) = f(z)$.

Finally, it remains to prove the equivalence of (b) and (c). To see this, consider the closed path γ in Figure 3.26 containing the point z_1.

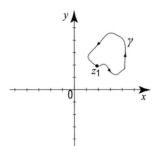

Fig. 3.26 A closed path starting and ending at the same point z_1.

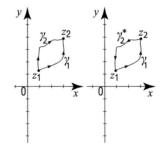

Fig. 3.27 γ_1 followed by γ_2^*.

If we start at z_1 and trace γ until, we return to z_1, by (3.3.3) we obtain

$$\int_\gamma f(z)\,dz = F(z_1) - F(z_1) = 0.$$

Hence, if f has an analytic antiderivative, its integral over any closed path is zero.

Conversely, suppose we know that the integral of f around every closed path is zero. Consider the two paths γ_1 and γ_2 joining z_1 to z_2 in Figure 3.27, and let Γ be the closed path consisting of γ_1 followed by γ_2^*, the reverse of γ_2. Then

$$0 = \int_\Gamma f(z)\,dz = \int_{\gamma_1} f(z)\,dz + \int_{\gamma_2^*} f(z)\,dz = \int_{\gamma_1} f(z)\,dz - \int_{\gamma_2} f(z)\,dz,$$

implying that $\int_{\gamma_1} f(z)\,dz = \int_{\gamma_2} f(z)\,dz$. Thus, the integral of f is independent of path. In conclusion, the integral of f is independent of path in a region Ω if and only if the integral of f over an arbitrary closed path in Ω is 0.   ∎

Thus, in view of Theorem 3.3.4, to compute a path integral of a function whose antiderivative is known, we may ignore the path and evaluate the integral using the endpoints. Here are some applications.

Example 3.3.5. (Integrals involving entire functions) Evaluate the path integrals

(a) $\displaystyle\int_\gamma e^z\,dz$, where $\gamma(t) = e^{it}$, $0 \le t \le \pi$ (Figure 3.28).

(b) $\displaystyle\int_\gamma ze^{z^2}\,dz$, where $\gamma(t) = -\frac{i}{2} + \frac{1}{2}e^{it}$, $-\frac{\pi}{2} \le t \le \frac{\pi}{2}$ (Figure 3.29).

(c) $\displaystyle\int_\gamma (z^3 + z^2 - 2)\,dz$, where $\gamma = [z_1, z_2, z_3, z_1]$ is the closed polygonal path with $z_1 = -1$, $z_2 = 1$, $z_3 = i$ (Figure 3.30).

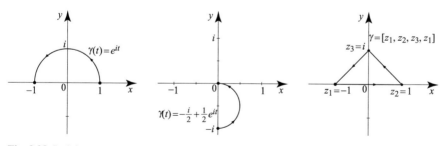

Fig. 3.28 Path in part (a).   **Fig. 3.29** Path in part (b).   **Fig. 3.30** Path in part (c).

Solution. (a) The function e^z is continuous in the entire plane, with an antiderivative e^z. The initial point of γ is $z_1 = \gamma(0) = 1$ and its terminal point is $z_2 = \gamma(\pi) = -1$. By Theorem 3.3.4,

$$\int_\gamma e^z\,dz = e^z\Big|_1^{-1} = e^{-1} - e^1 = -2\sinh(1).$$

(b) As one can easily verify by direct computation, an antiderivative of ze^{z^2} is $\frac{1}{2}e^{z^2}$. The initial point of γ is $z_1 = -i$ and its terminal point is $z_2 = 0$. By Theorem 3.3.4,

$$\int_\gamma ze^{z^2}\,dz = \frac{1}{2}e^{z^2}\Big|_{-i}^0 = \frac{1}{2}\left(1 - e^{-1}\right).$$

(c) An antiderivative of $z^3 + z^2 - 2$ is $\frac{z^4}{4} + \frac{z^3}{3} - 2z$. The initial and terminal points of the path are the same, $z_1 = -1$. By Theorem 3.3.4, we have

$$\int_{[z_1,z_2,z_3,z_1]} (z^3 + z^2 - 2)\,dz = \left[\frac{z^4}{4} + \frac{z^3}{3} - 2z\right]_{-1}^{-1} = 0. \qquad \square$$

In Example 3.3.2, the region that contained the paths was of little concern, because the integrands and their antiderivatives were entire. This is not the case in the next two examples, where the domain of the antiderivative must be carefully chosen.

Example 3.3.6. (Choosing an appropriate region) Evaluate

$$\int_{[z_1,z_2,z_3]} \frac{1}{z}\,dz$$

where $[z_1, z_2, z_3]$ is the polygonal path with $z_1 = 1$, $z_2 = 2 + i$, $z_3 = 3$ (Figure 3.31).

Solution. The function $1/z$ is continuous in $\mathbb{C}\setminus\{0\}$. An antiderivative of $1/z$ is $\operatorname{Log} z$ in the region $\Omega = \mathbb{C}\setminus(-\infty, 0]$. The path $[z_1, z_2, z_3]$ lies entirely in Ω, see Figure 3.31. We apply Theorem 3.3.4 to obtain

$$\int_{[z_1,z_2,z_3]} \frac{1}{z}\,dz = \operatorname{Log} z\Big|_1^3 = \ln 3. \qquad \square$$

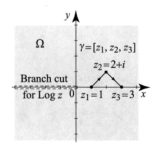

Fig. 3.31 The polygonal path $[z_1, z_2, z_3]$.

Example 3.3.7. (Choosing an appropriate antiderivative) Evaluate $\int_{[z_1,z_2,z_3]} \frac{1}{z}\,dz$, where $[z_1, z_2, z_3]$ is the polygonal path with $z_1 = -1$, $z_2 = -1 + i$, $z_3 = -4 - 4i$ (see Figure 3.32).

Solution. In order to apply Theorem 3.3.4, we must find an antiderivative of $\frac{1}{z}$ that is analytic in a region that contains the path $[z_1, z_2, z_3]$. We cannot use $\mathrm{Log}\,z$ as antiderivative, because it is not analytic in a region that contains the path $[z_1, z_2, z_3]$. Instead, we will use a different branch of the logarithm. We know from Example 3.3.2(d) that $\log_\alpha z$ is an antiderivative of $\frac{1}{z}$ in the region Ω_α ($\mathbb{C}$ minus the ray at angle α). Choose α in such a way that the branch cut of $\log_\alpha z$ does not intersect the path $[z_1, z_2, z_3]$.

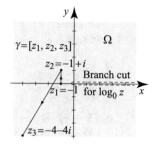

Fig. 3.32 Picture for Example 3.3.7.

Taking, for example, $\alpha = 0$ we write

$$\log_0 z = \ln|z| + i\arg_0 z,$$

where $0 < \arg_0 z \le 2\pi$. By Theorem 3.3.4 we have

$$\int_{[z_1, z_2, z_3]} \frac{1}{z}\, dz = \log_0(z_3) - \log_0(z_1)$$

$$= \frac{1}{2}\ln|-4-4i| + i\arg_0(-4-4i) - (\ln 1 + i\arg_0(-1))$$

$$= \frac{1}{2}\ln 32 + i\frac{5\pi}{4} - i\pi = \frac{5}{2}\ln 2 + i\frac{\pi}{4}.$$

Thus the value of the integral is equal to $\frac{5}{2}\ln 2 + i\frac{\pi}{4}$. $\square$

Integrals over Closed Paths

We now turn to applications of Theorem 3.3.4 related to integrals over closed paths. We start with some straightforward ones.

Example 3.3.8. (Integrals over closed paths)
(a) Since $z^2/2$ is an antiderivative of z on the plane, if γ is a closed path, then by Theorem 3.3.4,

$$\int_\gamma z\, dz = 0.$$

(b) Likewise,

$$\int_\gamma e^{2iz}\, dz = 0,$$

because e^{2iz} has an antiderivative $\frac{e^{2iz}}{2i}$ for all z in the plane.

(c) Let $C_{\frac{1}{2}}(0)$ denote the positively oriented circle with center at 0 and radius $\frac{1}{2}$.
Then, by Theorem 3.3.4,

$$\int_{C_{\frac{1}{2}}(0)} \frac{1}{1+z}\,dz = 0,$$

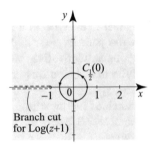

Fig. 3.33 Example 3.3.8(c).

because $\mathrm{Log}\,(1+z)$ is an antideriva-
tive of $\frac{1}{z+1}$ in the region $\mathbb{C} \setminus (-\infty, -1]$,
which contains $C_{\frac{1}{2}}(0)$. The branch cut of
$\mathrm{Log}\,(z+1)$ is obtained by translating to
the left by one unit the branch cut of
$\mathrm{Log}\,z$. See Figure 3.33.

(d) Let $C_{\frac{1}{2}}(z_0)$ be a positively oriented closed circle of radius $1/2$ centered at z_0 and
$n \neq 1$ an integer. Then, Theorem 3.3.4 implies that

$$\int_{C_{\frac{1}{2}}(z_0)} \frac{1}{(z-z_0)^n}\,dz = 0,$$

since $F(z) = \frac{-1}{(n-1)(z-z_0)^{n-1}}$ is an analytic antiderivative of $\frac{1}{(z-z_0)^n}$ on $\mathbb{C} \setminus \{z_0\}$.

(e) Let $z_0 \in \mathbb{C}$. The function $(z-z_0)^{-1}$ defined on $\Omega = B_1(z_0) \setminus \{z_0\}$ does not have
an analytic antiderivative. Indeed, if it did, then by Theorem 3.3.4, its integral over
any closed path in Ω would be zero. But we have shown in (3.2.16) that

$$\int_{C_{\frac{1}{2}}(z_0)} \frac{1}{z-z_0}\,dz = 2\pi i \neq 0,$$

which is a contradiction.   □

The next example is somewhat surprising. In calculus, if g is a continuous
function on a closed interval $[a, b]$, then it has an antiderivative; namely, $G(x) =
\int_a^x g(t)\,dt$. This is not the case in complex analysis, as the next example shows.

Example 3.3.9. (A continuous function with no antiderivative)
(a) Let R be a positive real number and $C_R(z_0)$ the positively oriented circle, centered
at z_0, with radius R. Evaluate

$$\int_{C_R(z_0)} \overline{z - z_0}\,dz.$$

(b) Prove that $f(z) = \bar{z}$ has no antiderivative on any region.

Solution. (a) Parametrize $C_R(z_0)$ by $z = z_0 + Re^{it}$, where $0 \leq t \leq 2\pi$. Then $dz =
Rie^{it}\,dt$ and

$$\overline{z - z_0} = \overline{z_0 + Re^{it} - z_0} = \overline{Re^{it}} = Re^{-it},$$

so we have

$$\int_{C_R(z_0)} \overline{z - z_0}\, dz = \int_0^{2\pi} Re^{-it} Rie^{it}\, dt = R^2 i \int_0^{2\pi} dt = 2\pi i R^2.$$

(b) The proof is by contradiction. Assume to the contrary that $f(z) = \bar{z}$ has an antiderivative F in some region Ω and let z_0 be a point in Ω. Then, we check directly that $F(z) - \overline{z_0} z$ is an antiderivative of $\bar{z} - \overline{z_0} = \overline{z - z_0}$. Let $R > 0$ be such that $C_R(z_0)$ is contained in Ω. By Theorem 3.3.4(c), the integral of $\overline{z - z_0}$ over $C_R(z_0)$ is 0, which contradicts the outcome of the calculation in part (a). Hence the assumption that there is an antiderivative of $\overline{z - z_0}$ must be false. ∎

While Theorem 3.3.4 is very useful, it has limited applications when it comes to the evaluation of general integrals. For example, while it is immediate that the integral of e^z over a closed path is zero (e^z is an antiderivative of itself), it is not clear whether the same is true for the function e^{z^2}, since we do not know whether it has an antiderivative. It is desirable to have a result that would enable us to compute integrals of analytic functions over closed paths without resorting to antiderivatives. In the next section we prove one such result, called Cauchy's theorem; this is at the heart of the theory of path integrals and indeed all of complex analysis.

Exercises 3.3

In Exercises 1–14, find an antiderivative for each of the listed functions and specify the region on which the antiderivatives are defined.

1. $z^2 + z - 1$

2. $ze^z - \sin z$

3. $\dfrac{\mathrm{Log}\, z}{z}$

4. $\dfrac{1}{z-1}$

5. $\dfrac{1}{(z-1)(z+1)}$

6. $\dfrac{1}{z \,\mathrm{Log}\, z}$

7. $\cos(3z+2)$

8. $ze^{z^2} - \dfrac{1}{z}$

9. $z \sinh z^2$

10. $e^z \cos z$

11. $z \,\mathrm{Log}\, z$

12. $\log_\alpha z$

13. $\log_0 z - \log_{\frac{\pi}{2}} z$

14. $z^{\frac{1}{3}}$ (principal branch)

In Exercises 15–27, evaluate the integral and explain clearly how you are applying Theorem 3.3.4.

15. $\displaystyle\int_{[z_1, z_2, z_3]} 3(z-1)^2\, dz$, where $z_1 = 1$, $z_2 = i$, $z_3 = 1 + i$.

16. $\displaystyle\int_{[z_1, z_2, z_3]} (z^2 - 1)^2 z\, dz$, where $z_1 = 0$, $z_2 = 1$, $z_3 = -i$.

17. $\displaystyle\int_\gamma z^2\, dz$, where $\gamma(t) = e^{it} + 3e^{2it}$, $0 \le t \le \frac{\pi}{4}$.

18. $\displaystyle\int_{C_1(0)} \left((z - 2 - i)^2 + \dfrac{i}{z - 2 - i} - \dfrac{3}{(z - 2 - i)^2}\right) dz$, where $C_1(0)$ is the unit circle.

19. $\displaystyle\int_{[z_1, z_2, z_3]} ze^z\, dz$, where $z_1 = \pi$, $z_2 = -1$, $z_3 = -1 - i\pi$.

20. $\displaystyle\int_{[z_1, z_2, z_3]} e^{i\pi z}\, dz$, where $z_1 = 2$, $z_2 = i$, $z_3 = 4$.

21. $\int_\gamma \sin z\,dz$, where $\gamma(t) = 2e^{it}$, $0 \leq t \leq \frac{\pi}{2}$.

22. $\int_\gamma \sin^2 z\,dz$, where γ is an arbitrary closed path.

23. $\int_\gamma \frac{1}{z}\,dz$, where γ is a path contained in $\{z \in \mathbb{C} : \operatorname{Im} z < 0\}$ joining $1 - i$ to $-i$.

24. $\int_{[z_1, z_2, \ldots, z_n]} dz$, where $z_1, z_2, \ldots, z_n$ are arbitrary.

25. $\int_{[z_1, z_2, z_3, z_1]} z \operatorname{Log} z\,dz$, where $z_1 = 1$, $z_2 = 1 + i$, $z_3 = -2 + 2i$.

26. $\int_{[z_1, z_2, z_3]} \frac{\operatorname{Log} z}{z}\,dz$, where $z_1 = -i$, $z_2 = 1$, $z_3 = i$.

27. $\int_\gamma \frac{1}{z^5}\,dz$, where γ is any closed path not containing 0.

28. Show that if $p(z)$ is a polynomial and γ a closed path, then $\int_\gamma p(z)\,dz = 0$.

29. (a) Recall from (2.5.13) that for a complex number α,

$$\frac{d}{dz} z^\alpha = \alpha z^{\alpha - 1},$$

where we define both complex powers using a single logarithm branch. Conclude that for $\alpha \neq -1$, an antiderivative of z^α is $\frac{1}{\alpha+1} z^{\alpha+1}$, where the same branch of the logarithm is used.

(b) Evaluate $\int_\gamma \frac{1}{\sqrt{z}}\,dz$ (principal branch), where $\gamma(t) = e^{it}$, $-\frac{\pi}{2} \leq t \leq \frac{\pi}{2}$.

30. Use the result of Exercise 29(a) to evaluate $\int_\gamma z^{i\pi}\,dz$ (use the branch $\log_\pi z$), where $\gamma(t) = e^{it}$, $-\frac{\pi}{2} \leq t \leq 0$.

31. Replacing the integrand. Consider the integral

$$\int_\gamma \operatorname{Log} z\,dz, \qquad \gamma(t) = e^{it}, \ 0 \leq t \leq \pi.$$

Since $\operatorname{Log} z$ is discontinuous at the point -1, it cannot be continuous in any region containing the path, and so we cannot apply Theorem 3.3.4 directly. The idea in this problem is to replace $\operatorname{Log} z$ by a different branch of the logarithm for which Theorem 3.3.4 does apply.

(a) Show that $\operatorname{Log} z = \log_{-\frac{\pi}{2}} z$ for all z on γ.

(b) Conclude that

$$\int_\gamma \operatorname{Log} z\,dz = \int_\gamma \log_{-\frac{\pi}{2}} z\,dz,$$

and evaluate the integral on the right side by using Theorem 3.3.4.

32. Evaluate $\int_{C_1(0)} \operatorname{Log} z\,dz$, where $C_1(0)$ is the positively oriented unit circle. First write the integral as the sum of two integrals over the upper and lower semicircles, then use the ideas of Exercise 31 to evaluate each integral in this sum by appealing to Theorem 3.3.4.

33. (a) Compute $\int_{C_R(z_0)} \operatorname{Im} z\,dz$, where $C_R(z_0)$ is the positively oriented circle with center at z_0 and radius $R > 0$.

(b) Show that the function $\operatorname{Im} z$ has no antiderivative in any open subset of $\mathbb{C}$.

(c) Show that the function $\operatorname{Re} z$ has no antiderivative in any open subset of $\mathbb{C}$.

3.4 Cauchy's Integral Theorem for Simple Paths

In this section, we prove Cauchy's theorem for analytic functions, assuming continuity of their derivatives. This version is not the most general available but is sufficient for concrete applications where the derivative is explicit. Versions of this result without assuming continuity of derivatives can be found in subsequent sections.

Simple Curves and Green's Theorem

Definition 3.4.1. A curve is called **simple** if it does not intersect itself except possibly at the endpoints. In other words, if γ is parametrized by $[a,b]$ and $a \leq t_1 < t_2 \leq b$, then γ is a simple curve if $\gamma(t_1) \neq \gamma(t_2)$ whenever $t_1 \neq t_2$. A **simple closed curve** is both simple and closed; i.e., it intersects itself only at the endpoints; that is, if $a \leq t_1 < t_2 \leq b$, then $\gamma(t_1) = \gamma(t_2)$ if and only if $t_1 = a$ and $t_2 = b$. A simple curve that is also a path (see Definition 3.1.10) is called a **simple path**.

A simple closed curve can loop around and look very complicated, but it cannot cross itself. A simple closed curve is also called a **Jordan curve**, after the French mathematician Camille Jordan (1838–1922), who showed that a simple closed curve γ divides the plane into two regions: a bounded region, called the **interior** of γ, and an unbounded region, called the **exterior** of γ.

In Figure 3.34 the region D is the interior and U is the exterior. This is the famous **Jordan curve theorem** from topology, which is easy to picture but quite difficult to prove. A proof can be found in *The Jordan curve theorem, formally and informally*, Amer. Math. Monthly **114** (10), (2007), 882–894, by Thomas C. Hales.

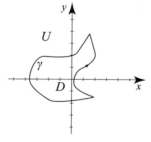

Fig. 3.34 A Jordan path γ and the interior domain D and the exterior domain U.

Since the interior and the exterior regions relative to a simple closed curve are identifiable, we may define positive and negative orientations of such a curve. The **positive orientation** of a simple closed curve is the one that places the interior region to our left when traced; see Figure 3.35. The **negative orientation** is the one that, when traced, places the interior region to our right; see Figure 3.36. Our definition of positive versus negative orientation thus generalizes our earlier idea of counterclockwise versus clockwise orientation of circles.

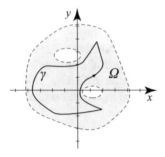

Fig. 3.35 A curve γ with positive orientation inside a region Ω.

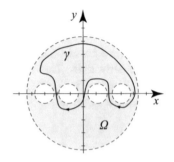

Fig. 3.36 A curve γ with negative orientation inside a region Ω.

We now recall an important result from advanced calculus concerning line integrals over simple paths. This theorem bears the name of the British Mathematician George Green (1793–1841).

Theorem 3.4.2. (Green's Theorem) *Let γ be a simple closed path with positive orientation and D be the region interior to γ. Suppose that $P(x, y)$ and $Q(x, y)$, along with their first partial derivatives, are continuous real-valued functions on $\overline{D} = D \cup \gamma$. Then*

$$\int_{\gamma} P\,dx + Q\,dy = \iint_{D} \left(\frac{\partial Q}{\partial x} - \frac{\partial P}{\partial y} \right) dx\,dy. \tag{3.4.1}$$

The integral on the right side of (3.4.1) is a double integral over the region D. To understand the notation in the line integral on the left side of (3.4.1), parametrize γ by $x = x(t)$ and $y = y(t)$, where $a \le t \le b$. Then the integral on the left side of (3.4.1) means

$$\int_{a}^{b} \left(P(x(t), y(t)) x'(t) + Q(x(t), y(t)) y'(t) \right) dt.$$

Example 3.4.3. Use Green's theorem to evaluate:

(a) $\displaystyle\int_{\gamma} x\,dx + y\,dy$ and (b) $\displaystyle\int_{\gamma} -y\,dx + x\,dy$, where γ is a simple closed path with positive orientation.

Solution. (a) Let $P(x, y) = x$ and $Q(x, y) = y$. It is clear that P, Q, and their first partial derivatives are continuous for all (x, y), and so we may apply Green's theorem. We have $\frac{\partial Q}{\partial x} = 0$ and $\frac{\partial P}{\partial y} = 0$. By Green's theorem,

$$\int_{\gamma} x\,dx + y\,dy = \iint_{D} 0\,dx\,dy = 0.$$

(b) Here $P(x, y) = -y$ and $Q(x, y) = x$, $\frac{\partial Q}{\partial x} = 1$ and $\frac{\partial P}{\partial y} = -1$, and so by Green's theorem

$$\int_\gamma -y\,dx + x\,dy = \iint_D 2\,dx\,dy = 2 \times (\text{Area of } D).$$

For example, if γ is the unit circle with positive orientation, then the integral is equal to $2 \times (\text{Area of unit disk}) = 2\pi$. $\qquad\square$

We apply Green's theorem to evaluate path integrals of the form $\int_\gamma f(z)\,dz$ where γ is a simple closed path and f is an analytic function on γ and the region interior to γ. For this purpose, parametrize γ by $\gamma(t) = x(t) + iy(t)$, where $a \le t \le b$, and write $f(z) = u(x, y) + iv(x, y)$, where u and v are the real and imaginary parts of f, respectively. Then, by the definition of path integrals, we write

$$\int_\gamma f(z)\,dz = \int_a^b \left[u(x(t), y(t)) + iv(x(t), y(t))\right] (x'(t) + iy'(t))\,dt$$

$$= \int_a^b \left(u(x(t), y(t))x'(t) - v(x(t), y(t))y'(t) \right) dt$$

$$+ i \int_a^b \left(v(x(t), y(t))x'(t) + u(x(t), y(t))y'(t) \right) dt$$

$$= \int_\gamma (u\,dx - v\,dy) + i \int_\gamma (v\,dx + u\,dy). \tag{3.4.2}$$

Thus the path integral $\int_\gamma f(z)\,dz$ is equal to a complex linear combination of two line integrals of real-valued functions. To remember (3.4.2), starting with $\int_\gamma f(z)\,dz$, we use $f(z) = u + iv$ and $dz = dx + i\,dy$ as follows:

$$\int_\gamma f(z)\,dz = \int_\gamma (u + iv)(dx + i\,dy) = \int_\gamma (u\,dx - v\,dy) + i \int_\gamma (v\,dx + u\,dy).$$

We now state the main result of this section.

Theorem 3.4.4. (Cauchy's Theorem for Simple Paths) *Suppose that f is an analytic function on a region U that contains a simple closed path γ and its interior. If f' is continuous on U, then we have*

$$\int_\gamma f(z)\,dz = 0. \tag{3.4.3}$$

Remark 3.4.5. We later prove that derivatives of analytic functions are also analytic, hence continuous (Corollary 3.8.9). Thus the assumption that f' is continuous in Theorem 3.4.4 is superfluous. The merit of Theorem 3.4.4 lies in the fact that all analytic functions we encounter in examples have continuous derivatives.

Proof. Recall that if $f = u + iv$, u, v real-valued, then $f' = u_x + iv_x = u_y + iv_y$ in view of (2.5.8), and the Cauchy-Riemann equations (2.5.7) hold: $u_x = v_y$ and $u_y = -v_x$. Using (3.4.2) and then applying Green's theorem along with the Cauchy-Riemann equations, we write

$$\int_\gamma f(z)\,dz = \int_\gamma (u\,dx - v\,dy) + i\int_\gamma (v\,dx + u\,dy)$$
$$= \iint_D \left(\frac{\partial}{\partial x}(-v) - \frac{\partial u}{\partial y}\right) dx\,dy + i\iint_D \left(\frac{\partial u}{\partial x} - \frac{\partial v}{\partial y}\right) dx\,dy$$
$$= \iint_D \overbrace{(-v_x - u_y)}^{0}\,dx\,dy + i\iint_D \overbrace{(u_x - v_y)}^{0}\,dx\,dy = 0.$$

This completes the proof of the theorem.   ■

Remark 3.4.6. It is very crucial in Theorem 3.4.4 for the domain U of f to contain all points in the interior of the curve γ. Indeed, the function $\frac{1}{z}$ is analytic and has a continuous derivative except at the origin. Thus the domain of $1/z$ does not contain the interior of the unit circle $C_1(0)$. And we showed in Example 3.2.10, that $\int_{C_1(0)} \frac{1}{z}\,dz = 2\pi i \neq 0$.

Example 3.4.7. Let γ be a simple closed path. Show that

$$\int_\gamma e^{z^2}\,dz = 0. \tag{3.4.4}$$

Solution. The function e^{z^2} is analytic and its derivative $2z e^{z^2}$ is continuous. Then identity (3.4.4) is a consequence of Theorem 3.4.4. You may recall that we were not able to compute this integral by the methods of the previous section.   □

Example 3.4.8. (Cauchy's integral theorem) Let $C_1(0)$ be the positively oriented unit circle. Evaluate

$$I = \int_{C_1(0)} \frac{\sin z}{z^2 - 4}\,dz.$$

Solution. The integrand is an analytic function with continuous derivatives everywhere except at $z = \pm 2$. Since these values are outside the path of integration, by Cauchy's theorem, $I = 0$.   □

Exercises 3.4

In Exercises 1–4, verify Green's theorem for the listed functions P and Q and path γ. That is, in each case, compute the integrals on both sides of (3.4.1) and show that they are equal.

1. $P(x, y) = xy$, $Q(x, y) = y$, γ is the positively oriented square path with vertices at $(0, 0)$, $(1, 0)$, $(1, 1)$, and $(0, 1)$.

2. $P(x, y) = y^2$, $Q(x, y) = x^2$, γ as in Exercise 1.

3. $P(x, y) = y$, $Q(x, y) = 1$, γ is the positively oriented triangular path with vertices at $(0, 0)$, $(1, 0)$, and $(1, 1)$.

4. $P(x, y) = \cos x$, $Q(x, y) = \sin y$, γ is the positively oriented unit circle.

5. Let γ be a positively oriented simple path and D be the region interior to γ. Show that the area of D is equal to the following integrals:

$$\int_\gamma -y\,dx, \qquad \int_\gamma x\,dy, \qquad \frac{1}{2}\int_\gamma -y\,dx + x\,dy.$$

6. Express the area of the ellipse $\frac{x^2}{a^2} + \frac{y^2}{b^2} = 1$ as a line integral and then find this area. [*Hint:* Use one of the integrals from the previous exercise.]

7. Let $u(x, y) = x^2 - y^2$ and let γ be a positively oriented simple path. Show that

$$\int_\gamma \frac{\partial u}{\partial y}\,dx - \frac{\partial u}{\partial x}\,dy = 0.$$

8. Let γ be the positively oriented triangle with vertices $(0,0)$, $(1,0)$, $(0,1)$. Show that

$$\int_\gamma \frac{e^{iz}\,dz}{z - \frac{1}{4} - \frac{4}{5}i} = 0.$$

9. Let $f(t) = t^2 \sin\frac{1}{t}$ if $t \neq 0$ and $f(0) = 0$. (a) Show that $f'(t)$ exists for all t.
(b) Let γ consists of the graph of f for $0 \leq t \leq \frac{1}{\pi}$ (initial point at $(0, 0)$ and terminal point at $(\frac{1}{\pi}, 0)$, followed by the line segment from $(\frac{1}{\pi}, 0)$ to $(0, 0)$. Argue that γ is a closed path that intersects itself an infinite number of times.

10. Show that the integrals of the functions $e^z(z-2)^{-1}$, $e^{-z}(z+2)^{-1}$, $(z-2i)^{-1}\cos z$, $(z+2i)^{-1}\sin z$ over the unit circle $C_1(0)$ are all vanishing.

11. Evaluate

$$\mathrm{Re}\int_{C_1(i)} \frac{z^6 - z^5 + iz}{z+i}\,dz + i\,\mathrm{Im}\int_{C_2(2i)} \frac{z^6 - z^5 + iz}{z+i}\,dz.$$

12. Let $r > 0$. Compute the integral of $\mathrm{Log}\,z$ over the circle $C_r(0)$ and over the circle $C_r(2r)$, both traced once in the positive orientation. Are the answers equal?

3.5 The Cauchy-Goursat Theorem

Our next goal is to prove Cauchy's theorem without assuming that the derivative of the analytic function is continuous. This requires an effort and occupies this and the next section. In this section we prove this theorem only for special kinds of sets, called star-shaped.

The fact that an analytic function has a complex derivative at a point means that near this point it can be approximated by a linear function. Since linear functions have antiderivatives and thus vanishing integral over closed paths, it follows that the integral of the analytic function over every closed triangle is small. This argument can be made precise and is the foundation of the proof of the general form of Cauchy's theorem. By a closed triangle we mean the interior and the perimeter of a triangle. For a closed triangle $\triangle$ we set $\mathrm{Diam}(\triangle)$ to be the diameter of a triangle, i.e., the length of its longest side, and we denote by $\partial\Delta$ its boundary, i.e., the union of the three line segments joining its vertices.

Lemma 3.5.1. *Let f be a continuous function on an open set U in $\mathbb{C}$ and assume that $f'(z_0)$ exists for some z_0 in U. Then for $\varepsilon > 0$ there is $\delta > 0$ such that for every closed triangle $\triangle$ that contains z_0 with $\triangle \subset B_\delta(z_0) \cap U$ we have*

$$\left| \int_{\partial\triangle} f(z)\,dz \right| < \varepsilon \left(Diam(\triangle) \right)^2 . \tag{3.5.1}$$

Proof. Let $\varepsilon > 0$. Since f has a complex derivative $f'(z_0)$ at z_0, there is $\delta > 0$ such that $B_\delta(z_0)$ is contained in U and for all $z \in B_\delta(z_0)$ we have

$$\left| \frac{f(z) - f(z_0)}{z - z_0} - f'(z_0) \right| < \frac{\varepsilon}{3} .$$

It follows that for an arbitrary $z \in B_\delta(z_0)$ we have

$$|f(z) - f(z_0) - f'(z_0)(z - z_0)| < \frac{\varepsilon}{3}|z - z_0| . \tag{3.5.2}$$

Let $\triangle$ be a closed triangle containing z_0 contained in $B_\delta(z_0)$. Then we have

$$\int_{\partial\triangle} f(z)\,dz = \int_{\partial\triangle} \left[f(z) - f(z_0) - f'(z_0)(z - z_0) \right] dz + \int_{\partial\triangle} \left[f(z_0) + f'(z_0)(z - z_0) \right] dz$$

$$= \int_{\partial\triangle} \left[f(z) - f(z_0) - f'(z_0)(z - z_0) \right] dz, \tag{3.5.3}$$

in view of the following consequence of Theorem 3.3.4(c)

$$\int_{\partial\triangle} \left[f(z_0) + f'(z_0)(z - z_0) \right] dz = 0, \tag{3.5.4}$$

given the fact that the integrand in (3.5.4) has an antiderivative in U, namely $f(z_0)z + \frac{1}{2}f'(z_0)(z - z_0)^2$. Since Δ is contained in $B_\delta(z_0)$ it follows that for $z \in \partial\Delta$ we have $|z - z_0| < \delta$; hence, using (3.5.3) and (3.5.2) we obtain

$$\left| \int_{\partial\triangle} f(z)\,dz \right| = \left| \int_{\partial\triangle} \left[f(z) - f(z_0) - f'(z_0)(z - z_0) \right] dz \right| \leq \frac{\varepsilon}{3} \operatorname{Perim}(\triangle) \max_{z \in \partial\triangle} |z - z_0| ,$$

where $\operatorname{Perim}(\triangle)$ is the perimeter of $\triangle$. Finally,

$$\frac{\varepsilon}{3} \operatorname{Perim}(\triangle) \max_{z \in \partial\triangle} |z - z_0| \leq \frac{\varepsilon}{3} \left(3 \operatorname{Diam}(\triangle) \right) \operatorname{Diam}(\triangle) = \varepsilon \left(Diam(\triangle) \right)^2 ,$$

and this yields (3.5.1).   ∎

We use this lemma to obtain the fundamental result that analytic functions have vanishing integral over boundaries of closed triangles contained in their domains.

This theorem is attributed to Cauchy and the French mathematician Édouard Goursat (1858–1936).

Theorem 3.5.2. (Cauchy-Goursat Theorem) *Suppose that* $\triangle$ *is a closed triangle contained in a region U and let f be an analytic function on U. Let* $\partial\triangle$ *denote the boundary of* $\triangle$ *with positive orientation. Then we have*

$$\int_{\partial\triangle} f(z)\,dz = 0. \tag{3.5.5}$$

Proof. We show that $\int_{\partial\triangle} f(z)\,dz$ is zero by showing that $\left|\int_{\partial\triangle} f(z)\,dz\right| < \varepsilon$ for all $\varepsilon > 0$. Thus, we fix $\varepsilon > 0$.

We begin by subdividing the triangle $\triangle$ into four closed triangles $\triangle^1, \triangle^2, \triangle^3, \triangle^4$ of size $1/4$ of that of $\triangle$ by joining the midpoints of the sides of $\triangle$. (See Figure 3.37). Assume that the boundaries of $\triangle^j$ are positively oriented. Then we have

$$\int_{\partial\triangle} f(z)\,dz = \int_{\partial\triangle^1} f(z)\,dz + \int_{\partial\triangle^2} f(z)\,dz + \int_{\partial\triangle^3} f(z)\,dz + \int_{\partial\triangle^4} f(z)\,dz,$$

since the contributions of the integrals over the line segments joining the midpoints of $\triangle$ cancel. Therefore we obtain

$$\left|\int_{\partial\triangle} f(z)\,dz\right| \leq \left|\int_{\partial\triangle^1} f(z)\,dz\right| + \left|\int_{\partial\triangle^2} f(z)\,dz\right| + \left|\int_{\partial\triangle^3} f(z)\,dz\right| + \left|\int_{\partial\triangle^4} f(z)\,dz\right|,$$

and so there is a triangle among $\triangle^1, \triangle^2, \triangle^3, \triangle^4$, which we call $\triangle_1$, such that

$$\left|\int_{\partial\triangle_1} f(z)\,dz\right| \geq \frac{1}{4}\left|\int_{\partial\triangle} f(z)\,dz\right|.$$

We now repeat the above procedure by replacing $\triangle$ by $\triangle_1$. Then we find a closed triangle $\triangle_2$ of size $1/4$ of that of $\triangle_1$ such that

$$\left|\int_{\partial\triangle_2} f(z)\,dz\right| \geq \frac{1}{4}\left|\int_{\partial\triangle_1} f(z)\,dz\right| \geq \frac{1}{4^2}\left|\int_{\partial\triangle} f(z)\,dz\right|.$$

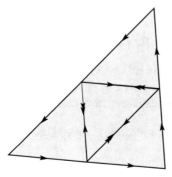

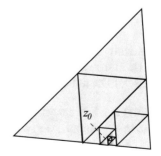

Fig. 3.37 The four triangles $\triangle^1$, $\triangle^2$, $\triangle^3$, $\triangle^4$ of the first step of the construction.

Fig. 3.38 The point z_0 lies in the intersection of $\triangle_n$ for all n.

Continuing this process by induction, for each positive integer n there is a triangle $\triangle_n$ of size 4^{-n} of that of $\triangle$ such that

$$\triangle = \triangle_1 \supset \triangle_2 \supset \triangle_3 \supset \cdots \supset \triangle_n \supset \triangle_{n+1} \supset \cdots$$

with

$$\left| \int_{\partial\triangle_n} f(z)\,dz \right| \geq \frac{1}{4} \left| \int_{\partial\triangle_{n-1}} f(z)\,dz \right| \geq \cdots \geq \frac{1}{4^n} \left| \int_{\partial\triangle} f(z)\,dz \right|. \tag{3.5.6}$$

The collection $\{\triangle_n\}_{n=1}^{\infty}$ is a nested sequence of nonempty closed and bounded (compact) sets and by the nonempty intersection property (Appendix, p. 484), there is a point z_0 in their intersection. This point is shown in Figure 3.38.

We now apply Lemma 3.5.1. Since f has a complex derivative at z_0, for the number $\varepsilon/(\text{Diam}(\triangle))^2$ there is a $\delta > 0$ such that $B_\delta(z_0) \subset U$ and such that the integral of f over the boundary of any closed triangle containing z_0 and contained in $B_\delta(z_0)$ is at most $\varepsilon/(\text{Diam}(\triangle))^2$. There is n large enough such that $\triangle_n$ is contained in $B_\delta(z_0)$; then

$$\left| \int_{\partial\triangle_n} f(z)\,dz \right| < \frac{\varepsilon}{(\text{Diam}(\triangle))^2} \left((\text{Diam}(\triangle_n))^2 \right).$$

But $\text{Diam}(\triangle) = 2^n \text{Diam}(\triangle_n)$, and thus, it follows that

$$\left| \int_{\partial\triangle_n} f(z)\,dz \right| < \frac{\varepsilon}{4^n}.$$

Combining this fact with the inequality in (3.5.6), we obtain $\left| \int_{\partial\triangle} f(z)\,dz \right| < \varepsilon$. Since $\varepsilon > 0$ was arbitrary, (3.5.5) holds. ∎

Cauchy's Theorem for Star-shaped Domains

A subset V of the complex plane $\mathbb{C}$ is called **convex** if for all $a, b \in V$ we have

$$(1-t)a + tb \in V \qquad \text{for all } t \in [0,1].$$

Geometrically speaking, convex sets contain all closed line segments whose endpoints lie in the set. For instance, interiors of open or closed squares and balls are convex sets. The notion of star-shaped sets extends that of convex sets.

Definition 3.5.3. A subset V of the complex plane $\mathbb{C}$ is called **star-shaped** about a point z_0 in V if for all $z \in V$ we have

$$(1-t)z_0 + tz \in V \qquad \text{for all } t \in [0,1].$$

In other words, star-shaped sets contain all closed line segments starting from a fixed point z_0 and ending at an arbitrary point in the set. See Figure 3.39. Convex sets are always star-shaped about any point in them, but the converse is not true.

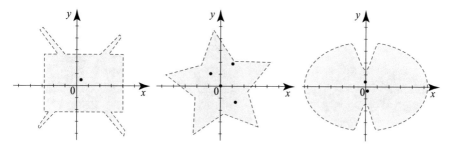

Fig. 3.39 These non-convex regions are star-shaped regions about the points indicated by dots.

Theorem 3.5.4. *Suppose that V is an open star-shaped subset of the plane and let f be an analytic function on V. Then for a simple closed path γ contained in V we have*

$$\int_\gamma f(z)\,dz = 0. \tag{3.5.7}$$

Proof. Fix a point z_0 in V such that $[z_0, z]$ lie in V for all $z \in V$. Define the function

$$F(z) = \int_{[z_0,z]} f(\zeta)\,d\zeta, \qquad \text{for } z \in V.$$

Since the line segment is contained in V, the function F is well defined. We show that F has a complex derivative equal to f. Let $z \in V$ and let $\delta > 0$ be such that $B_\delta(z)$ is contained in V. Then the closed triangle $[z_0, z, z+h, z_0]$ is contained in V and Theorem 3.5.2 yields

$$\int_{[z_0,z,z+h,z_0]} f(\zeta)\,d\zeta = 0.$$

In view of Properties (3.2.21) and (3.2.22) we have for $h \in B_\delta(0)$

$$\int_{[z_0,z]} f(\zeta)\,d\zeta + \int_{[z,z+h]} f(\zeta)\,d\zeta - \int_{[z_0,z+h]} f(\zeta)\,d\zeta = 0,$$

thus

$$\frac{F(z+h) - F(z)}{h} = \frac{1}{h}\int_{[z,z+h]} f(\zeta)\,d\zeta.$$

Letting $h \to 0$ and using Lemma 3.3.3, we conclude that F has a complex derivative at any point z in V and $F'(z) = f(z)$. Then by Theorem 3.3.4, the integral of f around an arbitrary closed path γ contained in V is zero. ∎

Exercises 3.5

In exercises 1–12 determine which sets are convex and which ones are star-shaped.

1.

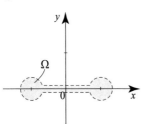

2.

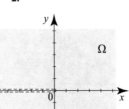

3.

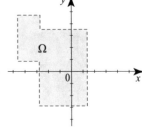

4.

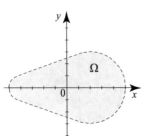

5.

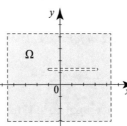

6.

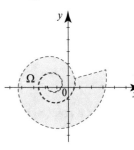

7.

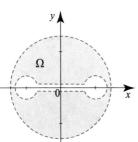

8.

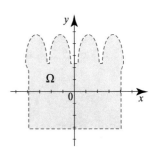

9. Evaluate the integral of the function $\frac{e^{z^2}-1}{z^2-4}$ over the square with vertices $1+i$, $-1+i$, $-1-i$, $1-i$ traced ten times counterclockwise.

10. Let g be an analytic function on a region U. Show that the integral of g over a closed polygonal path contained in U is zero. [*Hint:* Write the polygonal path as a union of triangular paths.]

11. Let Δ be the triangle with vertices $0, 1, i$. Without appealing to Theorem 3.5.2, evaluate the integrals $\int_{\partial\Delta} z\,dz$ and $\int_{\partial\Delta} \bar{z}\,dz$, where $\partial\Delta$ is traced once and is oriented counterclockwise.

3.6 Cauchy Integral Theorem For Simply Connected Regions

In this section we extend Cauchy's theorem from star-shaped sets to general regions. The approach we take is based on the geometric concept of deformation of paths that is easy to visualize, as illustrated by many examples. A precise definition is motivated by the following three cases:

- Suppose that α and β are points in a region Ω, and γ_0 and γ_1 are paths in Ω joining α to β. This means that the initial point of both γ_0 and γ_1 is α while the terminal point of both γ_0 and γ_1 is β. It may be possible to move γ_0 continuously so that it coincides with γ_1, keeping the ends fixed at α and β and without leaving Ω (Figure 3.40).
- If γ_0 and γ_1 are two closed paths in a region Ω, it may be possible to continuously move γ_0 without leaving Ω, in such a way that it coincides with γ_1 in position and direction (Figure 3.41).
- If γ_0 is a closed path in a region Ω, it may be the case that we can continuously shrink γ_0 to a point z_0 in Ω without leaving Ω (Figure 3.42). This situation is a special case of the preceding one, when the $\gamma_1(t) = z_0$ for all t, i.e., it is degenerated into a point.

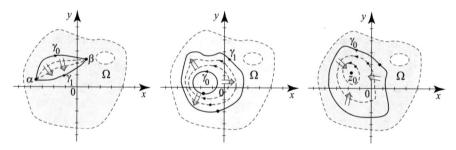

Fig. 3.40 Continuous deformation of γ_0 into γ_1.   **Fig. 3.41** Continuous deformation of γ_0 into γ_1.   **Fig. 3.42** Continuous deformation of γ_0 into a point z_0.

Definition 3.6.1. Let γ_0, γ_1 be paths in a region Ω, both defined on $[0, 1]$. Assume that $\gamma_0(0) = \gamma_1(0) = \alpha \in \Omega$ and $\gamma_0(1) = \gamma_1(1) = \beta \in \Omega$. We say that γ_0 is **homotopic** to γ_1 **(relative to Ω)** if there exists a continuous map H from the unit square $Q = [0, 1] \times [0, 1]$ to Ω with the following properties:

$$H(t, 0) = \gamma_0(t), \quad 0 \leq t \leq 1; \tag{3.6.1}$$
$$H(t, 1) = \gamma_1(t), \quad 0 \leq t \leq 1; \tag{3.6.2}$$
$$H(0, s) = \alpha, \quad 0 \leq s \leq 1; \tag{3.6.3}$$
$$H(1, s) = \beta, \quad 0 \leq s \leq 1. \tag{3.6.4}$$

The continuous mapping H is called a **homotopy** or a **continuous deformation**.

To understand this definition, examine Figure 3.43. Condition (3.6.1) tells us that the image of the lower side of Q is γ_0. Condition (3.6.2) says that the image of the upper side of Q is γ_1. Conditions (3.6.3) and (3.6.4) dictate that the vertical sides of Q are mapped to the endpoints of γ_0 and γ_1.

The fact that Q is mapped into Ω asserts that the image of any horizontal section in Q is a curve in Ω and because the vertical sides of Q are mapped to the endpoints of γ_0 and γ_1, this curve has the same endpoints as γ_0 and γ_1. As s varies from 0 to 1, the curve image of a horizontal section of the square Q varies continuously between the two extreme curves, from γ_0 to γ_1. For this reason a homotopy is also called a continuous deformation.

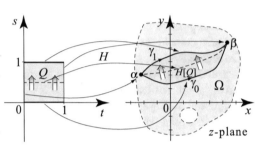

Fig. 3.43 Continuous deformation

The definition of a continuous deformation for closed paths is similar, but there is an additional requirement that all intermediate paths be closed.

Definition 3.6.2. Let γ_0, γ_1 be closed paths in a region Ω, both defined on $[0, 1]$. We say that γ_0 is **homotopic to** γ_1 **(relative to Ω)** if there exists a continuous map H from $Q = [0, 1] \times [0, 1]$ to Ω with the following properties:

$$H(t, 0) = \gamma_0(t), \qquad 0 \le t \le 1; \tag{3.6.5}$$
$$H(t, 1) = \gamma_1(t), \qquad 0 \le t \le 1; \tag{3.6.6}$$
$$H(0, s) = H(1, s), \qquad 0 \le s \le 1. \tag{3.6.7}$$

Such a map H is called a **homotopy** or a **continuous deformation** (Figure 3.44). If γ_1 is a constant path, then we say that γ_0 is **homotopic to a point** (Figure 3.45).

When γ_0 is **homotopic to a point** condition (3.6.7) states that the initial point $H(0, s)$ and the terminal point $H(1, s)$ are the same. In this case the images of all horizontal sections of the square Q are closed curves (Figure 3.44).

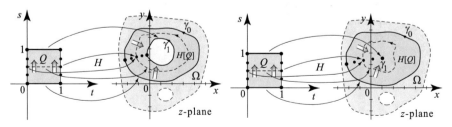

Fig. 3.44 Deformation of closed paths.   **Fig. 3.45** Deformation to a point.

It is geometrically clear that if γ_0 is homotopic to γ_1 relative to Ω, then γ_1 is homotopic to γ_0 relative to Ω. Also, if γ_0 is homotopic to γ_1 relative to Ω and γ_1 is homotopic to γ_2 relative to Ω, then γ_0 is homotopic to γ_2 relative to Ω. See Exercise 8. For this reason, we often say that γ_0 and γ_1 are homotopic to each other.

Example 3.6.3. (Continuous deformation of paths)
(a) Let $\Omega = \mathbb{C}$. The upper semicircle $\gamma(t) = e^{it}$, $0 \leq t \leq \pi$ in Figure 3.46 is homotopic to the directed line segment joining 1 to -1.
(b) In the open unit disk $B_1(0)$, the path $[z_1, z_2, z_3]$ is homotopic to the directed line segment $[z_1, z_3]$. See Figure 3.47.
(c) In the open unit disk $B_1(0)$, the arc γ_0 in Figure 3.48 is homotopic to the triangular path $[z_1, z_2, z_3, z_1]$.

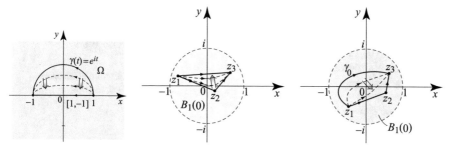

Fig. 3.46 Semicircle γ.   **Fig. 3.47** Path $[z_1, z_2, z_3]$.   **Fig. 3.48** Arc γ_0.

(d) In the annular region Ω (Figure 3.49) the circle γ_0 is homotopic to the circle γ_1.
(e) In the region Ω in Figure 3.50, the ellipse γ_0 is homotopic to a point.
(f) In Figure 3.51, the region Ω consists of a disk minus the points z_1, z_2, z_3. Relative to Ω, the circle γ_0 is homotopic to the curve γ_1 consisting of three circles centered at z_1, z_2, z_3 and connected by line segments traversed in both directions, as shown in Figure 3.51. $\qquad\square$

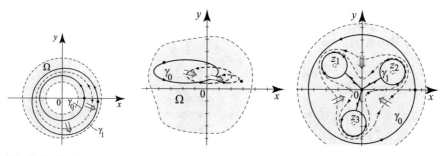

Fig. 3.49 Circle γ_0.   **Fig. 3.50** Ellipse γ_0.   **Fig. 3.51** Circle γ_0.

As one may imagine, it is often difficult to construct the homotopy H, and so in many situations we rely on intuition to decide whether a continuous deformation is possible. There are many interesting cases, however, where H can be constructed explicitly.

Example 3.6.4. (Continuous deformation of paths) Show that the positively oriented circle $C_1(0)$ is homotopic to the positively oriented circle $C_2(0)$, relative to the open disk of radius 3.

Solution. Intuitively, the result is obvious. To provide a rigorous proof, we construct a continuous mapping H of the unit square taking $C_1(0)$ continuously to $C_2(0)$, in the sense that (3.6.1), (3.6.2), and (3.6.7) hold. Parametrize $C_1(0)$ by $\gamma_0(t) = e^{2\pi i t}$, $0 \leq t \leq 1$, and $C_2(0)$ by $\gamma_1(t) = 2e^{2\pi i t}$, $0 \leq t \leq 1$. For a fixed t in $[0, 1]$, consider the corresponding points $\gamma_0(t)$ on $C_1(0)$ and $\gamma_1(t)$ on $C_2(0)$. We can move continuously from $\gamma_0(t)$ to $\gamma_1(t)$ along the line segment

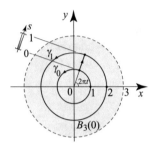

Fig. 3.52 Continuous deformation of $C_1(0)$ to $C_2(0)$.

$$(1-s)\gamma_0(t) + s\gamma_1(t), \qquad 0 \leq s \leq 1,$$

while staying in Ω (Figure 3.52).

Motivated by this idea, we let

$$H(t, s) = (1-s)\gamma_0(t) + s\gamma_1(t), \quad 0 \leq t \leq 1, \quad 0 \leq s \leq 1. \tag{3.6.8}$$

Then H is a continuous mapping of (t, s) in the unit square, because $\gamma_0(t)$ and $\gamma_1(t)$ are continuous. Moreover, it is easy to check that (3.6.1), (3.6.2), and (3.6.7) hold. Indeed,

$$H(t, 0) = \gamma_0(t) \quad \Rightarrow \quad (3.6.1) \text{ holds,}$$
$$H(t, 1) = \gamma_1(t) \quad \Rightarrow \quad (3.6.2) \text{ holds.}$$

Also, since $\gamma_0(0) = 1 = \gamma_0(1)$ and $\gamma_1(0) = 2 = \gamma_1(1)$, we obtain

$$H(0, s) = (1-s)\gamma_0(0) + s\gamma_1(0) = 1+s,$$
$$H(1, s) = (1-s)\gamma_0(1) + s\gamma_1(1) = 1+s,$$

and so $H(0, s) = H(1, s)$ implying (3.6.7). $\qquad \square$

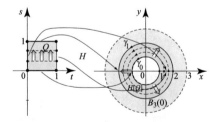

Fig. 3.53 H maps the lower side of Q onto the inner circle $C_1(0)$, the upper side onto $C_2(0)$, and all the in between horizontal segments in Q are mapped to in between circles $C_R(0)$, where $1 < R < 2$.

Integral Invariance for Homotopic Paths

Theorem 3.6.5. *Let f be an analytic function in a region Ω and let γ_0 and γ_1 be paths in Ω.*
(i) If $\gamma_0(0) = \gamma_1(0) = \alpha \in \Omega$, $\gamma_0(1) = \gamma_1(1) = \beta \in \Omega$, and γ_0 is homotopic to γ_1 according to Definition 3.6.1, then

$$\int_{\gamma_0} f(z)\,dz = \int_{\gamma_1} f(z)\,dz. \tag{3.6.9}$$

(ii) If γ_0 and γ_1 are closed homotopic paths then (3.6.9) holds.
(iii) If γ_0 is homotopic to a point in Ω, then

$$\int_{\gamma_0} f(z)\,dz = 0. \tag{3.6.10}$$

Proof. Note that the assertion in (*iii*) follows from that in (*ii*) because the integral of a function over a point is zero. Also, the statement in (*ii*) is a consequence of that in (*i*). Indeed, let $\alpha = \gamma_0(0) = \gamma_0(1)$ and $\beta = \gamma_1(0) = \gamma_1(1)$. Define a path $\gamma_{\alpha,\beta}(s) = H(0,s) = H(1,s)$ in Ω that connects α to β. The consider the path $\tilde{\gamma}_1 = \gamma_{\alpha,\beta} \cup \gamma_1 \cup (\gamma_{\alpha,\beta})^*$, which starts and ends at the point α just like γ_0. Then $\tilde{\gamma}_1$ is homotopic[2] to γ_0. Assuming the assertion in (*i*), (3.6.9) gives

$$\int_{\gamma_0} f(z)\,dz = \int_{\tilde{\gamma}_1} f(z)\,dz$$
$$= \int_{\gamma_{\alpha,\beta}} f(z)\,dz + \int_{\gamma_1} f(z)\,dz + \int_{(\gamma_{\alpha,\beta})^*} f(z)\,dz$$
$$= \int_{\gamma_1} f(z)\,dz,$$

since $\int_{(\gamma_{\alpha,\beta})^*} f(z)\,dz = -\int_{\gamma_{\alpha,\beta}} f(z)\,dz$. Thus (*ii*) holds assuming (*i*).

We now prove the claimed assertion in (*i*). Let γ_0 and γ_1 be paths in Ω joining α to β. If Ω is the entire plane, which is a star-shaped region, Theorem 3.5.4, yields the claimed assertion. If Ω is not the entire plane, its boundary $\partial\Omega$ is closed and nonempty. Let H denote the mapping of the unit square $Q = [0,\,1] \times [0,\,1]$ into Ω that continuously deforms γ_0 into γ_1, and let $K = H[Q]$. Since Q is closed and bounded, and H is continuous, it follows that K is closed and bounded. Since K is contained in the (open region) Ω, it is disjoint from the closed boundary of Ω. So if δ denotes the distance from $\partial\Omega$ to K, then δ is positive (Appendix, page 483).

Since H is continuous on Q and Q is closed and bounded, it follows that H is uniformly continuous on Q. So for the given $\delta > 0$ (which plays the role of $\varepsilon > 0$ in the definition of uniform continuity), there is a positive integer n such that

---

[2] The mapping $\mathcal{H}(t,s) = H(0,3ts)$ for $0 \le t \le \frac{1}{3}$, $\mathcal{H}(t,s) = H(3t-1,s)$ for $\frac{1}{3} \le t \le \frac{2}{3}$, $\mathcal{H}(t,s) = H(0,(3-3t)s)$ for $\frac{2}{3} \le t \le 1, 0 \le s \le 1$, continuously deforms γ_0 to $\tilde{\gamma}_1$.

$$|(t,s) - (t',s')| \le \frac{\sqrt{2}}{n} \Rightarrow |H(t,s) - H(t',s')| < \delta. \qquad (3.6.11)$$

Having chosen n, construct a grid over Q by subdividing the sides into n equal subintervals (Figure 3.54). This is a **network of simplices**, where each smaller square in this network is called a **simplicial square**. Label the points on the grid by $p_{jk} = \left(\frac{j}{n}, \frac{k}{n}\right)$, $j, k = 0, 1, \dots, n$, and let $z_{jk} = H(p_{jk})$.

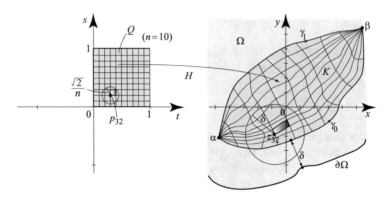

Fig. 3.54 A simplicial network on the square Q with a sample simplicial square with one corner at p_{32}. The simplicial squares are so small that the image of each is contained in a disk in Ω of radius $\delta > 0$, where δ is the distance between the boundary of Ω, $\partial\Omega$, and the closed and bounded set $H[Q]$.

Condition (3.6.11) guarantees that the image of each simplicial square lies in the open disk $B_\delta(z_{jk})$, and hence, the closed polygonal path

$$\sigma_{jk} = [z_{jk}, z_{(j+1)k}, z_{(j+1)(k+1)}, z_{j(k+1)}, z_{jk}] \qquad (3.6.12)$$

is also contained in $B_\delta(z_{jk})$ (see Figure 3.55). Since f is analytic on $B_\delta(z_{jk})$, which is obviously star-shaped, Theorem 3.5.4 yields

$$\int_{\sigma_{jk}} f(z)\,dz = 0.$$

Adding the integrals over all the σ_{jk}'s, we note that the integrals over the segments that correspond to internal sides of the simplicial squares cancel, since each internal segment is traversed twice in opposite directions. Thus the only integrals that do not cancel are those over line segments that correspond to external sides of the simplicial squares, that is, the boundary of Q.

Figure 3.55 shows a typical closed polygonal path obtained by joining the images of the four corners of one simplicial square. By construction, these polygonal paths are contained in disks of radius δ in Ω. Figure 3.56 displays the outcome of the cancelation; after summing all the path integrals and canceling those that are traversed in opposite direction, the only ones that remain are those that correspond to the boundary of Q.

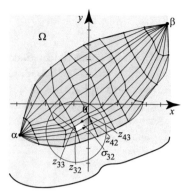

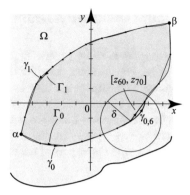

Fig. 3.55 A typical closed polygonal path.

Fig. 3.56 The outcome of cancellation.

The integrals on the line segments connecting z_{jk} going up the right side and down the left side of Q are each zero because the z_{jk} are fixed at α and β. The only remaining integrals are on the line segments corresponding to the bottom and top sides of Q. So we obtain

$$0 = \sum_{j,k=0}^{n-1} \int_{\sigma_{jk}} f(z)\,dz = \int_{\Gamma_0} f(z)\,dz - \int_{\Gamma_1} f(z)\,dz, \qquad (3.6.13)$$

where $\Gamma_0 = [z_{00}, z_{10}, \ldots, z_{n0}]$ and $\Gamma_1 = [z_{0n}, z_{1n}, \ldots, z_{nn}]$. Applying Theorem 3.5.4 for the disk $B_\delta(z_{j0})$, which is star-shaped, we write

$$\int_{[z_{j0}, z_{(j+1)0}]} f(z)\,dz = \int_{\gamma_{0,j}} f(z)\,dz, \qquad (3.6.14)$$

where $\gamma_{0,j}$ is the portion of the path γ_0 that joins the points z_{j0} and $z_{(j+1)0}$ (Figure 3.56). Adding equations (3.6.14) as j runs from 0 to $n-1$, we deduce

$$\int_{\Gamma_0} f(z)\,dz = \int_{\gamma_0} f(z)\,dz.$$

Similarly, we have

$$\int_{\Gamma_1} f(z)\,dz = \int_{\gamma_1} f(z)\,dz.$$

Comparing with (3.6.13), we find that

$$\int_{\gamma_0} f(z)\,dz = \int_{\Gamma_0} f(z)\,dz = \int_{\Gamma_1} f(z)\,dz = \int_{\gamma_1} f(z)\,dz,$$

which completes the proof of the assertion in (i) and thus of the theorem. ∎

Simply Connected Regions

It is important to be precise about the region Ω when talking about deformation of paths. For instance, consider the situations depicted in Figures 3.57 and 3.58. In Figure 3.57 the upper and lower semicircles joining the points -1 to 1 are homotopic in the disk of radius 2. However, in Figure 3.58, the upper and lower semicircles joining the points -1 to 1 are not homotopic in the punctured disk of radius 2. You can imagine a stick protruding at the origin, which will prevent us from continuously deforming the upper part of the unit circle onto the lower one.

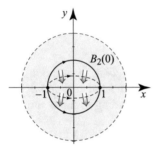

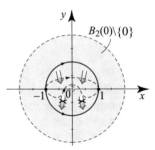

Fig. 3.57 Continuous deformation is possible.

Fig. 3.58 Continuous deformation is not possible.

The situation encountered in the preceding examples leads to the following important definition.

Definition 3.6.6. A region Ω is called **simply connected** if all paths joining α to β in Ω are homotopic relative to Ω. If a region is not simply connected, then it is called **multiply connected**.

Geometrically, a region is simply connected if and only if it has no holes in it. The regions in Figures 3.59 are simply connected but the ones in 3.60 are not simply connected.

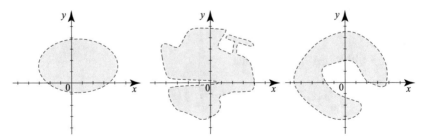

Fig. 3.59 Three simply-connected regions.

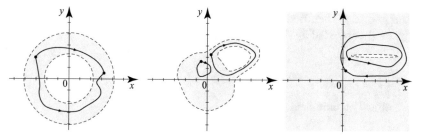

Fig. 3.60 Three multiply connected regions. In the figure on the left a path is shown that is not homotopic to a point. In the remaining two pictures two non-homotopic paths relative to the region are shown.

Suppose that Ω is a simply connected region, f is analytic on Ω, α and β are two points in Ω, and γ_0 and γ_1 are arbitrary paths in Ω joining α to β. Since γ_0 and γ_1 are homotopic relative to Ω, Theorem 3.6.5 implies that the integral of f over γ_0 is equal to its integral over γ_1. In other words, the integral of f is independent of path in Ω.

Theorem 3.6.7. (Cauchy's Theorem for Simply Connected Regions) *Let f be an analytic function on a simply connected region Ω. If γ is a closed path in Ω, then*

$$\int_\gamma f(z)\,dz = 0. \tag{3.6.15}$$

Proof. Equality (3.6.15) is an immediate consequence of Theorem 3.6.5 (ii), since every closed path in Ω is homotopic to a point. ∎

Remark 3.6.8. An informal restatement of Theorem 3.6.7 is that if f is analytic on a simple closed path[3] and its interior, then the integral of f around this path is zero.

Corollary 3.6.9. *Let f be an analytic function on a simply connected region Ω. Then the following are valid:*
(i) If γ_0 and γ_1 are two paths in Ω with the same initial and terminal point, then

$$\int_{\gamma_0} f(z)\,dz = \int_{\gamma_1} f(z)\,dz.$$

(ii) There is an analytic function F defined on Ω such that $F' = f$.

Proof. The claimed assertions (i) and (ii) follow by combining Theorem 3.6.7 with Theorem 3.3.4. ∎

Remark 3.6.10. We note that the antiderivative F of f in part (ii) of Corollary 3.6.9 is

$$F(z) = \int_{\gamma(z_0,z)} f(\zeta)\,d\zeta,$$

where z_0 is a fixed point in Ω and the integral is taken over a path $\gamma(z_0,z)$ in Ω joining z_0 to z.

---

[3] analytic "on" a path means analytic on a neighborhood of a path

Our geometric intuition that the punctured unit disk is not simply connected is analytically expressed by the identity

$$\int_{C_1(0)} \frac{dz}{z} = 2\pi i \neq 0,$$

which directly contradicts (3.6.15).

Exercises 3.6

In Exercises 1–6, figures are shown describing two paths γ_0 and γ_1 in a region Ω. Decide whether the two paths are homotopic relative to Ω. Justify your answer based on the picture.

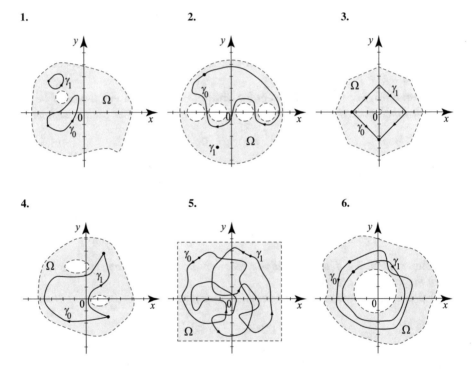

1.   **2.**   **3.**

4.   **5.**   **6.**

7. Of the regions shown in Figures 1–6, determine the ones that are simply connected. Justify your answer with arguments based on the figure.

8. Let Ω be a region in the complex plane.
(a) Suppose that γ_0 is homotopic to γ_1 relative to Ω. Show that γ_1 is homotopic to γ_0 relative to Ω.
(b) Show that if γ_0 is homotopic to γ_1 relative to Ω and γ_1 is homotopic to γ_2 relative to Ω, then γ_0 is homotopic to γ_2 relative to Ω.

9. Show that every non-closed path in $\mathbb{C}$ is homotopic to a point.

In Exercises 10–12, describe the mapping $H(t, s)$ of the unit square to Ω that continuously deforms γ_0 to γ_1 relative to Ω.

10.   **11.**   **12.**

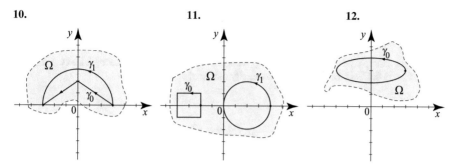

13. Project Problem: Convex sets. Recall that a subset S of the complex plane is **convex** if whenever z_0 and z_1 are in S, then the closed line segment joining z_0 to z_1 lies entirely in S. In other words, S is convex if whenever z_0 and z_1 are in S, then the point

$$(1-s)z_0 + sz_1, \qquad 0 \le s \le 1, \qquad\qquad (3.6.16)$$

also lies in S.
(a) Determine which of the sets in the figures below are convex.

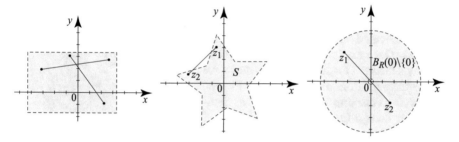

(b) Let Ω be a convex region and let γ_0 and γ_1 be paths in Ω with the same initial and terminal points. Suppose that γ_0 and γ_1 are parametrized by the interval $[0, 1]$. Consider the mapping

$$H(t, s) = (1-s)\gamma_0(t) + s\gamma_1(t), \quad 0 \le t \le 1, \quad 0 \le s \le 1. \qquad (3.6.17)$$

Show that H is a homotopy from the unit square into Ω that continuously deforms γ_0 to γ_1. Thus, any two paths γ_0 and γ_1 in Ω, joining two points α and β in Ω, are homotopic relative to Ω.
(c) Show that two arbitrary closed paths γ_0 and γ_1 in Ω are homotopic relative to Ω.
(d) Show that a closed path γ_0 in Ω is homotopic to a point in Ω. [*Hint*: Consider the homotopy H in (3.6.17) with $\gamma_1(t) = z_1$ for all $t \in [0, 1]$.]
(e) In the region Ω_0 consisting of the open disk centered at the origin with radius 4, consider the closed paths

$$\gamma_0(t) = e^{2\pi i t} + e^{4\pi i t}, \qquad 0 \le t \le 1,$$

and

$$\gamma_1(t) = 3e^{2\pi i t}, \qquad 0 \le t \le 1.$$

Describe the homotopy that continuously deforms γ_0 to γ_1 in Ω_0. See the adjacent figure.

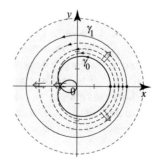

14. Give an example of a closed and bounded subset K of the real line and a bounded subset S of the real line such that K and S are disjoint but the distance from K to S is 0. [*Hint*: The set S is necessarily not closed.]

15. Give an example of two closed and disjoint subsets K and S of the plane such that the distance from K to S is 0. [*Hint*: Both sets are necessarily unbounded.]

3.7 Cauchy's Theorem for Multiply Connected Regions

For subsequent applications, we prove a version of Cauchy's theorem that involves several simple paths arranged as follows: C is a simple closed path and $C_1, C_2 \ldots, C_n$ are disjoint simple closed paths, contained in the interior of C and such that the interior regions of any two C_j's have no common points. To simplify our study, we require that C and all $C'_j s$ have the same positive orientation.

An interesting situation arises when the C_j's contain in their interior parts of the complement of a region Ω and C contains all C_j. For instance Ω could be the shaded region in Figure 3.61. Our goal in this section is to prove Cauchy's theorem for such paths. To do so, we need a theorem from topology which, like the Jordan curve theorem, although it is very intuitive, it is quite difficult to prove. For a proof of the next theorem we refer to *Selected Topics in the Classical Theory of Functions of a Complex Variable* by Maurice Heins, Dover (2015), republication of the edition by Holt, Rinehart & Winston, New York, 1962.

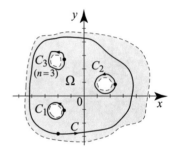

Fig. 3.61 The domain Ω contains the region interior to C and exterior to all C_j's.

Theorem 3.7.1. (Jordan Arc Theorem) *Suppose that U is a region that contains a simple non-closed path γ. Then $U \setminus \gamma$ is also a region, i.e., it is open and connected.*

The main result of this section is the following extension of Theorem 3.6.7 to the setting of multiply connected domains.

Theorem 3.7.2. (Cauchy's Theorem for Multiply Connected Regions) *Let $n \in \mathbb{N}$ and $\Omega_1, \ldots, \Omega_n$ be pairwise disjoint simply connected regions whose union is contained in another simply connected region Ω_0. Let $\Omega = \Omega_0 \setminus (\overline{\Omega_1} \cup \cdots \cup \overline{\Omega_n})$. For $j = 1, \ldots, n$, let C_j be disjoint simple closed paths with positive orientation whose interior regions contain $\overline{\Omega_j}$ and assume that C_j are pairwise disjoint. Let C be a simple closed path with positive orientation in Ω whose interior region contains all C_j. Suppose that f is an analytic function on Ω. Then we have*

$$\int_C f(z)\, dz = \sum_{j=1}^{n} \int_{C_j} f(z)\, dz. \tag{3.7.1}$$

Remark 3.7.3. We may informally say that in the hypothesis of the preceding theorem the function f is required to be analytic on a path[4] C and on its interior minus a few simply connected 'holes' denoted by Ω_j.

Proof. Fix a point z_0 on the outer path C. Join z_0 to a point w_1 in C_1 via a simple polygonal path L_1. Pick z_1 on C and let P_1 be the part of C_1 from w_1 to z_1 traversed in the orientation of C_1. By Theorem 3.7.1, $\Omega \setminus (L_1 \cup P_1)$ is connected, and thus, there is a simple polygonal path L_2 disjoint from $L_1 \cup P_1$ that joins z_1 to a point w_2 in C_2.

Pick $z_2 \in C_2$ and let P_2 be the part of C_2 from w_2 to z_2 traversed in the orientation of C_2. By Theorem 3.7.1, $\Omega \setminus (L_1 \cup P_1 \cup L_2 \cup P_2)$ is connected, and thus, there is simple polygonal path L_3 disjoint from $L_1 \cup P_1 \cup L_2 \cup P_2$ that joins z_2 to a point w_3 in C_3.

Continuing in this fashion, we find points w_n and z_n on C_n and we let P_n be the part of C_n from w_n to z_n traversed in the same orientation. At the end, join z_n to a point w_{n+1} in C via a simple polygonal path L_{n+1} that does not intersect the previously selected path from z_0 to z_n, passing through $w_1, z_1, w_2, z_2, \ldots, w_n$. In the selection of these points we defined

$$P_j = \text{part of } C_j \text{ from } w_j \text{ to } z_j \text{ traversed in the orientation of } C_j$$

for $j = 1, \ldots, n$, and now also define

$$Q_j = \text{part of } C_j \text{ from } z_j \text{ to } w_j \text{ traversed in the orientation of } C_j.$$

Also let P be the part of C from z_0 to w_{n+1} and let Q be the part of C from w_{n+1} to z_0 both traversed in the orientation inherited by C.

This construction yields two simple closed paths Γ_1 and Γ_2, as illustrated in Figure 3.62, precisely defined as follows:

$$\Gamma_1 = \left[P, L_{n+1}^*, Q_n^*, L_n^*, Q_{n-1}^*, \ldots, Q_1^*, L_1^* \right]$$

$$\Gamma_2 = \left[L_1, P_1^*, L_2, P_2^*, L_3, \ldots, P_n^*, L_{n+1}, Q \right]$$

for $j = 1, \ldots, n$. (Recall that γ^* is the reverse of a path γ.)

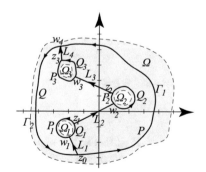

Fig. 3.62 The construction of Γ_1 and Γ_2.

Moreover, we have arranged so that all pieces of the complement of Ω in the interior of C do not lie in the interior of Γ_1 or Γ_2. Thus the interior regions of Γ_1 and Γ_2 are simply connected and f is analytic in a slightly larger simply connected

---

[4] analytic on C means analytic in a neighborhood of C

neighborhood of either Γ_1 and Γ_2. Theorem 3.6.7 is now applicable and yields that

$$\int_{\Gamma_1} f(z)\,dz = 0 \qquad \text{and} \qquad \int_{\Gamma_2} f(z)\,dz = 0.$$

Adding these two equalities, we obtain

$$\int_{\Gamma_1} f(z)\,dz + \int_{\Gamma_2} f(z)\,dz = 0,$$

or

$$\int_P f(z)\,dz + \int_Q f(z)\,dz + \sum_{j=1}^{n} \left(\int_{P_j^*} f(z)\,dz + \int_{Q_j^*} f(z)\,dz \right) = 0,$$

using Proposition 3.2.12(iii). Proposition 3.2.12(ii) now yields

$$\int_{P_j^*} f(z)\,dz + \int_{Q_j^*} f(z)\,dz = -\int_{P_j} f(z)\,dz - \int_{Q_j} f(z)\,dz = -\int_{C_j} f(z)\,dz,$$

hence, we conclude that

$$\int_P f(z)\,dz + \int_Q f(z)\,dz - \sum_{j=1}^{n} \int_{C_j} f(z)\,dz = 0,$$

which is equivalent to (3.7.1). ∎

Next, in the fundamental integral of Example 3.2.10, we replace the circle by an arbitrary positively oriented, simple, closed path C that does not contain a fixed point z_0.

Example 3.7.4. Let C be a positively oriented, simple, closed path, and z_0 be a point not on C. Then

$$\int_C \frac{1}{z - z_0}\,dz = \begin{cases} 0 & \text{if } z_0 \text{ lies in the exterior of } C; \\ 2\pi i & \text{if } z_0 \text{ lies in the interior of } C. \end{cases}$$

Moreover, for $n \neq 1$,

$$\int_C \frac{1}{(z - z_0)^n}\,dz = 0. \tag{3.7.2}$$

Solution. If z_0 lies in the exterior of C, then $\frac{1}{z-z_0}$ is analytic inside and on C and hence the integral is 0, by Theorem 3.6.7 (Instead of Theorem 3.6.7 we may also apply Theorem 3.4.4 since $\frac{1}{z-z_0}$ has continuous derivatives.)

To deal with the case where z_0 lies in the interior of C, choose $R > 0$ such that $C_R(z_0)$ is contained in the interior of C. The function $\frac{1}{z-z_0}$ is analytic in $\mathbb{C} \setminus \{z_0\}$. Applying conclusion (3.7.1) of Theorem 3.7.2, we see that

$$\int_C \frac{1}{z-z_0} \, dz = \int_{C_R(z_0)} \frac{1}{z-z_0} \, dz = 2\pi i,$$

where the second integral follows from Example 3.2.10. The same argument applies for (3.7.2) but in this case the outcome is zero in view of Example 3.2.10. □

In what follows, we illustrate the applications of the formula in Example 3.7.4.

Example 3.7.5. (Integral of a rational function) Let C be a positively oriented, simple, closed path containing i in its interior and $-i$ in its exterior; see Figure 3.63. Evaluate

$$\int_C \frac{dz}{z^2 + 1}.$$

Solution. The first thing to check is whether $f(z) = \frac{1}{z^2+1}$ is analytic inside and on C. If it is, then we can apply Cauchy's theorem and be done. Clearly f is not analytic at the points $z = \pm i$, where the denominator $z^2 + 1$ vanishes. Since one of these values is inside C, we apply Cauchy's theorem for multiple paths, as in the previous example or, better yet, use the result of the previous example. The method uses partial fractions and goes as follows. Since $z^2 + 1 = (z-i)(z+i)$, we have the partial fraction decomposition

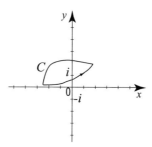

Fig. 3.63 The curve C

$$\frac{1}{z^2 + 1} = \frac{A}{z-i} + \frac{B}{z+i}. \tag{3.7.3}$$

Thus

$$\int_C \frac{1}{z^2+1} dz = A \overbrace{\int_C \frac{dz}{z-i}}^{2\pi i} + B \overbrace{\int_C \frac{dz}{z+i}}^{0} = 2\pi i A,$$

where both integrals after the first equality follow from Example 3.7.4. To complete the evaluation of the integral, we must determine A. To this end, multiply both sides of (3.7.3) by $(z-i)$; then evaluate the equation at $z = i$:

$$\frac{1}{z+i}\Big|_{z=i} = \Big[A + \frac{B}{z+i}(z-i) \Big]\Big|_{z=i};$$

$$\frac{1}{2i} = A + 0; \ A = -\frac{i}{2}.$$

Using this value of A, we get

$$\int_C \frac{dz}{z^2+1} = 2\pi i \left(-\frac{i}{2} \right) = \pi. \qquad \square$$

Example 3.7.6. (Path integrals of rational functions) Compute

$$\int_C \frac{dz}{(z-1)(z+i)(z-i)},$$

where C is a simple closed path in the following three cases.
(a) The point 1 is in the interior of C, and the points $\pm i$ are in the exterior of C.
(b) The points 1 and i are in the interior of C, and the point $z = -i$ is in the exterior of C.
(c) All three points 1, $\pm i$ are in the interior of C.

Solution. We start by finding the partial fractions decomposition of the integrand,

$$\frac{1}{(z-1)(z+i)(z-i)} = \frac{a}{z-1} + \frac{b}{z+i} + \frac{c}{z-i},$$

where a, b, and c are complex numbers. Combining, we obtain

$$\frac{1}{(z-1)(z+i)(z-i)} = \frac{a(z+i)(z-i) + b(z-1)(z-i) + c(z-1)(z+i)}{(z-1)(z+i)(z-i)},$$

so

$$1 = a(z+i)(z-i) + b(z-1)(z-i) + c(z-1)(z+i). \qquad (3.7.4)$$

Taking $z = 1$, we get

$$1 = a(1+i)(1-i) \quad \Rightarrow \quad a = \frac{1}{(1+i)(1-i)} = \frac{1}{2}.$$

Similarly, setting $z = -i$ yields $b = -\frac{1}{4} - \frac{i}{4}$. Setting $z = i$ yields $c = -\frac{1}{4} + \frac{i}{4}$. Thus the partial fractions decomposition

$$\frac{1}{(z-1)(z+i)(z-i)} = \frac{1}{2(z-1)} - \frac{1+i}{4(z+i)} + \frac{-1+i}{4(z-i)}.$$

So we have

$$\int_C \frac{dz}{(z-1)(z+i)(z-i)}$$
$$= \frac{1}{2} \int_C \frac{1}{z-1} dz - \left(\frac{1}{4} + \frac{i}{4}\right) \int_C \frac{1}{z+i} dz + \left(-\frac{1}{4} + \frac{i}{4}\right) \int_C \frac{1}{z-i} dz. \quad (3.7.5)$$

We can now compute the desired integrals using Example 3.7.4.
(a) Since 1 is the only point in the interior of C, the last two integrals on the right side of (3.7.5) are zero, and hence

$$\int_C \frac{dz}{(z-1)(z+i)(z-i)} = \frac{1}{2} \int_C \frac{1}{z-1} = \frac{1}{2} 2\pi i = \pi i.$$

(b) Since 1 and i are in the interior of C and $-i$ is in the exterior of C, the middle integral on the right side of (3.7.5) is zero, and the desired integral in this case is equal to

$$\frac{1}{2}\int_C \frac{1}{z-1} + \left(-\frac{1}{4}+\frac{i}{4}\right)\int_C \frac{1}{z-i}\,dz = \left(\frac{1}{2}+\left(-\frac{1}{4}+\frac{i}{4}\right)\right)2\pi i = \left(\frac{1}{2}+\frac{i}{2}\right)\pi i.$$

(c) Since 1 and $\pm i$ are all in the interior of C, all the integrals on the right side of (3.7.5) must be accounted for. The answer in this case is

$$\left[\frac{1}{2}+\left(\frac{-1}{4}+\frac{i}{4}\right)+\left(-\frac{1}{4}-\frac{i}{4}\right)\right]2\pi i = 0.$$

See the related more general result of Exercise 36. ☐

Example 3.7.7. (A path that is not simple) Evaluate the integral

$$\int_C \frac{z}{(z-i)(z-1)}\,dz, \tag{3.7.6}$$

where C is the figure-eight path shown in Figure 3.64.

The path C in Figure 3.64 is not simple because it intersects itself. To use Cauchy's integral theorem for simple paths, we break up C into two simple paths, as shown in Figure 3.65.

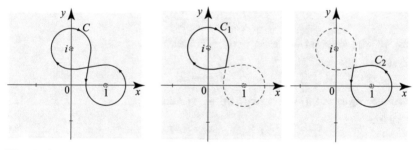

Fig. 3.64 The path C is not simple.   **Fig. 3.65** Breaking up C as a union of two simple paths.

Solution. Start by breaking up the path into two simple closed paths, C_1 and C_2, as shown in Figure 3.65. Note that the orientation of C_1 is negative, while the orientation of C_2 is positive. The integral (3.7.6) becomes

$$\int_{C_1} \frac{z}{(z-i)(z-1)}\,dz + \int_{C_2} \frac{z}{(z-i)(z-1)}\,dz = I_1 + I_2. \tag{3.7.7}$$

You can verify that the partial fractions decomposition of the integrand is

$$\frac{z}{(z-i)(z-1)} = \frac{\frac{1}{2}-\frac{i}{2}}{z-i} + \frac{\frac{1}{2}+\frac{i}{2}}{z-1}.$$

Since $\frac{1}{z-1}$ is analytic inside and on C_1, its integral along C_1 is 0, by Theorem 3.6.7 (Remark 3.6.8). Also $\int_{C_1} \frac{dz}{z-i} = -2\pi i$, by Example 3.7.4. Hence

$$I_1 = \left(\frac{1}{2} - \frac{i}{2}\right) \int_{C_1} \frac{dz}{z-i} + \left(\frac{1}{2} + \frac{i}{2}\right) \int_{C_1} \frac{dz}{z-1} = \left(\frac{1}{2} - \frac{i}{2}\right)(-2\pi i) + 0 = -\pi - i\pi.$$

Arguing similarly to evaluate the integral along C_2, we find

$$I_2 = \left(\frac{1}{2} - \frac{i}{2}\right) \overbrace{\int_{C_2} \frac{dz}{z-i}}^{0} + \left(\frac{1}{2} + \frac{i}{2}\right) \overbrace{\int_{C_2} \frac{dz}{z-1}}^{2\pi i} = \left(\frac{1}{2} + \frac{i}{2}\right) 2\pi i = -\pi + i\pi.$$

Adding the two integrals, we find the answer $I_1 + I_2 = -2\pi$.   □

The idea of Example 3.7.7 can be used to evaluate integrals over closed paths that intersect themselves finitely many times.

Exercises 3.7

In Exercises 1–8, C is a simple, closed, and positively oriented path.

1. Evaluate

$$\int_C \frac{2i}{z-i} dz$$

in the following cases: (a) i is inside C; (b) i is outside C.

2. Evaluate

$$\int_C \left[\frac{z}{2+i} + \frac{4i}{z+i}\right] dz$$

in the following cases: (a) $-i$ is inside C; (b) $-i$ is outside C.

3. Evaluate

$$\int_C \frac{z}{z+1} dz$$

in the following cases: (a) -1 is inside C; (b) -1 is outside C.

4. Evaluate

$$\int_C \left[(3-2i)z^2 + \frac{2i}{z+1+i}\right] dz$$

in the following cases: (a) $-1-i$ is inside C; (b) $-1-i$ is outside C.

5. Evaluate

$$\int_C \left[\frac{2i}{z-2} - \frac{3+2i}{z+i}\right] dz$$

in the following cases:
(a) 2 is inside C and $-i$ is outside C;
(b) 2 and $-i$ are inside C;
(c) 2 and $-i$ are outside C.
Have we covered all possibilities? If not, evaluate the integral in the overlooked cases.

6. Evaluate

$$\int_C \frac{z-1}{z+1} dz$$

in the following cases: (a) -1 is inside C; (b) -1 is outside C.

7. Let C be a positively oriented closed simple path, let z_0 be a complex number in the interior of C, and let $p(z) = \sum_{j=1}^{n-1} a_j z^j$ be a polynomial of degree $n - 1$ for some integer $n \geq 1$. Show that

$$\int_C \frac{p(z - z_0)}{(z - z_0)^n} \, dz = 2\pi i a_{n-1}.$$

8. Evaluate

$$\int_C \left[\frac{i}{z - 1} + \frac{6}{(z - 1)^2} \right] dz$$

in the following cases: (a) 1 is inside C; (b) 1 is outside C.

In Exercises 9–12 evaluate the path integral over the curve γ parametrized by $\gamma(t) = 1 + i + 2e^{it}$, $0 \leq t \leq 2\pi$.

9. $\displaystyle \int_\gamma \frac{dz}{z - 1}$

10. $\displaystyle \int_\gamma \frac{dz}{(z - i)(z - 1)}$

11. $\displaystyle \int_\gamma \frac{dz}{(z - 3i)(z - 1)}$

12. $\displaystyle \int_\gamma \frac{dz}{z^2 + 9}$

In Exercises 13–32, evaluate the integrals. Indicate clearly how you are applying previously established results.

12. $\displaystyle \int_{[z_1, z_2, z_3, z_1]} \sin(z^2) \, dz$, where $z_1 = 0$, $z_2 = -i$, $z_3 = 1$.

13. $\displaystyle \int_{C_1(0)} \frac{e^z}{z + 2} \, dz$.

14. $\displaystyle \int_\gamma (z^2 + 2z + 3) \, dz$, where γ is an arbitrary path joining 0 to 1.

15. $\displaystyle \int_{C_1(i)} \left(\frac{z - 1}{z + 1} \right)^2 z \, dz$.

16. $\displaystyle \int_\gamma \frac{3i}{z - 2i} \, dz$, where $\gamma(t) = e^{it} + \frac{e^{2it}}{2}$, $0 \leq t \leq 2\pi$.

17. $\displaystyle \int_\gamma \frac{e^z}{z + i} \, dz$, where $\gamma(t) = i + e^{it}$, $0 \leq t \leq 2\pi$.

18. $\displaystyle \int_{C_1(0)} \frac{1}{z - \frac{1}{2}} \, dz$.

19. $\displaystyle \int_{C_1(0)} \frac{1}{(z - \frac{1}{2})^2} \, dz$.

20. $\displaystyle \int_{C_4(0)} \left((z - 2 + i)^2 + \frac{i}{z - 2 + i} - \frac{3}{(z - 2 + i)^2} \right) dz$.

21. $\displaystyle \int_{[z_1, z_2, z_3, z_1]} z^2 \operatorname{Log} z \, dz$, where $z_1 = 1$, $z_2 = 1 + i$, $z_3 = -1 + i$.

22. $\displaystyle \int_{C_1(i)} \frac{1}{(z - i)(z + i)} \, dz$.

23. $\displaystyle \int_{C_3(i)} \frac{1}{(z - i)(z + i)} \, dz$.

24. $\displaystyle \int_{C_{\frac{3}{2}}(1+i)} \frac{1}{(z - 1)(z - i)(z + i)} \, dz$.

25. $\displaystyle \int_{C_2(0)} \frac{z}{z^2 - 1} \, dz$.

26. $\displaystyle \int_{C_2(0)} \frac{1}{z^2 + 1} \, dz$.

27. $\displaystyle \int_{C_{\frac{3}{2}}(0)} \frac{z^2 + 1}{(z - 2)(z + 1)} \, dz$.

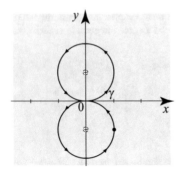

Fig. 3.66 Exercise 28.   **Fig. 3.67** Exercise 29.

28. $\int_\gamma \dfrac{z}{(z-i)(z+i)}\, dz$, where γ is the path that consists of the two circles in Figure 3.66.

29. $\int_\gamma \dfrac{1}{(z+1)^2(z^2+1)}\, dz$, where γ is the path that consists of the two circles in Figure 3.67.

30. $\int_\gamma \dfrac{1}{z+1}\, dz$, where γ is the path in Figure 3.68.

31. $\int_\gamma \dfrac{z+1}{z-i}\, dz$, where γ is the path in Figure 3.69.

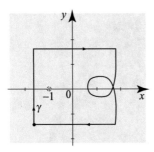

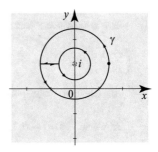

Fig. 3.68 Exercise 30.   **Fig. 3.69** Exercise 31.

32. Let $n \geq 1$ be an integer. (a) Verify the identity

$$\frac{z^n}{z-z_0} = z^{n-1} + z_0 \frac{z^{n-1}}{z-z_0}.$$

(b) Suppose that C is a simple, closed, positively oriented path, and z_0 a complex number inside C. Use induction and part (a) to show that

$$\frac{1}{2\pi i} \int_C \frac{z^n}{z-z_0}\, dz = z_0^n.$$

33. Let p be a polynomial with complex coefficients, C a simple, closed, positively oriented path, and z_0 be in the interior of C. Show that

$$\frac{1}{2\pi i}\int_C \frac{p(z)}{z-z_0}\,dz = p(z_0).$$

This is a special case of **Cauchy's integral formula** (Section 3.8). [*Hint*: Use Exercise 33.]

34. Suppose that C is a simple closed path with positive orientation, and α and β are complex numbers not on C. What are the possible values of

$$\int_C \frac{1}{(z-\alpha)(z-\beta)}\,dz?$$

$\left[\textit{Hint}: \text{Distinguish all possible locations of the points } \alpha \text{ and } \beta \text{ relative to the path } C.\right]$

35. (a) Let C be a simple, closed, positively oriented path. Show that

$$\int_C \bar{z}\,dz = 2iA,$$

where A is the area of the region inside C.

(b) Use (a) to evaluate $\displaystyle\int_C \operatorname{Re}(z)\,dz$ and $\displaystyle\int_C \operatorname{Im}(z)\,dz$.

36. (a) Let $z_1, z_2, \ldots, z_n$ be distinct complex numbers ($n \geq 2$). Show that in the partial fractions decomposition

$$\frac{1}{(z-z_1)(z-z_2)\cdots(z-z_n)} = \frac{A_1}{z-z_1} + \frac{A_2}{z-z_2} + \cdots + \frac{A_n}{z-z_n}$$

we must have $A_1 + A_2 + \cdots + A_n = 0$.

(b) Suppose that C is a simple closed path that contains the points $z_1, z_2, \ldots, z_n$ in its interior. Use the result in part (a) to prove that

$$\int_C \frac{1}{(z-z_1)(z-z_2)\cdots(z-z_n)}\,dz = 0.$$

37. Show that $\displaystyle\int_{C_2(0)} \frac{dz}{z^4+1} = 0$, where $C_2(0)$ is the positively oriented circle, with center at 0 and radius 2.

38. (a) Let $p(z) = a_n z^n + a_{n-1}z^{n-1} + \cdots + a_1 z + a_0$, $(a_n \neq 0)$ be a polynomial of degree $n \geq 2$. Show that, for all z with $|z| = R > 0$, we have

$$|p(z)| \geq |a_n|R^n - |a_{n-1}|R^{n-1} - \cdots - |a_1|R - |a_0|.$$

Conclude that for z on $C_R(0)$, the circle centered at 0 with radius R sufficiently large, we have

$$\frac{1}{|p(z)|} \leq \frac{1}{\left||a_n|R^n - |a_{n-1}|R^{n-1} - \cdots - |a_1|R - |a_0|\right|}.$$

(b) Use (a) to prove that

$$\left|\int_{C_R(0)} \frac{1}{p(z)}\,dz\right| \to 0, \quad \text{as} \quad R \to \infty.$$

(c) Let C be a simple closed path that contains all the roots of $p(z)$ in its interior. Show that

$$\int_C \frac{1}{p(z)}\,dz = 0.$$

[*Hint*: Take C to be positively oriented and choose R so large that $C_R(0)$ contains C in its interior. Explain why $\int_C \frac{1}{p(z)} dz = \int_{C_R(0)} \frac{1}{p(z)} dz$, where the right side is in fact independent of R. Then let $R \to \infty$ and use (b).]

3.8 Cauchy Integral Formula

In this section we establish the Cauchy integral formula and derive some applications related to the evaluation of integrals and properties of analytic functions.

We have managed to compute integrals of the form

$$\int_C \frac{1}{\zeta - z} d\zeta$$

where C is a simple closed curve and z is a point in the interior of the curve. In this section we discuss how to compute integrals of the form

$$\int_C \frac{e^{\zeta^2}}{\zeta - z} d\zeta$$

or similar integrals for other analytic functions in place of e^{ζ^2}.

Theorem 3.8.1. (Cauchy Integral Formula) *Suppose that f is an analytic function defined on a region U that contains a simple closed path C with positive orientation and its interior. If z is a point in the interior of C, then*

$$f(z) = \frac{1}{2\pi i} \int_C \frac{f(\zeta)}{\zeta - z} d\zeta. \tag{3.8.1}$$

Proof. Given a point z in the interior of C, pick $R > 0$ such that the closed disk $\overline{B_R(z)} = \{w : |w - z| \le R\}$ is contained inside C. The function

$$\zeta \mapsto \frac{f(\zeta)}{\zeta - z}$$

is analytic in the region inside C and outside the circle $C_r(z)$, where $0 < r \le R$ (see Figure 3.70). Applying Theorem 3.7.2 (Cauchy's theorem for multiply connected regions) with the variable of integration being ζ, we obtain

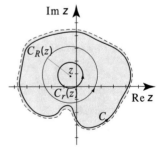

Fig. 3.70 Picture of the proof.

$$\frac{1}{2\pi i} \int_C \frac{f(\zeta)}{\zeta - z} d\zeta = \frac{1}{2\pi i} \int_{C_r(z)} \frac{f(\zeta)}{\zeta - z} d\zeta,$$

where C and $C_r(z)$ are positively oriented. Since this is true for all $0 < r \leq R$, the equality remains valid if we let $r \to 0$. So

$$\frac{1}{2\pi i} \int_C \frac{f(\zeta)}{\zeta - z} d\zeta = \lim_{r \to 0} \frac{1}{2\pi i} \int_{C_r(z)} \frac{f(\zeta)}{\zeta - z} d\zeta.$$

To finish the proof, we show that the limit on the right is $f(z)$. We need the *ML*-inequality (Theorem 3.2.19) and the following observations:

- Since z lies inside $C_r(z)$ and we are integrating with respect to ζ, by Example 3.7.4 we have

$$2\pi i = \int_{C_r(z)} \frac{1}{\zeta - z} d\zeta \;\Rightarrow\; f(z) = \frac{1}{2\pi i} \int_{C_r(z)} \frac{f(z)}{\zeta - z} d\zeta.$$

- The perimeter of $C_r(z)$ is $\ell(C_r(z)) = 2\pi r$, and since z is the center of $C_r(z)$, for all ζ on $C_r(z)$, we have $|\zeta - z| = r$.
- Since f is continuous at z, $M_r(z) = \max_{\zeta \in C_r(z)} |f(z) - f(\zeta)|$ tends to 0 as $r \to 0$.

We can now complete the proof:

$$\left| \frac{1}{2\pi i} \int_{C_r(z)} \frac{f(\zeta)}{\zeta - z} d\zeta - f(z) \right| = \left| \frac{1}{2\pi i} \int_{C_r(z)} \frac{f(\zeta) - f(z)}{\zeta - z} d\zeta \right|$$

$$\leq \frac{1}{2\pi} \ell(C_r(z)) M_r(z) \frac{1}{r}$$

$$= M_r(z) \to 0$$

as $r \to 0$, which establishes the desired limit. ∎

If z is a point outside C, then the function $\zeta \mapsto \frac{f(\zeta)}{\zeta - z}$ is analytic inside and on C, and so its integral along C is zero, by Cauchy's theorem. Let us combine this observation with (3.8.1) in one convenient formula in which the variable of integration is denoted z:

$$\frac{1}{2\pi i} \int_C \frac{f(z)}{z - z_0} dz = \begin{cases} f(z_0) & \text{if } z_0 \text{ is inside } C, \\ 0 & \text{if } z_0 \text{ is outside } C. \end{cases} \tag{3.8.2}$$

Example 3.8.2. (Cauchy integral formula) Let $C_R(z_0)$ denote the positively oriented circle with center at z_0 and radius $R > 0$. Compute the following integrals.

(a) $\displaystyle \int_{C_2(0)} \frac{e^z}{z+1} dz,$ (b) $\displaystyle \int_{C_2(1)} \frac{z^2 + 3z - 1}{(z+3)(z-2)} dz.$

Solution. (a) Write the integral as $\int_{C_2(0)} \frac{e^z}{z-(-1)} dz$. Since -1 is inside the circle $C_2(0)$, Cauchy's integral formula (3.8.2) with $f(z) = e^z$ and $z_0 = -1$ implies

$$\int_{C_2(0)} \frac{e^z}{z - (-1)} dz = 2\pi i e^{-1}.$$

(b) In evaluating $\int_{C_2(1)} \frac{z^2+3z-1}{(z+3)(z-2)} dz$, we first note that the integrand is not analytic at the points $z = -3$ and $z = 2$. Only the point $z = 2$ is inside the curve $C_2(1)$. So if we let $f(z) = \frac{z^2+3z-1}{z+3}$ the integral takes the form

$$\int_{C_2(1)} \frac{f(z)}{z-2} dz = 2\pi i f(2) = \frac{18\pi}{5}i,$$

by the Cauchy integral formula, applied at $z_0 = 2$.   ☐

Some integrals require multiple applications of Cauchy's formula along with applications of Cauchy's theorem. We illustrate this situation with an example.

Example 3.8.3. Compute

$$\int_{C_2(0)} \frac{e^{\pi z}}{z^2+1} dz.$$

Solution. Since

$$\frac{1}{z^2+1} = \frac{1}{(z+i)(z-i)},$$

the integral cannot be computed directly from Cauchy's formula, since the path contains both $\pm i$ in its interior. To overcome this difficulty, draw small nonintersecting circles inside $C_2(0)$ around $\pm i$, say $C_{1/4}(i)$ and $C_{1/4}(-i)$, as illustrated in Figure 3.71. Since $\frac{e^{\pi z}}{z^2+1}$ is analytic in a region containing the interior of $C_2(0)$ and the exterior of the smaller circles, by Cauchy's theorem for multiple connected domains (Theorem 3.7.2), we have

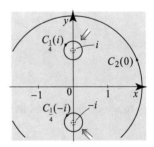

Fig. 3.71 Fig. 2 The integral over the outer path C is equal to the sum of the integrals over the inner non-overlapping circles.

$$\int_{C_2(0)} \frac{e^{\pi z}}{z^2+1} dz = \int_{C_{1/4}(i)} \frac{e^{\pi z}}{z^2+1} dz + \int_{C_{1/4}(-i)} \frac{e^{\pi z}}{z^2+1} dz.$$

Now, the two integrals on the right can be evaluated with the help of Cauchy's integral formula (3.8.2). For the first one, we apply Cauchy's formula (3.8.2) with $f(z) = \frac{e^{\pi z}}{z+i}$ and $z_0 = i$, and obtain

$$\int_{C_{1/4}(i)} \frac{e^{\pi z}}{(z-i)(z+i)} dz = \int_{C_{1/4}(i)} \frac{f(z)}{z-i} dz = 2\pi i f(i).$$

Since $f(i) = \frac{e^{i\pi}}{2i} = \frac{-1}{2i} = \frac{i}{2}$, we get

$$\int_{C_{1/4}(i)} \frac{f(z)}{z-i}\, dz = -\pi.$$

For the second integral, we have

$$\int_{C_{1/4}(-i)} \frac{e^{\pi z}}{(z-i)(z+i)}\, dz = \int_{C_{1/4}(-i)} \frac{g(z)}{z+i}\, dz = 2\pi i g(-i),$$

where $g(z) = \frac{e^{\pi z}}{z-i}$, and so $g(-i) = \frac{e^{-i\pi}}{-2i} = \frac{1}{2i} = -\frac{i}{2}$. Hence

$$\int_{C_{1/4}(-i)} \frac{g(z)}{z-i}\, dz = \pi.$$

Adding the two integrals together, we find that

$$\int_{C_2(0)} \frac{e^{\pi z}}{(z-i)(z+i)}\, dz = 0. \qquad \square$$

Cauchy's integral formula (3.8.1) shows that the values of $f(z)$, for z inside the path C, are determined by the values of f on the curve C, and the way to recapture the values inside C is to integrate $f(\zeta)$ against the function $1/[2\pi i(\zeta - z)]$ on C. Something analogous is valid for the derivatives of f. To achieve this we need to know how to differentiate under the integral sign.

Differentiation Under the Integral Sign

We focus on the analyticity (and also continuity) of a function of the form

$$g(z) = \int_C \phi(z, \zeta)\, d\zeta,$$

ss where ζ lies on a simple closed curve C and z lies in some open set. For instance in Theorem 3.8.1 we had $\phi(z, \zeta) = \frac{1}{2\pi i} \frac{f(\zeta)}{\zeta - z}$. We begin with a lemma.

Lemma 3.8.4. *Suppose that f is analytic on an open set containing the closed disk $\overline{B_R(z_0)}$ and satisfies $|f(z)| \le M$ for all $z \in \overline{B_R(z_0)}$. Then for $0 < |z - z_0| < \frac{R}{2}$ the following are valid*

$$\left| \frac{f(z) - f(z_0)}{z - z_0} - f'(z_0) \right| \le 2M \frac{|z - z_0|}{R^2} \tag{3.8.3}$$

and

$$f'(z_0) = \frac{1}{2\pi i} \int_{C_R(z_0)} \frac{f(\zeta)}{(\zeta - z_0)^2}\, d\zeta. \tag{3.8.4}$$

Proof. Using (3.8.1) for $0 < |z - z_0| < \frac{R}{2}$ we write

$$f(z) = \frac{1}{2\pi i} \int_{C_R(z_0)} \frac{f(\zeta)}{\zeta - z} \, d\zeta \quad \text{and} \quad f(z_0) = \frac{1}{2\pi i} \int_{C_R(z_0)} \frac{f(\zeta)}{\zeta - z_0} \, d\zeta.$$

Combining these integrals and simplifying, we obtain

$$\frac{f(z) - f(z_0)}{z - z_0} = \frac{1}{2\pi i} \int_{C_R(z_0)} \frac{f(\zeta)}{(\zeta - z)(\zeta - z_0)} \, d\zeta. \tag{3.8.5}$$

For ζ on $C_R(z_0)$ and $0 < |z - z_0| < \frac{R}{2}$ we have $|\zeta - z| > \frac{R}{2}$, and so $\frac{1}{|\zeta - z|} < \frac{2}{R}$. Thus

$$\left| \frac{f(\zeta)}{(\zeta - z)(\zeta - z_0)} - \frac{f(\zeta)}{(\zeta - z_0)^2} \right| = \left| \frac{f(\zeta)(z - z_0)}{(\zeta - z)(\zeta - z_0)^2} \right| \leq M \frac{2}{R} \frac{|z - z_0|}{R^2} = 2M \frac{|z - z_0|}{R^3}.$$

It follows from the *ML*-inequality that

$$\left| \frac{1}{2\pi i} \int_{C_R(z_0)} \left[\frac{f(\zeta)}{(\zeta - z)(\zeta - z_0)} - \frac{f(\zeta)}{(\zeta - z_0)^2} \right] d\zeta \right| \leq \frac{2\pi R}{|2\pi i|} \frac{2M|z - z_0|}{R^3} = 2M \frac{|z - z_0|}{R^2}.$$

Separating the integrals and using identity (3.8.5) we obtain

$$\left| \frac{f(z) - f(z_0)}{z - z_0} - \frac{1}{2\pi i} \int_{C_R(z_0)} \frac{f(\zeta)}{(\zeta - z_0)^2} \, d\zeta \right| \leq 2M \frac{|z - z_0|}{R^2}. \tag{3.8.6}$$

Letting $z \to z_0$ in (3.8.6) we deduce (3.8.4). Combining (3.8.4) and (3.8.6) yields (3.8.3). ∎

Theorem 3.8.5. *Let C be a path and let U be an open set. Let $\phi(z, \zeta)$ be a function defined for $z \in U$ and $\zeta \in C$. Suppose that $\phi(z, \zeta)$ is continuous in $\zeta \in C$ and analytic in $z \in U$ and that the complex derivative $\frac{d\phi}{dz}(z, \zeta)$ is continuous in $\zeta \in C$. Then the function*

$$g(z) = \int_C \phi(z, \zeta) \, d\zeta \tag{3.8.7}$$

is analytic in U and its derivative is

$$g'(z) = \int_C \frac{d\phi}{dz}(z, \zeta) \, d\zeta. \tag{3.8.8}$$

Proof. For a fxied $z_0 \in U$ choose $R > 0$ so that $\overline{B_R(z_0)}$ is contained in U. Let

$$M = \max_{z \in B_R(z_0), \zeta \in C} |\phi(z, \zeta)|.$$

Since continuous functions attain a maximum on compact (i.e., closed and bounded) sets, we have $M < \infty$. For each ζ, by assumption $\phi(z, \zeta)$ is an analytic function of z in U. For z such that $0 < |z - z_0| < \frac{R}{2}$ and $\zeta \in C$, Lemma 3.8.4 implies

$$\left| \frac{\phi(z,\zeta) - \phi(z_0,\zeta)}{z - z_0} - \frac{d\phi}{dz}(z_0,\zeta) \right| \leq 2M \frac{|z - z_0|}{R^2}. \tag{3.8.9}$$

Integrating the expression inside the absolute value on the left in (3.8.9) and using (3.8.7) we find

$$\left| \frac{g(z) - g(z_0)}{z - z_0} - \int_C \frac{d\phi}{dz}(z_0,\zeta)d\zeta \right| = \left| \frac{\int_C [\phi(z,\zeta) - \phi(z_0,\zeta)]d\zeta}{z - z_0} - \int_C \frac{d\phi}{dz}(z_0,\zeta)d\zeta \right|$$

$$= \left| \int_C \left[\frac{\phi(z,\zeta) - \phi(z_0,\zeta)}{z - z_0} - \frac{d}{dz}\phi(z_0,\zeta) \right] d\zeta \right|$$

$$\leq \ell(C) 2M \frac{|z - z_0|}{R^2},$$

where we have used the *ML*-inequality and (3.8.9). As $z \to z_0$, the difference quotient on the left side of the inequality approaches $g'(z_0)$, while the right side of the inequality tends to 0. This proves (3.8.8). ∎

Generalized Cauchy Integral Formula

We now use Theorem 3.8.5 to deduce the generalized Cauchy integral formula.

Theorem 3.8.6. (Generalized Cauchy Integral Formula) *Suppose that f is analytic on a region Ω that contains a positively oriented simple closed path C and its interior. Then f has derivatives of any order at all points z in the interior of C given by*

$$f^{(n)}(z) = \frac{n!}{2\pi i} \int_C \frac{f(\zeta)}{(\zeta - z)^{n+1}} d\zeta. \tag{3.8.10}$$

Proof. When $n = 0$, $f^{(0)} = f$ and $0! = 1$; in this case identity was proved in Theorem 3.8.1. Let U be the interior of C. Assuming by induction that (3.8.10) holds for a natural number n, for $z \in U$ and $\zeta \in C$ define

$$\phi(z,\zeta) = \frac{n!}{2\pi i} \frac{f(\zeta)}{(\zeta - z)^{n+1}},$$

and note that $\phi(z,\zeta)$ is analytic in z in U and continuous in $\zeta \in C$. Moreover, for $z \in U$ we have

$$\frac{d\phi}{dz}(z,\zeta) = \frac{(n+1)!}{2\pi i} \frac{f(\zeta)}{(\zeta - z)^{n+2}},$$

which is continuous in $\zeta \in C$. Applying Theorem 3.8.5 we obtain that $f^{(n+1)}$ is analytic in U and

$$f^{(n+1)}(z) = \frac{(n+1)!}{2\pi i} \int_C \frac{f(\zeta)}{(\zeta - z)^{n+2}} d\zeta$$

for all $z \in U$. This completes the induction and the proof. ∎

Example 3.8.7. (Generalized Cauchy integral formula) Let γ be the ellipse in Figure 3.72. Compute the following integrals:

(a) $\displaystyle \frac{1}{2\pi i} \int_{C_2(0)} \frac{z^{10}}{(z-1)^{11}}\, dz$, (b) $\displaystyle \int_{\gamma} \frac{e^{iz}}{(z-\pi)^3}\, dz$.

Solution. (a) By (3.8.10), we have

$$\frac{10!}{2\pi i} \int_{C_2(0)} \frac{z^{10}}{(z-1)^{11}}\, dz = \frac{d^{10}}{dz^{10}} z^{10} \bigg|_{z=1} = 10!.$$

Dividing by 10!, we find the desired integral to be 1.

(b) By (3.8.10), we have

$$\frac{2!}{2\pi i} \int_{\gamma} \frac{e^{iz}}{(z-\pi)^3}\, dz = \frac{d^2}{dz^2} e^{iz} \bigg|_{z=\pi} = -e^{i\pi} = 1.$$

Hence the desired integral is πi. ☐

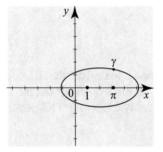

Fig. 3.72 Ellipse with foci at $1, \pi$.

In the following example, one should pay attention to the orientation of the paths as we decompose a figure-eight path into two simple paths.

Example 3.8.8. (A path that intersects itself) Let Γ be as in Figure 3.73 Compute

$$\int_{\Gamma} \frac{z}{(z-i)(z^2+1)}\, dz.$$

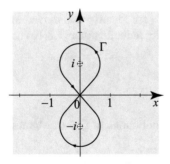

Fig. 3.73 The figure-eight path Γ is not a simple path.

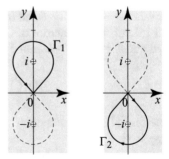

Fig. 3.74 Decomposition of Γ into two simple paths.

Solution. Since the path Γ intersects itself, it is not simple. As we cannot directly appeal to Cauchy's formulas, we decompose Γ into two simple paths Γ_1 and Γ_2, as shown in Figure 3.74. Noting that $(z-i)(z^2+1) = (z-i)^2(z+i)$, we have

$$\int_\Gamma \frac{z}{(z-i)(z^2+1)}\,dz = \int_{\Gamma_1} \frac{z}{(z-i)^2(z+i)}\,dz + \int_{\Gamma_2} \frac{z}{(z-i)^2(z+i)}\,dz.$$

The integrals on the right can now be evaluated with the help of Cauchy's generalized integral formula (3.8.10). We must be careful with the orientation of the paths: The orientation of Γ_1 is positive, while the orientation of Γ_2 is negative. On Γ_1, we apply (3.8.10) with $n = 1$, $f(z) = \frac{z}{z+i}$ at $z = i$, and get

$$\int_{\Gamma_1} \frac{z}{z+i}\frac{dz}{(z-i)^2} = 2\pi i\frac{d}{dz}\frac{z}{z+i}\Big|_{z=i} = 2\pi i\frac{i}{(z+i)^2}\Big|_{z=i} = \frac{\pi}{2}.$$

On Γ_2, apply (3.8.10) with $n = 0$, $f(z) = \frac{z}{(z-i)^2}$ at $z = -i$, and recall that the orientation of Γ_2 is negative. Then

$$\int_{\Gamma_2} \frac{z}{(z-i)^2}\frac{dz}{(z+i)} = -2\pi i\frac{z}{(z-i)^2}\Big|_{z=-i} = \frac{\pi}{2}.$$

Thus the desired integral is π, the sum of the integrals along Γ_1 and Γ_2. $\qquad\square$

We end the section with some important theoretical applications of Cauchy's formula. The first result is already contained in Theorem 3.8.6 but deserves a separate statement because of its importance.

Corollary 3.8.9. *Suppose that f is analytic in an open set Ω. Then f has derivatives of all orders $f', f'', f^{(3)}, f^{(4)}, \ldots$ which are analytic functions in Ω.*

Proof. For z in Ω, let $\overline{B_R(z)}$ be a closed disk contained in Ω. Pick C to be the boundary of this closed disk. By Theorem 3.8.6, all derivatives of f exist in $\Omega \setminus C$, and in particular at the given point z. These derivatives are shown in (3.8.10). $\blacksquare$

Consider the function $h(t) = t^{\frac{5}{3}}$, $-\infty < t < \infty$. Its derivative, $f'(t) = \frac{5}{3}t^{\frac{2}{3}}$, exists and is continuous for all real t; however, $f''(t)$ does not exist at $t = 0$. Thus Corollary 3.8.9 has no analog in the theory of functions of a real variable.

The following is a converse to Cauchy's theorem. It is named after the Italian mathematician Giacinto Morera (1856–1907).

Theorem 3.8.10. (Morera's Theorem) *Let f be a continuous complex-valued function on a region Ω. Suppose that for all closed disks $\overline{B}$ contained in Ω and for all closed paths γ in $\overline{B}$ we have*

$$\int_\gamma f(z)\,dz = 0. \tag{3.8.11}$$

Then the function f is analytic on Ω.

Proof. For every $z \in \Omega$ there is a disk $\overline{B_r(z)}$ contained in Ω. Since (3.8.11) holds for all closed paths γ contained in $B_r(z)$, applying Theorem 3.3.4, we obtain that there is an analytic function F on $B_r(z)$ such that $F' = f$. By Corollary 3.8.9, f is analytic in $B_r(z)$. Since z is arbitrary in Ω, it follows that f is analytic in Ω. $\blacksquare$

Exercises 3.8

In Exercises 1–20, evaluate the integrals. State clearly which theorem you are using and justify its application. Plot the path in each problem and describe exactly the points of interest. As usual, $C_R(z_0)$ denotes the positively oriented circle centered at z_0 with radius $R > 0$.

1. $\displaystyle\int_{C_1(0)} \frac{\cos z}{z} \, dz$

2. $\displaystyle\int_{C_3(0)} \frac{e^{z^2} \cos z}{z - i} \, dz$

3. $\displaystyle\frac{1}{2\pi i} \int_{C_2(1)} \frac{1}{z^2 - 5z + 4} \, dz$

4. $\displaystyle\frac{1}{2\pi i} \int_{C_3(1)} \frac{\cos z}{(z - \pi)^4} \, dz$

5. $\displaystyle\int_{C_{\frac{1}{2}}(i)} \frac{\operatorname{Log} z}{-z + i} \, dz$

6. $\displaystyle\frac{1}{2\pi i} \int_{C_2(1)} \frac{z^5 - 1}{(z + 3i)(z - 2)} \, dz$

7. $\displaystyle\int_{[z_1, z_2, z_3, z_1]} \frac{z^{19}}{(z - 1)^{19}} \, dz$, where $z_1 = 0$, $z_2 = -i$, $z_3 = 3 + i$.

8. $\displaystyle\int_{[z_1, z_2, z_3, z_1]} \frac{z^{19}}{(z - 1)^{20}} \, dz$, where $z_1 = 0$, $z_2 = -i$, $z_3 = 3 + i$.

9. $\displaystyle\int_\gamma \frac{\sin z}{(z - \pi)^3} \, dz$, where γ is as in Figure 3.75.

10. $\displaystyle\int_\gamma \frac{\sin z}{(z^2 - \pi^2)^2} \, dz$, where γ is as in Figure 3.75.

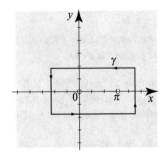

Fig. 3.75 Exercises 9, 10, 11.

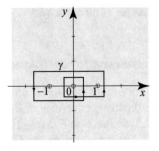

Fig. 3.76 Exercise 12.

11. $\displaystyle\int_\gamma \frac{e^z \sin z}{z^2(z - \pi)} \, dz$, where γ is as in Figure 3.75.

12. $\displaystyle\int_\gamma \frac{dz}{z^2(z - 1)^3(z + 3)}$, where γ is as in Figure 3.76.

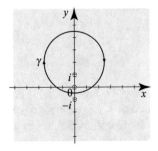

Fig. 3.77 Exercise 13.

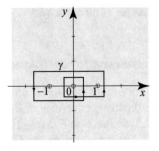

Fig. 3.78 Exercise 14.

13. $\displaystyle\int_\gamma \frac{z+\cos(\pi z)}{z(z^2+1)}\,dz$, where γ is the negatively oriented path of Figure 3.77.

14. $\displaystyle\int_\gamma \frac{1}{z(z-1)^2(z^2-1)}\,dz$, where γ is as in Figure 3.78.

15. $\displaystyle\int_{C_2(0)} \frac{z^2+z+1}{z^2-1}\,dz.$
16. $\displaystyle\frac{1}{2\pi i}\int_{C_2(1)} \frac{1}{z^2-z}\,dz.$

17. $\displaystyle\int_{C_{\frac32}(0)} \frac{1}{z^3-3z+2}\,dz.$
18. $\displaystyle\frac{1}{2\pi i}\int_{C_{\frac52}(1)} \frac{1}{z^3+2z^2-z-2}\,dz.$

19. $\displaystyle\int_{C_{\frac32}(1)} \frac{1}{z^4-1}\,dz.$
20. $\displaystyle\int_{C_2(0)} \frac{1}{z^4-1}\,dz.$

21. For $|z|<1$, let $F(z)=\dfrac{1}{2\pi}\displaystyle\int_0^{2\pi} \frac{e^{it}}{e^{it}-z}\,dt$. Show that $F(z)=1$ for all $|z|<1$. [*Hint:* Express the integral as a path integral.]

22. Compute $\dfrac{1}{2\pi}\displaystyle\int_0^{2\pi} \frac{1}{2+e^{it}}\,dt$. [*Hint:* See the hint in Exercise 21.]

23. Show that $\dfrac{1}{2\pi}\displaystyle\int_0^{2\pi} e^{e^{int}}\,dt = 1$ for $n=\pm1,\pm2,\dots.$

24. Show that $\dfrac{1}{2\pi}\displaystyle\int_0^{2\pi}\cos(e^{it})\,dt=1$ and $\dfrac{1}{2\pi}\displaystyle\int_0^{2\pi}\sin(e^{it})\,dt=0.$

25. Explain why the function $z\mapsto\displaystyle\int_0^1\cos(zt)\,dt$ is entire and find an explicit formula for it.

26. Explain why the function $z\mapsto\displaystyle\int_0^1 e^{z^2t}\,dt$ is entire and find an explicit formula for it.

27. Write the functions as integrals of the form (3.8.7) for appropriate $\phi(z,\zeta)$. Then use Theorem 3.8.5 to conclude that these functions are analytic at 0, when defined appropriately there.
(a) $\dfrac{1-e^z}{z}$ and (b) $\dfrac{\cos z-1}{z^2}.$

28. (a) Compute $\dfrac{1}{2\pi i}\displaystyle\int_{C_1(0)} \frac{e^z}{z}\,dz.$
(b) Use your answer in (a) to show that $\displaystyle\int_0^\pi e^{\cos t}\cos(\sin t)\,dt=\pi.$

29. Let f and g be analytic inside and on a simple closed nonconstant path C. Suppose that $f=g$ on C. Show that $f=g$ inside C.

30. Suppose that f is analytic inside and on $C_1(0)$. For $|z|<1$, show that

$$\int_{C_1(0)} \frac{f(\zeta)}{\zeta-\frac{1}{z}}\,d\zeta = 0.$$

31. (a) Suppose that f is analytic inside and on $C_R(0)$ for some $R>0$. For $|z|<R$, show that

$$\frac{1}{2\pi i}\int_{C_R(0)} \frac{f(\zeta)}{(\zeta-z)\zeta}\,d\zeta = \begin{cases} f'(0) & \text{if } z=0, \\ \frac{f(z)-f(0)}{z} & \text{if } z\neq 0. \end{cases}$$

(b) Let ϕ be the function defined by the formula on the right side of the equality in part (a). Show that ϕ is analytic for $|z|<R$. [*Hint:* Use Theorem 3.8.5.]

32. (a) From the previous exercise, show that if f is analytic at 0 and $f(0) = 0$, then the function ϕ defined by $\phi(z) = \frac{f(z)}{z}$ if $z \neq 0$ and $\phi(0) = f'(0)$ is analytic at 0.

(b) Show that the function ϕ defined by $\phi(z) = \frac{\sin z}{z}$ if $z \neq 0$ and $\phi(0) = 1$ is entire.

33. Based on the previous exercise, show that if f is analytic at $z = z_0$ then the function ϕ defined by $\phi(z) = \frac{f(z) - f(z_0)}{z - z_0}$ if $z \neq z_0$ and $\phi(z_0) = f'(z_0)$ is analytic at z_0.

34. Project Problem: Wallis' formulas. (a) If n is an integer recall the useful identity

$$\frac{1}{2\pi i} \int_{C_1(0)} \frac{1}{z^n} \, dz = \begin{cases} 1 & \text{if } n = 1, \\ 0 & \text{if } n \neq 1. \end{cases}$$

(b) Parametrize the circle $C_1(0)$ to show that

$$\frac{1}{2\pi i} \int_{C_1(0)} \left(z + \frac{1}{z} \right)^n \frac{dz}{z} = \frac{2^n}{2\pi} \int_0^{2\pi} \cos^n t \, dt.$$

(c) Expand $\left(z + \frac{1}{z} \right)^n$ using the binomial formula and use part (a) to prove that

$$\frac{1}{2\pi} \int_0^{2\pi} \cos^{2k} t \, dt = \frac{(2k)!}{2^{2k}(k!)^2} \quad \text{and} \quad \int_0^{2\pi} \cos^{2k+1} t \, dt = 0,$$

where $k = 0, 1, 2, \ldots$. These are some of **Wallis' formulas**.

(d) Show that for $k = 0, 1, 2, \ldots$ we have

$$\frac{1}{2\pi} \int_0^{2\pi} \sin^{2k} t \, dt = \frac{(2k)!}{2^{2k}(k!)^2} \quad \text{and} \quad \int_0^{2\pi} \sin^{2k+1} t \, dt = 0.$$

35. Generalized Cauchy formula for multiple simple paths. Under the hypothesis of Theorem 3.3.4, show that for all $n = 1, 2, \ldots$ and all points z interior to C and exterior to all C_j we have

$$f^{(n)}(z) = \frac{n!}{2\pi i} \int_C \frac{f(\zeta)}{(\zeta - z)^{n+1}} \, d\zeta - \sum_{j=1}^n \frac{n!}{2\pi i} \int_{C_j} \frac{f(\zeta)}{(\zeta - z)^{n+1}} \, d\zeta. \qquad (3.8.12)$$

3.9 Bounds for Moduli of Analytic Functions

In this section, we present several fundamental results about analytic functions, including the famous Liouville theorem, and the maximum modulus principle. We also give a simple proof of the fundamental theorem of algebra.

Theorem 3.9.1. (Cauchy Estimates) *Suppose that f is analytic on an open disk $B_R(z_0)$ with center at z_0 and radius $R > 0$. Suppose that $|f(z)| \leq M$ for all z in $B_R(z_0)$. Then*

$$\left| f^{(n)}(z_0) \right| \leq M \frac{n!}{R^n}, \qquad n = 1, 2, \ldots. \qquad (3.9.1)$$

Proof. Since we are not assuming that f is analytic on a domain that contains the

circle $C_R(z_0)$, we fix $0 < r < R$ and work on
a disk of radius r on which f is analytic (see
Figure 3.79). Applying Theorem 3.8.6, for any
$n \in \mathbb{N}$, we write

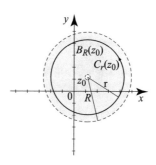

$$f^{(n)}(z_0) = \frac{n!}{2\pi i} \int_{C_r(z_0)} \frac{f(\zeta)}{(\zeta - z_0)^{n+1}} \, d\zeta. \quad (3.9.2)$$

For ζ on $C_r(z_0)$, we have

$$|\zeta - z_0| = r$$

and so it follows that

Fig. 3.79 Picture of the proof.

$$\left| \frac{f(\zeta)}{(\zeta - z_0)^{n+1}} \right| = \frac{|f(\zeta)|}{|\zeta - z_0|^{n+1}} = \frac{|f(\zeta)|}{r^{n+1}} \leq \frac{M}{r^{n+1}}. \quad (3.9.3)$$

Applying the *ML*-inequality [i.e., estimate (3.2.31)] to the integral on the right side
of (3.9.2) and using (3.9.3) we find that

$$\left| f^{(n)}(z_0) \right| \leq \frac{n!}{2\pi} \frac{M}{r^{n+1}} \ell(C_r(z_0)) = M \frac{n!}{r^n}.$$

Since this holds for all $0 < r < R$, letting $r \to R$ we deduce (3.9.1). ∎

The following surprising application of Cauchy's estimate was proved by the
French Mathematician Joseph Liouville (1809–1882). Recall that a function is
called entire if it is analytic on all of $\mathbb{C}$.

Theorem 3.9.2. (Liouville's Theorem) *A bounded entire function must be constant.*

Proof. Let f be an entire function satisfying $|f(z)| \leq M$ for all $z \in \mathbb{C}$. Applying
Cauchy's estimate to f' on a disk of radius $R > 0$ around a point $z_0 \in \mathbb{C}$, we obtain
$|f'(z_0)| \leq \frac{M}{R}$. Letting $R \to \infty$, we obtain that $f'(z_0) = 0$. Since z_0 is arbitrary, it
follows that $f'(z) = 0$ for all z, and hence, f is constant as a consequence of Theorem 2.5.7. ∎

Here is one useful application of Liouville's theorem.

Corollary 3.9.3. *If f is entire and $|f(z)| \to \infty$ as $|z| \to \infty$, then f must have at least
one zero.*

Proof. Suppose to the contrary that f has no zeros in $\mathbb{C}$. Then $g = 1/f$ is also entire
and $|g(z)| \to 0$ as $|z| \to \infty$. This property implies that g is bounded on $\mathbb{C}$ (Exercise
14). By Theorem 3.9.2, g is constant and consequently f is constant. ∎

The exponential function e^z is entire and never equals to 0. As a consequence of
Corollary 3.9.3, we deduce that $|e^z|$ does not tend to infinity as $|z| \to \infty$, although
$e^{|z|}$ tends to infinity as $|z| \to \infty$.

We now turn our attention to the fundamental theorem of algebra, first proved by Gauss in 1799 (with a topological gap) and then again by him in 1816 in two other ways different from the first one.

Theorem 3.9.4. (Fundamental Theorem of Algebra) *Every polynomial of degree $n \geq 1$ has exactly n zeros counted according to multiplicity.*

Proof. It is enough to show that every polynomial p of degree $n \geq 1$ has at least one zero. For, if we know that z_0 is a zero of p, then we write $p(z) = (z - z_0)q(z)$, where q is a polynomial of degree $n - 1$ (Exercise 13). We continue factoring until we have written p as the product of n linear terms times a constant, which shows that p has exactly n roots. So let us show that p has at least one zero. If $a_n \neq 0$, write

$$p(z) = a_n z^n + a_{n-1} z^{n-1} + \cdots + a_1 z + a_0 = z^n \left(a_n + \frac{a_{n-1}}{z} + \cdots + \frac{a_1}{z^{n-1}} + \frac{a_0}{z^n} \right).$$

As $z \to \infty$, the quantity inside the parentheses approaches a_n, while $|z^n| \to \infty$, and hence $|p(z)| \to \infty$. We complete the proof by applying Corollary 3.9.3. ∎

Maximum and Minimum Principles

Suppose that f is analytic in a region Ω and let $\overline{B_R(z)}$ be a closed disk contained in Ω. Parametrize the circle $C_R(z)$ by $\zeta(t) = z + Re^{it}$, $0 \leq t \leq 2\pi$, $d\zeta = Rie^{it} dt$. The Cauchy integral formula implies that

$$f(z) = \frac{1}{2\pi i} \int_{C_R(z)} \frac{f(\zeta)}{\zeta - z} d\zeta = \frac{1}{2\pi} \int_0^{2\pi} \frac{f(z + Re^{it})}{Re^{it}} Re^{it} dt,$$

and after simplifying this yields

$$f(z) = \frac{1}{2\pi} \int_0^{2\pi} f(z + Re^{it}) dt. \tag{3.9.4}$$

The integral on the right is a Riemann integral of a complex-valued function of t. Recalling that the integral is an average, this formula shows that the value of an analytic function at a point z in Ω is equal to the average value of f over a circle centered at z and contained in Ω. This important property is expressed by saying that an analytic function f has the **mean value property**. If we take absolute values on both sides of (3.9.4), we obtain

$$|f(z)| = \frac{1}{2\pi} \left| \int_0^{2\pi} f(z + Re^{it}) dt \right| \leq \frac{1}{2\pi} \int_0^{2\pi} |f(z + Re^{it})| dt, \tag{3.9.5}$$

expressing that the absolute value of an analytic function has the **sub-mean value property** on Ω.

The following lemma about real functions states an obvious fact: If the values of a function are less than or equal to some constant M and if the average of the function is equal to M, then the function must be identically equal to M.

Lemma 3.9.5. (i) *Suppose that h is a continuous real-valued function such that $h(t) \geq 0$ for all t in $[a, b]$ $(a < b)$. If*

$$\int_a^b h(t)\, dt = 0,$$

then $h(t) = 0$ for all t in $[a, b]$.
(ii) *Suppose that h is a continuous real-valued function such that $h(t) \leq M$ (alternatively, $h(t) \geq M$) for all t in $[a, b]$. If*

$$\frac{1}{b-a} \int_a^b h(t)\, dt = M,$$

then $h(t) = M$ for all t in $[a, b]$.

Proof. (i) The proof is by contradiction. Assume that $h(t_0) = \delta > 0$ for some t_0 in (a, b). Since h is continuous, we can find an interval (c, d) that contains t_0 with $(c, d) \subset (a, b)$ and $h(t) > \delta/2$ for all t in (c, d). Since the integral of a nonnegative function increases if we increase the interval of integration, we conclude that

$$\int_a^b h(t)\, dt \geq \int_c^d h(t)\, dt \geq \frac{\delta}{2}(d - c) > 0,$$

which is a contradiction. Hence $h(t) = 0$ for all t in (a, b) and, by continuity of h, h is also 0 at the endpoints a and b.
(ii) We consider the case $h \leq M$ only as the other case is similar. Let $k = M - h$. Then k is a continuous and nonnegative function. Note that

$$\int_a^b k(t)\, dt = 0 \Leftrightarrow \frac{1}{b-a} \int_a^b (M - h(t))\, dt = 0 \Leftrightarrow \frac{1}{b-a} \int_a^b h(t)\, dt = M.$$

As the last equation holds, k must have vanishing integral, and from part (i) it follows that k is identically equal to zero; thus, h is identically equal to M. ∎

Theorem 3.9.6. (Maximum Modulus Principle) *Suppose that f is analytic on a region Ω. If $|f|$ attains a maximum in Ω, then f is constant in Ω.*

Proof. The connectedness property of Ω is crucial in the proof. Suppose that $|f|$ attains a maximum in Ω. If we show that $|f|$ is constant, then by Exercise 36, Section 2.5, it will follow that f is constant. Let $M = \max_{z \in \Omega} |f(z)|$,

$$\Omega_0 = \{z \in \Omega : |f(z)| < M\}$$
$$\Omega_1 = \{z \in \Omega : |f(z)| = M\}.$$

Clearly, $\Omega = \Omega_0 \cup \Omega_1$, and Ω_0 and Ω_1 are disjoint, and Ω_1 is nonempty because $|f|$ is assumed to attain its maximum in Ω. The set Ω_0 is open because $|f|$ is continuous (Exercise 41, Section 2.2). If we show that Ω_1 is also open, then, as Ω is open and connected, it cannot be written as the union of two disjoint open nonempty sets (Proposition 2.1.7). This will force Ω_0 to be empty. Consequently, $\Omega = \Omega_1$, implying that $|f| = M$ is constant in Ω.

So let us prove that Ω_1 is open. Pick z in Ω_1. Since Ω is open, we can find an open disk $B_\delta(z)$ in Ω, centered at z with radius $\delta > 0$. We will show that $B_\delta(z)$ is contained in Ω_1. This will imply that Ω_1 is open. Let $0 < r < \delta$ as shown in Figure 3.80. Using (3.9.5) and the fact that $|f(z)| = M$, we obtain

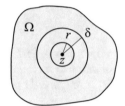

Fig. 3.80 We have that $B_r(z) \subset B_\delta(z) \subset \Omega$.

$$M = |f(z)| \le \frac{1}{2\pi} \int_0^{2\pi} \overbrace{\left|f(z+re^{it})\right|}^{\le M} dt \le \frac{1}{2\pi} \int_0^{2\pi} M\, dt = M.$$

Hence

$$\frac{1}{2\pi} \int_0^{2\pi} \left|f(z+re^{it})\right| dt = M,$$

and Lemma 3.9.5(*ii*) implies that

$$\left|f(z+re^{it})\right| = M$$

for all t in $[0, 2\pi]$. This shows that $C_r(z)$, the circle of radius r and center at z, is contained in Ω_1. But this is true for all r satisfying $0 < r < \delta$, and this implies that $B_\delta(z)$ is contained in Ω_1. ∎

Suppose that f is analytic in Ω and continuous on the boundary of Ω. By Theorem 3.9.6, $|f|$ cannot attain its maximum inside Ω unless f is constant. This leads us to the following two questions.

- Does $|f|$ attain its maximum on the boundary of Ω?
- If $|f(z)| \le M$ on the boundary of Ω, can we infer that $|f(z)| \le M$ for all z in Ω?

The next example shows that in general the answers to both questions are negative.

Example 3.9.7. (Failure of the maximum principle on unbounded regions)
Let $\Omega = \{z : \operatorname{Re} z > 0, \ \operatorname{Im} z > 0\}$ be the first quadrant, bounded by the semi-infinite nonnegative x- and y-axes.

Let $f(z) = e^{-iz^2}$ which can also be written as $f(x + iy) = e^{-i(x^2-y^2+2ixy)} = e^{-i(x^2-y^2)}e^{2xy}$ if $z = x + iy$. We have

$$|f(z)| = \left|e^{i(x^2-y^2)}e^{2xy}\right| = e^{2xy}.$$

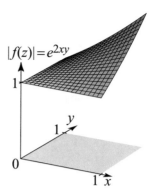

On the boundary, we have $x = 0$ or $y = 0$ and so $|f(z)| = 1$; however, it is clear that $|f(z)| = e^{2xy}$ is not bounded in Ω. To see this, take $x = y$ and let $x, y \to \infty$; then $|f(x + iy)| = e^{2x^2} \to \infty$ as shown in Figure 3.81.   □

Fig. 3.81 The maximum of $|f|$ in Example 3.9.7 does not occur on the boundary of the first quadrant.

Example 3.9.7 shows that the modulus of an analytic function need not attain its maximum on the boundary, and the maximum value of the modulus on the boundary may not be the maximum value inside the region. As we now show, the situation is different if Ω is bounded. In this case, the answers to both questions above are affirmative.

Corollary 3.9.8. (Maximum Modulus Principle) *Suppose that Ω is a bounded region and f is analytic on Ω and continuous on its closure $\overline{\Omega}$. Then*
(i) $|f|$ attains its maximum M on the boundary of Ω, and
(ii) either f is constant or $|f(z)| < M$ for all z in Ω.

Proof. If f is constant then (i) and (ii) hold obviously. Suppose f is not constant. The set consisting of Ω and its boundary is closed and bounded. Since $|f|$ is continuous on this set, it attains its maximum M on this set. But Theorem 3.9.6 says $|f|$ cannot attain its maximum in Ω, so $|f|$ attains its maximum on the boundary of Ω and $|f| < M$ in Ω.   ■

The modulus of a nonconstant analytic function can attain its minimum on a region Ω. Consider, for example, the function z on the open disk $|z| < 1$. Then the minimum of $|z|$ is 0 and it is attained at $z = 0$. However, if the function never vanishes in Ω, then we have the following useful principle.

Theorem 3.9.9. *Suppose that f is nonvanishing and analytic on a region Ω. If $|f|$ attains a minimum in Ω, then f is constant in Ω.*

Proof. Apply the maximum modulus principle to $g = 1/f$.   ■

Combining the previous two results, we obtain the following principle.

Corollary 3.9.10. *Suppose that Ω is a bounded region and f is analytic and non-vanishing on Ω and continuous on $\overline{\Omega}$. Then*
(i) $|f|$ attains a maximum M and minimum m on the boundary of Ω, and
(ii) either f is constant or $m < |f(z)| < M$ for all z in Ω.

Example 3.9.11. (Maximum and minimum values) Let $f(z) = \frac{e^z}{z}$, where z takes values in the annulus $\frac{1}{2} \leq |z| \leq 1$. Find the points where the maximum and minimum values of $|f(z)|$ occur and determine these values.

Solution. By Corollary 3.9.10, we must look for the maximum and minimum values on the boundary of the annular region, which consists of two circles $|z| = \frac{1}{2}$ and $|z| = 1$ (Figure 3.82). On the inner boundary, $|z| = \frac{1}{2}$, we have

$$|f(z)| = \frac{|e^z|}{|z|} = \frac{e^x}{1/2} = 2e^x,$$

where we have used the fact that $|e^z| = e^x$. For z on the inner circle, x varies from $-1/2$ to $1/2$. The minimum value of $2e^x$ occurs at $x = -1/2$ (or $z = -1/2$) and its maximum value occurs when $x = 1/2$ (or $z = 1/2$).

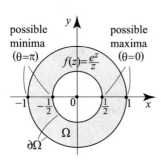

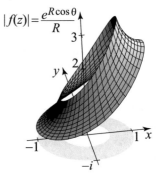

Fig. 3.82 The domain of $f(z) = e^z/z$ is the annulus $1/2 < |z| < 1$.

Fig. 3.83 The minimum value of $|f(z)|$ is attained at $z = -1$ and the maximum value at $z = \frac{1}{2}$.

Thus the minimum, respectively, maximum, values of $|f(z)|$ on the inner boundary are

$$|f(-1/2)| = \frac{e^{-1/2}}{1/2} = \frac{2}{\sqrt{e}} \approx 1.2 \quad \text{and} \quad |f(1/2)| = \frac{e^{1/2}}{1/2} = 2\sqrt{e} \approx 3.3.$$

Similarly, on the outer boundary $|z| = 1$, we have

$$|f(z)| = \frac{|e^z|}{|z|} = \frac{e^x}{1} = e^x.$$

For z on the outer circle, since x varies from -1 to 1, e^x varies from e^{-1} to e. Thus the minimum, respectively, maximum values of $|f(z)|$ on the outer boundary are $|f(-1)| = e^{-1} = 1/e \approx 0.4$ and $|f(1)| = e \approx 2.7$. Comparing the extreme values of $|f(z)|$, we find that its maximum value is $2\sqrt{e} \approx 3.3$ and it occurs at $z = \frac{1}{2}$, and its minimum value is $\frac{1}{e} \approx 0.4$ and it occurs at $z = -1$ (Figure 3.83).   □

In Example 3.9.11, the maximum value of the modulus was attained at precisely one point on the boundary, and similarly for the minimum value. This is not always be the case. Consider the function $g(z) = 1/z$ for $1 \leq |z| \leq 2$. The maximum value of $|g(z)|$ is attained at all points on the inner circle $|z| = 1$, and the minimum value is attained at all points of the outer circle $|z| = 2$.

Exercises 3.9

1. Find the maximum and minimum values of the modulus of the function $|z|$, defined for $|z| \leq 1$, and determine the points where these values occur. Explain your answers in view of Corollary 3.9.10.

2. Consider $f(z) = 2z + 3$, where z is in the closed square area with vertices at $1 \pm i$ and $-1 \pm i$. Find the maximum and minimum values of $|f|$ and determine where these values occur.

3. Consider $f(z) = e^{-z^2}$, where $1 \leq |z| \leq 2$. Find the maximum and minimum values of $|f|$ and determine where these values occur.

4. Consider the rectangle R with vertices at 0, π, i, and $\pi + i$. For z in R, let $f(z) = \frac{\sin z}{z}$ if $z \neq 0$, and $f(0) = 1$. Find the maximum and minimum values of $|f|$ and determine where these values occur.

5. Consider $f(z) = \frac{z}{z^2+2}$, where $2 \leq |z| \leq 3$. Find the maximum and minimum values of $|f|$ and determine where these values occur.

6. Consider $f(z) = \frac{3z}{1-z^2}$, where $|z| \leq \frac{1}{2}$. Find the maximum and minimum values of $|f|$ and determine where these values occur.

7. Consider $f(z) = \frac{2z-1}{-z+2}$, where $|z| \leq 1$. Find the maximum and minimum values of $|f|$ and determine where these values occur.

8. Consider $f(z) = \frac{2z-i}{iz+2}$, where $|z| \leq 1$. Find the maximum and minimum values of $|f|$ and determine where these values occur.

9. Consider $f(z) = \text{Log} z$, where $1 \leq |z| \leq 2$ and $0 \leq \text{Arg} z \leq \frac{\pi}{4}$. Find the maximum and minimum values of $|f|$ and determine where these values occur.

10. Consider $f(z) = \text{Log} z$, where $\frac{1}{2} \leq |z| \leq 2$ and $-\frac{\pi}{4} \leq \text{Arg} z \leq \frac{\pi}{4}$. Find the maximum and minimum values of $|f|$ and determine where these values occur.

11. Consider $f(z) = e^{e^z}$, where z belongs to the infinite horizontal strip $-\frac{\pi}{2} \leq \text{Im} z \leq \frac{\pi}{2}$. Show that $|f(z)| = 1$ for all z on the boundary, $\text{Im} z = \pm\frac{\pi}{2}$. Is f bounded inside the region? Does this contradict Theorem 3.9.6? Explain.

12. Suppose that $p(z) = a_n z^n + a_{n-1} z^{n-1} + \cdots + a_1 z + a_0$ $(a_n \neq 0)$ is a polynomial of degree $n \geq 1$.
(a) Show that $a_j = \frac{p^{(j)}(0)}{j!}$. (b) Suppose that $|p(z)| \leq M$ for all $|z| \leq R$. Show that $|a_j| \leq \frac{M}{R^j}$ for $j = 0, 1, \ldots, n$. Can we just assume that $|p(z)| \leq M$ for all $|z| < R$ and still get that $|a_j| \leq \frac{M}{R^j}$?

13. Factoring roots. (a) Verify the algebraic identity for complex numbers z and w and positive integers $n \geq 2$,

$$z^n - w^n = (z-w)(z^{n-1} + z^{n-2}w + z^{n-3}w^2 + \cdots + zw^{n-2} + w^{n-1}).$$

(b) Show that if $p(z) = p_n z^n + p_{n-1} z^{n-1} + \cdots + p_1 z + p_0$ is a polynomial of degree $n \geq 2$, and if $p(z_0) = 0$, then

$$p(z) = p(z) - p(z_0) = p_n(z^n - z_0^n) + p_{n-1}(z^{n-1} - z_0^{n-1}) + \cdots + p_1(z - z_0) = (z - z_0)q(z),$$

where $q(z)$ is a polynomial of degree $n - 1$.

14. Suppose that f is continuous on $\mathbb{C}$ and $\lim_{z\to\infty} |f(z)| = c$ exists and is finite. Show that f is bounded. [*Hint*: Make the following argument rigorous. For large values of $|z|$, say $|z| > M$, $|f(z)|$

is near c and so it is bounded. For $|z| \leq M$, $|f|$ is bounded because it is a continuous function on a closed and bounded set.]

15. Suppose that f is entire and $\lim_{z \to \infty} f(z) = 0$. Show that f is identically 0.

16. (a) Suppose that f is entire and f' is bounded in $\mathbb{C}$. Show that $f(z) = az + b$ for all $z \in \mathbb{C}$.
(b) Show that if f is entire and $f^{(n)}$ is bounded, then f is a polynomial of degree at most n.

17. Suppose that f is entire and omits an nonempty open set, i.e., there is an open disk $B_R(w_0)$ with $R > 0$ in the w-plane such that $f(z)$ does not lie in $B_R(w_0)$ for all z. Show that f is constant. [*Hint*: Consider $g(z) = \frac{1}{f(z) - w_0}$ and show that you can apply Liouville's theorem.] (A deep result in complex analysis known as Picard's theorem asserts that an entire nonconstant function can omit at most one value.)

18. Suppose that f is entire. Show that if either Re f or Im f are bounded, then f is constant. [*Hint*: Use Exercise 17.]

19. Suppose that f is entire and $\lim_{z \to \infty} f(z)/z = 0$. Show that f is constant. [*Hint*: Use Cauchy's estimate to show that $f'(z) = 0$.]

20. Suppose that f is entire and $\lim_{z \to \infty} \frac{f(z)}{z} = c$, where c is a constant. Show that $f(z) = cz + b$. [*Hint*: Apply the result of the previous exercise $g(z) = f(z) - cz$.]

21. A function $f(z) = f(x + iy)$ is called **doubly periodic** if there are real numbers $T_1 > 0$ and $T_2 > 0$ such that $f(x + T_1 + iy) = f(x + i(y + T_2)) = f(x + iy)$ for all $z = x + iy$ in $\mathbb{C}$. Show that if a function is entire and doubly periodic then it is constant. Can an entire function f be periodic in one of x or y without being constant?

22. What conclusion do you draw from Corollary 3.9.3 about the function e^{z^2}?

23. (a) Suppose that f is analytic in a bounded region Ω and continuous on the boundary of Ω. Suppose that $|f|$ is constant on the boundary of Ω. Show that either f has a zero in Ω or f is constant in Ω.
(b) Find all analytic functions f on the unit disk such that $|z| < |f(z)|$ for all $|z| < 1$ and $|f(z)| = 1$ for all $|z| = 1$. Justify your answer.

24. Let f and g be analytic functions on the open unit disk $B_1(0)$ and continuous and nonvanishing on the closed disk $\overline{B_1(0)}$. Suppose that $|f(z)| = |g(z)|$ for all $|z| = 1$. Show that $f(z) = A g(z)$ for all $|z| \leq 1$, where A is a constant such that $|A| = 1$.

25. Suppose that f is analytic on $|z| < 1$ and continuous on $|z| \leq 1$. Suppose that $f(z)$ is real-valued for all $|z| = 1$. Show that f is constant for all $|z| \leq 1$. [*Hint*: Consider $g(z) = e^{if(z)}$.]

26. Suppose that f and g are analytic in a bounded region Ω and continuous on the boundary of Ω. Suppose that g does not vanish in Ω and $|f(z)| \leq |g(z)|$ for all z on the boundary of Ω. Show that $|f(z)| \leq |g(z)|$ for all z in Ω.

Chapter 4
Series of Analytic Functions and Singularities

> *If, for every increasing value of n, the sum S_n indefinitely approaches a certain limit S, the series will be called convergent, and the limit in question will be called the sum of the series. If, on the contrary, while n increases indefinitely, the sum S_n does not approach a fixed limit, the series will be divergent and will no longer have a sum.*
>
> -Augustin-Louis Cauchy (1789–1857)
> (Cauchy's definition of a convergent series from his *Cours d' analyse.*

In calculus, we use Taylor series to represent functions on intervals centered at fixed points with a radius of convergence that could be positive, infinite, or zero, depending on the remainder associated with the function. For example, $\cos x$, e^x, $\frac{1}{1+x^2}$, and the function defined by e^{-1/x^2} for $x \neq 0$ and 0 if $x = 0$ are all infinitely differentiable for all real x. The radius of convergence of the Taylor series representation around zero is ∞ for the first two, 1 for the third one, and 0 for the last one. However in complex analysis, Taylor series are much nicer, in the sense that the remainder will play no role in determining their convergence. If a function is analytic on a disk of radius R centered at z_0, then it has a Taylor series representation centered at z_0 with radius at least R. For example the function $\frac{1}{1+z^2}$ is analytic on the disk $|z| < 1$ but, as it is not differentiable at $z = \pm i$, we do not expect the series to have a radius of convergence larger than 1.

Taylor series, and more generally Laurent series, are used to study important properties of analytic functions, concerning their zeros and isolated problem points (singularities). They are also useful in studying properties of special functions, such as Bessel functions.

The theory of power series as described in this chapter owes a lot to the German mathematician Karl Weierstrass (1815–1897). Weierstrass introduced to analysis the (ε, δ)-notation in proofs, replacing Cauchy's terminology, such as "indefinitely approaches a certain limit" and "increases indefinitely." Weierstrass' contributions to analysis are evidenced by the number of fundamental results we encounter here that bear his name.

4.1 Sequences and Series of Functions

In Section 1.5 we considered sequences and series of complex numbers. Now we turn our attention to sequences and series of functions.

Suppose that f_n and f are complex-valued functions defined on a subset $E \subset \mathbb{C}$. We say that f_n **converges pointwise** to f on E, if $\lim_{n \to \infty} f_n(z) = f(z)$ for every z in E.

© Springer International Publishing AG, part of Springer Nature 2018
N. H. Asmar and L. Grafakos, *Complex Analysis with Applications*,
Undergraduate Texts in Mathematics, https://doi.org/10.1007/978-3-319-94063-2_4

Hence f_n converges pointwise to f on E if for every z in E and every $\varepsilon > 0$ we can find $N > 0$ (depending on z and ε) such that for all $n \geq N$, we have $|f(z) - f_n(z)| < \varepsilon$. It is important to notice that the integer N depends in general on both z and ε.

Suppose u_n is defined on a set $E \subset \mathbb{C}$. The series of functions $\sum_{n=1}^{\infty} u_n$ is said to **converge pointwise** if the sequence of partial sums $s_n = \sum_{k=1}^{n} u_k$ converges pointwise on E.

In general, pointwise convergence does not preserve some desirable properties of the functions f_n. For example, the pointwise limit of continuous functions may not be continuous; and the pointwise limit of analytic functions may not be analytic. These properties are preserved, however, by the following stronger mode of convergence.

Definition 4.1.1. Let $\{f_n\}_{n=1}^{\infty}$ be a sequence of functions. We say that f_n **converges uniformly** to f on E, and we write $\lim_{n \to \infty} f_n = f$ uniformly on E, if for every $\varepsilon > 0$ we can find $N > 0$ such that for all $n \geq N$ and all z in E, we have $|f_n(z) - f(z)| < \varepsilon$.

A series of functions $\sum_{n=1}^{\infty} u_n$ is said to **converge uniformly** on E if the sequence of partial sums $s_n = \sum_{k=1}^{n} u_k$ converges uniformly on E.

The key words in Definition 4.1.1 are "for all z in E." These require that $f_n(z)$ be close to $f(z)$ for all z in E simultaneously. Let $M_n = \max|f_n(z) - f(z)|$, where the maximum is taken over all z in E. If no maximum is attained, we set[1] $M_n = \sup\{|f_n(z) - f(z)| : z \in E\}$. An equivalent formulation of uniform convergence is

$$f_n \to f \text{ uniformly on } E \quad \Longleftrightarrow \quad M_n \to 0 \text{ as } n \to \infty. \tag{4.1.1}$$

As a first example, we take E to be a real interval.

Example 4.1.2. (Pointwise versus uniform convergence) For $0 \leq x \leq 1$ and $n = 1, 2, \ldots$ define

$$f_n(x) = \frac{2nx}{1 + n^2 x^2}.$$

(a) Does the sequence converge pointwise on $[0, 1]$?
(b) Does it converge uniformly on $[0, 1]$?
(c) Does it converge uniformly on $[0.1, 1]$?

Solution. (a) We have $f_n(0) = 0$ for all n, and so $f_n(x) \to 0$ if $x = 0$. For all $x \neq 0$, we have

$$\lim_{n \to \infty} f_n(x) = \lim_{n \to \infty} \frac{2nx}{1 + n^2 x^2} = \lim_{n \to \infty} \frac{1}{n} \frac{2x}{x^2 + \frac{1}{n^2}} = 0.$$

So for all x in $[0, 1]$, the sequence $\{f_n(x)\}_{n=1}^{\infty}$ converges pointwise to $f(x) = 0$.
(b) Figure 4.1 suggests that the sequence does not converge to 0 uniformly on $[0, 1]$. To confirm this, let us see how large $|f_n(x)|$ can get on the interval $[0, 1]$. For this purpose, we compute the derivative

---

[1] The supremum of a bounded set is the least upper bound of the set. Its existence follows from the completeness of the real number system.

$$f_n'(x) = \frac{2n(1-n^2x^2)}{(1+n^2x^2)^2}.$$

Thus, for $0 < x \le 1$,

$$f_n'(x) = 0 \quad \Leftrightarrow \quad -n^2x^2 + 1 = 0 \quad \Leftrightarrow \quad x = \frac{1}{n}.$$

Plugging this value into $f_n(x)$ and simplifying, we find $f_n\left(\frac{1}{n}\right) = 1$. Thus, no matter how large n is, we can always find x in $[0, 1]$, namely $x = \frac{1}{n}$, with $f_n(x) = 1$. This shows that $M_n = \max_{x \in [0,1]} |f_n(x)| \ge 1$ (in fact, $M_n = 1$), and so f_n does not converge to 0 uniformly over $[0, 1]$. To see what is going on, note that $f_n(x) = f_1(nx)$. That is, $f_n(x)$ is merely a horizontally shrunken version of the curve $f_1(x)$ and has maximum value of 1. This maximum value moves left as n increases, but it never leaves the interval $[0, 1]$.

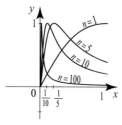

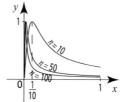

Fig. 4.1 Graphs of $f_n(x)$ for $n = 1, 5, 10, 100$. $f_n(x)$ has a maximum at $x = \frac{1}{n}$ in the interval $[0, 1]$, and $f_n\left(\frac{1}{n}\right) = 1$.

Fig. 4.2 Graphs of $f_n(x)$ for $n = 10, 50, 100$. Note how $f_n(x)$ attains its maximum value in $[0.1, 1]$ at $x = 0.1$.

(c) The situation here is quite different than that in part (b). For $n > 10$, we have $\frac{1}{n} < 0.1$, and so the maximum value of $f_n(x)$, which we found to be 1 in (b), is not attained in the interval $[0.1, 1]$; it is attained at $\frac{1}{n}$ outside this interval. So how large is $f_n(x)$ for x in $[0.1, 1]$? Since $f_n'(x) \le 0$ on $[\frac{1}{n}, 1]$, $f_n(x)$ is decreasing on $[\frac{1}{n}, 1]$ and so the maximum of $f_n(x)$ for x in the interval $[0.1, 1]$ occurs at the left endpoint, $x = \frac{1}{10}$ (Figure 4.2). We have $M_n = f_n(\frac{1}{10})$, and we know that $f_n(\frac{1}{10}) \to 0$ as $n \to \infty$ from part (a). Thus $M_n \to 0$, implying that $f_n \to 0$ uniformly on $[0.1, 1]$. □

There is an important point to be made about part (c) of Example 4.1.2. Clearly we could have replaced the left endpoint $x = 0.1$ by any number a with $0 < a < 1$ and still had uniform convergence on $[a, 1]$. While the sequence failed to converge uniformly on $[0, 1]$, it does converge uniformly on any proper closed subinterval $[a, 1]$ where $0 < a < 1$. This is a common phenomenon that one encounters with many sequences or series. They may fail to converge uniformly on a whole region, but they may converge uniformly on all closed (and bounded) proper subregions.

We now examine the preservation of continuity and integrability in the context of uniform limits of functions.

Theorem 4.1.3. *(i) Suppose that $f_n \to f$ uniformly on E and f_n is continuous on E for every n. Then f is continuous on E.*
(ii) Suppose $u = \sum_{n=1}^{\infty} u_n$ converges uniformly on E and u_n is continuous on E for every n. Then u is continuous on E.

Proof. (i) Fix z_0 in E. Given $\varepsilon > 0$, by uniform convergence we can find f_N such that $|f_N(z) - f(z)| < \frac{\varepsilon}{3}$ for all z in E. Since f_N is continuous at z_0 there is a $\delta > 0$ such that $|f_N(z_0) - f_N(z)| < \frac{\varepsilon}{3}$ for all $z \in E$ with $|z - z_0| < \delta$. Putting these two inequalities together and using the triangle inequality, we find for $|z - z_0| < \delta$

$$|f(z_0) - f(z)| \leq |f(z_0) - f_N(z_0)| + |f_N(z_0) - f_N(z)| + |f_N(z) - f(z)|$$
$$< \frac{\varepsilon}{3} + \frac{\varepsilon}{3} + \frac{\varepsilon}{3} = \varepsilon,$$

which establishes the continuity of f at z_0. Part (ii) follows from (i) by taking $f_n = \sum_{k=1}^{n} u_k$ and noting that each f_n is continuous, being a finite sum of continuous functions. $\qquad \square$

Sometimes we use Theorem 4.1.3 to prove the failure of uniform convergence.

Example 4.1.4. (Failure of uniform convergence) A sequence of functions is defined on the closed unit disk by

$$f_n(z) = \begin{cases} n|z| & \text{if } |z| \leq \frac{1}{n}, \\ 1 & \text{if } \frac{1}{n} \leq |z| \leq 1. \end{cases}$$

Does the sequence converge uniformly on $|z| \leq 1$?

Solution. The function f_n with $n = 5$ is depicted in Figure 4.3. It is clear that each f_n is continuous on $\{z \in \mathbb{C} : |z| \leq 1\}$ and

$$\lim_{n \to \infty} f_n(z) = \begin{cases} 1 & \text{if } 0 < |z| \leq 1, \\ 0 & \text{if } z = 0. \end{cases}$$

Since the limit function is not continuous at $z = 0$, we conclude from Theorem 4.1.3 that $\{f_n\}_{n=1}^{\infty}$ cannot converge uniformly on any set containing 0; in particular, it does not converge uniformly in $|z| \leq 1$. $\square$

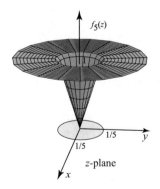

Fig. 4.3 Figure 3. Graph of $f_n(z)$ (with $n = 5$) in Example 4.1.4.

If a sequence of continuous functions converges uniformly to a limit, then the limit is continuous and thus it makes sense to integrate it.

Theorem 4.1.5. *Let* $\{f_n\}_{n=1}^{\infty}$ *be a sequence of continuous functions on a region* Ω, *and let* γ *be a path in* Ω. *If* $f_n \to f$ *uniformly on* γ, *then*

$$\lim_{n\to\infty} \int_{\gamma} f_n(z)\,dz = \int_{\gamma} f(z)\,dz.$$

Proof. Let $M_n = \max_{z\in\gamma}|f_n(z) - f(z)|$; then $M_n \to 0$. Using the integral inequality, (3.2.31) in Theorem 3.2.19 we obtain

$$\left| \int_{\gamma} f_n(z)\,dz - \int_{\gamma} f(z)\,dz \right| = \left| \int_{\gamma} (f_n(z) - f(z))\,dz \right| \le \ell(\gamma)\,M_n \to 0,$$

which proves the theorem.   ∎

Applying Theorem 4.1.5 to the partial sums of a uniformly convergent series, we obtain the following important corollary.

Corollary 4.1.6. *Suppose that* $\{u_n\}_{n=1}^{\infty}$ *is a sequence of continuous functions on a region* Ω, *and let* γ *be a path in* Ω. *Suppose that* $\{\sum_{n=1}^{N} u_n\}_{N=1}^{\infty}$ *converges uniformly on* γ *to a (continuous) function* u. *Then the series* $\sum_{n=1}^{\infty} \int_{\gamma} u_n(z)\,dz$ *converges and*

$$\int_{\gamma} u(z)\,dz = \sum_{n=1}^{\infty} \left(\int_{\gamma} u_n(z)\,dz \right).$$

We now discuss a very useful test for uniform convergence.

Theorem 4.1.7. (Weierstrass M-test) *Let* $\{u_n\}_{n=1}^{\infty}$ *be a sequence of functions on a subset* E *of* $\mathbb{C}$ *and let* $\{M_n\}_{n=1}^{\infty}$ *be a sequence of numbers such that for all* n
(i) $|u_n(z)| \le M_n$ *for all* z *in* E, *and*
(ii) $\sum_{n=1}^{\infty} M_n < \infty$.
Then $\sum_{n=1}^{\infty} u_n$ *converges uniformly and absolutely on* E.

Proof. The absolute convergence of $\sum_{n=1}^{\infty} u_n$ follows from (*i*) and (*ii*) by comparison to the series $\sum_{n=1}^{\infty} M_n$. Absolute convergence implies convergence, so we set $\sum_{n=1}^{\infty} u_n(z) = s(z)$ for all z in E. We next prove the uniform convergence. Let $s_m = \sum_{k=1}^{m} u_k(z)$. For $n > m \ge 1$, using the triangle inequality we obtain

$$|s_n(z) - s_m(z)| = \left| \sum_{j=m+1}^{n} u_j(z) \right| \le \sum_{j=m+1}^{n} |u_j(z)| \le \sum_{j=m+1}^{n} M_j \le \sum_{j=m+1}^{\infty} M_j. \quad (4.1.2)$$

Letting $n \to \infty$ and using the fact that $s_n(z) \to s(z)$ for all z in E, we obtain from (4.1.2) that $|s(z) - s_m(z)| \le \sum_{m+1}^{\infty} M_j$ for all z in E. This means that

$$\max_{z\in E} |s(z) - s_m(z)| \le \sum_{j=m+1}^{\infty} M_j.$$

But the tail of a convergent series tends to zero as $m \to \infty$ (Proposition 1.5.18). Thus s_m converges to s uniformly on E.   ∎

Example 4.1.8. (Weierstrass M-test) Establish the uniform convergence of the series on the given set.

(a) $\displaystyle\sum_{n=1}^{\infty} \frac{e^{inx}}{n^2}, \quad -\infty < x < \infty$ (b) $\displaystyle\sum_{n=1}^{\infty} \frac{z^n}{n^2}, \quad |z| \le 1$

(c) $\displaystyle\sum_{n=1}^{\infty} \left(\frac{z}{2}\right)^n, \quad |z| \le 1.9$ (d) $\displaystyle\sum_{n=1}^{\infty} \frac{1}{(1-z)^n}, \quad 1.01 \le |1-z|$

Solution. We use the notation of the Weierstrass M-test.

(a) We have $E = (-\infty, \infty)$, $u_n(x) = \frac{e^{inx}}{n^2}$. For all x in E, we have

$$|u_n(x)| = \left| \frac{e^{inx}}{n^2} \right| = \frac{1}{n^2} = M_n.$$

Since $\sum_{n=1}^{\infty} M_n = \sum \frac{1}{n^2}$ is convergent, we conclude from the Weierstrass M-test that $\sum_{n=1}^{\infty} \frac{e^{inx}}{n^2}$ converges uniformly for all x in E.

(b) Here E is the set $|z| \le 1$ and $u_n(z) = \frac{z^n}{n^2}$. For all $|z| \le 1$, we have

$$|u_n(z)| = \left| \frac{z^n}{n^2} \right| \le \frac{1}{n^2} = M_n.$$

Since $\sum_{n=1}^{\infty} \frac{1}{n^2}$ is convergent, we conclude from the Weierstrass M-test that $\sum_{n=1}^{\infty} \frac{z^n}{n^2}$ converges uniformly for all $|z| \le 1$.

(c) Here E is the set $|z| \le 1.9$ and $u_n(z) = \left(\frac{z}{2}\right)^n$. For all $|z| \le 1.9$, we have

$$|u_n(z)| = \left| \frac{z}{2} \right|^n \le \left(\frac{1.9}{2}\right)^n = r^n = M_n,$$

where $r = \frac{1.9}{2} < 1$. Since $\sum_{n=1}^{\infty} M_n = \sum_{n=1}^{\infty} r^n$ is convergent, we conclude that $\sum_{n=1}^{\infty} \left(\frac{z}{2}\right)^n$ converges uniformly for all $|z| \le 1.9$.

(d) Here E is the set $1.01 \le |1-z|$ and $u_n(z) = \frac{1}{(1-z)^n}$. For all z in E, we have

$$|u_n(z)| = \frac{1}{|1-z|^n} \le r^n = M_n,$$

where $r = \frac{1}{1.01} < 1$. Since $\sum_{n=1}^{\infty} M_n = \sum_{n=1}^{\infty} r^n$ is convergent, we conclude that $\sum_{n=1}^{\infty} \frac{1}{(1-z)^n}$ converges uniformly for all $1.01 \le |1-z|$.   □

Example 4.1.9. (Uniform convergence and the geometric series)

(a) Show that the geometric series $\sum_{n=0}^{\infty} z^n$ converges uniformly to $\frac{1}{1-z}$ on all closed subdisks $|z| \le r$, $r < 1$, of the open unit disk $|z| < 1$.

(b) Show that the geometric series $\sum_{n=0}^{\infty} z^n$ does not converge uniformly on the open disk $|z| < 1$.

Solution. (a) We refer to Example 1.5.13 for results about the geometric series. To establish the uniform convergence of the series for $|z| \le r < 1$, we apply the Weierstrass M-test. We have $|u_n(z)| = |z^n| \le r^n = M_n$ for all $|z| \le r$. Since $\sum_{n=0}^{\infty} M_n = \sum_{n=0}^{\infty} r^n$ is convergent if $0 \le r < 1$, we conclude that the series $\sum_{n=0}^{\infty} z^n$ converges uniformly for $|z| \le r$.

(b) The series $\sum_{n=0}^{\infty} z^n$ converges to $s(z) = 1/(1-z)$ in $|z| < 1$. Its nth partial sum is $s_n(z) = \frac{1-z^{n+1}}{1-z}$. Take $z = x$ to be a real number with $0 < x < 1$. Then

$$|s(x) - s_n(x)| = \frac{|x^{n+1}|}{|1-x|} = \frac{x^{n+1}}{1-x},$$

which tends to ∞ as x increases to 1. Thus for every n we can find a z with $|z| < 1$ and $|s(z) - s_n(z)|$ as large as we wish; this shows that s_n does not converge to s uniformly in $|z| < 1$. $\qquad\square$

Sequences and Series of Analytic Functions

The remaining results of this section concern limits of analytic functions. Based on the behavior of the geometric series (Example 4.1.9), we study the limit under the assumption of uniform convergence on closed subdisks. This hypothesis is less restrictive than uniform convergence. The following is a central result in the theory of analytic functions. Its proof is a nice application of Morera's theorem (Theorem 3.8.10).

Theorem 4.1.10. *Suppose that $\{f_n\}_{n=1}^{\infty}$ is a sequence of analytic functions on a region Ω such that $f_n \to f$ uniformly on every closed disk contained in Ω. Then*
(i) f is analytic on Ω, and
(ii) for an integer $k \ge 1$, $f_n^{(k)}(z) \to f^{(k)}(z)$ for all z in Ω. Thus, the limit of the kth derivative is the kth derivative of the limit.

The proof will be facilitated by the following auxiliary result.

Lemma 4.1.11. *(i) Suppose that $f_n \to f$ uniformly on a closed and bounded subset E of $\mathbb{C}$, and g is a continuous function on E. Then $f_n g \to f g$ uniformly on E.*
(ii) Suppose that the series $u = \sum_{n=1}^{\infty} u_n$ converges uniformly on a closed and bounded set E, and g is a continuous function on E. Then the series $gu = \sum_{n=1}^{\infty} g u_n$ converges uniformly on E.

Proof. (i) Since g is continuous on E and E is closed and bounded, it follows that g is bounded on E. Let $M = \max_{z \in E} |g(z)|$. For all z in E, we have

$$|f_n(z)g(z) - f(z)g(z)| = |f_n(z) - f(z)||g(z)| \le M|f_n(z) - f(z)|.$$

Thus

$$\max |f_n(z)g(z) - f(z)g(z)| \leq M \max |f_n(z) - f(z)| \to 0,$$

because f_n converges uniformly to f on E. Thus $f_n g \to f g$ uniformly on E. To prove (ii), apply (i) to the sequence of partial sums $u_1 + \cdots + u_n$ which converges to u. ∎

Proof (Theorem 4.1.10). (i) The function f is continuous by Theorem 4.1.3(i). To prove that f is analytic, we apply Morera's theorem (Theorem 3.8.10). Let γ be an arbitrary closed triangular path lying in a closed disk in Ω. It is enough to show that $\int_\gamma f(z)\,dz = 0$. We have $\int_\gamma f_n(z)\,dz = 0$ for all n, by Cauchy's theorem (Theorem 3.5.4), because f_n is analytic inside and on γ, and by Theorem 4.1.5, $\int_\gamma f_n(z)\,dz \to \int_\gamma f(z)\,dz$ as $n \to \infty$. So $\int_\gamma f(z)\,dz = 0$ and (i) follows.

(ii) Let $z_0 \in \Omega$ and let $\overline{B_R(z_0)}$ be a closed disk contained in Ω, centered at z_0 with radius $R > 0$, with positively oriented boundary $C_R(z_0)$. Since $f_n \to f$ uniformly on $C_R(z_0)$ and $\frac{1}{(z-z_0)^{k+1}}$ is continuous on $C_R(z_0)$, it follows from Lemma 4.1.11(i) that

$$\frac{f_n(z)}{(z-z_0)^{k+1}} \to \frac{f(z)}{(z-z_0)^{k+1}}$$

uniformly for all z on $C_R(z_0)$. Applying Theorem 4.1.5 and using the generalized Cauchy integral formula, we deduce

$$f_n^{(k)}(z_0) = \frac{k!}{2\pi i} \int_{C_R(z_0)} \frac{f_n(z)}{(z-z_0)^{k+1}}\,dz \to \frac{k!}{2\pi i} \int_{C_R(z_0)} \frac{f(z)}{(z-z_0)^{k+1}}\,dz = f^{(k)}(z_0),$$

which proves (ii). ∎

Theorem 4.1.10 may fail if we replace analytic functions by differentiable functions of a real variable. That is, if E is a subset of the real line and $f_n(x) \to f(x)$ uniformly on E, it does not follow in general that f_n' converges to f', as the next example shows.

Example 4.1.12. (Failure of termwise differentiation) For $0 \leq x \leq 2\pi$ and $n = 1, 2, \ldots$ define $f_n(x) = \frac{e^{inx}}{in}$. It is clear that $f_n \to 0$ uniformly on $[0, 2\pi]$. But $f_n'(x) = e^{inx}$ and this sequence does not converge except at $x = 0$ or $x = 2\pi$. (See Example 1.5.9) Consequently, f_n' does not converge to 0. Can we understand how this occurs within the larger framework of complex functions? Replace x by z and consider the sequence functions $f_n(z) = \frac{e^{inz}}{in}$. We cannot find a complex neighborhood of the real interval $[0, 2\pi]$ where f_n converges, as such a neighborhood would contain z with $\operatorname{Im} z < 0$. Thus Theorem 4.1.10 does not apply. □

Corollary 4.1.13. *Suppose that $\{u_n\}_{n=1}^\infty$ is a sequence of analytic functions on a region Ω and that $u = \sum_{n=1}^\infty u_n$ converges uniformly on every closed disk in Ω. Then u is analytic on Ω. Moreover, for all integers $k \geq 1$, the series may be differentiated term by term k times to yield*

$$u^{(k)}(z) = \sum_{n=1}^{\infty} u_n^{(k)}(z) \quad \text{for all } z \text{ in } \Omega.$$

Proof. Apply Theorem 4.1.10 to the sequence of partial sums $u_1 + \cdots + u_n$. ∎

Example 4.1.14. (Term-by-term differentiation of the geometric series) The geometric series $\sum_{n=0}^{\infty} z^n$ converges uniformly to $\frac{1}{1-z}$ on a disk $|z| \leq r < 1$, by Example 4.1.9. It will then converge uniformly on all closed disks contained in the open disk $|z| < 1$. Applying Corollary 4.1.13, we may differentiate term by term in the open disk $|z| < 1$ and obtain

$$\frac{1}{(1-z)^2} = \frac{d}{dz}\frac{1}{1-z} = \frac{d}{dz}\sum_{n=0}^{\infty} z^n = \sum_{n=0}^{\infty}\frac{d}{dz} z^n = \sum_{n=1}^{\infty} n z^{n-1}, \qquad |z| < 1. \qquad \square$$

There is an important technique in analysis that is manifested in Theorem 4.1.10, Corollary 4.1.13, and Example 4.1.14. We only have uniform convergence of a series on closed and bounded subsets of a region Ω, but we infer that the series can be differentiated termwise on the entire open set Ω.

Example 4.1.15. (Termwise differentiation and integration)
(a) Let $\rho > 0$ be a positive real number. Show that $u(z) = \sum_{n=0}^{\infty} \frac{1}{n! z^n}$ converges uniformly for all $|z| \geq \rho$.
(b) Conclude that u is analytic in $\mathbb{C} \setminus \{0\}$.
(c) Let $C_R(0)$ denote a circle of radius $R > 0$ centered at 0, with positive orientation. Evaluate $\int_{C_R(0)} u(z)\, dz$.

Solution. (a) For $|z| \geq \rho$, we have $\left|\frac{1}{z^n}\right| \leq \frac{1}{\rho^n}$, and so

$$\left|\frac{1}{n! z^n}\right| \leq \frac{1}{n! \rho^n} \qquad \text{for all } \rho \leq |z|.$$

The series $\sum_{n=0}^{\infty} \frac{1}{n! \rho^n}$ is convergent by the ratio test, since

$$\lim_{n \to \infty} \frac{a_{n+1}}{a_n} = \lim_{n \to \infty} \frac{n! \rho^n}{(n+1)! \rho^{n+1}} = \lim_{n \to \infty} \frac{1}{(n+1)\rho} = 0 < 1.$$

In fact, one observes that

$$\sum_{n=0}^{\infty} \frac{1}{n! \rho^n} = \sum_{n=0}^{\infty} \frac{\rho^{-n}}{n!} = e^{-\rho} < \infty.$$

By the Weierstrass M-test, it follows that

$$u(z) = \sum_{n=0}^{\infty} \frac{1}{n!z^n}$$

is uniformly convergent for all $|z| \geq \rho$.

(b) Since S is closed and 0 is not in S, then we can find $\rho > 0$ so that S is contained in the annulus $|z| \geq \rho$. From part (a) note that the series will be uniformly convergent on all closed and bounded subsets of the punctured plane $\mathbb{C} \setminus \{0\}$, since any such set is contained in an annular region $|z| \geq \rho$ (Figure 4.4). Applying Corollary 4.1.13, it follows that u is analytic on $\mathbb{C} \setminus \{0\}$.

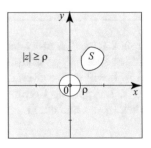

Fig. 4.4 The set S and the disc $B_\rho(0)$

(c) Pick ρ such that $0 < \rho < R$. Since $\sum_{n=0}^{\infty} \frac{1}{n!z^n}$ converges uniformly on $|z| \geq \rho$, and since $C_R(0)$ is contained in the region $|z| \geq \rho$, it follows that $\sum_{n=0}^{\infty} \frac{1}{n!z^n}$ converges uniformly on $C_R(0)$. Hence, by Corollary 4.1.6, the series may be integrated term by term; this yields

$$\int_{C_R(0)} u(z)\,dz = \int_{C_R(0)} \left\{ \sum_{n=0}^{\infty} \frac{1}{n!z^n} \right\} dz = \sum_{n=0}^{\infty} \frac{1}{n!} \int_{C_R(0)} \frac{1}{z^n}\,dz.$$

The integral $\int_{C_R(0)} \frac{1}{z^n}\,dz$ is quite familiar (Example 3.2.10). Its value is 0 if $n \neq 1$ and $2\pi i$ if $n = 1$. Thus,

$$\int_{C_R(0)} \left\{ \sum_{n=0}^{\infty} \frac{1}{n!z^n} \right\} dz = 2\pi i.$$

Notice what was achieved: Our function is expanded in powers of z, and its integral around the origin is precisely $2\pi i$ times the coefficient of $\frac{1}{z}$. This is part of a general technique that falls under the scope of Chapter 5 and will be revisited then. □

Exercises 4.1

In Exercises 1–8, (a) find the pointwise limit of the sequences. (b) Determine if the sequences converge uniformly on the given intervals. (c) If the sequence does not converge uniformly on the corresponding interval, describe some subintervals on which uniform convergence takes place.

1. $f_n(x) = \dfrac{\sin nx}{n}$, $0 \leq x \leq \pi$

2. $f_n(x) = \dfrac{\sin nx}{nx}$, $0 < x \leq \pi$

3. $f_n(x) = \dfrac{x}{nx+1}$, $0 \leq x \leq 1$

4. $f_n(x) = nx^n$, $0 \leq x < .99$

5. $f_n(x) = \dfrac{nx}{n^2x^2 - x + 1}$, $0 \leq x \leq 1$

6. $f_n(x) = \dfrac{nx}{n^2x^2 - nx + 1}$, $0 \leq x \leq 1$

7. $f_n(x) = \dfrac{nx}{n^2x^2 + 2}$, $0 \leq x \leq 1$

8. $f_n(x) = \displaystyle\int_0^x e^{-n\sqrt{t}}\,dt$, $0 \leq x \leq 1$

In Exercises 9–12, (a) determine whether or not the sequence of functions converges pointwise and find its limit if it converges. (b) Determine if the sequence converges uniformly on the given subset Ω of $\mathbb{C}$. (c) If uniform convergence fails on Ω, describe some closed and bounded subsets of Ω on which uniform convergence takes place.

9. $f_n(z) = \dfrac{nz+1}{z+2n^2}$, $|z| \le 1$ **10.** $f_n(z) = \dfrac{z^n+z}{n+1}$, $|z| \le 1$

11. $f_n(z) = \dfrac{\sin nz}{n^2}$, $|z| \le 1$ **12.** $f_n(z) = \dfrac{z^2+nz+1}{n^2 z+1}$, $2 < |z|$

In Exercises 13–22, use the Weierstrass M-test to show that the series converge uniformly on the indicated regions.

13. $\displaystyle\sum_{n=1}^{\infty} \frac{z^n}{n(n+1)}$, $|z| \le 1$ **14.** $\displaystyle\sum_{n=1}^{\infty} \frac{(3z)^n}{n(n+1)}$, $|z| \le \frac{1}{3}$

15. $\displaystyle\sum_{n=0}^{\infty} \left(\frac{3z}{4}\right)^n$, $|z| \le 1.1$ **16.** $\displaystyle\sum_{n=0}^{\infty} \left(\frac{z^2-1}{4}\right)^n$, $|z| \le 1$

17. $\displaystyle\sum_{n=0}^{\infty} \left(\frac{z+2}{5}\right)^n$, $|z| \le 2$ **18.** $\displaystyle\sum_{n=0}^{\infty} \frac{1}{(5-z)^n}$, $|z| \le \frac{7}{2}$

19. $\displaystyle\sum_{n=0}^{\infty} \frac{(z+1-3i)^n}{4^n}$, $|z-3i| \le .5$ **20.** $\displaystyle\sum_{n=0}^{\infty} \frac{(z-1)^n}{4^n}$, $|z| \le 2$

21. $\displaystyle\sum_{n=0}^{\infty} \left\{\frac{(z-2)^n}{3^n} + \frac{2^n}{(z-2)^n}\right\}$, $2.01 \le |z-2| \le 2.9$

22. $\displaystyle\sum_{n=1}^{\infty} \left\{\left(\frac{z}{5}\right)^n + \frac{1}{z^n}\right\}$, $1.001 \le |z| \le 4.9$

23. (a) The nth partial sum of a series is $s_n(z) = \frac{z^n}{n}$ for $|z| \le 1$. Does the series converge uniformly on $|z| \le 1$?
(b) Construct a series with partial sum the function s_n in part (a).

24. The nth partial sum of a series is $s_n(z) = \frac{e^{inz}}{n}$ for $|z| \le 1$. Does the series converge uniformly on $|z| \le 1$? Does it converge pointwise?

25. (a) Does $\sum_{n=0}^{\infty} z^n$ converge uniformly on $|z - \frac{1}{2}| < \frac{1}{6}$?
(b) Does $\sum_{n=0}^{\infty} z^n$ converge uniformly on $|z - \frac{1}{2}| < \frac{1}{2}$? Justify your answers.

26. (a) If $\sum_{n=0}^{\infty} u_n$ converges absolutely on a set E, does this imply that $\sum_{n=0}^{\infty} u_n$ converges uniformly on E? If not, indicate this via an example.
(b) If $\sum_{n=0}^{\infty} u_n$ converges uniformly on E, does this imply that $\sum_{n=0}^{\infty} u_n$ converge absolutely for all z in E? [*Hint:* Consider $\sum_{n=1}^{\infty} (-1)^{n+1} \frac{x^n}{n} = \log(1+x)$ for $0 \le x \le 1$.]

27. Derivative of the exponential function. Show that $\frac{d}{dz} e^z = e^z$ by differentiating the series term by term. Justify this process via Theorem 4.1.10.

28. (a) Show that $u(z) = \sum_{n=0}^{\infty} \frac{z^n}{1+z^{2n}}$ is analytic for all $|z| < 1$ and all $|z| > 1$. [*Hint:* Treat separately the uniform convergence on $|z| \le r < 1$ and on $|z| \ge R > 1$.]
(b) What is $u'(z)$ for $|z| < 1$ or $|z| > 1$?

29. Riemann zeta function. The Riemann zeta function is defined by

$$\zeta(z) = \sum_{n=1}^{\infty} \frac{1}{n^z} \qquad \text{(principal branch of } n^z\text{)}.$$

(a) Let $\delta > 1$ be a positive real number. Show that the series converges uniformly on every half-plane $H_\delta = \{z : \operatorname{Re} z \geq \delta > 1\}$.

(b) Conclude that $\zeta(z)$ is analytic on the half-plane $H = \{z : \operatorname{Re} z > 1\}$.

(c) What is $\zeta'(z)$?

Riemann used the zeta function to study the distribution of the prime numbers. Although this function was previously known to Euler, Riemann was the first one to consider it over the complex numbers. An important thesis, conjectured by Riemann, is the **Riemann hypothesis**, which states that the analytic continuation of the zeta function has infinitely many nonreal roots that lie on the line $\operatorname{Re} z = \frac{1}{2}$.

30. Uniformly Cauchy sequence. A sequence of functions $\{f_n\}_{n=1}^\infty$ is **uniformly Cauchy** on $E \subset \mathbb{C}$ if for every $\varepsilon > 0$ we can find a positive integer N so that

$$m, n \geq N \quad \Rightarrow \quad |f_n(z) - f_m(z)| < \varepsilon \quad \text{for all } z \text{ in } E.$$

Prove the following **Cauchy criterion**: $\{f_n\}_{n=1}^\infty$ converges uniformly on $E \subset \mathbb{C}$ if and only if $\{f_n\}_{n=1}^\infty$ is uniformly Cauchy on E.

31. Sums and products of uniformly convergent sequences. Suppose that $f_n \to f$ uniformly on E and $g_n \to g$ uniformly on E. Show that

(a) $f_n + g_n \to f + g$ uniformly on E;

(b) if f and g are bounded on E, then $f_n g_n \to fg$ uniformly on E.

$\left[\textit{Hint: } f_n g_n - fg = f_n g_n - f_n g + f_n g - fg.\right]$

32. Suppose that $\sum_{n=1}^\infty u_n$ converges uniformly on $E \subset \mathbb{C}$. Show that $u_n \to 0$ uniformly on E. [*Hint:* Let s_n denote the nth partial sum of u_n. What can you say about $s_{n+1} - s_n$?]

33. Boundary criterion for uniform convergence. Suppose that C is a simple closed path (for simplicity, take C to be a circle) and let Ω denote the region interior to C. Let f_n be analytic on Ω and continuous on C for $n = 1, 2, \ldots$ and suppose that f_n converges uniformly to some function f on C. Show that $\{f_n\}_{n=1}^\infty$ converges uniformly on Ω. [*Hint:* Show that $\{f_n\}_{n=1}^\infty$ is a uniformly Cauchy sequence on Ω.]

4.2 Power Series

Definition 4.2.1. A **power series** is a series of the form

$$\sum_{n=0}^\infty c_n (z - z_0)^n,$$

where z is a complex variable, z_0 is a fixed complex number called the **center**, and c_n are complex numbers called the **coefficients**.

We adhere to the usual notational convention that $(z - z_0)^0 = 1$ for all z. A power series always converges at the center $z = z_0$ because all the terms vanish for $n \geq 1$. The term with $n = 0$ may not vanish but produces the constant c_0. For $z \neq z_0$ the power series may converge or diverge. Let us consider a few examples.

Example 4.2.2. Study the convergence of the power series:

(a) $\displaystyle\sum_{n=0}^{\infty} \frac{(z-1+i)^n}{(n!)^2}$,   (b) $\displaystyle\sum_{n=1}^{\infty} \frac{n^n (z-i)^n}{n}$,   (c) $\displaystyle\sum_{n=0}^{\infty} (-2)^n \frac{z^n}{n+1}$.

Solution. (a) We apply the ratio test (Theorem 1.5.23). For $z \neq 1 - i$, we have

$$\rho = \lim_{n\to\infty} \left| \frac{(z-1+i)^{n+1} (n!)^2}{(z-1+i)^n ((n+1)!)^2} \right| = \lim_{n\to\infty} \frac{|z-1+i|(n!)^2}{(n+1)^2 (n!)^2} = \lim_{n\to\infty} \frac{|z-1+i|}{(n+1)^2} = 0.$$

Since $\rho < 1$, we conclude from the ratio test that the series converges for all $z \in \mathbb{C}$.
(b) We use the root test (Theorem 1.5.25). For $z \neq i$, we have

$$\rho = \lim_{n\to\infty} \sqrt[n]{|a_n|} = \lim_{n\to\infty} \left(\frac{n^n |z-i|^n}{n} \right)^{\frac{1}{n}} = \lim_{n\to\infty} \frac{n|z-i|}{n^{\frac{1}{n}}} = \infty,$$

because $n|z-i| \to \infty$ and $n^{\frac{1}{n}} \to 1$ as $n \to \infty$ (Exercise 10, Section 1.5). Since $\rho > 1$ for all $z \neq i$, we conclude from the root test that the series diverges for all $z \neq i$.
(c) We use the ratio test (Theorem 1.5.23). For $z \neq 0$, we have

$$\rho = \lim_{n\to\infty} \left| \frac{a_{n+1}}{a_n} \right| = \lim_{n\to\infty} \left| \frac{2^{n+1} z^{n+1} (n+1)}{2^n z^n (n+2)} \right| = \lim_{n\to\infty} 2|z| \frac{n+1}{n+2} = 2|z|.$$

By the ratio test, the series converges if $2|z| < 1$ and diverges if $2|z| > 1$. Equivalently, the series converges if $|z| < \frac{1}{2}$ and diverges if $|z| > \frac{1}{2}$.   □

For the rest of this section, the following notion will be useful.

Definition 4.2.3. For a sequence of real numbers $\{x_n\}_{n=1}^{\infty}$, define the **limit superior** of $\{x_n\}_{n=1}^{\infty}$, denoted $\limsup_{n\to\infty} x_n$, or simply $\limsup x_n$, to be the limit of the decreasing sequence $\{\sup_{m\geq n} x_m\}_{n=1}^{\infty}$, i.e.,

$$\limsup_{n\to\infty} x_n = \lim_{n\to\infty} \left(\sup_{m\geq n} x_m \right). \tag{4.2.1}$$

If $\sup_{m\geq n} x_m = \infty$ for all $n \geq 1$, then we set $\limsup x_n = \infty$.

The existence of a limit superior follows from the completeness of the real number system. It is easy to show that if $L = \lim x_n$ exists, then $L = \limsup x_n$. Moreover, $\limsup x_n$ coincides with the largest limit of a convergent subsequence[2] of the sequence $\{x_n\}_{n=1}^{\infty}$. Since $\{\sup_{m\geq n} x_m\}_{n=1}^{\infty}$ is a convergent decreasing sequence, in explicit calculations, it is convenient to find the limit superior as the limit of a subsequence of $\{\sup_{m\geq n} x_m\}_{n=1}^{\infty}$ that can be easily calculated.

Example 4.2.4. (a) Let $x_n = 1 + \frac{1}{n}$ if n is even and $x_n = \frac{1}{n}$ if n is odd. Then $\{x_n\}_{n=1}^{\infty}$ does not converge, but one has that $\limsup x_n = 1$, since $\sup\{x_m : m \geq 2n\} = 1 + \frac{1}{2n}$ which converges to 1.

---

[2] A subsequence of a sequence $\{a_n\}_{n=1}^{\infty}$ is a sequence of the form $\{a_{n_k}\}_{k=1}^{\infty}$, where $n_1 < n_2 < \cdots$ is a strictly increasing sequence of natural numbers.

(b) Let $y_n = \frac{3k+4}{k+1}$ if $n = 3k$, $y_n = \frac{4k+5}{k+1}$ if $n = 3k+1$, and $y_n = \frac{2k+3}{k+1}$ if $n = 3k+2$.
Then $\sup\{y_m : m \geq 3k+1\} = \frac{4k+5}{k+1}$ which converges to 4, hence $\limsup y_n = 4$.
(c) Let $u_n = \frac{3k+4}{k+1}$ if $n = 3k$, $u_n = (-1)^k \frac{4k+5}{k+1}$ if $n = 3k+1$, $u_n = \frac{2k+3}{k+1}$ if $n = 3k+2$.
Then $\sup\{u_m : m \geq 6k+1\} = \frac{4k+5}{k+1}$ which converges to 4, hence $\limsup u_n = 4$. $\square$

Let $\{s_n\}_{n=1}^\infty$ be a sequence of real numbers such that $s = \limsup_{n\to\infty} s_n$ is a real number. The following two properties are consequences of Definition 4.2.3.

For every $t > s$ there is n_0 such that for all $n \geq n_0$ we have $s_n < t$.   (4.2.2)

For any $u < s$ we have $s_n > u$ for infinitely many n.   (4.2.3)

Theorem 4.2.5. *Associate with a power series $\sum_{n=0}^\infty c_n(z-z_0)^n$ the number*

$$\frac{1}{R} = \limsup \sqrt[n]{|c_n|}, \tag{4.2.4}$$

with the convention $\frac{1}{0} = \infty$ and $\frac{1}{\infty} = 0$. Then
(i) if $R = 0$ the series converges only at $z = z_0$.
(ii) if $R = \infty$ the series converges absolutely for all z and uniformly on all closed disks $|z - z_0| \leq r$.
(iii) if $0 < R < \infty$ the series converges absolutely if $|z - z_0| < R$, diverges if $|z - z_0| > R$, and converges uniformly on a closed disk $|z - z_0| \leq r$ for all $r < R$.

Identity (4.2.4) that relates the number R having properties (i), (ii), and (iii) with the coefficients of the power series is called the **Cauchy-Hadamard formula**.

Proof. (iii): We first show that $\sum_{n=0}^\infty c_n(z-z_0)^n$ converges for $|z-z_0| < R$. To prove this, choose ρ such that $|z-z_0| < \rho < R$ which implies $\frac{1}{R} < \frac{1}{\rho}$. Then there is an n_0 such that $\sqrt[n]{|c_n|} < \frac{1}{\rho}$ for all $n \geq n_0$. Hence $|c_n||z-z_0|^n < \left(\frac{|z-z_0|}{\rho}\right)^n$ for all $n \geq n_0$, and $\sum_{n=0}^\infty c_n(z-z_0)^n$ converges absolutely by comparison to the convergent geometric series $\sum_{n=0}^\infty \left(\frac{|z-z_0|}{\rho}\right)^n$. Next we prove that for $r < R$ the series converges uniformly on the closed disk $\overline{B_r(z_0)}$. Indeed, pick ρ such that $r < \rho < R$. Then $1/\rho > 1/R$, and as before there is an n_0 such that for $n \geq n_0$ we have $\sqrt[n]{|c_n|} < \frac{1}{\rho}$. Then $|c_n||z-z_0|^n < \left(\frac{|z-z_0|}{\rho}\right)^n$ for all $n \geq n_0$, and thus $|c_n(z-z_0)^n| \leq \left(\frac{r}{\rho}\right)^n$. We conclude that the series $\sum_{n=n_0}^\infty c_n(z-z_0)^n$ converges uniformly on $\overline{B_r(z_0)}$ by the Weierstrass M-test.

Next we claim that $\sum_{n=0}^\infty c_n(z-z_0)^n$ diverges for $|z-z_0| > R$. Note that $\frac{1}{|z-z_0|} < \frac{1}{R}$ implies that $\sqrt[n]{|c_n|} > \frac{1}{|z-z_0|}$ for infinitely many n's, by definition of R. So, for $|z-z_0| > R$, we have $|c_n(z-z_0)^n| > 1$ for infinitely many n's, and thus the series diverges by the nth term test.

(ii): The proof of the assertion that $\sum_{n=0}^\infty c_n(z-z_0)^n$ converges for $|z-z_0| < R$ in case (iii) also applies when R is replaced by ∞ and $1/R$ by 0. The same is valid for uniform convergence on closed disks $|z-z_0| \leq r$ for all $r < \infty$.

(i): The proof of the assertion that $\sum_{n=0}^\infty c_n(z-z_0)^n$ diverges for $|z-z_0| > R$ in case (iii) also applies when R is replaced by 0 and $1/R$ by ∞. $\blacksquare$

Theorem 4.2.5 illustrates a typical dichotomy of power series. This behavior is determined by the number $R > 0$ with the property that the power series converges for $|z - z_0| < R$ and diverges for $|z - z_0| > R$.

Definition 4.2.6. The number R in all cases of Theorem 4.2.5 is called the **radius of convergence** of the series. As a convention, the radius of convergence is ∞ in case (ii) and 0 in case (i). In case (iii), the open disk $|z - z_0| < R$ is called the **disk of convergence** and the circle $|z - z_0| = R$ the **circle of convergence**.

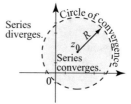

Fig. 4.5 The circle of convergence of a power series.

Remark 4.2.7. The radius of convergence of a power series $\sum_{n=0}^{\infty} c_n (z - z_0)^n$ is unique. Indeed, if there were two radii of convergence R and R', then for any point z with $R < |z - z_0| < R'$ the power series would have to both converge and diverge at z, which is impossible.

Example 4.2.8. (Cauchy-Hadamard formula) For the series $\sum_{n=0}^{\infty} c_n z^n$, where $c_n = 2^n$ if n is even and 2^{-n} if n is odd, formula (4.2.4) gives

$$\frac{1}{R} = \limsup \sqrt[n]{|c_n|} = \limsup \left\{ 2, \frac{1}{2}, 2, \frac{1}{2}, \dots \right\} = 2.$$

Thus the radius of convergence of this power series is $R = \frac{1}{2}$. □

Corollary 4.2.9. Let $\sum_{n=0}^{\infty} c_n (z - z_0)^n$ be a power series with radius of convergence $R > 0$. Define the function

$$f(z) = \sum_{n=0}^{\infty} c_n (z - z_0)^n \tag{4.2.5}$$

on the disk $B_R(z_0) = \{ z : |z - z_0| < R \}$. Then f is analytic on $B_R(z_0)$ and

$$f^{(k)}(z) = \sum_{n=k}^{\infty} n(n-1) \cdots (n-k+1) c_n (z - z_0)^{n-k}. \tag{4.2.6}$$

Proof. Since the power series (4.2.5) converges uniformly on all closed subdisks of $B_R(z_0)$, it follows from Corollary 4.1.13 that f is analytic on $B_R(z_0)$ and that the power series can be differentiated term by term as many times as necessary. Thus, for all integers $k \geq 1$ and all z in $B_R(z_0)$, (4.2.6) holds. ∎

If we evaluate (4.2.6) at $z = z_0$, all the terms in the series vanish except when $n = k$ and we get $f^{(k)}(z_0) = \frac{k!}{(k-k)!} c_k = k! c_k$. Solving for c_k, we get

$$c_k = \frac{f^{(k)}(z_0)}{k!} \qquad (k = 0, 1, 2, \ldots), \tag{4.2.7}$$

with the usual convention: $f^{(0)}(z) = f(z)$ and $0! = 1$. This formula relates the coefficients of the power series to the derivatives of f at the center of the series and is known as the **Taylor's formula**. A consequence of this formula is the following useful result.

Theorem 4.2.10. (Uniqueness of Power Series Expansions) *Let $R > 0$. If for all z satisfying $|z - z_0| < R$ we have*

$$\sum_{n=0}^{\infty} a_n(z - z_0)^n = f(z) = \sum_{n=0}^{\infty} b_n(z - z_0)^n,$$

then $a_n = b_n$ for all n. In particular, if $\sum_{n=0}^{\infty} c_n(z - z_0)^n = 0$ for all $|z - z_0| < R$, then $c_n = 0$ for all $n = 0, 1, 2, \ldots$.

Proof. By Taylor's formula, we have $a_n = \frac{f^{(n)}(z_0)}{n!} = b_n$. The second assertion of the theorem follows by setting $b_n = 0$. ∎

Power series can be integrated term by term over paths contained in their disk of convergence.

Example 4.2.11. (Term-by-term integration) Start with the geometric series

$$\frac{1}{1 + \zeta} = \sum_{n=0}^{\infty} (-1)^n \zeta^n \qquad |\zeta| < 1.$$

Since the integral of an analytic function on a disk is independent of path, integrating both sides from 0 to z, where $|z| < 1$, we obtain

$$\int_{[0,z]} \frac{1}{1 + \zeta} \, d\zeta = \sum_{n=0}^{\infty} (-1)^n \int_{[0,z]} \zeta^n \, d\zeta.$$

Using the independence of paths (Theorem 3.3.4) and the fact that $\frac{d}{dz} \mathrm{Log}\,(1 + z) = \frac{1}{1+z}$ and $\mathrm{Log}\, 1 = 0$, we obtain

$$\mathrm{Log}\,(1 + z) = \sum_{n=0}^{\infty} (-1)^n \frac{z^{n+1}}{n + 1} \qquad |z| < 1. \qquad \square$$

Example 4.2.12. (Term-by-term differentiation) Find the sum $\sum_{n=1}^{\infty} nz^n$ and determine its radius of convergence.

Solution. The series looks like the derivative of the geometric series

$$\frac{1}{1-z} = \sum_{n=0}^{\infty} z^n, \qquad |z| < 1,$$

Indeed, differentiating this series term by term we obtain the power series

$$\frac{d}{dz}\frac{1}{1-z} = \frac{1}{(1-z)^2} = \sum_{n=1}^{\infty} nz^{n-1}, \qquad |z| < 1.$$

Multiplying both sides by z,

$$\frac{z}{(1-z)^2} = \sum_{n=1}^{\infty} nz^n, \qquad |z| < 1.$$

In particular, the radius of convergence is 1. □

Example 4.2.13. (Matching series) Find the sum $\sum_{n=2}^{\infty} \frac{n}{(n-2)!} z^n$ and determine its radius of convergence.

Solution. The factorial in the denominator of the coefficients suggests that we look at the exponential series

$$e^z = \sum_{n=0}^{\infty} \frac{z^n}{n!}, \qquad z \in \mathbb{C}.$$

Differentiating the series twice term by term, we obtain

$$e^z = \sum_{n=2}^{\infty} \frac{n(n-1)}{n!} z^{n-2} = \sum_{n=2}^{\infty} \frac{z^{n-2}}{(n-2)!}, \qquad z \in \mathbb{C}.$$

Multiplying both sides by z^2, we deduce

$$z^2 e^z = \sum_{n=2}^{\infty} \frac{z^n}{(n-2)!}, \qquad z \in \mathbb{C}.$$

Differentiating term by term to get n in the numerator of the coefficients, and then multiplying by z, we obtain

$$\frac{d}{dz}\left(z^2 e^z\right) = 2ze^z + z^2 e^z = \sum_{n=2}^{\infty} \frac{n}{(n-2)!} z^{n-1}$$

$$(2z^2 + z^3)e^z = \sum_{n=2}^{\infty} \frac{nz^n}{(n-2)!}$$

for all complex number z. In particular, the radius of convergence is ∞. □

In the following example, we introduce the **Bessel functions** using power series and derive some of their properties.

Example 4.2.14. (Bessel functions of integer order) For $n = 0, 1, 2, \ldots$ the **Bessel function of order** n is defined by

$$J_n(z) = \sum_{k=0}^{\infty} \frac{(-1)^k}{k!(k+n)!} \left(\frac{z}{2}\right)^{2k+n} \qquad \text{for all } z.$$

For negative integers n, we set $J_n(z) = (-1)^n J_{-n}(z)$, $z \in \mathbb{C}$.
(a) Show that J_n is entire.
(b) Verify the identity $\dfrac{d}{dz} [z^n J_n(z)] = z^n J_{n-1}(z)$ for all complex numbers z.
(c) Prove the recurrence relation: $zJ_n'(z) + nJ_n(z) = zJ_{n-1}(z)$ for all z.

Solution. (a) Since $J_n(z)$ is defined by a power series, to prove that it is entire, it suffices by Theorem 4.2.10 to show that the power series converges for all z. Using the ratio test, we have for $z \neq 0$

$$\rho = \lim_{k \to \infty} \left| \frac{a_{k+1}}{a_k} \right| = \lim_{k \to \infty} \left| \frac{z^{2(k+1)+n}}{2^{2(k+1)+n}(k+1)!(k+1+n)!} \frac{2^{2k+n}k!(k+n)!}{z^{2k+n}} \right|$$

$$= \lim_{k \to \infty} 2^2 \frac{|z|^2}{(k+1)(k+1+n)} = 0.$$

Since $\rho < 1$, the series converges for all z.
(b) We use the series definition of J_n, differentiate term by term, and get

$$\frac{d}{dz} \left[z^n J_n(z) \right] = \frac{d}{dz} \sum_{k=0}^{\infty} \frac{(-1)^k 2^n}{k!(k+n)!} \left(\frac{z}{2}\right)^{2k+2n}$$

$$= \sum_{k=0}^{\infty} \frac{(-1)^k 2^n (k+n)}{k!(k+n)!} \left(\frac{z}{2}\right)^{2k+2n-1}$$

$$= z^n \sum_{k=0}^{\infty} \frac{(-1)^k}{k!(k+n-1)!} \left(\frac{z}{2}\right)^{2k+n-1}$$

$$= z^n J_{n-1}(z).$$

(c) The identity is true if $z = 0$. For $z \neq 0$, expand the left side of the identity in (b) using the product rule and get

$$z^n J_n'(z) + nz^{n-1} J_n(z) = z^n J_{n-1}(z).$$

Now (c) follows upon dividing through by $z^{n-1} \neq 0$.   □

Exercises 4.2

In Exercises 1–12, find the radius, disk, and circle of convergence of the power series.

1. $\displaystyle\sum_{n=0}^{\infty} (-1)^n \frac{z^n}{2n+1}$
2. $\displaystyle\sum_{n=0}^{\infty} \frac{n! z^n}{(2n)!}$
3. $\displaystyle\sum_{n=0}^{\infty} (2)^n \frac{(z-i)^n}{n!}$

4. $\displaystyle\sum_{n=0}^{\infty} \frac{(2z+1-i)^{2n}}{n^2+i}$ **5.** $\displaystyle\sum_{n=0}^{\infty} \frac{(4iz-2)^n}{2^n}$ **6.** $\displaystyle\sum_{n=0}^{\infty} (n-2)! \frac{z^n}{n^2}$

7. $\displaystyle\sum_{n=0}^{\infty} (n+i)^4 (z+6)^n$ **8.** $\displaystyle\sum_{n=0}^{\infty} \left(\frac{z+1}{3-i}\right)^{2n}$ **9.** $\displaystyle\sum_{n=0}^{\infty} (1-e^{in\frac{\pi}{4}})^n z^n$

10. $\displaystyle\sum_{n=0}^{\infty} z^{n!}$ **11.** $\displaystyle\sum_{n=0}^{\infty} (1+i^n)^n z^n$ **12.** $\displaystyle\sum_{n=0}^{\infty} (2+2i^n)^n z^n$

In Exercises 13–18, find the sum of the series and their radii of convergence. [Hint: Relate each series to a geometric series using differentiation, integration, or some other operation.]

13. $\displaystyle\sum_{n=1}^{\infty} 2nz^{n-1}$ **14.** $\displaystyle\sum_{n=2}^{\infty} \frac{z^n}{n(n-1)}$ **15.** $\displaystyle\sum_{n=1}^{\infty} (-1)^n \frac{z^{n-1}}{n(n+1)}$

16. $\displaystyle\sum_{n=1}^{\infty} \frac{(1-(-1)^n)}{n} z^n$ **17.** $\displaystyle\sum_{n=0}^{\infty} \frac{(3z-i)^n}{3^n}$ **18.** $\displaystyle\sum_{n=1}^{\infty} \frac{(z+1)^n}{n(n+1)}$

19. Cauchy's estimate for Taylor coefficients. Suppose that $f(z) = \sum_{n=0}^{\infty} c_n(z-z_0)^n$ is a power series with radius of convergence R, $0 < R < \infty$, and $|f(z)| \leq M$ for all $|z-z_0| < R$. Show that the coefficients satisfy Cauchy's estimate

$$|c_n| \leq \frac{M}{R^n} \qquad (n = 0, 1, 2, \ldots). \qquad (4.2.8)$$

[*Hint*: Use (4.2.7) and Theorem 3.9.1.]

20. Suppose that $f(z) = \sum_{n=0}^{\infty} c_n z^n$ converges for all z and that $|f(z)| \leq A + B|z|^p$ for some non-negative real numbers A, B, and p. Show that f is a polynomial of degree $\leq p$. (Compare with Liouville's theorem.)

21. Cauchy products of power series. Suppose that $\sum_{n=0}^{\infty} a_n(z-z_0)^n$ has radius of convergence $R_1 > 0$ and $\sum_{n=0}^{\infty} b_n(z-z_0)^n$ has radius of convergence $R_2 > 0$. Show that their Cauchy product is

$$\left(\sum_{n=0}^{\infty} a_n(z-z_0)^n\right)\left(\sum_{n=0}^{\infty} b_n(z-z_0)^n\right) = \sum_{n=0}^{\infty} c_n(z-z_0)^n, \qquad |z-z_0| < R, \qquad (4.2.9)$$

where

$$c_n = \sum_{k=0}^{n} a_k b_{n-k} = a_0 b_n + a_1 b_{n-1} + \cdots + a_{n-1} b_1 + a_n b_0, \qquad (4.2.10)$$

and R is at least as large as the smallest of R_1 and R_2. [*Hint*: Use Theorem 1.5.28.]

22. (a) Compute the Cauchy product of the series $\sum_{n=0}^{\infty} z^n$ and $\sum_{n=0}^{\infty} \frac{z^n}{n!}$. What is the radius of convergence of the Cauchy product series?
(b) Compute the Cauchy product of the series $\sum_{n=0}^{\infty} z^n$ and $1 - z$. What is the radius of convergence of the Cauchy product series? Notice how it is larger than the smaller of the radii of convergence of $\sum_{n=0}^{\infty} z^n$ and $1 - z$.
(c) Compute the Cauchy product of the series $\sum_{n=0}^{\infty} z^n$ with itself. What is the radius of convergence of the product series?

23. Show that the radius of convergence of a power series is equal to the radius of convergence of the k-times term-by-term differentiated power series. [*Hint*: It is enough to prove the result with $k = 1$ (why?). Use the comparison test or the Cauchy-Hadamard formula.]

24. Project Problem: The gamma function. For $\mathrm{Re}\, z > 0$, define

$$\Gamma(z) = \int_0^\infty e^{-t} t^{z-1}\, dt. \qquad (4.2.11)$$

This integral is improper on both ends and should be interpreted as $\lim_{A\downarrow 0, B\uparrow\infty} \int_A^B e^{-t} t^{z-1}\, dt$. Let Ω denote the half-plane $\Omega = \{z: \operatorname{Re} z > 0\}$.

(a) Show that $|e^{-t} t^{z-1}| \leq e^{-t} t^{\operatorname{Re} z - 1}$ and conclude that the integral in (4.2.11) converges absolutely for all z in Ω.

(b) Write $\Gamma(z) = \sum_{n=0}^{\infty} I_n(z)$, where $I_n(z) = \int_n^{n+1} e^{-t} t^{z-1}\, dt$. Show that I_n is analytic on Ω. [*Hint*: Use Theorem 3.8.5.]

(c) Let S be a bounded subset of Ω, and let $0 < \varepsilon \leq \operatorname{Re} z \leq K < \infty$ for all z in S. Show that for all z in S we have $|I_0(z)| \leq \varepsilon^{-1} = M_0$ and $|I_n(z)| \leq (n+1)^K e^{-n} = M_n$ for $n \geq 1$. Apply the Weierstrass M-test and use Corollary 4.1.13 to conclude that Γ is analytic on Ω.

(d) Use integration by parts to prove the **basic property** of the gamma function

$$\Gamma(z+1) = z\Gamma(z), \qquad \operatorname{Re} z > 0. \tag{4.2.12}$$

(e) Show by direct computation that $\Gamma(1) = 1$. Then use part (d) to prove that for all positive integers n we have $\Gamma(n) = (n-1)!$. For this reason, the gamma function is sometimes called the **generalized factorial function**.

25. Project Problem: The gamma function (continued). Follow the steps to derive the formula

$$\frac{\Gamma(z_1)\Gamma(z_2)}{\Gamma(z_1+z_2)} = 2\int_0^{\frac{\pi}{2}} (\cos^{2z_1-1}\theta)(\sin^{2z_2-1}\theta)\, d\theta, \qquad \operatorname{Re} z_1, \operatorname{Re} z_2 > 0.$$

All complex powers here are principal branches.

(a) Apply the formula with $z_1 = \frac{1}{2} = z_2$ and show that $\Gamma(\frac{1}{2}) = \sqrt{\pi}$.

(b) Make the change of variables $u^2 = t$ in (4.2.11) and obtain

$$\Gamma(z) = 2\int_0^{\infty} e^{-u^2} u^{2z-1}\, du, \qquad \operatorname{Re} z > 0.$$

(c) Use (b) to show that for $\operatorname{Re} z_1, \operatorname{Re} z_2 > 0$,

$$\Gamma(z_1)\Gamma(z_2) = 4\int_0^{\infty}\int_0^{\infty} e^{-(u^2+v^2)} u^{2z_1-1} v^{2z_2-1}\, du\, dv.$$

(d) Use polar coordinates, $u = r\cos\theta$, $v = r\sin\theta$, $du\, dv = r\, dr\, d\theta$ and get

$$\Gamma(z_1)\Gamma(z_2) = 2\Gamma(z_1+z_2)\int_0^{\pi/2} (\cos^{2z_1-1}\theta)(\sin^{2z_2-1}\theta)\, d\theta.$$

[*Hint*: Recall (a) as you compute the integral in r.]

4.3 Taylor Series

In the previous section, we showed that a power series is analytic on its disk of convergence. In this section, we prove essentially the converse that every analytic function can be expressed as a power series when restricted to a disk contained in its domain of analyticity. We are going to achieve this goal by reducing matters to the expansion of the geometric series

$$\frac{1}{1-w} = \sum_{n=0}^{\infty} w^n \qquad (|w| < 1). \tag{4.3.1}$$

Theorem 4.3.1. *Suppose that f is analytic in a region Ω and z_0 is in Ω. Let $B_R(z_0)$ be the largest open disk centered at z_0 and contained in Ω. Then f has the following* **Taylor series** *expansion around z_0*

$$f(z) = \sum_{n=0}^{\infty} \frac{f^{(n)}(z_0)}{n!}(z - z_0)^n, \qquad |z - z_0| < R. \tag{4.3.2}$$

Moreover, the expansion is unique in the following sense: If the uniformly convergent [on the closed subdisks of $B_R(z_0)$] series $\sum_{n=0}^{\infty} a_n(z - z_0)^n$ equals $f(z)$ for all z in $B_R(z_0)$, then we must have $a_n = \frac{f^{(n)}(z_0)}{n!}$ for all $n = 0, 1, 2, \ldots$.

Remark 4.3.2. The coefficient $\frac{f^{(n)}(z_0)}{n!}$ is called the nth **Taylor coefficient** of f. Taylor series are named after the English mathematician Brook Taylor (1685–1731), who wrote about them in 1715. The idea of expanding a function by a power series was known to mathematicians before Taylor and appeared in the work of Sir Isaac Newton (1643–1727), the Scottish mathematician James Gregory (1638–1675), and the Swiss mathematician Johann Bernoulli (1667–1748). Maclaurin series are named after the Scottish mathematician Colin Maclaurin (1698–1746), who used them and acknowledged that they are special cases of Taylor series. When $z_0 = 0$ in Theorem 4.3.1, the series is called the **Maclaurin series** of f.

Proof (Theorem 4.3.1). We show that f has a power series expansion that converges uniformly on a closed subdisk $\overline{B_\rho(z_0)}$ of $B_R(z_0)$. Let $0 < \rho < R$. Pick $r > 0$ such that $\rho < r < R$. Since f is analytic inside and on $C_r(z_0)$, Cauchy's formula implies that

$$f(z) = \frac{1}{2\pi i} \int_{C_r(z_0)} \frac{f(\zeta)}{\zeta - z} d\zeta \qquad \text{for all } |z - z_0| < r. \tag{4.3.3}$$

See Figure 4.6. We have

$$\frac{1}{\zeta - z} = \frac{1}{(\zeta - z_0) - (z - z_0)}$$
$$= \frac{1}{(\zeta - z_0)} \frac{1}{1 - \frac{z - z_0}{\zeta - z_0}}$$
$$= \frac{1}{(\zeta - z_0)} \sum_{n=0}^{\infty} \left(\frac{z - z_0}{\zeta - z_0}\right)^n$$
$$= \sum_{n=0}^{\infty} \frac{(z - z_0)^n}{(\zeta - z_0)^{n+1}},$$

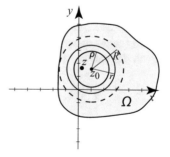

Fig. 4.6 Picture of the proof.

where we have used the geometric series identity (4.3.1) with $w = \frac{z - z_0}{\zeta - z_0}$ and $|w| < 1$, which is a consequence of the fact that $|z - z_0| \leq \rho < r = |\zeta - z_0|$. The function f is continuous and bounded on $C_r(z_0)$. For all ζ on $C_r(z_0)$, we have

$$\left| (z - z_0)^n \frac{f(\zeta)}{(\zeta - z_0)^{n+1}} \right| \le M \frac{\rho^n}{r^{n+1}} = \frac{M}{r} \left(\frac{\rho}{r} \right)^n = M_n.$$

Since $\rho/r < 1$, the series $\sum M_n$ converges and it follows from the Weierstrass M-test that the series

$$\frac{f(\zeta)}{\zeta - z} = \sum_{n=0}^{\infty} (z - z_0)^n \frac{f(\zeta)}{(\zeta - z_0)^{n+1}}$$

converges uniformly for all ζ on $C_r(z_0)$. Integrating term by term both sides of the equality

$$\frac{1}{2\pi i} \frac{f(\zeta)}{\zeta - z} = \frac{1}{2\pi i} \sum_{n=0}^{\infty} (z - z_0)^n \frac{f(\zeta)}{(\zeta - z_0)^{n+1}}$$

and using Cauchy's generalized integral formula, we deduce

$$f(z) = \frac{1}{2\pi i} \int_{C_r(z_0)} \frac{f(\zeta)}{\zeta - z} d\zeta$$

$$= \sum_{n=0}^{\infty} (z - z_0)^n \overbrace{\frac{1}{2\pi i} \int_{C_r(z_0)} \frac{f(\zeta)}{(\zeta - z_0)^{n+1}} d\zeta}^{\frac{f^{(n)}(z_0)}{n!}}$$

$$= \sum_{n=0}^{\infty} \frac{f^{(n)}(z_0)}{n!} (z - z_0)^n,$$

which completes the first part of the proof. The uniqueness of the power series is a consequence of Theorem 4.2.10. ∎

Remark 4.3.3. Since there is no guarantee that Ω is the largest region on which the function f in Theorem 4.3.1 is defined, the series (4.3.2) may converge on an open disk $B_{R_1}(z_0)$ larger than $B_R(z_0)$. Since a Taylor series is a power series, it follows from Theorem 4.2.5(*iii*) that the series (4.3.2) converges to $f(z)$ absolutely on $B_{R_1}(z_0)$ and uniformly on all closed subdisks of $B_R(z_0)$ (in fact all closed subdisks of $B_{R_1}(z_0)$.) Also, in view of Corollary 4.2.9, the series may be differentiated term by term in $B_R(z_0)$ as many times as we wish. This yields the Taylor series representation of the nth derivative

$$f^{(n)}(z) = \sum_{j=n}^{\infty} \frac{f^{(j)}(z_0)}{(j-n)!} (z - z_0)^{j-n}, \qquad |z - z_0| < R. \tag{4.3.4}$$

Example 4.3.4. (Maclaurin series of e^z, $\cos z$, and $\sin z$) Find the Maclaurin series expansions of   (a) e^z,   (b) $\cos z$,   (c) $\sin z$.

Solution. We first note that all three functions are entire, so the Maclaurin series will converge for all z; that is, $R = \infty$ in all three cases.

(a) Recall that the exponential function was defined as the power series: $e^z = \sum_{n=0}^{\infty} \frac{z^n}{n!}$ which converges for all complex numbers z. So this is the Maclaurin series expansion of e^z. Let us confirm this using Theorem 4.3.1. If $f(z) = e^z$, then $f^{(n)}(z) = e^z$, so $f^{(n)}(0) = 1$ for all n. Therefore, as expected, the Maclaurin series is

$$\sum_{n=0}^{\infty} \frac{f^{(n)}(0)}{n!} z^n = \sum_{n=0}^{\infty} \frac{z^n}{n!}, \qquad \text{for all } z.$$

(b) We have

$$f(z) = \cos z \Rightarrow f(0) = 1;$$
$$f'(z) = -\sin z \Rightarrow f'(0) = 0;$$
$$f''(z) = -\cos z \Rightarrow f''(0) = -1;$$
$$f'''(z) = \sin z \Rightarrow f'''(0) = 0;$$
$$f^{(4)}(z) = \cos z \Rightarrow f^{(4)}(0) = 1;$$

and so on. The values of the derivatives at 0 will repeat with period 4. Thus

$$\cos z = f(0) + \frac{f'(0)}{1!}z + \frac{f''(0)}{2!}z^2 + \frac{f'''(0)}{3!}z^3 + \frac{f^{(4)}(0)}{4!}z^4 + \cdots$$
$$= 1 - \frac{z^2}{2!} + \frac{z^4}{4!} + \cdots = \sum_{n=0}^{\infty}(-1)^n \frac{z^{2n}}{(2n)!} \qquad \text{for all } z.$$

(c) Using the relation $\frac{d}{dz}\cos z = -\sin z$ and the series for $\cos z$,

$$\sin z = -\frac{d}{dz}\cos z = -\frac{d}{dz}\left(1 - \frac{z^2}{2!} + \frac{z^4}{4!} + \cdots\right)$$
$$= \frac{2z}{2!} - \frac{4z^3}{4!} + \cdots = \frac{z}{1!} - \frac{z^3}{3!} + \cdots = \sum_{n=0}^{\infty}(-1)^n \frac{z^{2n+1}}{(2n+1)!}. \qquad \square$$

Example 4.3.4 shows the advantage of complex Taylor series over their real counterparts. A real Taylor series converges if and only if a certain remainder goes to zero. In the complex case, the remainder is irrelevant; the Taylor series will converge in the largest disk that you can fit inside the domain of definition of the analytic function.

As $\cos x = \sum_{n=0}^{\infty}(-1)^n \frac{x^{2n}}{(2n)!}$, $\sin x = \sum_{n=0}^{\infty}(-1)^n \frac{x^{2n+1}}{(2n+1)!}$ for $-\infty < x < \infty$, isn't it reasonable to expect that the series for $\cos z$ and $\sin z$ are obtained by merely replacing x by z? The answer is affirmative and we have the following useful result, which will allow us to turn well-known Taylor series from calculus into Taylor series of complex-valued functions.

Proposition 4.3.5. *Suppose that $f(x) = \sum_{n=0}^{\infty} c_n x^n$ is a Taylor series that converges for all $x \in (-R,R)$. Suppose that g is an analytic function on $|z| < \rho$ and $g(x) = f(x)$ for all real numbers x in some open interval that contains 0. Then the Taylor series of g is $g(z) = \sum_{n=0}^{\infty} c_n z^n$ for all $|z| < \rho$.*

Proof. Since g is analytic in a neighborhood of 0, it has a Taylor series expansion $g(z) = \sum_{n=0}^{\infty} \frac{g^{(n)}(0)}{n!} z^n$. We can compute the derivatives $g^{(n)}(0)$ by taking limits as $z \to 0$ with $z = x$ real, and since $g = f$ in an open interval that contains 0, we must have $g^{(n)}(0) = f^{(n)}(0)$ for all n. So the coefficients in the Taylor series of f and g are the same. $\blacksquare$

Example 4.3.6. (Manipulating Taylor series) Find the Maclaurin series expansions of the functions and determine their radii of convergence:

(a) ze^{z^2},   (b) $\frac{1}{1+z^2}$.

Solution. (a) Since $e^z = \sum_{n=0}^{\infty} \frac{z^n}{n!}$ is valid for all z, replacing z by z^2, we obtain

$$e^{z^2} = \sum_{n=0}^{\infty} \frac{(z^2)^n}{n!} = \sum_{n=0}^{\infty} \frac{z^{2n}}{n!}, \qquad \text{for all } z.$$

Multiplying both sides by z, we get the desired Maclaurin series

$$ze^{z^2} = \sum_{n=0}^{\infty} \frac{z^{2n+1}}{n!}, \qquad \text{for all } z.$$

(b) Start with the geometric series $\frac{1}{1-w} = \sum_{n=0}^{\infty} w^n$, which is valid if and only if $|w| < 1$. Replace w by $-z^2$, note that $|w| < 1 \Leftrightarrow |-z^2| < 1 \Leftrightarrow |z| < 1$ and get

$$\frac{1}{1+z^2} = \sum_{n=0}^{\infty} (-z^2)^n = \sum_{n=0}^{\infty} (-1)^n z^{2n}, \qquad \text{for } |z^2| < 1, \text{ equivalently, } |z| < 1. \quad \Box$$

Example 4.3.6(b) takes us back to a question that we mentioned in the introduction of Section 4.1. The function $f(x) = \frac{1}{1+x^2}$ is infinitely differentiable for all x. Yet its Maclaurin series $\sum_{n=0}^{\infty} (-1)^n x^{2n}$ converges only for $|x| < 1$. While it is difficult to explain this from a real analysis point of view, the justification is immediate in complex analysis: $\frac{1}{1+x^2}$ is the restriction to the real line of $\frac{1}{1+z^2}$, and since $\frac{1}{1+z^2}$ is not analytic at $z = \pm i$, its Maclaurin series cannot converge outside the disk $|z| < 1$. So in view of Proposition 4.3.5, the Maclaurin series of $\frac{1}{1+x^2}$ cannot converge outside $|x| < 1$.

If we wish to expand the function $\frac{a}{b-z}$ as a power series centered at a points other than 0, we can still use (4.3.1), as the next example illustrates.

Example 4.3.7. (Expanding about points other than 0) Find the Taylor series representation of $\frac{1}{1-z}$ about $z_0 = i$ and determine its radius of convergence.

Solution. Before computing the series, we can determine its radius of convergence from Theorem 4.3.1. Indeed the Taylor series of $f(z) = \frac{1}{1-z}$ about $z_0 = i$ converges in the largest disk centered at $z_0 = i$ on which f is analytic. Since f is analytic for all $z \neq 1$, the Taylor series has radius of convergence $R = \sqrt{2}$, which is the distance form i to 1 (Figure 4.7). To find the Taylor series about $z_0 = i$, we start with $\frac{1}{1-z}$, introduce the expression $z - i$ by adding and subtracting i to the denominator, and then use (4.3.1) with $w = \frac{z-i}{1-i}$. The details follow:

$$\frac{1}{1-z} = \frac{1}{1-i-(z-i)}$$

$$= \frac{1}{1-i} \frac{1}{1-\frac{z-i}{1-i}}$$

$$= \frac{1}{1-i} \frac{1}{1-w}$$

$$= \frac{1}{1-i} \sum_{n=0}^{\infty} w^n$$

$$= \frac{1}{1-i} \sum_{n=0}^{\infty} \left(\frac{z-i}{1-i} \right)^n$$

$$= \sum_{n=0}^{\infty} \frac{(z-i)^n}{(1-i)^{n+1}}.$$

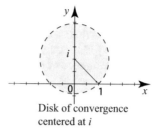

Disk of convergence
centered at i

Fig. 4.7 The function $\frac{1}{1-z}$ is not analytic at $z = 1$. Its Taylor series converges in the largest disk around i that does not contain 1. This disk has radius $\sqrt{2}$.

We have used (4.3.1), with $w = \frac{z-i}{1-i}$. The series converges if and only if $|w| < 1$, equivalently, $|z - i| < |1 - i| = \sqrt{2}$. This confirms that the radius of convergence of the series is $\sqrt{2}$, as we surmised earlier.   □

Example 4.3.8. (Expanding about points other than 0) Find the Taylor series representation of $\frac{1}{4-iz}$ about $z_0 = 3$ and determine its radius of convergence.

Solution. The function is analytic for all z except when $4 - iz = 0$ or $z = -4i$. Its Taylor series about $z_0 = 3$ converges in the disk centered at 3 and reaching as far as $-4i$. Since the distance from 3 to $-4i$ is 5, we conclude the radius of convergence of the series should be 5. To find the Taylor series, we use ideas similar to Example 4.3.7:

$$\frac{1}{4-iz} = \frac{1}{i(-4i-z)} = \frac{-i}{-4i-z}$$

$$= \frac{-i}{-4i-3-(z-3)} = \frac{-i}{-3-4i} \frac{1}{1-\frac{z-3}{-3-4i}}$$

$$= \frac{-i}{-3-4i} \frac{1}{1-w} = \frac{-i}{-3-4i} \sum_{n=0}^{\infty} w^n = \frac{-i}{-3-4i} \sum_{n=0}^{\infty} \left(\frac{z-3}{-3-4i} \right)^n$$

$$= -i \sum_{n=0}^{\infty} \frac{(z-3)^n}{(-3-4i)^{n+1}} = -i \sum_{n=0}^{\infty} \frac{(-3+4i)^{n+1}}{25^{n+1}}(z-3)^n.$$

In the last step, we multiplied and divided the terms of the series by the conjugate of the denominator and used

$$(-3-4i)^{n+1}(-3+4i)^{n+1} = [(-3-4i)(-3+4i)]^{n+1} = 25^{n+1}.$$

To obtain the Taylor series, we used (4.3.1) with $w = \frac{z-3}{-3-4i}$. The series converges if and only if $|w| < 1$, equivalently, $|z-3| < |-3-4i| = 5$, which confirms that the radius of convergence of the series is 5. $\qquad\square$

Example 4.3.9. (Expanding about points other than 0) Find the Taylor series representation of $\sin z$ and $\cos z$ about a point z_0.

Solution. Begin by writing

$$\sin(z) = \sin(z - z_0 + z_0) = \cos(z_0)\sin(z - z_0) + \sin(z_0)\cos(z - z_0) \quad (4.3.5)$$
$$\cos(z) = \cos(z - z_0 + z_0) = \cos(z_0)\cos(z - z_0) - \sin(z_0)\sin(z - z_0). \quad (4.3.6)$$

It follows from (4.3.5) and (4.3.6) that

$$\sin(z) = \cos(z_0) \sum_{n=0}^{\infty} \frac{(-1)^n}{(2n+1)!}(z - z_0)^{2n+1} + \sin(z_0) \sum_{n=0}^{\infty} \frac{(-1)^n}{(2n)!}(z - z_0)^{2n}$$

and

$$\cos(z) = \cos(z_0) \sum_{n=0}^{\infty} \frac{(-1)^n}{(2n)!}(z - z_0)^{2n} - \sin(z_0) \sum_{n=0}^{\infty} \frac{(-1)^n}{(2n+1)!}(z - z_0)^{2n+1}.$$

Thus the Taylor coefficients of the power series of $\sin(z)$ centered at z_0 are $a_k = \cos(z_0)\frac{(-1)^n}{(2n+1)!}$ if $k = 2n+1$ and $a_k = \sin(z_0)\frac{(-1)^n}{(2n)!}$ if $k = 2n$. Also the Taylor coefficients of the power series of $\cos(z)$ centered at z_0 are $a_k = -\sin(z_0)\frac{(-1)^n}{(2n+1)!}$ if $k = 2n+1$ and $a_k = \cos(z_0)\frac{(-1)^n}{(2n)!}$ if $k = 2n$. $\qquad\square$

Factoring Zeros and Extending Analytic Functions

When a polynomial p has a zero at z_0, we can write $p(z) = (z - z_0)g(z)$, where g is another polynomial. In particular, the function $p(z)/(z - z_0)$ (which equals $g(z)$ for all $z \neq z_0$) can be made analytic even at z_0 by defining it to be $g(z_0)$ at z_0. The same is true for analytic functions in general.

Example 4.3.10. (Extending $\dfrac{\sin z}{z}$ to an entire function) Show that the function

$$
g(z) = \begin{cases} \dfrac{\sin z}{z} & \text{if } z \neq 0, \\ 1 & \text{if } z = 0, \end{cases}
$$

is entire.

Solution. We have

$$
\sin z = z - \frac{z^3}{3!} + \frac{z^5}{5!} - \frac{z^7}{7!} + \cdots = z\left(1 - \frac{z^2}{3!} + \frac{z^4}{5!} - \frac{z^6}{7!} + \cdots\right) = z g_1(z),
$$

where $g_1(z) = 1 - \frac{z^2}{3!} + \frac{z^4}{5!} - \frac{z^6}{7!} + \cdots$. The Taylor series that defines $g_1(z)$ converges for all z. Hence $g_1(z)$ is entire by Corollary 4.2.9. Clearly, $g_1(z) = \frac{\sin z}{z}$ if $z \neq 0$ and $g_1(0) = 1$; thus it coincides with g and hence g is entire. $\qquad\square$

In Example 4.3.10, observe that $g(0) = \frac{d}{dz}\sin z\big|_{z=0} = \cos 0 = 1$. With this in mind, we have the following useful result.

Theorem 4.3.11. *Suppose that f is an analytic function on a region Ω and $z_0 \in \Omega$. Define*

$$
g(z) = \begin{cases} \dfrac{f(z)-f(z_0)}{z-z_0} & \text{if } z \neq z_0, \\ f'(z_0) & \text{if } z = z_0. \end{cases} \tag{4.3.7}
$$

Then g is analytic in Ω.

Proof. Clearly g is analytic for all $z \neq z_0$ in Ω. To show that g is analytic at z_0, let $B_R(z_0)$ be an open disk contained in Ω. We know from Theorem 4.3.1 that f has a Taylor series expansion about z_0 that converges in $B_R(z_0)$. So

$$
f(z) = f(z_0) + f'(z_0)(z-z_0) + \frac{f''(z_0)}{2!}(z-z_0)^2 + \cdots
$$

$$
f(z) - f(z_0) = f'(z_0)(z-z_0) + \frac{f''(z_0)}{2!}(z-z_0)^2 + \cdots
$$

$$
= (z-z_0)\left[f'(z_0) + \frac{f''(z_0)}{2!}(z-z_0) + \cdots\right] = (z-z_0)g(z),
$$

where $g(z) = f'(z_0) + \frac{f''(z_0)}{2!}(z-z_0) + \cdots$ for all z in $B_R(z_0)$. Hence g is analytic in $B_R(z_0)$ by Corollary 4.2.9, and g is the function in (4.3.7). $\qquad\blacksquare$

Theorem 4.3.11 is one of several interesting properties of zeros of analytic functions. Such results are derived in Section 4.6. We end the section with an example that deals with an important family of numbers.

Example 4.3.12. (Bernoulli numbers) Let $f(z) = \frac{z}{e^z - 1}$, $z \neq 0$, $f(0) = 1$.
(a) Show that f is analytic at 0.
(b) By Theorem 4.3.1, f has a Maclaurin series expansion. Show that its radius of convergence is $R = 2\pi$.
(c) Write the Maclaurin series in the form

$$f(z) = \sum_{n=0}^{\infty} \frac{B_n}{n!} z^n, \qquad |z| < 2\pi.$$

The number B_n is called the nth **Bernoulli number**. Show that $B_0 = 1$, and derive the recurrence relation

$$B_n = -\frac{1}{n+1} \sum_{k=0}^{n-1} \binom{n+1}{k} B_k, \qquad n \geq 1. \tag{4.3.8}$$

(d) Find $B_0, B_1, B_2, \ldots, B_{12}$, with the help of the recursion formula and a calculator.
(e) Show that $B_{2n+1} = 0$ for $n \geq 1$. Here $\binom{n}{k} = \frac{n!}{k!(n-k)!}$ is the binomial coefficient.

Solution. (a) Consider $g(z) = \frac{1}{f(z)} = \frac{e^z - 1}{z}$ for $z \neq 0$, and $g(0) = 1$. By Theorem 4.3.11, g is analytic at 0. Since $g(0) \neq 0$, $\frac{1}{g} = f$ is therefore analytic at $z = 0$.
(b) The Maclaurin series of f converges in the largest disk around $z_0 = 0$ on which f is defined and is analytic. Away from zero, f is analytic on the set where $e^z - 1 \neq 0$. Since $e^z = 1$ precisely when z is an integer multiple of $2\pi i$, we see that the Maclaurin series converges for all $|z| < 2\pi$, and the radius of convergence is 2π.

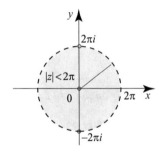

Fig. 4.8 The domain of $\frac{z}{e^z - 1}$.

(c) Multiplying both sides of the Maclaurin series expansion of $\frac{z}{e^z - 1}$ by $e^z - 1$ and using the Maclaurin series $e^z - 1 = z + \frac{z^2}{2!} + \frac{z^3}{3!} + \cdots = \sum_{n=1}^{\infty} \frac{z^n}{n!}$, we obtain

$$z = (e^z - 1) \sum_{n=0}^{\infty} \frac{B_n}{n!} z^n = \sum_{n=1}^{\infty} \overbrace{\frac{z^n}{n!}}^{a_n z^n} \sum_{n=0}^{\infty} \frac{B_n}{n!} z^n = \sum_{n=1}^{\infty} c_n z^n, \qquad |z| < 2\pi, \tag{4.3.9}$$

where c_n will be computed from the Cauchy product formula (Theorem 1.5.28). Note that because we are multiplying by the power series of $e^z - 1$ whose first term is z, the first term in the Cauchy product will have degree greater than or equal to 1 (thus $c_0 = 0$). We have for each $n \geq 1$

$$c_n = \sum_{k=0}^{n-1} \frac{B_k}{k!} \frac{1}{(n-k)!} \qquad \text{(because } a_0 = 0)$$

$$= \frac{1}{n!} \sum_{k=0}^{n-1} \frac{n!}{k!(n-k)!} B_k = \frac{1}{n!} \sum_{k=0}^{n-1} \binom{n}{k} B_k.$$

By the uniqueness of the power series expansion, (4.3.9) implies that $c_1 = 1$, and $c_n = 0$ for all $n \geq 2$. Thus,

$$c_1 = 1 \quad \Rightarrow \quad \frac{1}{1!} B_0 = 1 \quad \Rightarrow \quad B_0 = 1;$$

$$c_n = 0, \quad n \geq 2 \quad \Rightarrow \quad \frac{1}{n!} \sum_{k=0}^{n-1} \binom{n}{k} B_k = 0, \quad n \geq 2.$$

Changing n to $n+1$ in the last identity, we see that, for $n \geq 1$,

$$\frac{1}{(n+1)!} \sum_{k=0}^{n} \binom{n+1}{k} B_k = 0,$$

which implies that

$$\frac{1}{(n+1)!} \sum_{k=0}^{n-1} \binom{n+1}{k} B_k + \frac{1}{(n+1)!} \binom{n+1}{n} B_n = 0.$$

Now, realizing that $\binom{n+1}{n} = n+1$, we deduce (4.3.8).

(d) With the aid of a computer and the recurrence relation (4.3.8) we generated the Bernoulli numbers shown in Table 1.

n	0	1	2	3	4	5	6	7	8	9	10	11	12
B_n	1	$-\dfrac{1}{2}$	$\dfrac{1}{6}$	0	$-\dfrac{1}{30}$	0	$\dfrac{1}{42}$	0	$-\dfrac{1}{30}$	0	$\dfrac{5}{66}$	0	$-\dfrac{691}{2730}$

Table 2. Bernoulli numbers.

(e) As the table suggests, $B_{2n+1} = 0$ for $n \geq 1$. This is clearly a fact about an even function. Consider $f(z) - B_1 z$, i.e., eliminate the linear term of the Maclaurin series:

$$\frac{z}{e^z - 1} + \frac{z}{2} = \frac{z + ze^z}{2(e^z - 1)} = \frac{z(e^{\frac{z}{2}} + e^{-\frac{z}{2}})}{2(e^{\frac{z}{2}} - e^{-\frac{z}{2}})} = \frac{z}{2} \coth\left(\frac{z}{2}\right). \tag{4.3.10}$$

This is an even function. Hence all the odd-numbered coefficients in its Maclaurin series are zero (Exercise 29). This implies that $B_{2n+1} = 0$ for all $n \geq 1$. □

Using (4.3.10) and the Maclaurin series of $\frac{z}{e^z-1}$, we see that for $|z| < 2\pi$

$$\frac{z}{2}\coth\left(\frac{z}{2}\right) = \frac{z}{2} + 1 - \frac{z}{2} + \sum_{n=2}^{\infty}\frac{B_n}{n!}z^n = 1 + \sum_{n=1}^{\infty}\frac{B_{2n}}{(2n)!}z^{2n} = \sum_{n=0}^{\infty}\frac{B_{2n}}{(2n)!}z^{2n};$$

and upon replacing z by $2z$,

$$z\coth z = \sum_{n=0}^{\infty}\frac{2^{2n}B_{2n}}{(2n)!}z^{2n}, \qquad |z| < \pi. \qquad (4.3.11)$$

In the exercises, important Taylor series are investigated in terms of Bernoulli numbers.

Exercises 4.3

In Exercises 1–6, determine the radii of convergence of the Taylor series of the functions centered at the points shown, without explicitly writing the series.

1. e^{z-1}, $\quad z_0 = 0$ 2. $\dfrac{\sin z}{e^z}$, $\quad z_0 = 1 + 7i$ 3. $\dfrac{z}{z-3i}$, $\quad z_0 = 0$

4. $\sin\left(\dfrac{z+1}{z-1}\right)$, $\quad z_0 = 0$ 5. $\dfrac{z+1}{z-i}$, $\quad z_0 = 2 + i$ 6. $\dfrac{\sin z}{z^2 + 4}$, $\quad z_0 = 3$

In Exercises 7–12, find the Taylor series expansion of the function at the indicated point. Determine the radius of convergence.

7. $\dfrac{1}{1-z}$, $\quad z_0 = 1 + i$ 8. $\dfrac{3}{i-2z}$, $\quad z_0 = i$ 9. $\dfrac{2i}{3-iz}$, $\quad z_0 = -1$

10. $\dfrac{1}{1+i-(2+i)z}$, $\quad z_0 = 0$ 11. $\dfrac{2}{1-z^2}$, $\quad z_0 = i$ 12. $\dfrac{4}{z^2+2z}$, $\quad z_0 = 1$

In Exercises 13–18, use a known Taylor series to derive the Taylor series of the shown function around the indicated point z_0. Determine the radius of convergence in each case.

13. $\dfrac{z}{1-z}$, $\quad z_0 = 0$ 14. $\dfrac{z^2+1}{z-1}$, $\quad z_0 = 0$ 15. $\dfrac{2z}{(z+i)^3}$, $\quad z_0 = 0$

16. ze^{3z^2}, $\quad z_0 = 0$ 17. ze^z, $\quad z_0 = 1$ 18. $\cos^2 z$, $\quad z_0 = 0$

19. Use the Taylor series of ze^{z^2} from Example 4.3.6(a) to compute $\frac{d^n}{dz^n}\left(ze^{z^2}\right)\big|_{z=0}$ for $n = 0, 1, 2, \ldots$.

20. Compute $\frac{d^n}{dz^n}\sin(z^2)\big|_{z=0}$ for $n = 0, 1, 2, \ldots$.

21. Find the Maclaurin series of $\frac{1}{(1-z)(2-z)}$ in two different ways, as indicated.
(a) Prove the partial fractions decomposition $\frac{1}{(1-z)(2-z)} = \frac{1}{1-z} - \frac{1}{2-z}$, and then use a geometric series expansion of each term in the partial fraction decomposition.
(b) Use a geometric series to expand $\frac{1}{1-z}$ and $\frac{1}{2-z}$ separately, then form their Cauchy product.
(c) Verify that your answers are the same in (a) and (b) and give the radius of convergence of the Maclaurin series of $f(z)$.

22. Let $z_1 \neq z_2$ be two complex numbers and suppose that $0 < |z_1| \leq |z_2|$. Show that, for $|z| < |z_1|$,

$$\frac{1}{(z_1-z)(z_2-z)} = \frac{1}{z_1-z_2}\sum_{n=0}^{\infty}\frac{(z_1^{n+1} - z_2^{n+1})}{(z_1z_2)^{n+1}}z^n.$$

23. Find the Maclaurin series of $f(z) = \frac{1}{1+z+z^2}$ and determine its radius of convergence. You may use the result of Exercise 22.

24. Let $z_1 \neq 0$. (a) Show that

$$\frac{1}{z_1 - z} = \frac{1}{z_1} \sum_{n=0}^{\infty} \left(\frac{z}{z_1} \right)^n, \qquad |z| < |z_1|.$$

(b) For a positive integer k, show that, for $|z| < |z_1|$,

$$\frac{1}{(z_1 - z)^{k+1}} = \frac{1}{z_1^{k+1}} \sum_{n=k}^{\infty} \binom{n}{k} \left(\frac{z}{z_1} \right)^{n-k} = \frac{1}{z_1^{k+1}} \sum_{n=0}^{\infty} \binom{n+k}{k} \left(\frac{z}{z_1} \right)^n.$$

In Exercises 25–26, find the Maclaurin series of the functions and determine their radii of convergence. You may use the result of Exercise 24.

25. $f(z) = \dfrac{1}{(z - 2i)^3}.$

26. $f(z) = \dfrac{1}{(2z - i + 1)^6}.$

In Exercises 27–28, show that the function is analytic at $z_0 = 0$ and find its Maclaurin series. What is the radius of convergence of the series?

27. $f(z) = \begin{cases} \frac{\cos z - 1}{z} & \text{if } z \neq 0, \\ 0 & \text{if } z = 0. \end{cases}$

28. $f(z) = \begin{cases} \frac{e^z - 1}{z} & \text{if } z \neq 0, \\ 1 & \text{if } z = 0. \end{cases}$

29. Even and odd functions. Recall that a function f is even if $f(-z) = f(z)$ and is odd if $f(-z) = -f(z)$. Suppose that f is analytic on $|z| < R$ and write $f(z) = \sum_{n=0}^{\infty} c_n z^n$, $|z| < R$, $R > 0$. Show that

(a) f is even if and only if $c_{2n+1} = 0$ for all $n = 0, 1, 2, \ldots$. [*Hint:* $f(z) - f(-z) = 0$.]

(b) f is odd if and only if $c_{2n} = 0$ for all $n = 0, 1, 2, \ldots$.

30. Let z_0, z_1, z_2 be distinct complex numbers such that

$$|z_1 - z_0| < |z_2 - z_0|.$$

(See the adjacent figure.) Show that

$$\frac{1}{z_2 - z_1} = \sum_{n=0}^{\infty} \frac{(z_1 - z_0)^n}{(z_2 - z_0)^{n+1}}.$$

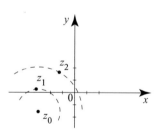

31. Maclaurin series of the tangent, cotangent, and cosecant.

(a) Replace z by iz in (4.3.11) and simplify to obtain

$$z \cot z = \sum_{n=0}^{\infty} (-1)^n \frac{2^{2n} B_{2n}}{(2n)!} z^{2n}, \qquad |z| < \pi.$$

(b) Derive the Maclaurin series of the tangent:

$$\tan z = \sum_{n=1}^{\infty} (-1)^{n-1} \frac{2^{2n}(2^{2n} - 1)B_{2n}}{(2n)!} z^{2n-1}, \qquad |z| < \frac{\pi}{2}.$$

[*Hint:* $\tan z = \cot z - 2\cot(2z)$.]

(c) Prove

$$z \csc z = \sum_{n=0}^{\infty} (-1)^{n-1} \frac{(2^{2n} - 2)B_{2n}}{(2n)!} z^{2n}, \qquad |z| < \pi.$$

[*Hint:* $\cot z + \tan z = 2\csc 2z$.]

32. Fibonacci numbers. Discovered in the early thirteenth century by the great mathematician Leonardo of Pisa (1180–1250), who wrote under the name of Fibonacci, this sequence of integers is defined inductively by $c_0 = c_1 = 1$, and $c_n = c_{n-1} + c_{n-2}$ for $n \geq 2$.

(a) Find $c_0, c_1, \ldots, c_{10}$.

In 1843, the French mathematician Jacques-Philippe-Marie Binet (1786–1856) found an explicit formula for c_n in terms of n. This identity, known as **Binet's formula**, states that

$$c_n = \frac{1}{\sqrt{5}}\left[\left(\frac{1+\sqrt{5}}{2}\right)^{n+1} - \left(\frac{1-\sqrt{5}}{2}\right)^{n+1}\right].$$

We use complex analysis to derive Binet's formula.

(b) Suppose for a moment that there is a function f, analytic at 0 and whose Maclaurin coefficients are the Fibonacci numbers: $f(z) = \sum_{n=0}^{\infty} c_n z^n$, $|z| < R$. Show that f is a solution of the equation $f(z) = 1 + zf(z) + z^2 f(z)$.

(c) Conclude that $f(z) = \frac{1}{1-z-z^2}$ and find the radius of convergence of its Maclaurin series.

(d) Compute the Maclaurin series and derive Binet's formula.

[*Hint*: $\frac{1}{1-z-z^2} = \frac{1}{\sqrt{5}}\left(\frac{1}{-z+\frac{-1+\sqrt{5}}{2}} + \frac{1}{z+\frac{1+\sqrt{5}}{2}}\right)$. Now use geometric series expansions.]

33. The Lucas numbers. This sequence of integers $\{l_n\}$, named after the French mathematician Edouard Lucas (1842–1891), is defined by a recurrence relation similar to the Fibonacci sequence $l_n = l_{n-1} + l_{n-2}$ but with $l_0 = 1$ and $l_1 = 3$. Take an approach similar to Exercise 32 and prove the following.

(a) Let $f(z) = \sum_{n=0}^{\infty} l_n z^n$, $|z| < R$. Show that f is a solution of the equation $f(z) = 1 + 2z + zf(z) + z^2 f(z)$, and conclude that $f(z) = \frac{1+2z}{1-z-z^2}$.

(b) Compute the Maclaurin series of f and derive the following formula for the Lucas sequence:

$$l_n = \left(\frac{1+\sqrt{5}}{2}\right)^{n+1} + \left(\frac{1-\sqrt{5}}{2}\right)^{n+1}, \quad n \geq 0.$$

34. Project Problem: Euler numbers. Like the Bernoulli numbers, the Euler numbers can be generated from the coefficients of special Maclaurin series.

(a) Show that $\sec z$ is analytic at $z_0 = 0$ and its Maclaurin series has radius of convergence $R = \frac{\pi}{2}$.

(b) Show that the odd-numbered coefficients in the Maclaurin series are all 0. Then write

$$\sec z = \sum_{n=0}^{\infty} (-1)^n \frac{E_{2n}}{(2n)!} z^{2n}, \quad |z| < \frac{\pi}{2}.$$

The numbers E_n are called the **Euler numbers**.

(c) Derive the recursion formula

$$E_0 = 1, \quad \sum_{k=0}^{n}\binom{2n}{2k} E_{2k} = 0, \quad n > 0.$$

(d) Use (c) to derive $E_2 = -1$, $E_4 = 5$, $E_6 = -61$, $E_8 = 1385$.

35. Project Problem: Log series and the inverse tangent. In this exercise, we will derive several useful series involving the principal branch of the logarithm and the series of the inverse tangent.

(a) Show that

$$\frac{1+z}{1-z} = \frac{1-|z|^2}{|1-z|^2} + 2i\frac{\text{Im}\,z}{|1-z|^2}.$$

(b) Conclude from (a) that if $|z| < 1$, then $\text{Re}\left(\frac{1+z}{1-z}\right) > 0$ and hence $\text{Log}\left(\frac{1+z}{1-z}\right)$ is analytic on $|z| < 1$.

(c) Show that for $|z| < 1$, $\text{Log}\left(\frac{1+z}{1-z}\right) = \text{Log}(1+z) - \text{Log}(1-z)$. [*Hint*: Show that the derivatives are equal.]

In parts (d)-(g), starting with the Log series in Example 4.2.11, derive the Maclaurin series of the functions in the open disk $|z| < 1$:

(d) $\text{Log}(1+z) = \sum_{n=1}^{\infty} \frac{(-1)^{n-1}}{n} z^n$;

(e) $-\text{Log}(1-z) = \sum_{n=1}^{\infty} \frac{z^n}{n}$;

(f) $\text{Log}\left(\dfrac{1+z}{1-z}\right) = 2\sum_{n=0}^{\infty}\dfrac{z^{2n+1}}{2n+1}$; (g) $\dfrac{1}{2i}\text{Log}\left(\dfrac{1+iz}{1-iz}\right) = \sum_{n=0}^{\infty}\dfrac{(-1)^n}{2n+1}z^{2n+1}$.

(h) Let $\phi(z) = \frac{1}{2i}\text{Log}\left(\frac{1+iz}{1-iz}\right)$, $|z| < 1$. Using the definition of $\tan z$, verify that $\tan(\phi(z)) = z$. Thus, $\phi(z)$ is the inverse function of $\tan z$. We call $\phi(z)$ the **principal branch of the inverse tangent** and denote it by $\text{Arctan}\, z$. Thus,

$$\text{Arctan}\, z = \frac{1}{2i}\text{Log}\left(\frac{1+iz}{1-iz}\right) = \sum_{n=0}^{\infty}\frac{(-1)^n}{2n+1}z^{2n+1}, \quad |z| < 1.$$

(i) Using term-by-term differentiation, show that

$$\frac{d}{dz}\text{Arctan}\, z = \frac{1}{1+z^2}, \quad |z| < 1.$$

36. Project Problem: Binomial series. In this exercise we study the binomial series with complex exponents. Let α be a complex number. Define the **generalized binomial coefficient** by

$$\binom{\alpha}{0} = 1, \quad \binom{\alpha}{n} = \frac{\alpha(\alpha-1)\cdots(\alpha-n+1)}{n!}, \quad n \geq 1. \tag{4.3.12}$$

Use Theorem 4.3.1 to show that for $|z| < 1$

$$(1+z)^{\alpha} = \sum_{n=0}^{\infty}\binom{\alpha}{n}z^n, \tag{4.3.13}$$

where $(1+z)^{\alpha}$ is the principal branch, $(1+z)^{\alpha} = e^{\alpha\,\text{Log}(1+z)}$. Note that if $\alpha = m$ is a positive integer, then from (4.3.12), $\binom{m}{n} = 0$ for all $n > m$ (why?) and the series (4.3.13) reduces to a finite sum

$$(1+z)^m = \sum_{n=0}^{m}\binom{m}{n}z^n,$$

which is the familiar **binomial theorem**. The finite sum clearly converges for all z, and so the restriction $|z| < 1$ is not necessary if α is a positive integer.

37. Derive the power series expansion

$$(1+z)^{\frac{1}{2}} = \sum_{n=0}^{\infty}\frac{(-1)^{n-1}}{2^{2n}(2n-1)}\binom{2n}{n}z^n, \quad |z| < 1.$$

38. Project Problem: The Catalan numbers. The Catalan numbers are named after the French mathematician Eugène Charles Catalan (1814–1894). They arise in many combinatorics problems, such as the following:

- The number of ways in which parentheses can be placed in a sequence of numbers to be multiplied, two at a time.
- The number of ways a polygon with $n+1$ sides can be cut into $n-1$ triangles by nonintersecting diagonals.
- The number of paths of length $2(n-1)$ through an n by n grid that connect two diagonally opposite vertices that stay below or on the main diagonal.

Denote the nth Catalan number by c_n. Take the first interpretation of these numbers, and say we have a sequence of three numbers a, b, c. We can multiply them two at a time in exactly two ways: $((ab)c)$ and $(a(bc))$. Thus $c_3 = 2$. With a sequence of four numbers, we can have $(((ab)c)d)$ or $((ab)(cd))$ or $((a(bc))d)$ or $(a((bc)d))$ or $(a(b(cd)))$. Thus $c_4 = 5$. In general, if we have a sequence of n numbers, we can break it into two subsequences of k and $n-k$ numbers. Then each arrangement of parentheses from the k sequence can be combined with an arrangement of parentheses of the $n-k$ sequence to yield an arrangement of parentheses of the n-sequence. Since we have c_k possibilities for the k-sequence and c_{n-k} possibilities for the $(n-k)$-sequence, we get this way $c_k \times c_{n-k}$ possibilities for the n-sequence. But this can be done for $k = 1, 2, \ldots, n-1$, and so we have the recursion relation

$$c_n = \sum_{k=1}^{n-1} c_k c_{n-k}, \quad \text{for } n \geq 2.$$

We have $c_2 = 1$, and we will take $c_1 = 1$ (for reference, $c_0 = 0$). Our goal in this exercise is to derive a formula for c_n in terms of n.

(a) Suppose that there is an analytic function $f(z) = \sum_{n=0}^{\infty} c_n z^n$, where c_n is the nth Catalan number. From the recursion relation, show that f satisfies the equation $f(z) = z + \left(f(z)\right)^2$. [*Hint*: Use Cauchy products.]

(b) Solve for f and choose the solution that gives $f(0) = 0$. You should get $f(z) = \frac{1 - \sqrt{1-4z}}{2}$.

(c) Expand f in a power series around 0, using Exercise 37. You should get

$$f(z) = \frac{1}{2} + \frac{1}{2} \sum_{n=0}^{\infty} \frac{1}{2n-1} \binom{2n}{n} z^n \qquad |z| < \frac{1}{4}.$$

(d) Conclude that for $n \geq 2$,

$$c_n = \frac{1}{2(2n-1)} \binom{2n}{n}.$$

(e) Verify your answer by computing c_3 and c_4 from the formula. What is c_5? Exhibit all c_5 arrangements of parentheses in a sequence of 5 numbers.

(f) Show that $\lim_{n \to \infty} \frac{c_{n+1}}{c_n} = 4$. Thus the Catalan numbers grow at a rate of 4.

39. Weierstrass double series theorem. Suppose that $F(z) = \sum_{k=0}^{\infty} f_k(z)$ converges uniformly for $|z - z_0| \leq \rho$ for every $\rho < R$, and that, for all k, $f_k(z)$ is analytic in $|z - z_0| < R$. Write $f_k(z) = \sum_{n=0}^{\infty} a_n^{(k)} (z - z_0)^n$. Prove that F is analytic on $|z - z_0| < R$ and has a power series expansion, $F(z) = \sum_{n=0}^{\infty} A_n (z - z_0)^n$, where $A_n = \sum_{k=0}^{\infty} a_n^{(k)}$. [*Hint*: $A_n = \frac{F^{(n)}(z_0)}{n!}$. Compute the derivatives of F using term-by-term differentiation.]

40. (a) Show that the series $f(z) = \sum_{n=1}^{\infty} \frac{z^n}{1+z^{2n}}$ converges uniformly on every closed disk of the form $|z| \leq r < 1$.

(b) Show that for $|z| < 1$ we have

$$\sum_{n=1}^{\infty} \frac{z^n}{1+z^{2n}} = \sum_{k=0}^{\infty} (-1)^k \frac{z^{2k+1}}{1 - z^{2k+1}}.$$

[*Hint*: Use Exercise 39.]

41. Analytic continuation. We briefly visit discuss the topic of analytic continuation. By constructing a Taylor series of a function f about z_0, we create a function g analytic on some disk centered at z_0, such that g and f coincide near z_0. If g is defined in a region where f is not, or if g has different values than f away from z_0, g is said to be an **analytic continuation of** f.

As a somewhat contrived example, suppose that e^z is defined only for $|z| < 1$. Can we find a function that coincides with e^z on the unit disk but is analytic elsewhere? It is clear that we can (just define the exponential function on $\mathbb{C}$), but let us see how we can get this from series expansions. The Maclaurin series of e^z is $g(z) = \sum_{n=0}^{\infty} \frac{z^n}{n!}$, which converges for all z. Thus $g(z) = e^z$ for all z is an analytic continuation of $f(z)$. We now present a more interesting case of analytic continuation of logarithm branches.

(a) Show that the Taylor expansion of the principal branch of the logarithm $f(z) = \text{Log}\, z$ about the point $z_0 = -1 + i$ is

$$g(z) = \text{Log}\,(-1+i) + \sum_{n=1}^{\infty} \frac{(-1)^{n-1}(z+1-i)^n}{n(-1+i)^n}.$$

(b) Show that the radius of convergence of the Taylor series is $\sqrt{2}$. Thus g is analytic in the disk

$$|z+1-i| < \sqrt{2}.$$

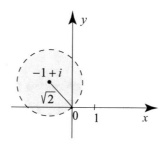

Fig. 4.9 The Taylor series expansion of $\text{Log}\, z$ around $-1+i$ has radius of convergence $\sqrt{2}$.

See Figure 4.9. it is important to note that this crosses the branch cut of f, but not the branch point (the origin). On the upper side of the branch cut, g coincides with f, but on the lower side of the branch cut, g does not coincide with f.

To see this, consider the branch $f_1(z) = \log_{-\frac{\pi}{4}} z$. Since $f = f_1$ in a neighborhood of z_0, their Taylor expansions about z_0 will be the same, and since f_1 is analytic on the disk $|z+1-i| < \sqrt{2}$, Theorem 4.3.1 says that g must equal f_1 in this disk. Thus, for z in the part of the disk above the branch cut (quadrants 1 and 2) we have $g(z) = f_1(z) = f(z)$, but for z in the part of the disk below the branch cut (quadrant 3) we have $g(z) = f_1(z) = f(z) + 2\pi i$. The Taylor expansion of f analytically continues the function into quadrant 3 where $\arg z > \pi$, and the analytic continuation is different from the original function.

4.4 Laurent Series

We saw in the previous section that if a function is analytic in a disk centered at a point z_0, then it has a Taylor series representation in that disk. There is a similar series representation in terms of both positive and negative powers of $(z - z_0)$ for functions that are analytic in annular regions around a point z_0. These series are known as **Laurent series**, after the French engineer and mathematician Pierre Alphonse Laurent (1813–1854), who discovered them around 1842.

For $0 \leq R_1 < R_2 \leq \infty$ we define the annular region (or annulus)

$$A_{R_1,R_2}(z_0) = \{z : R_1 < |z - z_0| < R_2\}.$$

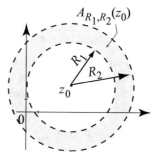

See Figure 4.10 for an illustration where R_1 and R_2 are nonzero and finite. Note that the annulus $A_{R_1,R_2}(z_0)$ degenerates into a punctured disk with z_0 removed when $R_1 = 0$ and $R_2 < \infty$, a punctured plane with z_0 removed when $R_1 = 0$, $R_2 = \infty$, or a plane with a disk centered at z_0 cut out of it when $0 < R_1$ and $R_2 = \infty$. These sets still count as annuli by our definition.

Fig. 4.10 The open annulus $A_{R_1,R_2}(z_0)$ centered at z_0 with radii R_1 and R_2.

Theorem 4.4.1. *Suppose that* f *is analytic on the annulus* $A_{R_1,R_2}(z_0)$ *where* $0 \leq R_1 < R_2 \leq \infty$. *Then* f *has a unique representation as a series of the form*

$$f(z) = \sum_{n=0}^{\infty} a_n(z - z_0)^n + \sum_{n=1}^{\infty} \frac{a_{-n}}{(z - z_0)^n}, \qquad R_1 < |z - z_0| < R_2, \qquad (4.4.1)$$

which converges absolutely for all z *in* $A_{R_1,R_2}(z_0)$ *and uniformly on every closed annulus* $\rho_1 \leq |z - z_0| \leq \rho_2$ *where* $R_1 < \rho_1$ *and* $\rho_2 < R_2$. *The series in* (4.4.1) *is called the* **Laurent series representation** *of* f. *The numbers* a_m *in* (4.4.1) *are*

$$a_m = \frac{1}{2\pi i} \int_{C_R(z_0)} \frac{f(\zeta)}{(\zeta - z_0)^{m+1}} \, d\zeta \qquad (m = 0, \pm 1, \pm 2, \dots), \qquad (4.4.2)$$

where R *is a number satisfying* $R_1 < R < R_2$, *and are called the* **Laurent coefficients** *of* f.

Proof. We show that the series (4.4.1) converges absolutely and uniformly on closed annuli $\rho_1 \leq |z - z_0| \leq \rho_2$, where $R_1 < \rho_1$ and $\rho_2 < R_2$. Fixing such ρ_1 and ρ_2, pick r_1 and r_2 such that

$$R_1 < r_1 < \rho_1 < \rho_2 < r_2 < R_2$$

and find a ρ so that $C_\rho(z)$ is contained in $A_{r_1,r_2}(z_0)$ (see Figure 4.11). The function

$$\zeta \longmapsto \frac{f(\zeta)}{\zeta - z}$$

is analytic inside and on the boundary of the region outside $C_\rho(z)$ and inside $A_{r_1,r_2}(z_0)$. So by Cauchy's theorem for multiply connected regions (Theorem 3.7.2), we have

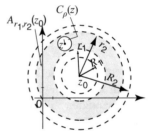

Fig. 4.11 The shaded region is the annulus $A_{r_1,r_2}(z_0)$.

$$\frac{1}{2\pi i} \int_{C_{r_2}(z_0)} \frac{f(\zeta)}{\zeta - z} \, d\zeta = \frac{1}{2\pi i} \int_{C_\rho(z)} \frac{f(\zeta)}{\zeta - z} \, d\zeta + \frac{1}{2\pi i} \int_{C_{r_1}(z_0)} \frac{f(\zeta)}{\zeta - z} \, d\zeta,$$

where all circular paths are positively oriented. By Cauchy's integral formula (Theorem 3.8.1), the first integral on the right is equal to $f(z)$, because f is analytic inside and on a small neighborhood of $C_\rho(z)$. So

$$f(z) = \frac{1}{2\pi i} \int_{C_{r_2}(z_0)} \frac{f(\zeta)}{\zeta - z} \, d\zeta - \frac{1}{2\pi i} \int_{C_{r_1}(z_0)} \frac{f(\zeta)}{\zeta - z} \, d\zeta. \qquad (4.4.3)$$

For $\zeta \in C_{r_2}(z_0)$ and z satisfying $\rho_1 \leq |z - z_0| \leq \rho_2$ we have $\frac{|z-z_0|}{|\zeta-z_0|} \leq \frac{\rho_2}{r_2} < 1$ and thus (as in the proof of Theorem 4.3.1) we write

$$\frac{1}{\zeta - z} = \frac{1}{\zeta - z_0 - (z - z_0)} = \frac{\frac{1}{\zeta - z_0}}{1 - \frac{z - z_0}{\zeta - z_0}} = \sum_{n=0}^{\infty} \frac{(z - z_0)^n}{(\zeta - z_0)^{n+1}} \tag{4.4.4}$$

and the series converges absolutely and uniformly in ζ satisfying $|\zeta - z_0| = r_2$ by the Weierstrass M-test, since the nth term of the series in (4.4.4) is bounded by $\frac{1}{r_2}\left(\frac{\rho_2}{r_2}\right)^n$ for all $z \in \overline{A_{\rho_1,\rho_2}(z_0)}$. Then we multiply (4.4.4) by $\frac{1}{2\pi i}f(\zeta)$ and we integrate the outcome over the circle $C_{r_2}(z_0)$ using Corollary 4.1.6 (that allows us to interchange summation and integration) to obtain

$$\frac{1}{2\pi i}\int_{C_{r_2}(z_0)} \frac{f(\zeta)}{\zeta - z}d\zeta = \sum_{n=0}^{\infty} a_n(z - z_0)^n$$

where

$$a_n = \frac{1}{2\pi i}\int_{C_{r_2}(z_0)} \frac{f(\zeta)}{(\zeta - z_0)^{n+1}}d\zeta. \tag{4.4.5}$$

For the second integral on the right of (4.4.3), we have $\zeta \in C_{r_1}(z_0)$ and for z satisfying $\rho_1 \le |z - z_0| \le \rho_2$ we have $\frac{|\zeta - z_0|}{|z - z_0|} \le \frac{r_1}{\rho_1} < 1$ and thus we write

$$\frac{1}{\zeta - z} = \frac{1}{\zeta - z_0 - (z - z_0)} = \frac{-\frac{1}{z - z_0}}{1 - \frac{\zeta - z_0}{z - z_0}} = \sum_{n=0}^{\infty} \frac{-(\zeta - z_0)^n}{(z - z_0)^{n+1}} = -\sum_{n=1}^{\infty} \frac{(\zeta - z_0)^{n-1}}{(z - z_0)^n} \tag{4.4.6}$$

and the series converges absolutely and uniformly in ζ satisfying $|\zeta - z_0| = r_1$ by the Weierstrass M-test, since the nth term of the series in (4.4.6) is bounded by $\frac{1}{r_1}\left(\frac{r_1}{\rho_1}\right)^n$ for all $z \in \overline{A_{\rho_1,\rho_2}(z_0)}$.

Multiplying both sides of (4.4.6) by $\frac{1}{2\pi i}f(\zeta)$ and then integrating the outcome over the circle $C_{r_1}(z_0)$ using Corollary 4.1.6 we obtain

$$-\frac{1}{2\pi i}\int_{C_{r_1}(z_0)} \frac{f(\zeta)}{\zeta - z}d\zeta = \sum_{n=1}^{\infty}\left\{\frac{1}{2\pi i}\int_{C_{r_1}(z_0)} f(\zeta)(\zeta - z_0)^{n-1}d\zeta\right\}\frac{1}{(z - z_0)^n}$$

$$= \sum_{n=1}^{\infty} \frac{a_{-n}}{(z - z_0)^n},$$

where

$$a_{-n} = \frac{1}{2\pi i}\int_{C_{r_1}(z_0)} f(\zeta)(\zeta - z_0)^{n-1}d\zeta. \tag{4.4.7}$$

As for $R \in (R_1, R_2)$ the circles $C_{r_1}(z_0)$, $C_{r_2}(z_0)$, and $C_R(z_0)$ are homotopic in $A_{R_1,R_2}(z_0)$, Theorem 3.6.5 yields that the expressions in (4.4.5) and (4.4.7) are equal to that in (4.4.2).

Finally we show uniqueness of the Laurent expansion (4.4.1). Suppose that on the annulus $R_1 < |z - z_0| < R_2$ the function f has two Laurent series:

$$f(z) = \sum_{n \in \mathbb{Z}} a_n(z - z_0)^n = \sum_{n \in \mathbb{Z}} b_n(z - z_0)^n$$

which converge uniformly on all closed subannuli of $A_{R_1,R_2}(z_0)$. Multiply both sides of preceding identity by $(z - z_0)^{-k-1}$ for some arbitrary integer k and integrate over the circle $C_R(z_0)$. We obtain

$$\frac{1}{2\pi i}\int_{C_R(z_0)}\sum_{n\in\mathbb{Z}}a_n(z-z_0)^{n-k-1}dz = \frac{1}{2\pi i}\int_{C_R(z_0)}\sum_{n\in\mathbb{Z}}b_n(z-z_0)^{n-k-1}dz.$$

But both series above converge uniformly on all closed annuli containing the circle $C_R(z_0)$ and contained in the annulus $A_{R_1,R_2}(z_0)$, so the integration and summation can be interchanged via Corollary 4.1.6. We obtain

$$\sum_{n\in\mathbb{Z}}\frac{1}{2\pi i}\int_{C_R(z_0)}a_n(z-z_0)^{n-k-1}dz = \sum_{n\in\mathbb{Z}}\frac{1}{2\pi i}\int_{C_R(z_0)}b_n(z-z_0)^{n-k-1}dz$$

and noting that all terms of the series vanish except for the terms with $n = k$, in view of the identity

$$\frac{1}{2\pi i}\int_{C_R(z_0)}(z-z_0)^{-1}dz = 1$$

we deduce that $a_k = b_k$ for all k. ∎

As with power series, often in computing Laurent series, we can avoid using (4.4.2) by resorting to known series.

Example 4.4.2. (Laurent series centered at 0) The function $e^{\frac{1}{z}}$ is analytic in the (degenerated) annulus $0 < |z|$, with center at $z_0 = 0$. Find its Laurent series expansions in this annulus.

Solution. Start with the exponential series $e^z = \sum_{n=0}^{\infty}\frac{z^n}{n!}$, which is valid for all z. In particular, if $z \neq 0$, putting $\frac{1}{z}$ into this series, we obtain

$$e^{\frac{1}{z}} = \sum_{n=0}^{\infty}\frac{1}{n!z^n} = 1 + \sum_{n=1}^{\infty}\frac{1}{n!z^n} = 1 + \frac{1}{1!z} + \frac{1}{2!z^2} + \frac{1}{3!z^3} + \cdots.$$

By the uniqueness of the Laurent series representation in the annulus $0 < |z|$, we have thus found the Laurent series of $e^{\frac{1}{z}}$ in the annulus $0 < |z|$. Note that the series has infinitely many negative powers of z. □

Example 4.4.3. (Laurent series centered at 0) The function $\frac{1}{1-z}$ is analytic in the annulus $1 < |z|$, with center at $z_0 = 0$. Find its Laurent series expansions in this annulus.

Solution. Here we use the geometric series $\sum_{n=0}^{\infty}w^n = \frac{1}{1-w}$ for $|w| < 1$. Factor z from the denominator and use the fact that $1 < |z|$, then

$$\frac{1}{1-z} = \frac{1}{z}\frac{1}{\frac{1}{z}-1} = \frac{-1}{z}\frac{1}{1-\frac{1}{z}} = \frac{-1}{z}\sum_{n=0}^{\infty}\left(\frac{1}{z}\right)^n = \sum_{n=0}^{\infty}\frac{-1}{z^{n+1}}, \qquad 1 < |z|.$$

We could stop here, but if we want to match the Laurent series form (4.4.1), we change $n + 1$ to n in the series, adjust the summation limit, and get

$$\frac{1}{1-z} = \sum_{n=1}^{\infty} \frac{-1}{z^n} = -\frac{1}{z} - \frac{1}{z^2} - \frac{1}{z^3} - \cdots, \qquad 1 < |z|.$$

Here again, the Laurent series has infinitely many negative powers of z. $\square$

Combining Example 4.4.3 with the geometric series, we obtain the following useful identities:

$$\frac{1}{1-w} = \begin{cases} \displaystyle\sum_{n=0}^{\infty} w^n & \text{if } |w| < 1, \\[4mm] \displaystyle-\sum_{n=1}^{\infty} \frac{1}{w^n} & \text{if } 1 < |w|. \end{cases} \tag{4.4.8}$$

We can use these identities to find Laurent series centered at points other than 0.

Example 4.4.4. (A Laurent series centered at a point other than 0) The function $\frac{1}{z-6}$ is analytic in the annulus $2 < |z - 4|$. Find its Laurent series in this annulus (Figure 4.12).

Solution. Since we are expanding about 4, we need to see the expression $(z - 4)$ in the denominator of f. For this purpose, and in order to apply (4.4.8), let us write

$$\frac{1}{z-6} = \frac{1}{(z-4)-2}$$
$$= -\frac{1}{2} \frac{1}{1 - \frac{z-4}{2}}$$
$$= -\frac{1}{2} \frac{1}{1-w},$$

where $w = \frac{z-4}{2}$. If $2 < |z - 4|$ then $|w| = \left|\frac{z-4}{2}\right| > 1$ and so the second identity in (4.4.8) implies that

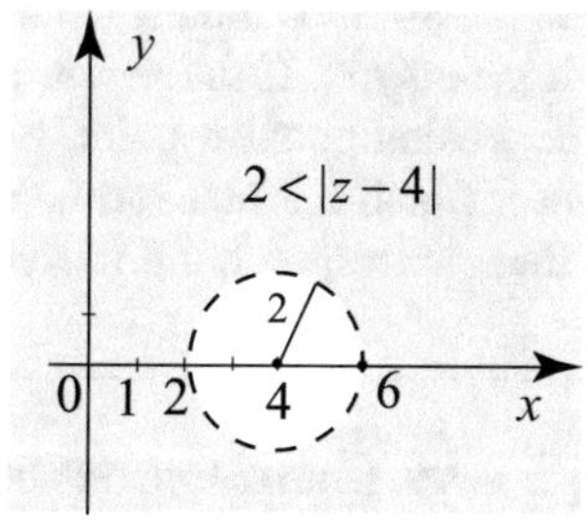

Fig. 4.12 The function $\frac{1}{z-6}$ is analytic in the annulus $2 < |z - 4|$ and has a Laurent series expansion there.

$$\frac{1}{z-6} = -\frac{1}{2} \frac{1}{1-w} = \frac{1}{2} \sum_{n=1}^{\infty} \frac{1}{w^n} = \frac{1}{2} \sum_{n=1}^{\infty} \frac{2^n}{(z-4)^n} = \sum_{n=1}^{\infty} \frac{2^{n-1}}{(z-4)^n},$$

which is the Laurent series of $\frac{1}{z-6}$ in the annulus $2 < |z - 4|$. $\square$

Example 4.4.5. (Laurent series of a rational function) Find the Laurent series expansion of $f(z) = \frac{3z^2 - 2z + 4}{z - 6}$ in the annulus $2 < |z - 4|$ of the previous example.

Solution. Since the degree of the numerator is larger than the degree of the denominator, the first step is to divide the numerator by the denominator:

$$\frac{3z^2 - 2z + 4}{z - 6} = 3z + 16 + \frac{100}{z - 6}.$$

The quotient $3z + 16$ has a simple expression in terms of powers of $(z - 4)$: This is $3z + 16 = 3(z - 4) + 28$. The next step is to compute the Laurent series of the remainder $\frac{100}{z-6}$. But this part follows from Example 4.4.4: for $2 < |z - 4|$,

$$\frac{100}{z - 6} = 100 \sum_{n=1}^{\infty} \frac{2^{n-1}}{(z - 4)^n}.$$

So, in the annulus $2 < |z - 4|$, we have the Laurent series

$$\frac{3z^2 - 2z + 4}{z - 6} = 3z + 16 + \frac{100}{z - 6} = 3(z - 4) + 28 + 100 \sum_{n=1}^{\infty} \frac{2^{n-1}}{(z - 4)^n}.$$

This Laurent series has positive and negative powers of $(z - 4)$. □

In the previous examples, we were given the function and the annular region on which it is analytic. If the region is not known and we are asked to find a Laurent series expansion about a point, our first step would be to determine the annular regions on which the function is analytic. As the next example illustrates, to each region there corresponds a different Laurent series.

Example 4.4.6. (Laurent series of a rational function) Find all the Laurent series expansions of $\dfrac{3}{(1+z)(2-z)}$ about the point $z_0 = 0$.

Solution. This problem has two parts. First we must determine how many different Laurent series expansions the function has around 0. Then we must find these Laurent series. To answer the first question, we look for the largest disjoint annular regions around $z_0 = 0$ on which the function is analytic. Obviously, it is analytic at all points except at $z = -1$ and $z = 2$. So $\frac{3}{(1+z)(2-z)}$ is analytic in the annular regions $|z| < 1$, $1 < |z| < 2$, and $2 < |z|$ (Figure 4.13). Note that the first region $|z| < 1$ is a disk, and so we use a power series expansion there, which is really a special case of a Laurent series expansion without negative powers (Theorem 4.3.1).

So the function has three different Laurent series expansions around 0; one of them being a power series. Let us find them. We will need the partial fraction decomposition

$$\frac{3}{(1+z)(2-z)} = \frac{1}{1+z} + \frac{1}{2-z},$$

which can be easily verified.

Next we undertake the task to find the Laurent expansions in the three regions.

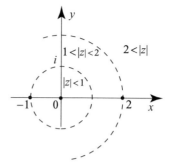

Fig. 4.13 The rational function $\frac{3}{(1+z)(2-z)}$ has three Laurent series expansions around 0.

For $|z| < 1$, we have from the geometric series expansion

$$\frac{1}{1+z} = \frac{1}{1-(-z)} = \sum_{n=0}^{\infty} (-1)^n z^n, \quad |z| < 1,$$

and

$$\frac{1}{2-z} = \frac{1}{2} \frac{1}{1-\left(\frac{z}{2}\right)} = \frac{1}{2} \sum_{n=0}^{\infty} \left(\frac{z}{2}\right)^n = \sum_{n=0}^{\infty} \frac{z^n}{2^{n+1}}, \quad \left|\frac{z}{2}\right| < 1, \text{ or } |z| < 2. \quad (4.4.9)$$

Adding the two series over their common region of convergence, we obtain

$$\frac{3}{(1+z)(2-z)} = \sum_{n=0}^{\infty} \left((-1)^n + \frac{1}{2^{n+1}}\right) z^n, \quad |z| < 1,$$

which is the Taylor series expansion in $|z| < 1$. To find the Laurent series in the annulus $1 < |z| < 2$ we use (4.4.8) or reason as in Example 4.4.3. We have

$$\frac{1}{1+z} = \frac{1}{z} \frac{1}{1-\left(-\frac{1}{z}\right)} = \frac{1}{z} \sum_{n=0}^{\infty} \left(\frac{-1}{z}\right)^n = \sum_{n=1}^{\infty} \frac{(-1)^{n-1}}{z^n}, \quad 1 < |z|. \quad (4.4.10)$$

For the term $\frac{1}{2-z}$ we can use the previous expansion (4.4.9). Adding the two, we obtain for $1 < |z| < 2$,

$$\frac{3}{(1+z)(2-z)} = \sum_{n=1}^{\infty} \frac{(-1)^{n-1}}{z^n} + \sum_{n=0}^{\infty} \frac{z^n}{2^{n+1}}$$

$$= \frac{1}{2} + \frac{z}{2^2} + \frac{z^2}{2^3} + \cdots + \frac{1}{z} - \frac{1}{z^2} + \frac{1}{z^3} - \cdots,$$

which is the Laurent series in the annulus $1 < |z| < 2$. Finally, let us consider the annulus $2 < |z|$. Since $2 < |z|$, then clearly $1 < |z|$, and for the term $\frac{1}{1+z}$ we can use (4.4.10). Also, if $2 < |z|$, then $\left|\frac{2}{z}\right| < 1$, and so

$$\frac{1}{2-z} = \frac{1}{z}\frac{1}{\left(\frac{2}{z}\right)-1} = \frac{-1}{z}\frac{1}{1-\left(\frac{2}{z}\right)} = \frac{-1}{z}\sum_{n=0}^{\infty}\left(\frac{2}{z}\right)^n = -\sum_{n=1}^{\infty}\frac{2^{n-1}}{z^n}, \qquad 2 < |z|.$$

Adding the two series, we obtain for $2 < |z|$,

$$\frac{3}{(1+z)(2-z)} = \sum_{n=1}^{\infty}\frac{(-1)^{n-1}}{z^n} - \sum_{n=1}^{\infty}\frac{2^{n-1}}{z^n} = \sum_{n=1}^{\infty}\frac{(-1)^{n-1}-2^{n-1}}{z^n}$$

$$= -\frac{3}{z^2} - \frac{3}{z^3} - \frac{9}{z^4} - \cdots.$$

This is the desired Laurent series in the annulus $2 < |z|$.   □

Example 4.4.7. (Determining the annulus of convergence) Determine the largest annulus around $z_0 = i$ of the form $R_1 < |z-i| < R_2 < \infty$ on which the function $\frac{1}{1+z^2}$ has a Laurent series and then find this Laurent series.

Solution. The function $\frac{1}{1+z^2}$ is analytic at all points z except $z = \pm i$. It has a Laurent series in the largest annulus around i on which it is analytic. In order to avoid the singularity at $-i$, we take the annulus $0 < |z-i| < 2$ (Figure 4.14). We have

$$\frac{1}{1+z^2} = \frac{1}{z-i}\frac{1}{z+i}.$$

Since we are expanding in terms of $(z-i)$, the factor $\frac{1}{z-i}$ is already in a desirable form, and so we keep it as is and work on the factor $\frac{1}{z+i}$.

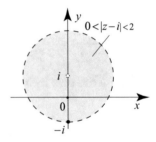

Fig. 4.14 The largest annulus around i on which $\frac{1}{1+z^2}$ has a Laurent series expansion.

To make $z-i$ appear in the latter factor, we add and subtract i from the denominator. Thus

$$\frac{1}{1+z^2} = \frac{1}{z-i}\frac{1}{z+i} = \frac{1}{z-i}\frac{1}{(z-i)+2i} = \frac{1}{2i}\frac{1}{z-i}\frac{1}{1+\left(\frac{z-i}{2i}\right)}.$$

Since $|z-i| < 2$, then $\left|\frac{z-i}{2i}\right| < 1$, and so we can use a geometric series expansion:

$$\frac{1}{1+z^2} = \frac{1}{2i}\frac{1}{z-i}\frac{1}{1-\left(-\frac{z-i}{2i}\right)} = \frac{1}{2i}\frac{1}{z-i}\sum_{n=0}^{\infty}(-1)^n\left(\frac{z-i}{2i}\right)^n$$

We conclude that on the annulus the annulus $0 < |z-i| < 2$ we have

$$\frac{1}{1+z^2} = \frac{1}{2i}\left[\frac{1}{z-i} - \frac{1}{2i} - \frac{z-i}{2^2} + \frac{(z-i)^2}{2^3i} + \cdots\right]. \qquad □$$

Differentiation and Integration of Laurent Series

Consider a Laurent series (4.4.1) in the annulus $A_{R_1,R_2}(z_0)$. Since each term of the series, $a_n(z - z_0)^n$, is analytic on the annulus $A_{R_1,R_2}(z_0)$, and the series converges uniformly on all subannuli of $A_{R_1,R_2}(z_0)$, we can integrate or differentiate term by term the series as many times as we want in the annulus $A_{R_1,R_2}(z_0)$. So, for example, differentiating the Laurent series (4.4.1) once, for $R_1 < |z - z_0| < R_2$, we obtain

$$f'(z) = \sum_{n=1}^{\infty} na_n(z - z_0)^{n-1} - \sum_{n=1}^{\infty} \frac{na_{-n}}{(z - z_0)^{n+1}}. \tag{4.4.11}$$

We could continue to differentiate term by term to find $f''(z)$, $f'''(z)$, and so forth. Also, if γ is an arbitrary path contained in $A_{R_1,R_2}(z_0)$, then

$$\int_\gamma f(z)\,dz = \sum_{n=0}^{\infty} a_n \int_\gamma (z - z_0)^n\,dz + \sum_{n=1}^{\infty} a_{-n} \int_\gamma \frac{dz}{(z - z_0)^n}. \tag{4.4.12}$$

Example 4.4.8. (Differentiating to find a Laurent series) Find the Laurent series for $f(z) = \frac{1}{(1-z)^3}$ in the annulus $1 < |z|$ (Figure 4.15).

Solution. Starting with the Laurent series

$$\frac{1}{1 - z} = -\sum_{n=1}^{\infty} \frac{1}{z^n}, \qquad 1 < |z|,$$

if we differentiate both sides we get the Laurent series

$$\frac{1}{(1 - z)^2} = \sum_{n=1}^{\infty} \frac{n}{z^{n+1}}, \qquad 1 < |z|.$$

Differentiating a second time, we get

$$\frac{2}{(1 - z)^3} = -\sum_{n=1}^{\infty} \frac{n(n+1)}{z^{n+2}}, \qquad 1 < |z|.$$

So the desired Laurent series is

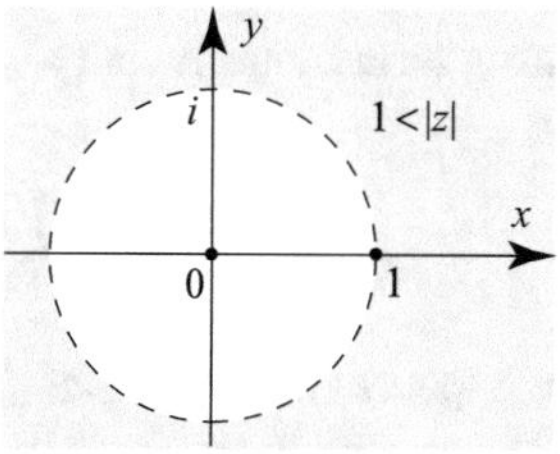

Fig. 4.15 The annulus $1 < |z|$ in Example 6.

$$\frac{1}{(1 - z)^3} = -\frac{1}{2} \sum_{n=1}^{\infty} \frac{n(n+1)}{z^{n+2}}, \qquad 1 < |z|. \qquad \square$$

Example 4.4.9. (Term-by-term integration of a Laurent series) Let $C_1(0)$ denote the positively oriented circle of radius 1, centered at 0. Evaluate the following integrals:

(a) $\displaystyle\int_{C_1(0)} \frac{e^{\frac{1}{z}}}{z}\, dz,$ (b) $\displaystyle\int_{C_1(0)} e^{z+\frac{1}{z}}\, dz.$

Solution. Note that the integrands are continuous on the path of integration; so the integrals do exist. However, they cannot be computed using Cauchy's theorem because of the singularity of the function $e^{1/z}$ at the point $z_0 = 0$ which lies inside the path. The idea is to expand the integrand (or part of the integrand) in a Laurent series in an annulus that contains the path, and then integrate term by term.

(a) In the annulus $0 < |z|$, we have

$$\frac{e^{\frac{1}{z}}}{z} = \frac{1}{z}e^{\frac{1}{z}} = \frac{1}{z}\sum_{n=0}^{\infty} \frac{1}{n!z^n} = \sum_{n=0}^{\infty} \frac{1}{n!z^{n+1}}.$$

So

$$\int_{C_1(0)} \frac{e^{\frac{1}{z}}}{z}\, dz = \int_{C_1(0)} \left(\sum_{n=0}^{\infty} \frac{1}{n!z^{n+1}} \right) dz = \sum_{n=0}^{\infty} \frac{1}{n!} \int_{C_1(0)} \frac{dz}{z^{n+1}} = 2\pi i,$$

where only the $n = 0$ term is nonzero.

(b) Here again, we work on the annulus $0 < |z|$. With an eye on Cauchy's integral formula, we do not expand e^z. For $0 < |z|$, we have

$$e^{z+\frac{1}{z}} = e^z e^{\frac{1}{z}} = e^z \sum_{n=0}^{\infty} \frac{1}{n!z^n} = e^z + \frac{e^z}{1!z} + \frac{e^z}{2!z^2} + \frac{e^z}{3!z^3} + \cdots = e^z + \sum_{n=0}^{\infty} \frac{e^z}{(n+1)!z^{n+1}}.$$

Integrating term by term yields

$$\int_{C_1(0)} e^{z+\frac{1}{z}}\, dz = \int_{C_1(0)} e^z\, dz + \sum_{n=0}^{\infty} \frac{1}{(n+1)!} \int_{C_1(0)} \frac{e^z}{z^{n+1}}\, dz.$$

Cauchy's theorem implies that $\int_{C_1(0)} e^z\, dz = 0$ because e^z is analytic on and inside $C_1(0)$. Cauchy's generalized integral formula tells us that

$$\int_{C_1(0)} \frac{e^z}{z^{n+1}}\, dz = \frac{2\pi i}{n!} f^{(n)}(0),$$

where $f(z) = e^z$. Hence $f^{(n)}(0) = e^0 = 1$, and so

$$\int_{C_1(0)} e^{z+\frac{1}{z}}\, dz = \sum_{n=0}^{\infty} \frac{1}{(n+1)!} \int_{C_1(0)} \frac{e^z}{z^{n+1}}\, dz = 2\pi i \overbrace{\sum_{n=0}^{\infty} \frac{1}{n!(n+1)!}}^{\approx 1.59} \approx 9.99\, i.$$

We should also mention that $\sum_{n=0}^{\infty} \frac{1}{n!(n+1)!} = -iJ_1(2i)$, where $J_1(z)$ is the Bessel function of order 1 (see Example 4.2.14). $\square$

The ideas of the previous example are the basis for the techniques of Chapter 5: To compute integrals around circles, we find Laurent expansions and integrate term by term. Only the term involving $\frac{1}{z}$ survives.

Exercises 4.4

In Exercises 1–18, find the Laurent series of the function in the indicated annulus.

1. $\dfrac{1}{1+z}$, $1 < |z|$

2. $\dfrac{1}{2+iz}$, $2 < |z|$

3. $\dfrac{1}{3+2iz}$, $\dfrac{5}{2} < |z+i|$

4. $\dfrac{1}{1+i-z}$, $\sqrt{2} < |z-2|$

5. $\dfrac{1}{1+z^2}$, $1 < |z|$

6. $\dfrac{1}{1-z^2}$, $1 < |z-2| < 3$

7. $\dfrac{3+z}{2-z}$, $2 < |z|$

8. $\dfrac{1+z}{1-z}$, $1 < |z|$

9. $z + \dfrac{1}{z}$, $1 < |z-1|$

10. $z^{22}e^{\frac{1}{z}}$, $0 < |z|$

11. $\coth z$, $0 < |z| < \pi$

12. $\cot z$, $0 < |z| < \pi$

13. $\dfrac{z}{(z+2)(z+3)}$, $2 < |z| < 3$

14. $\dfrac{-2}{(2z-1)(2z+1)}$, $\dfrac{1}{2} < |z|$

15. $\dfrac{1}{(3z-1)(2z+1)}$, $\dfrac{1}{3} < |z| < \dfrac{1}{2}$

16. $\dfrac{1}{2z^2 - 3z + 1}$, $1 < |z|$

17. $\dfrac{z^2 + (1-i)z + 2}{(z-i)(z+2)}$, $1 < |z| < 2$

18. $\dfrac{4z-5}{(z-2)(z-1)}$, $1 < |z-2|$

In Exercises 19–22, find all Laurent series expansions of the function at the indicated point z_0.

19. $\dfrac{1}{z+i}$, $z_0 = 1$

20. $\dfrac{1}{z^2-1}$, $z_0 = 2$

21. $\dfrac{1}{(z-1)(z+i)}$, $z_0 = -1$

22. $\dfrac{1}{z^2+1}$, $z_0 = 1+i$

23. (a) Derive the Laurent series

$$\frac{1}{1+z} = \sum_{n=1}^{\infty} \frac{(-1)^{n-1}}{z^n} \qquad 1 < |z|.$$

Starting with this Laurent series, find the Laurent series of the following functions in the annulus $1 < |z|$:

(b) $\dfrac{1}{(1+z)^2}$;   (c) $\dfrac{z}{(1+z)^2}$;   (d) $\dfrac{z^2}{(1+z)^3}$.

24. Find the Laurent series of $\csc^2 z$ in the annulus $0 < |z| < \pi$.

In Exercises 25–30, evaluate the integral using an appropriate Laurent series. As usual, we denote by $C_R(z_0)$ the positively oriented circle of radius $R > 0$ and center z_0.

25. $\displaystyle\int_{C_1(0)} \sin\frac{1}{z}\, dz$

26. $\displaystyle\int_{C_1(0)} \frac{\cos\frac{1}{z^2}}{z}\, dz$

27. $\displaystyle\int_{C_1(0)} \cos z \sin\frac{1}{z}\, dz$

28. $\displaystyle\int_{C_1(0)} e^{z^2 + \frac{1}{z}}\, dz$

29. $\displaystyle\int_{C_4(0)} \mathrm{Log}\left(1 + \frac{1}{z}\right) dz$

30. $\displaystyle\int_{C_1(0)} z^{10} e^{\frac{1}{z}}\, dz$

31. Suppose that f is analytic on a region Ω. Let $\overline{B_R(z_0)}$ be a closed disk contained in Ω.
(a) For $n = 0, 1, 2, \ldots$, derive the Laurent series expansion

$$\frac{f(z)}{(z-z_0)^{n+1}} = \sum_{k=0}^{\infty} \frac{f^{(k)}(z_0)}{k!} \frac{1}{(z-z_0)^{n+1-k}}, \qquad 0 < |z-z_0| < R.$$

(b) Prove Cauchy's generalized integral formula using (a) and integration term by term.

32. Project Problem: Generating function for Bessel functions. We derive the generating function for Bessel functions of integer order (Example 4.2.14):

$$e^{\frac{z}{2}\left(\zeta - \frac{1}{\zeta}\right)} = \sum_{n=-\infty}^{\infty} J_n(z)\zeta^n, \qquad 0 < |\zeta|. \tag{4.4.13}$$

This formula generates the Bessel functions $J_n(z)$ as the Laurent coefficients of the function $f(\zeta) = e^{\frac{z}{2}\left(\zeta - \frac{1}{\zeta}\right)}$, which is clearly analytic in the annulus $0 < |\zeta|$.

(a) Show that for $n = 0, \pm 1, \pm 2, \ldots,$

$$J_n(z) = \frac{1}{2\pi i} \int_{C_1(0)} e^{\frac{z}{2}\left(\zeta - \frac{1}{\zeta}\right)} \frac{d\zeta}{\zeta^{n+1}}. \tag{4.4.14}$$

[*Hint*: Write $e^{\frac{z}{2}\left(\zeta - \frac{1}{\zeta}\right)} = e^{\frac{z}{2}\zeta} e^{-\frac{z}{2\zeta}}$ and expand $e^{-\frac{z}{2\zeta}}$ in a Laurent series in $0 < |\zeta|$.]

(b) We know that f has a Laurent series expansion in $0 < |\zeta|$. Write this series as $f(\zeta) = \sum_{n=-\infty}^{\infty} c_n(z)\zeta^n$. Express the coefficients $c_n(z)$ by using (4.4.2) and integrating over $C_1(0)$. Conclude that $c_n(z)$ is equal to the integral in (4.4.14), and hence $c_n(z) = J_n(z)$.

33. Project Problem: Cosine integral representation of Bessel functions.

(a) Parametrize the path integral in (4.4.14) and obtain the formula

$$J_n(z) = \frac{1}{2\pi} \int_{-\pi}^{\pi} e^{i(z\sin\theta - n\theta)} d\theta$$

$$= \frac{1}{\pi} \int_{0}^{\pi} \cos(z\sin\theta - n\theta) d\theta.$$

This formula is known as the **cosine integral representation** of $J_n(z)$.

(b) Show that for $z = x$ a real number $|J_n(x)| \le 1$. This property is illustrated in the adjacent figure.

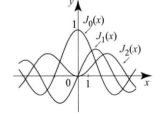

4.5 Zeros and Singularities

In this section, we use Taylor and Laurent series to study the zeros and singular points of analytic functions. To motivate the first result, suppose that p is a polynomial and $p(z_0) = 0$, then we can factor the highest power of $(z - z_0)$ out of $p(z)$ and write $p(z) = q(z)(z - z_0)^m$, where q is a polynomial with $q(z_0) \ne 0$. It is also straightforward to check that $p^{(j)}(z_0) = 0$ for all $j = 0, 1, \ldots, m-1$.

Definition 4.5.1. Let Ω be a region, let f be an analytic function defined on Ω, and let $z_0 \in \Omega$. Let $m \ge 1$ be an integer. The complex number z_0 is called a **zero of order** m of f if there is another analytic function g defined in a neighborhood of z_0 such that $g(z_0) \ne 0$ and

$$f(z) = (z - z_0)^m g(z)$$

for all z in this neighborhood. If $m = 1$, we call z_0 a **simple zero** of f. The zero z_0 is called **isolated** if there is a neighborhood of z_0 in Ω such that z_0 is the only zero of f in this neighborhood.

Theorem 4.5.2. *Suppose that f is analytic on a region Ω that contains a point z_0 and $f(z_0) = 0$. Then exactly one of the following two assertions holds:*
(i) f is identically zero in a neighborhood of z_0.
(ii) z_0 is an isolated zero of f.
Moreover, if (ii) holds, then there are an integer $m \geq 1$, a real number $r > 0$, and an analytic function λ on $B_r(z_0)$ such that $\lambda(z) \neq 0$ for all $z \in B_r(z_0)$ and

$$f(z) = (z - z_0)^m \lambda(z) \qquad \text{for all } z \in B_r(z_0).$$

Consequently, z_0 is a zero of order m and we have

$$f(z_0) = f'(z_0) = \cdots = f^{(m-1)}(z_0) = 0, \qquad f^{(m)}(z_0) \neq 0.$$

Proof. Since f is analytic at z_0, it has a Taylor series, $f(z) = \sum_{n=0}^{\infty} a_n (z - z_0)^n$, centered at z_0, and convergent in an open disk $B_R(z_0)$ contained in Ω. Theorem 4.3.1 gives that $a_k = f^{(k)}(z_0)/k!$. There are two possibilities: either (i) $a_n = 0$ for all $n = 0, 1, 2, \ldots$, or (ii) there exists an n such that $a_n \neq 0$. We show that these two possibilities correspond to the mutually exclusive assertions (i) and (ii) of the theorem. Clearly, if $a_n = 0$ for all n, then f is identically equal to zero on the disk $B_R(z_0)$. Now if $a_n \neq 0$ for some n, then there is a coefficient a_m with a least index m such that $a_m \neq 0$; then $a_0 = a_1 = \cdots = a_{m-1} = 0$. Thus we obtain

$$f(z_0) = f'(z_0) = \cdots = f^{(m-1)}(z_0) = 0$$

and

$$f^{(m)}(z_0) \neq 0.$$

For $|z - z_0| < R$, write

$$\begin{aligned}
f(z) &= a_m(z - z_0)^m + a_{m+1}(z - z_0)^{m+1} + a_{m+2}(z - z_0)^{m+2} + \cdots \\
&= (z - z_0)^m \left(a_m + a_{m+1}(z - z_0) + a_{m+2}(z - z_0)^2 + \cdots \right) \\
&= (z - z_0)^m \lambda(z),
\end{aligned}$$

where

$$\lambda(z) = a_m + a_{m+1}(z - z_0) + a_{m+2}(z - z_0)^2 + \cdots,$$

and $a_m \neq 0$. Since $\lambda(z)$ is a convergent power series in $|z - z_0| < R$, it defines an analytic function in view of Corollary 4.2.9. Also, $\lambda(z_0) = a_m \neq 0$ and λ is continuous at z_0, so we can find a neighborhood $B_r(z_0)$ of z_0 such that $\lambda(z) \neq 0$ for all z in $B_r(z_0)$. ∎

Example 4.5.3. (Order of zeros) Find the order m of the zero of $\sin z$ at $z_0 = 0$; then express $\sin z = z^m \lambda(z)$, where λ is analytic and satisfies $\lambda(0) \neq 0$.

Solution. Clearly, 0 is a zero of $\sin z$. The order of the zero is equal to the order of the first nonvanishing derivative of $f(z) = \sin z$ at 0. Since $f'(z) = \cos z$ and $\cos 0 = 1 \neq 0$, we conclude that the order of the zero at 0 is 1. We have for all z

$$\sin z = z - \frac{z^3}{3!} + \frac{z^5}{5!} - \cdots = z\left(1 - \frac{z^2}{3!} + \frac{z^4}{5!} - \cdots\right) = z\lambda(z),$$

where $\lambda(z) = 1 - \frac{z^2}{3!} + \frac{z^4}{5!} - \cdots$. The function $\lambda(z)$ is entire (because it is a convergent power series for all z), and $\lambda(0) = 1$. Also, for $z \neq 0$, $\lambda(z) = \frac{\sin z}{z}$, which is entire by Example 4.3.10.   □

We now address the following question: Suppose that a function is analytic and not identically zero on a region. Can it have zeros that are not isolated? The answer is no, and this depends on the fact that the region is connected.

Theorem 4.5.4. *An nonzero analytic function defined on a region has isolated zeros.*

Proof. Let f be an analytic function defined on a region Ω. Define the subsets

$$\Omega_0 = \{z \in \Omega : f(z) = 0 \text{ and } z \text{ is not isolated}\}$$
$$\Omega_1 = \{z \in \Omega : f(z) \neq 0\} \cup \{z \in \Omega : f(z) = 0 \text{ and } z \text{ is isolated}\}.$$

Then Ω_0 and Ω_1 are disjoint sets whose union is Ω. We show that Ω_0 and Ω_1 are open; then by the connectedness of Ω (Proposition 2.1.7) we have either $\Omega = \Omega_0$ or $\Omega = \Omega_1$. If $\Omega = \Omega_0$, then every point in Ω is a zero of f; in this case f vanishes everywhere on Ω. If $\Omega = \Omega_1$ then every point of Ω is either not a zero or an isolated zero of f. In this case all zeros of f are isolated.

If $w \in \Omega_0$, then assertion (ii) in Theorem 4.5.2 does not hold, hence assertion (i) must hold and thus f vanishes identically in some neighborhood $B_r(w)$. Clearly, all the points in $B_r(w)$ are not isolated zeros of f, so $B_r(w)$ is contained in Ω_0. This shows that Ω_0 is open in this case. Now let $w \in \Omega_1$. If $f(w) \neq 0$, then there is a neighborhood of w on which f is nonvanishing; this neighborhood is contained in Ω_1. If $f(w) = 0$ and w is an isolated zero, then Theorem 4.5.2 guarantees the existence of a neighborhood $B_\delta(w)$ of w on which f has no zeros except the isolated one at w. This neighborhood is also contained in Ω_1, since $B'_\delta(w) = B_\delta(w) \setminus \{w\}$ is contained in $\{z \in \Omega : f(z) \neq 0\}$. Thus Ω_1 is open.   ■

Theorem 4.5.5. (Identity Principle) *Suppose that f, g are analytic functions on a region Ω. If $\{z_n\}_{n=1}^{\infty}$ is an sequence of distinct points in Ω with $f(z_n) = g(z_n)$ for all n and $z_n \to z_0 \in \Omega$, then $f(z) = g(z)$ for all $z \in \Omega$.*

Proof. If z_n is a zero of $f - g$ in Ω and $z_n \to z_0$ as $n \to \infty$, then $f(z_n) = g(z_n)$ for all n and by continuity it follows that

$$f(z_0) = f(\lim_{n \to \infty} z_n) = \lim_{n \to \infty} f(z_n) = \lim_{n \to \infty} g(z_n) = g(\lim_{n \to \infty} z_n) = g(z_0)$$

and thus z_0 is zero of $f - g$. But obviously z_0 is not an isolated zero. By Theorem 4.5.2, $f - g$ is identically zero on Ω. ∎

In many interesting applications of Theorem 4.5.5 the functions f and g are equal on an interval $[a, b]$ of the real line, or on a whole disk $B_R(z_0)$. Such sets clearly contain infinite sequences of points that converge to a point in the set itself.

Example 4.5.6. (Applications of the identity principle) (a) Recall the identity $\cos^2 z + \sin^2 z = 1$ valid for all complex numbers z; see (1.7.9) in Proposition 1.7.3.

We prove this identity using Theorem 4.5.5 and the fact that it holds for real numbers z. Let $f(z) = \cos^2 z + \sin^2 z$ and $g(z) = 1$. Clearly, f and g are entire, and for real $z = x$, we have $f(x) = \sin^2 x + \cos^2 x = 1 = g(x)$. Since $f(z) = g(z)$ for all z on the real line, which contains infinite converging sequences, we infer from Theorem 4.5.5 that $f(z) = g(z)$ for all z; that is, $\cos^2 z + \sin^2 z = 1$.

(b) Modifying the method in (a), we can prove identities involving two or more variables. As an illustration, let us prove that for all complex numbers z_1 and z_2,

$$\cos(z_1 + z_2) = \cos z_1 \cos z_2 - \sin z_1 \sin z_2. \tag{4.5.1}$$

In a first step, let $z_2 = x_2$ be an arbitrary real number. Let $f(z) = \cos(z + x_2)$ and $g(z) = \cos z \cos x_2 - \sin z \sin x_2$. Clearly, f and g are entire, and from the addition formula for the cosines, we have $f(x) = g(x)$ for all real x. Hence by Theorem 4.5.5, we have $f(z) = g(z)$ for all z; equivalently,

$$\cos(z + x_2) = \cos z \cos x_2 - \sin z \sin x_2 \qquad \text{for all complex } z. \tag{4.5.2}$$

Now fix z_1 and define $F(z_2) = \cos(z_1 + z_2)$ and $G(z_2) = \cos z_1 \cos z_2 - \sin z_1 \sin z_2$. Here again, F and G are entire (as functions of z_2) and (4.5.2) states that F and G agree on the real line. By Theorem 4.5.5, $F(z_2) = G(z_2)$ for all complex z_2, implying that (4.5.1) holds. □

To establish other useful consequences regarding the number of zeros of an analytic functions, we need a topological property of complex numbers known as the **Bolzano-Weierstrass theorem**. This states the following: Let S denote a closed and bounded subset of $\mathbb{C}$, and let $\{w_n\}_{n=1}^{\infty}$ be a sequence formed by elements of S. Then there is a subsequence $\{z_{n_k}\}_{k=1}^{\infty}$ of $\{w_n\}_{n=1}^{\infty}$ that converges to a point z_0 in S. See Appendix, page 484.

Corollary 4.5.7. *An analytic function defined on a bounded region that admits a continuous extension on its closure and is nonvanishing on the boundary of the region has at most finitely many zeros.*

Proof. The closure $\overline{\Omega}$ of a region Ω is the region Ω together with its boundary. If Ω is bounded, then $\overline{\Omega}$ is closed and bounded. Suppose that f is a function as in the statement of the theorem and suppose that it has infinitely many zeros in Ω; then by the Bolzano-Weierstrass theorem, there is an infinite sequence of zeros, $\{z_n\}_{n=1}^{\infty}$ in Ω that converges to a point z_0 in $\overline{\Omega}$. Since f is continuous in $\overline{\Omega}$, $f(z_0) = \lim_{n \to \infty} f(z_n) = 0$, and since f is nonvanishing on the boundary, we conclude that z_0 lies in Ω. Hence by Theorem 4.5.5, f is identically zero on Ω, and since f is continuous, f must be zero on the boundary, which is a contradiction. Hence f can have at most finitely many zeros in Ω.   ∎

Isolated Singularities

If a function is analytic in a neighborhood of a point z_0 except possibly at the point z_0, then z_0 is called an **isolated singularity** of the function. There are only three different types of isolated singularities, which are defined as follows.

Definition 4.5.8. Suppose that z_0 is an isolated singularity of an analytic function f defined in a deleted neighborhood of z_0. Then
(i) z_0 is a **removable singularity** of f if the function can be redefined at z_0 to be analytic there.
(ii) z_0 is a **pole** of f if $\lim_{z \to z_0} |f(z)| = \infty$.
(iii) z_0 is an **essential singularity** of f if it is neither a pole nor a removable singularity.

(a) Removable singularity.

(b) Pole.

(c) Essential singularity.

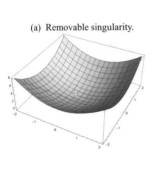

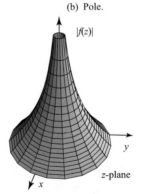

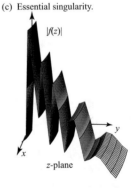

Fig. 4.16 Near a removable singularity, $|f|$ is bounded.

Fig. 4.17 Near a pole, $|f|$ tends to infinity.

Fig. 4.18 Near an essential singularity, $|f|$ is neither bounded nor tends to ∞. Its graph behaves erratically.

When redefining f at a removable singularity z_0, we set $f(z_0) = \lim_{z \to z_0} f(z)$; otherwise, f would not be continuous and hence could not be analytic at z_0. Note that when z_0 is a removable singularity, f must be bounded near z_0 (Figure 4.16).

The singularity is a pole if the graph of $|f|$ blows up to infinity as we approach z_0 (Figure 4.17). It is harder to explain the graph of an essential singularity, but as the function is neither bounded nor tends to infinity as we approach z_0 (Figure 4.18), its behavior is quite erratic. We provide equivalent characterizations of each type of singularity after we take a look at some examples.

Example 4.5.9. (Three types of singularities)

(a) The function $f(z) = \frac{z^2-1}{z-1}$ $(z \neq 1)$ is analytic everywhere, except at $z_0 = 1$. So $z_0 = 1$ is an isolated singularity. For $z \neq 1$, $f(z) = \frac{(z-1)(z+1)}{z-1} = z+1$. By defining $f(1) = 2$, we make f analytic at $z = 1$. This shows that the singularity at $z_0 = 1$ is removable.

(b) Consider the function

$$g(z) = \frac{z^2}{(z-i)^3}.$$

It has an isolated singularity at $z = i$. Unlike the previous example, the singularity here is not removable; it is a pole. Since

$$\lim_{z \to i} |g(z)| = \lim_{z \to i} \frac{|z|^2}{|z-i|^3} = \infty,$$

the graph of $|g|$ has a pole that blows up to infinity above the singularity $z_0 = i$.

(c) Consider

$$h(z) = e^{\frac{1}{z}}, \qquad z \neq 0.$$

We have an isolated singularity at $z_0 = 0$. We will show that it is an essential singularity by eliminating the possibility of the other two types. Suppose that $z = x$ is real and tends to 0 from the right. Then

$$\lim_{z=x\downarrow 0} \left|e^{\frac{1}{z}}\right| = \lim_{x\downarrow 0} e^{\frac{1}{x}} = \infty.$$

So 0 cannot be a removable singularity. Now suppose that $z = x$ is real and tends to 0 from the left. Then

$$\lim_{z=x\uparrow 0} \left|e^{\frac{1}{z}}\right| = \lim_{x\uparrow 0} e^{\frac{1}{x}} = 0;$$

hence 0 cannot be a pole, and so 0 is an essential singularity.   □

Removable singularities often occur when an analytic function $h(z)$ with a zero at z_0 is divided by a small enough power of $(z - z_0)$.

Proposition 4.5.10. *Let h be analytic on $B_R(z_0)$, $R > 0$, with a zero of order $m \geq 1$ at z_0. If $p \leq m$ is an integer, then $f(z) = \frac{h(z)}{(z-z_0)^p}$ has a removable singularity at z_0.*

Proof. By Theorem 4.5.2(ii), there is some $0 < r \leq R$ such that $h(z) = (z-z_0)^m \lambda(z)$ for $|z-z_0| < r$, where λ is analytic and nowhere vanishing on $B_r(z_0)$. So for $z \neq z_0$, $f(z) = (z-z_0)^{m-p}\lambda(z)$, where $m - p \geq 0$. Since in a neighborhood of z_0, $f(z)$ is equal to the analytic function $(z-z_0)^{m-p}\lambda(z)$, z_0 is a removable singularity. ∎

Example 4.5.11. (Dividing by powers) (a) The function $e^{z-2} - 1$ can be expanded in a Taylor series about $z_0 = 2$ as $e^{z-2} - 1 = (z-2) + \frac{1}{2}(z-2)^2 + \cdots$. Clearly, this function has a zero of order 1 at $z_0 = 2$, so by Proposition 4.5.10 (or just by dividing), $\frac{e^{z-2}-1}{z-2}$ has a removable singularity at $z_0 = 2$.

(b) The function $\cos z - 1$ can be expanded in a Maclaurin series as

$$\cos z - 1 = -\frac{1}{2}z^2 + \frac{1}{4!}z^4 - \cdots .$$

Clearly, this function has a zero of order 2 at 0, so from Proposition 4.5.10, each of $\frac{\cos z - 1}{z}$ and $\frac{\cos z - 1}{z^2}$ has a removable singularity at 0. □

We can now give several characterizations of a removable singularity.

Theorem 4.5.12. *Suppose that f is analytic on $0 < |z - z_0| < R$. The following are equivalent:*

(i) f has a removable singularity at z_0.

(ii) $f(z) = \displaystyle\sum_{n=0}^{\infty} a_n(z - z_0)^n$ for $0 < |z - z_0| < R$.

(iii) There is a complex number A such that $\displaystyle\lim_{z \to z_0} f(z)$ exists and equals A.

(iv) $\displaystyle\lim_{z \to z_0} |f(z)|$ exists and is finite.

(v) f is bounded in a neighborhood of z_0.

(vi) $\displaystyle\lim_{z \to z_0} (z - z_0)f(z) = 0$.

Moreover, if (iii) holds, then the analytic extension of f on the disk $|z - z_0| < R$ satisfies $f(z_0) = A$.

Proof. $(i) \Rightarrow (ii)$ follows from Theorem 4.3.1, i.e., the existence of a power series expansion for the analytic extension of f on the disk $|z - z_0| < R$.

$(ii) \Rightarrow (iii)$ is a consequence of the fact that power series are continuous; $A = a_0$.

$(iii) \Rightarrow (iv)$ follows from the continuity of the absolute value function $w \mapsto |w|$.

$(iv) \Rightarrow (v)$ is obtained from Theorem 2.2.7, which asserts that if a function has a limit as $z \to z_0$, then it is bounded in a neighborhood of z_0.

$(v) \Rightarrow (vi)$ is obtained by the squeeze theorem (Theorem 2.2.5).

$(vi) \Rightarrow (i)$ Define the function

$$h(z) = \begin{cases} (z - z_0)^2 f(z) & \text{when } 0 < |z - z_0| < R, \\ 0 & \text{when } z = z_0. \end{cases}$$

Then for $z \neq z_0$, $h'(z)$ exists and is equal to $(z - z_0)^2 f'(z) + 2(z - z_0)f(z)$. Also,

$$\lim_{z \to z_0} \frac{(z-z_0)^2 f(z)}{z-z_0} = \lim_{z \to z_0} (z-z_0)f(z) = 0,$$

hence $h'(z_0)$ exists and is equal to 0. Thus h is analytic on $B_R(z_0)$ and satisfies $h(z_0) = h'(z_0) = 0$. It follows that z_0 is a zero of h of order at least 2, since the first two terms of the power series expansion of h vanish. In view of Proposition 4.5.10, it follows that $f(z) = \frac{h(z)}{(z-z_0)^2}$ has a removable singularity at z_0.

Finally, if (iii) holds, then by continuity, the value of f at z_0 should be the limit of $f(z)$ as $z \to z_0$ which is A. ∎

Example 4.5.13. (Removable singularities) Show that the functions have removable singularities at the indicated points:

(a) $\dfrac{\sin z}{z}$ at $z_0 = 0$; (b) $\dfrac{e^{z-1} - 1}{z-1}$ at $z_0 = 1$.

Solution. (a) We have

$$\lim_{z \to 0} (z-0)f(z) = \lim_{z \to 0} z \frac{\sin z}{z} = \lim_{z \to 0} \sin z = 0.$$

Thus $\frac{\sin z}{z}$ has a removable singularity at $z = 0$, in view of Theorem 4.5.12(vi).
(b) Again we apply Theorem 4.5.12(vi). We have

$$\lim_{z \to 1} (z-1)f(z) = \lim_{z \to 1} (z-1) \frac{e^{z-1} - 1}{z-1} = \lim_{z \to 1} (e^{z-1} - 1) = 0.$$

Hence $z_0 = 1$ is a removable singularity. □

In order to characterize poles, we relate a pole of f to a zero of $\frac{1}{f}$. Let f be analytic in $0 < |z - z_0| < R$, so that z_0 is an isolated singularity. Suppose that z_0 is a pole of f. Since $\lim_{z \to z_0} |f(z)| = \infty$, we can find $\rho > 0$ such that $f(z) \neq 0$ for all $0 < |z - z_0| < \rho$. Consider the function

$$g(z) = \begin{cases} \frac{1}{f(z)} & \text{if } 0 < |z - z_0| < \rho, \\ 0 & \text{if } z = z_0. \end{cases} \tag{4.5.3}$$

Clearly, g is analytic and nonzero on $0 < |z - z_0| < \rho$. As $\lim_{z \to z_0} g(z) = 0 = g(z_0)$, it follows from Theorem 4.5.12 that g is analytic at $z = z_0$; and since g is not identically 0 in a neighborhood of z_0, it follows that z_0 is a zero of g. We are now able to define the order of the pole of f at z_0 as the order of the zero of g at z_0.

Definition 4.5.14. Let z_0 be a pole of an analytic function f on $0 < |z - z_0| < R$. The **order of the pole** of f at z_0 is defined as the order of the zero of $\frac{1}{f}$ at $z = z_0$.

We describe equivalent characterizations of poles.

Theorem 4.5.15. *Let $m \geq 1$ be an integer. Let $R > 0$ and suppose that f is analytic on the punctured disk $0 < |z - z_0| < R$. Then the following conditions are equivalent:*
(i) f has a pole of order m at z_0.
(ii) There is an $r > 0$ and there is a nowhere vanishing analytic function ϕ on $B_r(z_0)$ such that

$$f(z) = \frac{\phi(z)}{(z - z_0)^m}, \qquad \text{when } 0 < |z - z_0| < R.$$

(iii) There exists a complex number $\alpha \neq 0$ such that

$$\lim_{z \to z_0} (z - z_0)^m f(z) = \alpha.$$

(iv) There exist $r > 0$ and complex numbers a_n for $n \geq -m$ such that $a_{-m} \neq 0$ and

$$f(z) = \frac{a_{-m}}{(z - z_0)^m} + \frac{a_{-m+1}}{(z - z_0)^{m-1}} + \cdots + \frac{a_{-1}}{z - z_0} + a_0 + a_1(z - z_0) + a_2(z - z_0)^2 + \cdots$$

for all $z \in B_r(z_0) \setminus \{z_0\}$.

Proof. $(i) \Rightarrow (ii)$. If f has a pole of order m at z_0, then the function g in (4.5.3) has a zero of order m at z_0. By Theorem 4.5.2, there is a nowhere vanishing function λ on a neighborhood $B_r(z_0)$ of z_0 such that $g(z) = (z - z_0)^m \lambda(z)$. This implies assertion (ii) with $\phi = 1/\lambda$.

$(ii) \Rightarrow (iii)$. This is an easy consequence of the observation that $(z - z_0)^m f(z) = \phi(z)$ which converges to $\phi(z_0)$ as $z \to z_0$. Then $\alpha = \phi(z_0) \neq 0$ works.

$(iii) \Rightarrow (iv)$. Using Theorem 4.5.12(iii) we deduce that z_0 is a removable singularity of $(z - z_0)^m f(z)$ on the annulus $|z - z_0| < r$. Expanding $q(z) = (z - z_0)^m f(z)$ in a power series centered at z_0, we write

$$q(z) = \sum_{n=0}^{\infty} c_n(z - z_0)^n, \qquad |z - z_0| < r, \tag{4.5.4}$$

for some c_k with $c_0 = \alpha$. Dividing both sides of (4.5.4) by $(z - z_0)^m$ and using that $(z - z_0)^m f(z) = q(z)$ for $0 < |z - z_0| < r$, we obtain the claimed Laurent series of f. Note that $a_k = c_{k+m}$ for $k \geq -m$ and that $a_{-m} = \alpha \neq 0$.

$(iv) \Rightarrow (i)$. Factoring $\frac{1}{(z-z_0)^m}$ from the Laurent series of $f(z)$ in (iv) we write $f(z)$ as

$$\frac{1}{(z - z_0)^m} \left[a_{-m} + a_{-m+1}(z - z_0) + \cdots + a_{-1}(z - z_0)^{m-1} + a_0(z - z_0)^m + \cdots \right]$$

and we note that the expression inside the square brackets is an analytic function on $|z - z_0| < r$. This function, which we call h, satisfies $h(z_0) = a_{-m} \neq 0$; thus there is a neighborhood $B_\delta(z_0)$ of z_0 on which it is nowhere vanishing. We have

$$\lim_{z \to z_0} |f(z)| = \lim_{z \to z_0} \frac{1}{|z - z_0|^m} \lim_{z \to z_0} |h(z)| = |a_{-m}| \cdot \infty = \infty$$

which implies that z_0 is a pole of f. Since $1/h$ does not vanish on $B_\delta(z_0)$, we write

$$\frac{1}{f(z)} = (z - z_0)^m \frac{1}{h(z)}$$

and consequently the order of the zero of $1/f$ at z_0 is m; so the order of the pole z_0 of f is also m in view of Definition 4.5.14. ∎

Example 4.5.16. (Poles) Determine the order of the pole of the functions at the indicated points:

(a) $\dfrac{1}{z \sin z}$ at $z_0 = 0$;   (b) $\dfrac{e^{z^2} - 1}{z^4}$ at $z_0 = 0$.

Solution. (a) We use Theorem 4.5.15(*iii*) and our knowledge of the function $\sin z$ around 0. Since

$$\lim_{z \to 0} z^2 \frac{1}{z \sin z} = \lim_{z \to 0} \frac{z}{\sin z} = 1 \neq 0,$$

we conclude that 0 is a pole of order 2 of $f(z)$.

(b) We use Laurent series. Since

$$e^z = \sum_{n=0}^{\infty} \frac{z^n}{n!}, \qquad \text{all } z,$$

then

$$e^{z^2} = \sum_{n=0}^{\infty} \frac{z^{2n}}{n!} = 1 + \frac{z^2}{1!} + \frac{z^4}{2!} + \frac{z^6}{3!} + \cdots, \qquad \text{all } z,$$

hence

$$e^{z^2} - 1 = \frac{z^2}{1!} + \frac{z^4}{2!} + \frac{z^6}{3!} + \cdots, \qquad \text{all } z,$$

and so for $z \neq 0$

$$\frac{e^{z^2} - 1}{z^4} = \frac{z^2}{1!z^4} + \frac{z^4}{2!z^4} + \frac{z^6}{3!z^4} + \cdots = \frac{1}{1!z^2} + \frac{1}{2!} + \frac{z^2}{3!} + \frac{z^4}{4!} + \cdots.$$

Thus the order of the pole at 0 is 2. □

Having characterized removable singularities and poles in terms of Laurent series, this leaves one possibility for essential singularities. They must have infinitely many terms involving negative powers of $(z - z_0)$. For ease of reference, we list all three possibilities together.

Theorem 4.5.17. *Suppose that f is analytic in a region Ω except for an isolated singularity at z_0 in Ω. Let*

$$f(z) = \sum_{n=-\infty}^{\infty} a_n (z - z_0)^n$$

denote the Laurent series expansion of f about z_0, which is valid in some annulus $0 < |z - z_0| < R$. Then

(i) z_0 is a removable singularity $\Leftrightarrow a_n = 0$ for all $n = -1, -2, \ldots$.
(ii) z_0 is a pole of order $m \geq 1$ $\Leftrightarrow a_{-m} \neq 0$ for some $m > 0$ and $a_n = 0$ for all $n < -m$.
(iii) z_0 is an essential singularity $\Leftrightarrow a_n \neq 0$ for infinitely many $n < 0$.

Example 4.5.18. (Essential singularities) Classify the isolated singularities of the function $f(z) = e^{\frac{z}{\sin z}}$.

Solution. The function f is analytic at all points except where $\sin z = 0$; that is, except when $z = k\pi$, where k is an integer. As it is a bit complicated to find the Laurent series expansion of f, we use the characterizations of Theorems 4.5.12 and 4.5.15. When $z = 0$, we have $\lim_{z \to 0} \frac{z}{\sin z} = 1$, hence $\lim_{z \to 0} e^{\frac{z}{\sin z}} = e$, and so the function has a removable singularity at $z = 0$ by Theorem 4.5.12(iii). We claim that we have an essential singularity at $z = k\pi$, $k \neq 0$. To prove this, we eliminate the possibility of a removable singularity or a pole. Suppose that k is even. Then it is easy to see that if $z = x$ is real, then $\lim_{x \downarrow k\pi} \frac{x}{\sin x} = +\infty$ and $\lim_{x \uparrow k\pi} \frac{x}{\sin x} = -\infty$. So if $z = x$ is real, then

$$\lim_{x \downarrow k\pi} e^{\frac{x}{\sin x}} = \infty,$$

implying that $k\pi$ is not a removable singularity of $e^{\frac{z}{\sin z}}$. Also,

$$\lim_{x \uparrow k\pi} e^{\frac{x}{\sin x}} = 0,$$

implying that $k\pi$ is not a pole of $e^{\frac{z}{\sin z}}$. This leaves only one possibility: $k\pi$ is an essential singularity of $e^{\frac{z}{\sin z}}$. A similar argument works for odd k. $\square$

One way to determine whether an isolated singularity is an essential singularity is to rule out the possibility of the other two types of singularities. This can be achieved by showing that the function is unbounded and has different limits as we approach the isolated singularity in different ways. The following theorem is a useful characterization of essential singularities. The theorem was discovered independently by Weierstrass and the Italian mathematician Felice Casorati (1835–1890).

Theorem 4.5.19. (Casorati-Weierstrass) *Suppose that f is analytic on a punctured disk $B'_R(z_0)$. Then z_0 is an essential singularity of f if and only if the following two conditions hold:*

(i) There is a sequence $\{z_n\}_{n=1}^{\infty}$ in $B'_R(z_0)$ such that $z_n \to z_0$ and $|f(z_n)| \to \infty$ as $n \to \infty$.
(ii) For any complex number α, there is a sequence $\{z_n\}_{n=1}^{\infty}$ in $B'_R(z_0)$ (that depends on α) such that $z_n \to z_0$ and $f(z_n) \to \alpha$ as $n \to \infty$.

Remark 4.5.20. Notice that (i) is not saying that $\lim_{z \to z_0} |f(z)| = \infty$. It is just saying that you can approach z_0 in such a way that $|f(z)|$ will tend to infinity. Similarly, part (ii) says that you can approach z_0 in such a way that $f(z)$ comes arbitrarily close to any complex value α. In fact, a deep result in complex analysis, known as Picard's great theorem, states that, in a neighborhood of an essential singularity, a function takes on every complex value, with one possible exception, an infinite number of times.

Proof. If (i) holds, then f is not bounded near z_0, and so z_0 is not a removable singularity. If (ii) holds, then it is not the case that $\lim_{z \to z_0} |f(z)| = \infty$, and so z_0 is not a pole. Thus (i) and (ii) together imply that z_0 is an essential singularity.

Conversely, let z_0 be an essential singularity. Since f is not bounded near z_0 (otherwise z_0 would be a removable singularity), it follows that (i) holds. Now we prove (ii). Precisely, we show that

$$\forall n = 1, 2, \ldots \exists z_n \in B_{1/n}(z_0) \cap B'_R(z_0) \text{ such that } |f(z_n) - \alpha| < 1/n. \qquad (4.5.5)$$

Suppose that (4.5.5) does not hold. Then there is a natural number n such that for all $z \in B_{1/n}(z_0) \cap B'_R(z_0)$ we have $|f(z) - \alpha| \geq 1/n$. Consider the function $g(z) = \frac{1}{f(z) - \alpha}$ defined for $z \in B_{1/n}(z_0) \cap B'_R(z_0)$. Then $|g(z)| \leq n$ on its domain and so z_0 is a removable singularity for g, by Theorem 4.5.12(v). Also, since $\lim_{z \to z_0} |f(z)| \neq \infty$, we conclude that $g(z_0) \neq 0$. Solving for $f(z)$, we find $f(z) = \frac{1}{g(z)} + \alpha$, which is analytic in a neighborhood of z_0, since $g(z_0) \neq 0$. This is a contradiction. Hence statement (4.5.5) is valid. Consequently, there are z_n satisfying $0 < |z_n - z_0| < 1/n$ such that $|f(z_n) - \alpha| < 1/n$; then (ii) follows by letting $n \to \infty$. ∎

Example 4.5.21. (Casorati-Weierstrass) Suppose α is a complex number. Find a sequence $\{z_n\}_{n=1}^{\infty}$ that converges to 0 as $n \to \infty$ such that $e^{1/z_n} \to \alpha$ as $n \to \infty$. Also find a sequence $z_n \to 0$ such that $|e^{1/z_n}| \to \infty$ as $n \to \infty$.

Solution. If α is not a negative real number, choose $z_n = (\text{Log } \alpha + 2\pi i n)^{-1}$. If α is a negative real number, choose $z_n = (\text{Log } (i\alpha) + 2\pi i n - \frac{\pi i}{2})^{-1}$. The sequence $\{z_n\}_{n=1}^{\infty}$ tends to zero in each case but $e^{1/z_n} = \alpha$ for all n. If $\alpha = 0$ choose $z_n = -1/n$; then $e^{1/z_n} \to 0$ as $n \to \infty$. As already observed, for $\alpha = \infty$, $z_n = 1/n$ works. □

We end the section by extending the definitions of zeros and singularities to the point at infinity. If f is analytic on a neighborhood of infinity, i.e., f is analytic for all $|z| > R$, then $f\left(\frac{1}{z}\right)$ is analytic in the annulus $0 < |z| < \frac{1}{R}$ and hence it has an isolated singularity at 0. With this in mind, we make the following definitions.

Definition 4.5.22. Suppose that f is analytic for all $|z| > R$. Then f has

a **removable singularity at** ∞ if $f\left(\frac{1}{z}\right)$ has a removable singularity at 0;
a **pole of order** m **at** ∞ if $f\left(\frac{1}{z}\right)$ has a pole of order m at 0;
an **essential singularity at** ∞ if $f\left(\frac{1}{z}\right)$ has a essential singularity at 0.

When f has a removable singularity at ∞, $\lim_{z\to\infty} f(z)$ exists. When $\lim_{z\to\infty} f(z) = 0$, we say that f has a **zero** at ∞.

Example 4.5.23. (Singularities at ∞) Characterize all entire functions with a pole of order $m \geq 1$ at ∞.

Solution. Let f be an entire function with a pole of order $m \geq 1$ at ∞. This function has a Maclaurin series $f(z) = \sum_{n=0}^{\infty} c_n z^n$ that converges for all z. For $z \neq 0$, $f\left(\frac{1}{z}\right) = \sum_{n=0}^{\infty} \frac{c_n}{z^n}$. Appealing to Theorem 4.5.15(iv), we see that $f\left(\frac{1}{z}\right)$ has a pole of order $m \geq 1$ at 0 if and only if $c_m \neq 0$ and $c_n = 0$ for all $n > m$, in other words, if and only if f is a polynomial of degree m. Consequently, f has a pole of order $m \geq 1$ at ∞ if and only if f is a polynomial of degree m.   □

Exercises 4.5

In Exercises 1–8, find the isolated zeros of the functions. Also, find the order of each isolated zero.

1. $(1 - z^2)\sin z$ **2.** $z^3(e^z - 1)$ **3.** $\dfrac{z(z-1)^2}{z^2 + 2z - 1}$ **4.** $\sin\dfrac{1}{z}$

5. $\dfrac{\sin^7 z}{z^4}$ **6.** $(z-1)^3(e^{2z} - 1)^2$ **7.** $(z^2 + 2z - 1)^3$ **8.** $\sinh z$

In Exercises 9–12, find the order of the zero at $z_0 = 0$.

9. $1 - \dfrac{z^2}{2} - \cos z$ **10.** $z\,\mathrm{Log}\,(1+z)$ **11.** $z - \sin z$ **12.** $\tan z$

In Exercises 13–24, classify the isolated singularities of the functions. Do not include the case at ∞. At a removable singularity, redefine the functions to make them analytic. If it is a pole, determine its order.

13. $\dfrac{1 - z^2}{\sin z} + \dfrac{z - 1}{z + 1}$ **14.** $\dfrac{z - 1}{z - i} + \dfrac{z - i}{z - 1}$ **15.** $\dfrac{z(z-1)^2}{\sin(\pi z)\sin z}$

16. $e^{\frac{1}{1-z}} + \dfrac{1}{1 - z}$ **17.** $z\tan\dfrac{1}{z}$ **18.** $\dfrac{z}{e^z - 1}$

19. $\dfrac{z}{z^4 - 1} - \dfrac{\sin(2z)}{z^4}$ **20.** $z^2 \sin\dfrac{1}{z^2}$ **21.** $\dfrac{1}{z} - \sin\dfrac{1}{z}$

22. $\dfrac{1}{(e^z - e^{2z})^2}$ **23.** $\dfrac{\cot z}{(z - \frac{\pi}{2})^2}$ **24.** $\dfrac{z\sin z}{\cos z - 1}$

In Exercises 25–30, determine if the function has an isolated singularity at ∞, and determine its type. Does the function have a zero at ∞?

25. $\dfrac{1}{z + 1}$ **26.** $\dfrac{z^2 - 1}{z^2 + 2z + 3i}$ **27.** $\dfrac{z}{z^2 + 1} - \dfrac{1}{z}$

28. $e^z - \cos\dfrac{1}{z}$ **29.** $\sin\dfrac{1}{z}$ **30.** $\dfrac{e^z}{e^z - 1}$

31. Prove the identities using Theorem 4.5.5:

(a) $\sin^2 z = \dfrac{1 - \cos 2z}{2}$ (b) $\sin(2z) = 2\sin z\cos z$

(c) $\tan(2z) = \dfrac{2\tan z}{1 - \tan^2 z}$ $z \neq \dfrac{\pi}{2} + 2k\pi$ or $\dfrac{\pi}{4} + 2k\pi$

32. Prove the identities using Theorem 4.5.5.

(a) $e^{z_1 + z_2} = e^{z_1} e^{z_2}$ (b) $\sin(z_1 + z_2) = \sin z_1 \cos z_2 + \cos z_1 \sin z_2$

33. Give a new proof of Liouville's theorem (Theorem 3.9.2) based on Theorem 4.5.12 as follows:
(a) Suppose f is entire and bounded. Show that $f(\frac{1}{z})$ has a removable singularity at 0.
(b) Express the Laurent series expansion of $f(\frac{1}{z})$ around 0 by replacing z by $\frac{1}{z}$ in the Maclaurin series expansion of f. Using part (a), conclude that f must be constant.

34. Show that f has an essential singularity at z_0 if and only if $(z - z_0)^m f(z)$ has an essential singularity at z_0, where m is an integer.

35. Prove the following assertions concerning zeros and isolated singularities:
(a) If f has a zero of order $m \geq 0$ at z_0 and g has a zero of order $n \geq 0$ at z_0, then fg has a zero of order $m + n$ at z_0. (Here we take a zero of order 0 to mean analytic and nonvanishing in a neighborhood of z_0.)
(b) If f has a pole of order $m \geq 1$ at z_0 and g has a zero of order $n \geq 1$ at z_0, then at z_0, fg has a pole of order $m - n$ if $m > n$; a zero of order $n - m$ if $n > m$, and a removable singularity if $m = n$.
(c) If f has a removable singularity at z_0 and g is analytic at z_0, then fg has a removable singularity at z_0.
(d) If f has an essential singularity at z_0 and g is not identically 0 with a removable singularity or a pole at z_0, then fg has an essential singularity at z_0. [*Hint*: Multiply or divide by suitable $(z - z_0)^m$ so that $g(z)$ is analytic and nonvanishing in a neighborhood of z_0; use Exercise 34.] Does the assertion remain true if g has an essential singularity at z_0?

36. Suppose that f is analytic in a deleted neighborhood of z_0. Show that f has a removable singularity at z_0 if and only if either $\mathrm{Re}\, f$ or $\mathrm{Im}\, f$ is bounded in the deleted neighborhood of z_0. [*Hint*: One direction is easy. For the other direction, suppose that $\mathrm{Re}\, f$ is bounded and show that $g = e^f$ has a removable singularity at z_0. Compute g' in a deleted neighborhood of z_0 and conclude that f' is analytic at z_0. Conclude that f is analytic at z_0. If $\mathrm{Im}\, f$ is bounded, consider $g = e^{if}$.]

37. (a) Show that if f has a pole of order $m \geq 1$ at z_0 and n is a nonzero integer, then f^n has a pole at z_0 of order mn if $n > 0$ or a zero at z_0 of order $-mn$ if $n < 0$.
(b) Show that if f has an essential singularity at z_0 and n is a nonzero integer, then f^n has an essential singularity at z_0.

38. Suppose that f and g are entire functions such that $f \circ g$ is a polynomial. Show that both f and g must be polynomials. [*Hint*: If p is a nonconstant polynomial, then $\lim_{z \to \infty} |p(z)| = \infty$. Suppose that f is not a polynomial. Then f has an essential singularity at ∞. Use this and Theorem 4.5.19 to show that $\lim_{z \to \infty} |f \circ g(z)| \neq \infty$. Argue similarly if g is not a polynomial.]

39. Contrasts with the theory of functions of a real variable.
(a) Consider $f(x) = \sin \frac{1}{x}$. Show that f is bounded and differentiable for all $x \neq 0$.
(b) Show that there is no differentiable function $g(x)$ defined on $\mathbb{R}$ such that $f(x) = g(x)$ for all $x \neq 0$. Which aspect of the theory of analytic functions did we contrast with this example?
(c) Define $\phi(x) = x^2 \sin \frac{1}{x}$, $x \neq 0$, $\phi(0) = 0$. Show that ϕ is differentiable for all x. Is the zero of ϕ isolated at $x = 0$? Which aspect of the theory of analytic functions did we contrast?
(d) Define $\psi(x) = e^{-\frac{1}{x^2}}$, $x \neq 0$, $\psi(0) = 0$. Show that ψ has derivatives of all order at $x = 0$. Explain why ψ cannot possibly have a Maclaurin series representation. Which aspect of the theory of analytic functions does this behavior contrast?

40. Suppose that f and g are analytic in a region Ω and fg is identically zero in Ω. Show that either f or g is identically zero in Ω.

41. Determine all entire functions with zeros at $\frac{1}{n}$, $n = 1, 2, \ldots$.

42. Suppose that f is entire such that $f(z)f(\frac{1}{z})$ is bounded on $\mathbb{C} \setminus \{0\}$. Follow the outlined steps to show that $f(z) = az^n$ for some constant a and nonnegative integer n.
(a) Let $h(z) = f(z)f(\frac{1}{z})$. Show that h has a removable singularity at 0 and conclude that it is a constant . In what follows, suppose that the constant is nonzero, otherwise f is identically zero.
(b) Show that the only possible zero of f is at $z = 0$.
(c) If $f(0) \neq 0$, show that f is constant.

(d) If $f(0) = 0$, show that $f(z) = az^n$, where n is the order of the zero at 0. [*Hint*: Factor z^n and apply (c) to the entire function $\frac{f(z)}{z^n}$.]

4.6 Schwarz's Lemma

In this section we put together some of the material we have developed to prove a very elegant and useful lemma attributed to Karl Hermann Amandus Schwarz (1843–1921). This lemma reflects the rigidity of analytic functions from the unit disk to itself and has remarkable applications.

Lemma 4.6.1. (Schwarz's Lemma) *Suppose that f is analytic on the open unit disk $B_1(0)$ with $f(0) = 0$ and that f takes values in the closed unit disk $\overline{B_1}(0)$; i.e., it satisfies $|f(z)| \leq 1$ for all $|z| < 1$. Then we have*

$$|f(z)| \leq |z| \quad \text{for all } |z| < 1 \tag{4.6.1}$$

and

$$|f'(0)| \leq 1. \tag{4.6.2}$$

Moreover, if either (a) equality holds in (4.6.1) for some $z \neq 0$ (i.e., there is a $z_0 \neq 0$ with $|z_0| < 1$ such that $|f(z_0)| = |z_0|$) or (b) equality holds in (4.6.2), then there is a complex constant A with $|A| = 1$ such that

$$f(z) = Az$$

for all $|z| < 1$.

Proof. Consider the function

$$g(w) = \begin{cases} \frac{f(w)}{w} & \text{if } w \neq 0, \\ f'(0) & \text{if } w = 0. \end{cases} \tag{4.6.3}$$

It follows from Theorem 4.5.12(*iii*) that g has a removable singularity and hence is analytic on the open unit disk. Fix a point z in $B_1(0)$ and let R satisfy $|z| < R < 1$. Then the function g is analytic on the open disk $B_R(0)$ and is continuous on its boundary $C_R(0)$; moreover $|g(w)| \leq 1/R$ on the circle $|w| = R$ in view of the fact that $|f(w)| \leq 1$ for all w in $B_1(0)$. It follows from the maximum modulus principle (Theorem 3.9.6) that $|g(w)| \leq 1/R$ for all $|w| \leq R$, in particular $|g(0)| \leq 1/R$ and $|g(z)| \leq 1/R$. Letting $R \uparrow 1$ we obtain $|g(0)| \leq 1$ and $|g(z)| \leq 1$. Recasting these in terms of f, we obtain $|f'(0)| \leq 1$ and $|f(z)| \leq |z|$. Since $|z| < 1$ was an arbitrary point in the unit disk, (4.6.1) and (4.6.2) hold.

Now suppose that equality holds in (4.6.1) for some $z = z_0 \neq 0$ or equality holds in (4.6.2). Then g attains its maximum in the interior of $B_1(0)$ and it must be equal to a constant A by Theorem 3.9.6 Thus $f(w) = Aw$ for all $|w| < 1$. Since there is a $z_0 \neq 0$ in $B_1(0)$ such that $|f(z_0)| = |z_0|$ or $f'(0) = 1$, it follows that $|A| = 1$. ∎

Möbius Transformations on the Unit Disk

Maps of the form $\frac{az+b}{cz+d}$ are called **linear fractional transformations**. We would like to study a class of linear fractional transformations from the open unit disk to itself of the form

$$\phi_a(z) = \frac{z-a}{1-\bar{a}z}, \qquad |z| \le 1 \tag{4.6.4}$$

where a is a complex number with $|a| < 1$. As the denominator in (4.6.4) never vanishes when $|z| < 1$, we note that these maps are analytic on the unit disk. A map ϕ_a is called a **Möbius transformation** after the mathematician August Ferdinand Möbius (1790–1868).

Proposition 4.6.2. (Properties of Möbius Transformations) *Let a,b be complex numbers with $|a|, |b| < 1$. Then for all $|z| < 1$ we have*

(i) $\phi_a(0) = -a$ *and* $\phi_a(a) = 0$.

(ii) ϕ_a *is a one-to-one and onto map from* $B_1(0)$ *to* $B_1(0)$.

(ii) ϕ_a *is a one-to-one and onto map from* $C_1(0)$ *to* $C_1(0)$.

(iv) $\phi_a^{-1} = \phi_{-a}$.

(v) $\phi_a'(z) = \dfrac{1-|a|^2}{(1-\bar{a}z)^2}$.

(vi) $\phi_a(cz) = c\,\phi_{a\bar{c}}(z)$ *where c is a complex number with* $|c| = 1$.

(vii) *We have the identity*

$$\phi_a \circ \phi_b(z) = \frac{1+a\bar{b}}{1+\bar{a}b}\,\phi_{\frac{a+b}{1+\bar{a}b}}(z) = \frac{1+a\bar{b}}{1+\bar{a}b}\,\phi_{\phi_{-b}(a)}(z).$$

Proof. Assertion (i) is trivial. To prove (ii) and (iii) notice that for θ real we have

$$|\phi_a(e^{i\theta})| = \left| \frac{e^{i\theta} - a}{1 - \bar{a}e^{i\theta}} \right| = \left| \frac{1 - ae^{-i\theta}}{1 - \bar{a}e^{i\theta}} \right| = 1.$$

This shows that ϕ_a maps $C_1(0)$ to $C_1(0)$. It follows that ϕ_a maps $B_1(0)$ to $B_1(0)$, otherwise it would achieve its maximum in the interior of $B_1(0)$, contradicting the maximum modulus principle (Theorem 3.9.6).

To prove the one-to-one assertion in (ii) and (iii) we assume $\frac{z-a}{1-\bar{a}z} = \frac{z'-a}{1-\bar{a}z'}$ and we show that $z = z'$. Cross-multiplying gives $(z-a)(1-\bar{a}z') = (z'-a)(1-\bar{a}z)$ which implies $z - a - \bar{a}z'z + |a|^2 z' = z' - a - \bar{a}zz' + |a|^2 z$. This in turn implies $z - z' = |a|^2(z - z')$, hence $(z - z')(1 - |a|^2) = 0$, and so $z = z'$.

We prove the onto assertion in (ii) and (iii) by solving the equation $\frac{z-a}{1-\bar{a}z} = w$ in z for a given w in the unit disk or on the unit circle. Indeed, algebraic manipulations yield

$$z = \frac{w+a}{1+\bar{a}w} = \phi_{-a}(w)$$

and this z satisfies $|z| < 1$ if $|w| < 1$ and $|z| = 1$ if $|w| = 1$. This also proves (iv).

To prove (v) we use the quotient rule to write

$$\phi_a'(z) = \frac{(1 - \bar{a}z) - (z - a)(-\bar{a})}{(1 - \bar{a}z)^2} = \frac{1 - |a|^2}{(1 - \bar{a}z)^2}.$$

To prove (vi) we write

$$\phi_\alpha(cz) = \frac{cz - a}{1 - \bar{a}cz} = \frac{cz - c\bar{c}a}{1 - \bar{a}cz} = c\frac{z - a\bar{c}}{1 - \overline{a\bar{c}}z} = c\,\phi_{a\bar{c}}(z).$$

Finally, (vii) is proved by a direct calculation as follows:

$$\phi_a \circ \phi_b(z) = \frac{(1 + a\bar{b})z - (a + b)}{(1 + \bar{a}b) - (\bar{a} + \bar{b})z} = \frac{1 + a\bar{b}}{1 + \bar{a}b} \cdot \frac{z - \dfrac{a + b}{1 + a\bar{b}}}{1 - \dfrac{\bar{a} + \bar{b}}{1 + \bar{a}b}z} = \frac{1 + a\bar{b}}{1 + \bar{a}b}\,\phi_{\frac{a+b}{1+a\bar{b}}}(z).$$

Notice that the constant $\frac{1 + a\bar{b}}{1 + \bar{a}b}$ is unimodular and that $\frac{a+b}{1+a\bar{b}} = \phi_{-b}(a)$. ∎

Let us examine how the properties of ϕ_a interplay with the maximum and minimum modulus principles. Clearly, ϕ_a is not constant, but we showed that $|\phi_a(z)|$ equals 1 on $C_1(0)$. Studying Corollary 3.9.10, we conclude that $\phi_a(z)$ must vanish somewhere inside the unit disk. This is certainly the case since $\phi_a(a) = 0$, and in fact $z = a$ is the only zero of $\phi_a(z)$ inside the unit disk.

The Schwarz-Pick theorem

Let f be a map defined on the unit disk $B_0(1)$ that satisfies $|f(z)| \leq 1$ for all $|z| < 1$. We cannot directly apply Schwarz's Lemma on f since we may not have $f(0) = 0$. But we can compose f with two linear fractional transformations to achieve this.

Theorem 4.6.3. (Schwarz-Pick Theorem) *Let f be an analytic function that maps the open unit disk $B_1(0)$ to the closed unit disk $\overline{B_1(0)}$. Then for a number a in the disk $B_1(0)$ we have*

$$\left| \frac{f(z) - f(a)}{1 - \overline{f(a)}f(z)} \right| \leq \left| \frac{z - a}{1 - \bar{a}z} \right| \tag{4.6.5}$$

and also

$$|f'(a)| \leq \frac{1 - |f(a)|^2}{1 - |a|^2}. \tag{4.6.6}$$

Moreover, if equality holds in (4.6.5) for some $z \neq a$ or equality holds in (4.6.6), then f is equal to a unimodular constant times a Möbius transformation.

Proof. Let be f as in the statement of the theorem. Set $b = f(a)$ and consider the function

$$F = \phi_b \circ f \circ \phi_{-a}$$

defined on the unit disk $B_1(0)$; see Figure 4.19. Then F maps $B_1(0)$ to $\overline{B_1(0)}$ and we have

$$F(0) = \phi_b \circ f \circ \phi_{-a}(0) = \phi_b(f(a)) = 0,$$

and moreover $|F(z)| \le 1$ for all $|z| < 1$. It follows from Lemma 4.6.1 that

$$|F(w)| = |\phi_b \circ f \circ \phi_{-a}(w)| \le |w|,$$

so letting $\phi_{-a}(w) = z$ and $\phi_a(z) = w$ we get $|\phi_b \circ f(z)| \le |\phi_a(z)|$ for $z \in B_1(0)$. Using the definition of ϕ_b we deduce (4.6.5).

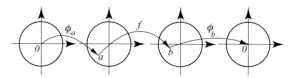

Fig. 4.19 The function $F = \phi_b \circ f \circ \phi_{-a}$.

Now consider $F'(0) = (\phi_b \circ f \circ \phi_{-a})'(0)$. Applying the chain rule, we write

$$F'(0) = \phi_b'(f(\phi_{-a}(0)))f'(\phi_{-a}(0))\phi_{-a}'(0).$$

Since $\phi_{-a}'(0) = 1 - |a|^2$ and $\phi_{-a}(0) = a$ we obtain

$$(\phi_b \circ f \circ \phi_{-a})'(0) = \frac{1 - |b|^2}{(1 - |b|^2)^2} f'(a)(1 - |a|^2) = f'(a)\frac{1 - |a|^2}{1 - |b|^2}.$$

By Lemma 4.6.1 this number should have modulus at most 1, thus we obtain (4.6.6).

Suppose now that equality holds in (4.6.5) for some $z \ne a$ or that equality holds in (4.6.6). Then by Lemma 4.6.1 we must have $\phi_b \circ f \circ \phi_{-a}(z) = cz$ for $|c| = 1$, so then $f \circ \phi_{-a}(z) = \phi_{-b}(cz)$ and leting $\phi_{-a}(z) = w$, we see $f(w) = \phi_{-b}(c\phi_a(w))$. Using Proposition 4.6.2 we write

$$f(w) = \phi_{-b}(c\phi_a(w)) = c\phi_{-b\bar{c}}(\phi_a(w)) = c\frac{1 - b\bar{a}c}{1 - \bar{b}ac}\, \phi_{\frac{a-b\bar{c}}{1-b\bar{a}c}}(z) = d\,\phi_q(z),$$

with

$$d = c\frac{1 - b\bar{a}c}{1 - \bar{b}ac}, \qquad q = -\phi_a(b\bar{c}), \quad (\text{note } |d| = 1, |q| < 1),$$

thus f must be a Möbius transformation times a unimodular constant. ∎

Example 4.6.4. Let f be an analytic map from the unit disk to itself.
(a) If f fixes a point a in the unit disk, i.e., $f(a) = a$, show that $|f'(a)| \le 1$.
(b) Show that $|f(0)|^2 + |f'(0)| \le 1$.

Solution. (a) Setting $f(a) = a$ in (4.6.6), we obtain $|f'(a)| \leq \frac{1-|a|^2}{1-|a|^2} = 1$.

(b) Setting $a = 0$ in (4.6.6) we obtain $|f(0)|^2 + |f'(0)| \leq 1$.   □

Example 4.6.5. Let f be an analytic map on the open unit disk such that $|f(z)| \leq 1$ for all $|z| < 1$.

(a) If for some $|a| < 1$ we have $f(a) = 0$, show that $|f(z)| \leq |\phi_a(z)|$ for all z in $B_1(0)$.

(b) Let $a_1, \ldots, a_m$ be (not necessarily distinct) zeros of f in $B_1(0)$. Prove that

$$|f(z)| \leq |\phi_{a_1}(z) \cdots \phi_{a_m}(z)| \tag{4.6.7}$$

for all z in $B_1(0)$.

(c) Let $a_1, \ldots, a_q$ be distinct zeros of f in $B_1(0)$ of orders $m_1, \ldots, m_q$, respectively. Show that

$$|f(z)| \leq |\phi_{a_1}(z)|^{m_1} \cdots |\phi_{a_q}(z)|^{m_q} \tag{4.6.8}$$

for all z in $B_1(0)$.

Solution. (a) Setting $f(a) = 0$ in (4.6.5) we obtain $|f(z)| \leq |\phi_a(z)|$ for all $z \in B_1(0)$.

(b) By part (a) we have $|f(z)| \leq |\phi_{a_1}(z)|$. The function $f(z)/\phi_{a_1}(z)$ is bounded by 1 and is analytic on $B_1(0)$, since the zero of $\phi_{a_1}(z)$ at $z = a_1$ is canceled by corresponding the zero of f. Applying the assertion in part (a) for $a = a_2$ to the function $f(z)/\phi_{a_1}(z)$ we obtain $|f(z)/\phi_{a_1}(z)| \leq |\phi_{a_2}(z)|$. Continuing in this way we deduce (4.6.7).

(c) Applying the result in part (b) to the list of repeated zeros (according to their order)

$$\underbrace{a_1, \ldots, a_1}_{m_1 \text{ times}}, \underbrace{a_2, \ldots, a_2}_{m_2 \text{ times}}, \ldots, \underbrace{a_q, \ldots, a_q}_{m_q \text{ times}},$$

we derive (4.6.8).   □

Exercises 4.6

1. Does there exist an analytic function g from the unit disk to itself such that $g(1/5) = 4/5$ and $g'(1/5) = 5/12$?

2. Does there exist an analytic function h from the unit disk to itself such that

$$7|h(\tfrac{1}{2}) - h(\tfrac{1}{4})| > 2|1 - \overline{h(\tfrac{1}{2})}h(\tfrac{1}{4})|?$$

3. Let f be an analytic function on $B_1(0)$ satisfying $|f(z)| \leq 1$ for all $|z| < 1$. Let $m \geq 1$. Assume that $z = 0$ is a zero of order m of f. Prove that for all $|z| < 1$ we have

$$|f(z)| \leq |z|^m$$

and also

$$|f^{(m)}(0)| \leq m!$$

4. Let f be as in Exercise 3. Suppose that either $|f^{(m)}(0)| = m!$ or $|f(z_0)| = |z_0|^m$ for some $z_0 \neq 0$. Show that there exists a complex number a with $|a| = 1$ such that $f(z) = az^m$.

5. Let f be an analytic one-to-one and onto map from the open unit disc to itself such that $f(0) = 0$. Suppose that the inverse map f^{-1} is also analytic. Show that $f(z) = cz$ for some constant c of modulus 1. [*Hint:* Apply Schwarz's lemma to both f and f^{-1} to obtain $|f'(0)| = 1$.]

6. Let f be an analytic one-to-one map from the open unit disc onto itself. Suppose that the inverse map f^{-1} is also analytic. Show that $f(z) = c \phi_q(z)$ for some constant c of modulus 1 and some $|q| < 1$. [*Hint:* Apply the preceding exercise to the function $F = \phi_b \circ f \circ \phi_{-a}$ where $b = f(a)$.]

7. Suppose that Ω is a region and f and g are two analytic one-to-one mappings from Ω onto the unit disk D such that f^{-1}, g^{-1} are also analytic. Show that there is a linear fractional transformation ϕ_α with $|\alpha| < 1$ such that $f(z) = c \phi_\alpha \circ g(z)$ for all $z \in \Omega$, where c is a unimodular constant. [*Hint:* Use Exercise 6. (The assumptions that f^{-1}, g^{-1} are analytic can be dropped by Corollary 5.7.18.)]

8. World's worst function. (a) For $z \neq 1$ let $I(z) = e^{\frac{z+1}{z-1}}$. Show that I is analytic for all $z \neq 1$ and that $|I(z)| = 1$ for all $|z| = 1$ and $z \neq 1$.

[*Hint:* $|I(z)| = e^{\mathrm{Re}\left(\frac{z+1}{z-1}\right)} = e^{\frac{|z|^2-1}{|z-1|^2}}$.]

(b) Show that if z is real and $z \to 1^-$, then $I(z) \to 0$.

(c) Show that $|I(z)| < 1$ for all $|z| < 1$.

9. Suppose that p is a polynomial such that $|p(z)| = 1$ for all $|z| = 1$. Show that $p(z) = Az^m$, where A is a unimodular constant and m is the number of zeros of $p(z)$ inside the unit disk counted according to multiplicity (that means a zero of order $k \geq 1$ is counted k times). [*Hint:* Recall that if f is analytic on $|z| \leq 1$, $f(z) \neq 0$ for all $|z| < 1$, and $|f(z)| = 1$ for $|z| = 1$, then f is constant. So if p has no zeros in the unit disk, we are done. Otherwise, let $a_1, a_2, \ldots, a_m$ denote the zeros of p inside the open unit disk, repeated according to multiplicity. Multiply $p(z)$ by a product of linear fractional transformations of the form $\phi_{a_j}(z) = \frac{1 - \bar{a}_j z}{a_j - z}$ to reduce to the case of a function without zeros in $|z| < 1$ and with modulus 1 on the circle $|z| = 1$. Then argue that $a_j = 0$ for all j and that $p(z) = Az^m$.]

10. Suppose that f is analytic for all $|z| \leq 1$ and $|f(z)| = 1$ for all $|z| = 1$. Suppose that f has n zeros inside the open unit disk $a_1, a_2, \ldots, a_n$ counted according to multiplicity. Show that $f = A \phi_{a_1} \cdots \phi_{a_n}$, where $\phi_{a_j}(z) = \frac{a_j - z}{1 - \bar{a}_j z}$ is a linear fractional transformation and A is unimodular.

11. Suppose that $f(z)$ is analytic on $|z| < 1$ and continuous on $|z| \leq 1$ such that $|f(z)| = 1$ for all $|z| = 1$. Show that f has finitely many zeros in $|z| \leq 1$ and conclude that $f = A \phi_{a_1} \cdots \phi_{a_n}$, where $\phi_{a_j}(z) = \frac{a_j - z}{1 - \bar{a}_j z}$ is a linear fractional transformation and A is unimodular. [*Hint:* The first part follows from Corollary 4.5.7. The second part follows from Exercise 10.]

12. Find all analytic functions f such that $|f(z)| \leq 1$ for all $|z| \leq 1$, $f(0) = 0$, and $f\left(\frac{i}{2}\right) = \frac{1}{2}$.

13. Suppose that f is analytic, $|f(z)| \leq 3$ for all $|z| = 1$, and $f(0) = 0$. Can $|f'(0)| > 3$?

14. Suppose that f is analytic on $|z| < 1$ satisfying $|f(z)| \leq 1$. Show that for all $|z| < 1$, we have $|f^{(n)}(z)| \leq \frac{n!}{(1-|z|)^n}$. [*Hint:* Apply Cauchy's estimate to $f^{(n)}(z)$. Consider the radius of the disk that is contained in $B_1(0)$ with center at z.]

Chapter 5
Residue Theory

After having thought about this subject, and brought together the diverse results mentioned above, I had the hope of establishing on a direct and rigorous analysis the passage from the real to the imaginary; and my research has lead me to this Memoire.

-Augustin-Louis Cauchy (1789–1857)
[Writing about his Memoire of 1814, which contained the residue theorem and several computations of real integrals by complex methods.]

In previous chapters, we introduced complex functions and studied properties of three essential tools: derivatives, integrals, and series of complex functions. In this chapter, we derive some exciting applications of complex analysis based on one formula, known as Cauchy's residue theorem. We have indirectly used this result when computing integrals in previous sections. Here we highlight the main ideas behind it and devise new techniques for computing Laplace transforms and integrals that arise in Fourier series. For instance, integrals like $\int_{-\infty}^{\infty} \frac{\sin x}{x}\, dx$ and $\int_0^{\infty}(\cos x)e^{-x^2}\, dx$ are often tedious to compute via real-variable techniques. With the residue theorem and some additional estimates with complex functions, the computations of these integrals become straightforward tasks.

In Section 5.7, we use residues to expand our knowledge of analytic functions. We use integrals to count the number of zeros of analytic functions and give a formula for the inverse of analytic functions. Theoretical results, such as the open mapping property, are derived and used to obtain a fresh and different perspective on concrete results such as the maximum modulus principle.

The residue theorem was discovered by Cauchy around 1814 but was explicitly stated in 1831. Cauchy's goal was to place under one umbrella the computations of certain special integrals, some of which involving complex substitutions, that were calculated by Euler, Laplace, Legendre, and others.

5.1 Cauchy's Residue Theorem

Suppose that f is an analytic function on a deleted neighborhood of a point z_0 and has an isolated singularity at z_0. We know from Theorem 4.4.1, that f has a Laurent series in an annulus around z_0: for $0 < |z - z_0| < R$,

$$f(z) = \cdots + \frac{a_{-2}}{(z-z_0)^2} + \frac{a_{-1}}{z-z_0} + a_0 + a_1(z-z_0) + a_2(z-z_0)^2 + \cdots.$$

© Springer International Publishing AG, part of Springer Nature 2018
N. H. Asmar and L. Grafakos, *Complex Analysis with Applications*,
Undergraduate Texts in Mathematics, https://doi.org/10.1007/978-3-319-94063-2_5

Furthermore, the series can be integrated term by term over any path that lies in the annulus $0 < |z - z_0| < R$. Let $C_r(z_0)$ be a positively oriented circle that lies in $0 < |z - z_0| < R$. If we integrate the Laurent series term by term over $C_r(z_0)$ and use the fact that $\int_{C_r(z_0)} (z - z_0)^n dz = 0$ if $n \neq -1$ and $\int_{C_r(z_0)} \frac{1}{z - z_0} dz = 2\pi i$ (Example 3.2.10), we find $\int_{C_r(z_{0s})} f(z) dz = a_{-1} 2\pi i$; hence,

$$a_{-1} = \frac{1}{2\pi i} \int_{C_r(z_0)} f(z) dz. \tag{5.1.1}$$

Definition 5.1.1. The coefficient a_{-1} in (5.1.1) is called the **residue of f at z_0** and is denoted by $\text{Res}(f, z_0)$ or simply $\text{Res}(z_0)$ when there is no risk of confusion.

With the concept of residue in hand, we state our main result of this section which reduces the evaluation of certain integrals to computations of residues.

Theorem 5.1.2. (Cauchy's Residue Theorem) *Let C be a simple closed positively oriented path. Suppose that f is analytic inside and on C, except at finitely many isolated singularities $z_1, z_2, \ldots, z_n$ inside C. Then*

$$\int_C f(z) dz = 2\pi i \sum_{j=1}^n \text{Res}(f, z_j). \tag{5.1.2}$$

Proof. We start by picking small circles $C_{r_j}(z_j)$ ($j = 1, 2, \ldots, n$) that do not intersect each other and are contained in the interior of C (Figure 5.1). Apply Cauchy's integral theorem for multiple simple paths (Theorem 3.7.2) to obtain

$$\int_C f(z) dz = \sum_{j=1}^n \int_{C_{r_j}} f(z) dz = 2\pi i \sum_{j=1}^n \text{Res}(z_j),$$

where the last equality follows from (5.1.1). So (5.1.2) holds. ∎

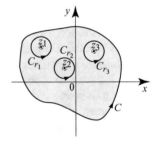

Fig. 5.1 The path C and the circles $C_{r_j}(z_j)$.

The computation of the residue will depend on the type of singularity of the function f, as we now illustrate.

Proposition 5.1.3. *(i) Suppose that z_0 is an isolated singularity of f. Then f has a simple pole at z_0 if and only if*

$$\text{Res}(f, z_0) = \lim_{z \to z_0} (z - z_0) f(z) \neq 0. \tag{5.1.3}$$

(ii) If $f(z) = \frac{p(z)}{q(z)}$, where p and q are analytic at z_0, $p(z_0) \neq 0$, and q has a simple zero at z_0, then

$$\text{Res}\left(\frac{p(z)}{q(z)}, z_0\right) = \frac{p(z_0)}{q'(z_0)}. \tag{5.1.4}$$

Proof. By Theorem 4.5.15, z_0 is a pole of order 1 if and only if the Laurent series of f at z_0 is of the form

$$f(z) = \frac{a_{-1}}{z - z_0} + a_0 + a_1(z - z_0) + a_2(z - z_0)^2 + \cdots = \frac{a_{-1}}{z - z_0} + h(z),$$

where $a_{-1} \neq 0$ and h is the analytic power series part of the Laurent series. Then, $(z - z_0)f(z) = a_{-1} + (z - z_0)h(z)$, and (i) follows upon taking the limit as $z \to z_0$. To prove (ii), note that f has a simple pole at z_0. Using (i) and $q(z_0) = 0$, we have

$$\text{Res}\left(\frac{p(z)}{q(z)}, z_0\right) = \lim_{z \to z_0}(z - z_0)\frac{p(z)}{q(z)} = \lim_{z \to z_0} p(z) \lim_{z \to z_0}\frac{z - z_0}{q(z) - q(z_0)} = \frac{p(z_0)}{q'(z_0)},$$

proving (5.1.4). ∎

Example 5.1.4. (An application of the residue theorem) Let C be a simple closed positively oriented path such that $1, -i$, and i are in the interior of C and -1 is in the exterior of C (Figure 5.2). Find

$$\int_C \frac{dz}{z^4 - 1}.$$

Solution. The function $f(z) = \frac{1}{z^4 - 1}$ has isolated singularities at $z = \pm 1$ and $\pm i$. Three of these are inside C, and according to (5.1.2) we have

$$\int_C \frac{dz}{z^4 - 1} = 2\pi i\Big(\text{Res}\,(1) + \text{Res}\,(i) + \text{Res}\,(-i)\Big). \tag{5.1.5}$$

We have $z^4 - 1 = (z - 1)(z + 1)(z - i)(z + i)$, so ± 1 and $\pm i$ are simple roots of the polynomial $z^4 - 1 = 0$. Hence $f(z) = \frac{1}{z^4 - 1}$ has simple poles at ± 1 and $\pm i$ by Theorem 4.5.15(ii). Let z_0 denote any one of the points $1, \pm i$. Because z_0 is a simple pole, we have from (5.1.3),

$$\lim_{z \to z_0}(z - z_0)f(z) = a_{-1} = \text{Res}\,(z_0).$$

Using the factorization

$$z^4 - 1 = (z - 1)(z + 1)(z - i)(z + i),$$

we have at $z_0 = 1$

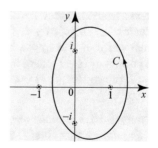

Fig. 5.2 The path C and the poles of $f(z)$ in Example 5.1.4.

$$\text{Res}(1) = \lim_{z \to 1}(z-1)\frac{1}{z^4-1} = \lim_{z \to 1}\frac{1}{(z+1)(z-i)(z+i)}$$

$$= \frac{1}{(z+1)(z-i)(z+i)}\bigg|_{z=1} = \frac{1}{4}.$$

Similarly, at $z_0 = i$, we have

$$\text{Res}(i) = \lim_{z \to i}(z-i)\frac{1}{z^4-1} = \frac{1}{(z-1)(z+1)(z+i)}\bigg|_{z=i} = \frac{i}{4},$$

and at $z = -i$,

$$\text{Res}(-i) = \lim_{z \to -i}(z+i)\frac{1}{z^4-1} = \frac{1}{(z-1)(z+1)(z-i)}\bigg|_{z=-i} = -\frac{i}{4}.$$

Plugging these values into (5.1.5), we obtain

$$\int_C \frac{dz}{z^4-1} = \frac{\pi i}{2}. \qquad \Box$$

Example 5.1.5. (Residue at a removable singularity) Let C be a simple closed positively oriented path such that 1 is in the interior of C and -1 is in the exterior of C (Figure 5.3). Evaluate

$$I = \int_C \frac{\sin(\pi z)}{z^2-1}\, dz.$$

Solution. The function $f(z) = \frac{\sin(\pi z)}{z^2-1}$ has isolated singularities at $z = \pm 1$; only 1 is inside C. Since

$$\lim_{z \to 1}(z-1)\frac{\sin(\pi z)}{z^2-1} = \lim_{z \to 1}\frac{\sin(\pi z)}{z+1} = 0,$$

it follows from Theorem 4.5.12(vi), that 1 is a removable singularity of f. Thus, the Laurent series of f at $z_0 = 1$ has no negative powers of $(z-1)$, and, in particular, $a_{-1} = 0$. Hence $\text{Res}(f, 1) = a_{-1} = 0$, and so $I = 0$. $\Box$

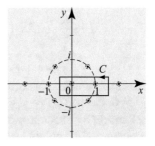

Fig. 5.3 The contour of Example 5.1.5.

Example 5.1.5 reminds us of the following simple observation: *If z_0 is a removable singularity, then $\text{Res}(f, z_0) = 0$.*

For poles of higher order the situation is more complicated.

Theorem 5.1.6. *Suppose that z_0 is a pole of order $m \geq 1$ of f. Then we have*

$$\text{Res}\,(f,z_0) = \lim_{z \to z_0} \frac{1}{(m-1)!} \frac{d^{m-1}}{dz^{m-1}} \left[(z-z_0)^m f(z) \right], \tag{5.1.6}$$

where as usual the derivative of order 0 of a function is the function itself.

Notice that formula (5.1.6) reduces to (5.1.3) when $m = 1$.

Proof. By the Laurent series characterization of poles (Theorem 4.5.15),

$$f(z) = \frac{a_{-m}}{(z-z_0)^m} + \cdots + \frac{a_{-1}}{z-z_0} + a_0 + a_1(z-z_0) + a_2(z-z_0)^2 + \cdots .$$

Multiply by $(z-z_0)^m$, then differentiate $(m-1)$ times to obtain

$$\frac{d^{m-1}}{dz^{m-1}} \left[(z-z_0)^m f(z) \right] = (m-1)!\, a_{-1} + m!\, a_0(z-z_0) + \frac{(m+1)!}{2} a_1(z-z_0)^2 + \cdots .$$

Take the limit as $z \to z_0$, and get

$$\lim_{z \to z_0} \frac{d^{m-1}}{dz^{m-1}} \left[(z-z_0)^m f(z) \right] = (m-1)!\, a_{-1} + 0,$$

which yields (5.1.6). ∎

Example 5.1.7. Let C be the simple closed path shown in Figure 5.4.
(a) Compute the residues of $f(z) = \frac{z^2}{(z^2+\pi^2)^2 \sin z}$ at all isolated singularities inside C.
(b) Evaluate the integral of the function in (a) over the curve C.

Solution. (a) Three steps are involved in answering this question.

Step 1: Determine the singularities of f inside C. The function $f(z) = \frac{z^2}{(z^2+\pi^2)^2 \sin z}$ is analytic except where $z^2 + \pi^2 = 0$ or $\sin z = 0$. Thus f has isolated singularities at $\pm i\pi$ and at $k\pi$ where k is an integer. Only 0 and $i\pi$ are inside C.

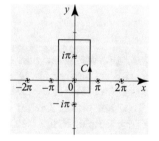

Fig. 5.4 The path C and the poles of $f(z)$ in Example 5.1.7.

Step 2: Determine the type of the singularities of f inside C. Let us start with the singularity at 0. Using $\lim_{z\to 0} \frac{\sin z}{z} = 1$, it follows that $\lim_{z\to 0} \frac{z}{\sin z} = 1$, and so

$$\lim_{z\to 0} f(z) = \lim_{z\to 0} \frac{z}{\sin z} \frac{z}{(z^2+\pi^2)^2} = 1 \cdot 0 = 0.$$

By Theorem 4.5.12(*iii*), f has a removable singularity at $z_0 = 0$. To treat the singularities at $i\pi$, we consider the function $\frac{1}{f(z)} = \frac{(z+i\pi)^2(z-i\pi)^2 \sin z}{z^2}$ which has a zero of order 2 at $i\pi$, and so by Definition 4.5.14, f has a pole of order 2 at $i\pi$.

Step 3: Determine the residues of f inside C. At 0, f has a removable singularity, so $a_{-1} = 0$, and hence the residue of f at 0 is 0. At $i\pi$, we apply Theorem 5.1.6, with $m = 2$, $z_0 = i\pi$. Then

$$
\begin{aligned}
\mathrm{Res}\,(i\pi) &= \lim_{z \to i\pi} \frac{d}{dz}\left[(z - i\pi)^2 f(z)\right] \\
&= \lim_{z \to i\pi} \frac{d}{dz}\left[\frac{z^2}{(z + i\pi)^2 \sin z}\right] \\
&= \lim_{z \to i\pi} \frac{2z(z + i\pi)\sin z - z^2((z + i\pi)\cos z + 2\sin z)}{(z + i\pi)^3 \sin^2 z} \\
&= \frac{2\sinh\pi + (-\pi\cosh\pi - \sinh\pi)}{-4\pi \sinh^2 \pi} = -\frac{1}{4\pi\sinh\pi} + \frac{\cosh\pi}{4\pi\sinh^2\pi},
\end{aligned}
$$

where the last line follows by plugging $z = i\pi$ into the previous line and using $\sin(i\pi) = i\sinh\pi$ and $\cos(i\pi) = \cosh\pi$.

(b) Using Theorem 5.1.2 and (a), we obtain

$$
\int_C \frac{z^2}{(z^2 + \pi^2)^2 \sin z}\,dz = 2\pi i\left(\mathrm{Res}\,(0) + \mathrm{Res}\,(i\pi)\right) = i\left(-\frac{1}{2\sinh\pi} + \frac{\cosh\pi}{2\sinh^2\pi}\right).\ \ \square
$$

Example 5.1.8. (Residues of the cotangent) (a) Let k be an integer. Show that $\mathrm{Res}\left(\cot(\pi z), k\right) = \frac{1}{\pi}$.

(b) Suppose that f is analytic at an integer k. Show that

$$
\mathrm{Res}\left(f(z)\cot(\pi z), k\right) = \frac{1}{\pi}f(k). \tag{5.1.7}
$$

(c) Let C be the positively oriented rectangular path in Figure 5.5. Evaluate

$$
\int_C \frac{\cot(\pi z)}{1 + z^4}\,dz.
$$

Solution. (a) We know that the zeros of $\phi(z) = \sin(\pi z)$ are precisely the integers. Also, since $\phi'(k) = \pi\cos(k\pi) \neq 0$, it follows from Theorem 4.5.2, that all the zeros of $\sin(\pi z)$ are simple. Hence $\cot(\pi z) = \frac{\cos(\pi z)}{\sin(\pi z)}$ has simple poles at the integers. In view of Proposition 5.1.3(ii), we write

$$
\mathrm{Res}\left(\cot(\pi z), k\right) = \mathrm{Res}\left(\frac{\cos(\pi z)}{\sin(\pi z)}, k\right)
$$

$$
= \frac{\cos(k\pi)}{\frac{d}{dz}\sin(\pi z)\big|_{z=k}} = \frac{1}{\pi}.
$$

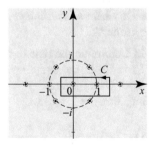

Fig. 5.5 The path C.

(b) This is immediate from (a) and Proposition 5.1.3(ii): Take $p(z) = f(z)\cos(\pi z)$ and $q(z) = \sin(\pi z)$.

(c) Since $1 + z^4$ is nonzero inside C and $\cot(\pi z)$ has simple poles at the integers, it follows that $\frac{\cot(\pi z)}{1+z^4}$ has two simple poles inside C at $z = 0$ and $z = 1$. Applying Theorem 5.1.2 and using (5.1.7) with $f(z) = \frac{1}{1+z^4}$ to compute the residues, we find

$$\int_C \frac{\cot(\pi z)}{1+z^4} dz = 2\pi i \left(\operatorname{Res}\left(\frac{\cot(\pi z)}{1+z^4}, 0 \right) + \operatorname{Res}\left(\frac{\cot(\pi z)}{1+z^4}, 1 \right) \right)$$

$$= 2\pi i \left(\frac{1}{\pi} \frac{1}{1+0^4} + \frac{1}{\pi} \frac{1}{1+1^4} \right) = 2i\left(1 + \frac{1}{2}\right) = 3i. \qquad \square$$

So far the examples that we treated involved residues at poles of finite order. There is no formula like (5.1.6) for computing the residue at an essential singularity. We have to rely on various tricks to evaluate the coefficient a_{-1} in the Laurent series expansion. We illustrate with several examples, starting with a useful observation.

Proposition 5.1.9. *Suppose that 0 is an isolated singularity of an even function f that is analytic on $\mathbb{C} \setminus \{0\}$. Then $\operatorname{Res}(f, 0) = 0$.*

Proof. We have to show that the a_{-1} the Laurent series coefficient of f at 0 is 0. Write $f(z) = \sum_{n=-\infty}^{\infty} a_n z^n$, where $0 < |z| < r$. Substitue $-z$ for z and use the fact that f is even, i.e., $f(-z) = f(z)$ for all z. Then $\sum_{n=-\infty}^{\infty} a_n z^n = \sum_{n=-\infty}^{\infty} (-1)^n a_n z^n$, and by the uniqueness of the Laurent series, it follows that $(-1)^n a_n = a_n$, which implies that $a_n = 0$ if $|n|$ is odd; in particular, $a_{-1} = 0$. $\blacksquare$

Example 5.1.10. (Residue at 0 of an even function) Compute $\operatorname{Res}\left(e^{-\frac{1}{z^2}} \cos\frac{1}{z}, 0\right)$.

Solution. The function $e^{-\frac{1}{z^2}} \cos\frac{1}{z}$ is even and has an isolated (essential) singularity at 0. By Proposition 5.1.9, $\operatorname{Res}\left(e^{-\frac{1}{z^2}} \cos\frac{1}{z}, 0\right) = 0$. $\square$

Multiplication of series is often useful in computing residues at a singularity, including essential singularities.

Example 5.1.11. (Multiplying series by a polynomial) Compute the residues of $z^2 \sin\frac{1}{z}$ at $z = 0$.

Solution. From the Laurent series

$$\sin\frac{1}{z} = \frac{1}{z} - \frac{1}{3!}\frac{1}{z^3} + \frac{1}{5!}\frac{1}{z^5} - \cdots,$$

we obtain

$$z^2 \sin\frac{1}{z} = z - \frac{1}{3!}\frac{1}{z} + \frac{1}{5!}\frac{1}{z^3} - \cdots,$$

and so $\operatorname{Res}\left(z^2 \sin\frac{1}{z}, 0\right) = -\frac{1}{3!}$. $\square$

Example 5.1.12. (Using Cauchy products) Find the residue of $\frac{e^{\frac{1}{z}}}{z^2+1}$ at $z=0$.

Solution. This function has an essential singularity at $z=0$. To compute the coefficient a_{-1} in its Laurent series around 0, we use two familiar Taylor and Laurent series as follows. We have, for $0 < |z| < 1$,

$$\frac{e^{\frac{1}{z}}}{z^2+1} = \frac{1}{z^2+1} e^{\frac{1}{z}} = \left(1 - z^2 + z^4 - \cdots\right)\left(1 + \frac{1}{z} + \frac{1}{2!}\frac{1}{z^2} + \frac{1}{3!}\frac{1}{z^3} + \cdots\right).$$

By properties of Taylor and Laurent series, both series are absolutely convergent in $0 < |z| < 1$. So we can multiply them term by term using a Cauchy product. Collecting all the terms in $\frac{1}{z}$, we find that

$$\text{Res}\left(\frac{e^{\frac{1}{z}}}{z^2+1}, 0\right) = a_{-1} = 1 - \frac{1}{3!} + \frac{1}{5!} - \cdots = \sin 1. \qquad \square$$

Exercises 5.1

In Exercises 1–12, find the residue of the functions at all their isolated singularities.

1. $\dfrac{1+z}{z}$

2. $\dfrac{1+z}{z^2+2z+2}$

3. $\dfrac{1+e^z}{z^2} + \dfrac{2}{z}$

4. $\dfrac{\sin(z^2)}{z^2(z^2+1)}$

5. $\left(\dfrac{z-1}{z+3i}\right)^3$

6. $\dfrac{1-\cos z}{z^3}$

7. $\dfrac{1}{z\sin z}$

8. $\dfrac{\cot(\pi z)}{z+1}$

9. $\csc(\pi z)\dfrac{z+1}{z-1}$

10. $z\sin\left(\dfrac{1}{z}\right)$

11. $e^{z+\frac{1}{z}}$

12. $\cos\left(\dfrac{1}{z}\right)\sin\left(\dfrac{1}{z}\right)$

In Exercises 13–26, evaluate the path integral. The path R in Exercises 15 and 20 is shown in the adjacent figure.

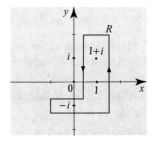

13. $\displaystyle\int_{C_1(0)} \frac{z^2+3z-1}{z(z^2-3)}\,dz.$

14. $\displaystyle\int_{C_{\frac{1}{10}}(1)} \frac{1}{z^5-1}\,dz$

15. $\displaystyle\int_R \frac{z+i}{(z-1-i)^3(z-i)}\,dz$

16. $\displaystyle\int_{C_3(0)} \frac{e^{iz^2}}{z^2+(3-3i)z-2-6i}\,dz$

17. $\displaystyle\int_{C_{\frac{3}{2}}(0)} \frac{dz}{z(z-1)(z-2)\cdots(z-10)}$

18. $\displaystyle\int_{C_3(0)} \frac{z^2+1}{(z-1)^2}\,dz$

19. $\displaystyle\int_{C_4(0)} z\tan z\,dz$

20. $\displaystyle\int_R \frac{dz}{1+e^{\pi z}}$

21. $\displaystyle\int_{C_1(0)} \frac{e^{z^2}}{z^6}\,dz$

22. $\displaystyle\int_{C_1(0)} \cos\left(\frac{1}{z^2}\right)e^{\frac{1}{z}}\,dz$

23. $\displaystyle\int_{C_1(0)} z^4\left(e^{\frac{1}{z}}+z^2\right)dz$

24. $\displaystyle\int_{C_{31/2}(0)} z^2\cot(\pi z)\,dz$

25. $\displaystyle\int_{C_1(0)} \frac{\sin z}{z^6}\,dz$

26. $\displaystyle\int_{C_{1/2}(0)} \frac{1}{z^2(e^z-1)}\,dz$

27. (a) Prove that if f has a simple pole at z_0 and g is analytic at z_0, then

$$\mathrm{Res}\,(f(z)g(z),z_0) = g(z_0)\,\mathrm{Res}\,(f(z),z_0).$$

(b) Use (a) to prove (5.1.7).

28. Show that $\mathrm{Res}\,(f+g,z_0) = \mathrm{Res}\,(f,z_0) + \mathrm{Res}\,(g,z_0)$.

29. Residues of the cosecant. (a) Show that $\csc(\pi z)$ has simple poles at the integers.
(b) For an integer k show that

$$\mathrm{Res}\,(\csc(\pi z),k) = \frac{(-1)^k}{\pi}.$$

(c) Suppose that f is analytic at an integer k. Show that

$$\mathrm{Res}\,(f(z)\csc(\pi z),k) = \frac{(-1)^k}{\pi}f(k).$$

30. Use Exercise 29 to compute

(a) $\displaystyle\int_{C_{25/2}(0)} z\csc(\pi z)\,dz,$ (b) $\displaystyle\int_{C_{25/2}(0)} \frac{\csc(\pi z)}{1+z^2}\,dz.$

31. Explain how the residue theorem implies Cauchy's theorem (Theorem 3.4.4) and Cauchy's integral formula (Theorem 3.8.1).

32. Suppose that f has an isolated singularity at z_0. Show that $\mathrm{Res}\,(f',z_0) = 0$.

33. Consider the Laurent series expansions in an annulus around z_0,

$$f(z) = \sum_{n=-\infty}^{\infty} a_n(z-z_0)^n \quad\text{and}\quad g(z) = \sum_{n=-\infty}^{\infty} b_n(z-z_0)^n.$$

Show that $\mathrm{Res}\,(fg,z_0) = \sum_{n=-\infty}^{\infty} a_n b_{-n-1}$.

34. Use Exercise 33 to compute $\mathrm{Res}\,\left(e^{\frac{1}{z^2}}\sin\frac{1}{z},0\right)$.

In Exercises 35–36, g and h are analytic at z_0. Compute a formula for the residue of $f = \frac{g}{h}$ at z_0 under the stated conditions.

35. If g and h have zeros of the same order at z_0, then $\mathrm{Res}\,\left(\frac{g}{h},z_0\right) = 0$.

36. If $g(z_0) \neq 0$ and h has a zero of order 2 at z_0, then

$$\mathrm{Res}\,\left(\frac{g}{h},z_0\right) = 2\frac{g'(z_0)}{h''(z_0)} - \frac{2}{3}\frac{g(z_0)h'''(z_0)}{[h''(z_0)]^2}.$$

37. Project Problem: Laplace transform of Bessel functions. In this project we use residues to derive the formula

$$\int_0^\infty J_n(t)e^{-st}\,dt = \frac{1}{\sqrt{s^2+1}}\left(\sqrt{s^2+1}-s\right)^n, \quad s>0, \tag{5.1.8}$$

where J_n is the Bessel function of integer order $n \geq 0$. This formula gives the Laplace transform of J_n. To illustrate the useful method that is involved in this project and for clarity's sake, we start with the case $n = 0$.

(a) Using the integral representation of J_0 from Exercise 32(a), Section 4.4, write

$$\int_0^\infty J_0(t)e^{-st}\,dt = \frac{1}{2\pi i}\int_0^\infty \int_{C_1(0)} e^{-t\left(s-\frac{1}{2}\left(\zeta-\frac{1}{\zeta}\right)\right)}\frac{d\zeta}{\zeta}\,dt.$$

(b) Show that for ζ on $C_1(0)$, the complex number $\zeta - \frac{1}{\zeta}$ is either 0 or purely imaginary and conclude that for real s and t

$$\left|e^{-t\left(s-\frac{1}{2}\left(\zeta-\frac{1}{\zeta}\right)\right)}\right| = e^{-ts}.$$

(c) Thus the integral in (a) is absolutely convergent. Although no details are required here, this fact is enough to justify interchanging the order of integration. Interchange the order of integration, evaluate the integral in t and obtain

$$\int_0^\infty J_0(t)e^{-st}\,dt = \frac{1}{\pi i}\int_{C_1(0)}\frac{d\zeta}{-\zeta^2+2s\zeta+1}, \quad s>0.$$

(d) Evaluate the integral using the residue theorem and derive

$$\int_0^\infty J_0(t)e^{-st}\,dt = \frac{1}{\sqrt{s^2+1}}, \quad s>0.$$

(e) Proceed as in (a)–(d), using the integral representation for J_n, and show that for $s>0$

$$\int_0^\infty J_n(t)e^{-st}\,dt = \frac{1}{\pi i}\int_{C_1(0)}\frac{d\zeta}{(-\zeta^2+2s\zeta+1)\zeta^n} = \frac{1}{\pi i}\int_{C_1(0)}\frac{\eta^n d\eta}{\eta^2+2s\eta-1}.$$

[*Hint*: To justify the second equality, let $\zeta = \frac{1}{\eta}$, $d\zeta = -\frac{d\eta}{\eta^2}$. As ζ runs through $C_1(0)$, η runs through the reverse of $C_1(0)$.]

(f) Derive (5.1.8) by computing the integral in (e) using residues.

5.2 Definite Integrals of Trigonometric Functions

We consider integrals of the form

$$\int_0^{2\pi} F\left(\cos\theta,\sin\theta\right)d\theta, \tag{5.2.1}$$

where $F\left(\cos\theta,\sin\theta\right)$ is a rational function of $\cos\theta$ and $\sin\theta$ with real coefficients and whose denominator does not vanish on the interval $[0, 2\pi]$. For example, the integrals

$$\int_0^{2\pi}\frac{d\theta}{2+\cos\theta} \quad \text{and} \quad \int_0^{2\pi}\frac{\cos^2(2\theta)}{4+2\sin\theta\cos\theta}\,d\theta$$

are of the form (5.2.1). Our goal is to transform the definite integral (5.2.1) into a contour integral that can be evaluated using the residue theorem. For this purpose, we recall the familiar identities

$$\cos\theta = \frac{1}{2}\left(e^{i\theta} + e^{-i\theta}\right) \quad \text{and} \quad \sin\theta = \frac{1}{2i}\left(e^{i\theta} - e^{-i\theta}\right). \qquad (5.2.2)$$

As θ varies in the interval $[0, 2\pi]$, the complex number $z = e^{i\theta}$ traces the unit circle $C_1(0)$ in the positive direction. This suggests transforming the integral (5.2.1) into a contour integral, where the variable $z = e^{i\theta}$ traces $C_1(0)$. Observing that

$$\frac{1}{z} = \frac{1}{e^{i\theta}} = e^{-i\theta},$$

we obtain from (5.2.2) the key substitutions

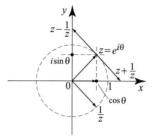

Fig. 5.6 Constructing $\cos\theta$ and $\sin\theta$.

$$\cos\theta = \frac{1}{2}\left(z + \frac{1}{z}\right) \quad \text{and} \quad \sin\theta = \frac{1}{2i}\left(z - \frac{1}{z}\right). \qquad (5.2.3)$$

This is represented graphically in Figure 5.6. Also, from $z = e^{i\theta}$, we have $dz = ie^{i\theta}\,d\theta = iz\,d\theta$ or

$$-i\frac{dz}{z} = d\theta. \qquad (5.2.4)$$

With (5.2.3) and (5.2.4) in hand, we now consider some examples.

Example 5.2.1. Evaluate $\displaystyle\int_0^{2\pi} \frac{d\theta}{10 + 8\cos\theta}$.

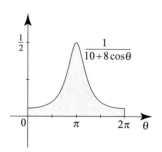

Fig. 5.7 The definite integral in Example 5.2.1.

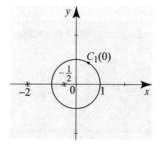

Fig. 5.8 The path and poles for the contour integral in Example 5.2.1.

Solution. Let $z = e^{i\theta}$. As θ varies from 0 to 2π, z traces the unit circle $C_1(0)$ in the positive direction. Using the substitutions (5.2.3) and (5.2.4), we transform the given integral into a path integral as follows

$$\int_0^{2\pi} \frac{d\theta}{10 + 8\cos\theta} = \int_{C_1(0)} \frac{-i\frac{dz}{z}}{10 + \frac{8}{2}\left(z + \frac{1}{z}\right)} = -i\int_{C_1(0)} \frac{dz}{4z^2 + 10z + 4}.$$

To compute the last integral using residues, we solve

$$4z^2 + 10z + 4 = 4(z+2)\left(z + \frac{1}{2}\right) = 0,$$

and get $z = -2$ and $z = -\frac{1}{2}$ (see Figure 5.8). So the only singularity in the unit disk is a simple pole at $-\frac{1}{2}$. Applying Proposition 5.1.3(ii) we find

$$\mathrm{Res}\left(\frac{1}{4z^2 + 10z + 4}, -\frac{1}{2}\right) = \frac{1}{\frac{d}{dz}(4z^2 + 10z + 4)\big|_{z=-\frac{1}{2}}} = \frac{1}{8(-1/2) + 10} \quad (5.2.5)$$

$$= \frac{1}{6}.$$

By the residue theorem, we conclude that

$$\int_{C_1(0)} \frac{dz}{4z^2 + 10z + 4} = \frac{2\pi i}{6} = \frac{\pi i}{3},$$

and so

$$\int_0^{2\pi} \frac{d\theta}{10 + 8\cos\theta} = -i\int_{C_1(0)} \frac{dz}{4z^2 + 10z + 4} = \frac{\pi}{3}. \qquad \square$$

Let n be an integer and $z = e^{i\theta}$. De Moivre's identity implies

$$z^n = e^{in\theta} = \cos n\theta + i\sin n\theta$$

and

$$\frac{1}{z^n} = e^{-in\theta} = \cos n\theta - i\sin n\theta.$$

Adding and subtracting the preceding equalities, we obtain the following pair of useful identities

$$\cos n\theta = \frac{1}{2}\left(z^n + \frac{1}{z^n}\right) \quad \text{and} \quad \sin n\theta = \frac{1}{2i}\left(z^n - \frac{1}{z^n}\right). \quad (5.2.6)$$

Example 5.2.2. Compute the definite integral (see Figure 5.9)

$$\int_0^{2\pi} \frac{\cos 2\theta}{2 + \cos\theta}\, d\theta.$$

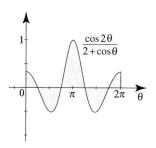

Fig. 5.9 The definite integral in Example 5.2.2.

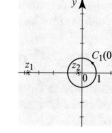

Fig. 5.10 The path and poles for the contour integral in Example 5.2.2.

Solution. Use (5.2.3), (5.2.4), and (5.2.6) with $n = 2$, and get

$$\int_0^{2\pi} \frac{\cos 2\theta}{2 + \cos \theta} \, d\theta = -i \int_{C_1(0)} \frac{\frac{1}{2}\left(z^2 + \frac{1}{z^2}\right)}{2 + \frac{1}{2}\left(z + \frac{1}{z}\right)} \frac{dz}{z} = -i \int_{C_1(0)} \frac{z^4 + 1}{z^2(z^2 + 4z + 1)} \, dz.$$

We now compute the last integral using residues. Since $1/z^2$ has a pole of order 2 at $z_0 = 0$, using Theorem 5.1.6 with $m = 2$ we obtain

$$\text{Res}\left(\frac{z^4 + 1}{z^2(z^2 + 4z + 1)}, 0\right) = \lim_{z \to 0} \frac{d}{dz} \frac{z^4 + 1}{z^2 + 4z + 1}$$

$$= \lim_{z \to 0} \frac{4z^3(z^2 + 4z + 1) - (z^4 + 1)(2z + 4)}{(z^2 + 4z + 1)^2} = -4.$$

The roots of $z^2 + 4z + 1 = 0$ are $z_1 = -2 - \sqrt{3} \approx -3.7$ and $z_2 = -2 + \sqrt{3} \approx -0.27$ (Figure 5.10). Only z_2 is inside the unit disk. Since z_2 is a simple pole, we can compute the residues at z_2 using Proposition 5.1.3(ii):

$$\text{Res}\left(\frac{z^4 + 1}{z^2} \frac{1}{z^2 + 4z + 1}, -2 + \sqrt{3}\right) = \frac{(-2 + \sqrt{3})^4 + 1}{(-2 + \sqrt{3})^2} \frac{1}{\frac{d}{dz}(z^2 + 4z + 1)\big|_{-2 + \sqrt{3}}}$$

$$= \frac{(-2 + \sqrt{3})^4 + 1}{(-2 + \sqrt{3})^2(2\sqrt{3})} = \frac{7}{\sqrt{3}}.$$

By the residue theorem, we conclude that

$$\int_{C_1(0)} \frac{z^4 + 1}{z^2(z^2 + 4z + 1)} \, dz = 2\pi i\left(-4 + \frac{7}{\sqrt{3}}\right),$$

and so

$$\int_0^{2\pi} \frac{\cos 2\theta}{2 + \cos \theta} \, d\theta = -i \int_{C_1(0)} \frac{z^4 + 1}{z^2(z^2 + 4z + 1)} \, dz = 2\pi\left(-4 + \frac{7}{\sqrt{3}}\right) \approx 0.26. \quad \square$$

The preceding examples dealt with integrals over the interval $[0, 2\pi]$. Integrals over the interval $[0, \pi]$ can be handled if the integrand is even and 2π-periodic. In this case,

$$\int_0^\pi f(\theta)\,d\theta = \frac{1}{2}\int_{-\pi}^\pi f(\theta)\,d\theta = \frac{1}{2}\int_0^{2\pi} f(\theta)\,d\theta. \qquad (5.2.7)$$

The first equality follows because the integrand is even and the second one is a consequence of the fact that the integrand is 2π-periodic, and so the integral is the same over all intervals of length 2π.

Example 5.2.3. Compute the definite integral (see Figure 5.11)

$$I = \int_0^{\frac{\pi}{2}} \frac{\cos 2\theta}{1+2\cos^2\theta}\,d\theta.$$

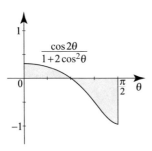

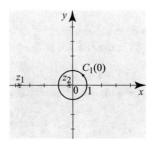

Fig. 5.11 The definite integral in Example 5.2.3.

Fig. 5.12 The path and poles for the contour integral in Example 5.2.3.

Solution. We first use the double angle identity $2\cos^2\theta = 1 + \cos 2\theta$ and then the change of variables $\theta' = 2\theta$ to obtain

$$I = \int_0^{\frac{\pi}{2}} \frac{\cos 2\theta}{2+\cos 2\theta}\,d\theta = \frac{1}{2}\int_0^\pi \frac{\cos\theta'}{2+\cos\theta'}\,d\theta'.$$

The integral over $[0, \pi]$ has an even integrand, so according to (5.2.7) we have

$$I = \frac{1}{4}\int_0^{2\pi} \frac{\cos\theta}{2+\cos\theta}\,d\theta = \frac{1}{4}\int_0^{2\pi} \frac{2+\cos\theta-2}{2+\cos\theta}\,d\theta$$

$$= \frac{1}{4}\int_0^{2\pi} d\theta + \frac{1}{4}\int_0^{2\pi} \frac{-2}{2+\cos\theta}\,d\theta = \frac{\pi}{2} - \frac{1}{2}\int_0^{2\pi} \frac{d\theta}{2+\cos\theta}.$$

We now evaluate this last integral by the residue method. Letting $z = e^{i\theta}$, $\cos\theta = \frac{1}{2}(z+\frac{1}{z})$, $d\theta = \frac{dz}{iz}$, we obtain

$$I = \frac{\pi}{2} - \frac{1}{i} \int_{C_1(0)} \frac{dz}{z^2 + 4z + 1}.$$

The integrand $\frac{1}{z^2+4z+1} = \frac{1}{(z-z_1)(z-z_2)}$ has simple poles at $z_1 = -2 - \sqrt{3}$ and $z_2 = -2 + \sqrt{3}$, and z_2 is the only pole inside $C_1(0)$; see Figure 5.12. The residue at z_2 is

$$\text{Res}\left(\frac{1}{(z-z_1)(z-z_2)}, z_2\right) = \lim_{z \to z_2}(z-z_2)\frac{1}{(z-z_1)(z-z_2)} = \frac{1}{z_2 - z_1} = \frac{1}{2\sqrt{3}},$$

and so

$$I = \frac{\pi}{2} - \frac{1}{i}2\pi i\frac{1}{2\sqrt{3}} = \pi\left(\frac{1}{2} - \frac{1}{\sqrt{3}}\right). \qquad \square$$

Although the integrals in this section are special, they have important applications, including the computation of certain Fourier series.

Exercises 5.2

In Exercises 1–10, use the method of this section to evaluate the integrals.

1. $\displaystyle\int_0^{2\pi} \frac{d\theta}{2 - \cos\theta}$

2. $\displaystyle\int_0^{2\pi} \frac{d\theta}{5 + 3\cos\theta}$

3. $\displaystyle\int_0^{2\pi} \frac{d\theta}{10 - 8\sin\theta}$

4. $\displaystyle\int_0^{2\pi} \frac{1}{\sin^2\theta + 2\cos^2\theta} d\theta$

5. $\displaystyle\int_0^{2\pi} \frac{\cos 2\theta}{5 + 4\cos\theta} d\theta$

6. $\displaystyle\int_0^{2\pi} \frac{\cos\theta - \sin^2\theta}{10 + 8\cos\theta} d\theta$

7. $\displaystyle\int_0^{\pi} \frac{d\theta}{9 + 16\sin^2\theta}$

8. $\displaystyle\int_0^{\pi} \frac{\cos\theta\sin^2\theta}{2 + \cos\theta} d\theta$

9. $\displaystyle\int_0^{2\pi} \frac{d\theta}{7 + 2\cos\theta + 3\sin\theta}$

10. $\displaystyle\int_0^{2\pi} \frac{d\theta}{7 - 2\cos^2\theta - 3\sin^2\theta}$

In Exercises 11–16, use the method of this section to derive the indicated formula, where a, b, c are real numbers.

11. $\displaystyle\int_0^{2\pi} \frac{d\theta}{1 + a\cos\theta} = \frac{2\pi}{\sqrt{1 - a^2}}, \quad 0 < |a| < 1.$

12. $\displaystyle\int_0^{2\pi} \frac{\sin^2\theta}{a + b\cos\theta} d\theta = \frac{2\pi}{b^2}\left(a - \sqrt{a^2 - b^2}\right), \quad 0 < |b| < a.$

13. (i) $\displaystyle\int_0^{2\pi} \frac{d\theta}{a + b\cos^2\theta} = \frac{2\pi}{\sqrt{a}\sqrt{a+b}}, \quad a > 0,\ b \geq 0.$

(ii) $\displaystyle\int_0^{2\pi} \frac{d\theta}{a + b\sin^2\theta} = \frac{2\pi}{\sqrt{a}\sqrt{a+b}}, \quad a > 0,\ b \geq 0.$

14. $\displaystyle\int_0^{2\pi} \frac{d\theta}{a\cos\theta + b\sin\theta + c} = \frac{2\pi}{\sqrt{c^2 - a^2 - b^2}}, \quad a^2 + b^2 < c^2.$

15. $\displaystyle\int_0^{2\pi} \frac{d\theta}{a\cos^2\theta + b\sin^2\theta + c} = \frac{2\pi}{\sqrt{(a+c)(b+c)}}, \quad 0 < c < a,\ c < b.$

5.3 Improper Integrals Involving Rational and Exponential Functions

In this section we present a useful technique to evaluate improper integrals involving rational and exponential functions. Let a and b be arbitrary real numbers. Consider the integrals

$$\int_{-\infty}^{b} f(x)\,dx \quad \text{and} \quad \int_{a}^{\infty} f(x)\,dx, \tag{5.3.1}$$

where in each case f is continuous in the closure of the interval of integration. These are called **improper integrals**, because the interval of integration is infinite. The integral $\int_{a}^{\infty} f(x)\,dx$ is **convergent** if $\lim_{b \to \infty} \int_{a}^{b} f(x)\,dx$ exists as a finite number. Similarly, $\int_{-\infty}^{b} f(x)\,dx$ is **convergent** if $\lim_{a \to -\infty} \int_{a}^{b} f(x)\,dx$ exists as a finite number.

The integral $\int_{-\infty}^{\infty} f(x)\,dx$ of a continuous function f on the real line is also improper since the interval of integration is infinite, but here it is infinite in both the positive and negative direction. Such an integral is said to be **convergent** if both $\int_{0}^{\infty} f(x)\,dx$ and $\int_{-\infty}^{0} f(x)\,dx$ are convergent. In this case, we set

$$\int_{-\infty}^{\infty} f(x)\,dx = \lim_{a \to -\infty} \int_{a}^{0} f(x)\,dx + \lim_{b \to \infty} \int_{0}^{b} f(x)\,dx. \tag{5.3.2}$$

This identity is shown pictorially in Figure 5.13.

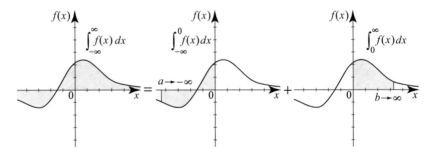

Fig. 5.13 Splitting an improper integral over the line.

Definition 5.3.1. We define the **Cauchy principal value** of the integral $\int_{-\infty}^{\infty} f(x)\,dx$ to be

$$\text{P.V.} \int_{-\infty}^{\infty} f(x)\,dx = \lim_{a \to \infty} \int_{-a}^{a} f(x)\,dx \tag{5.3.3}$$

if the limit exists. (See Figure 5.14.)

The Cauchy principal value of an integral may exist even though the integral itself is not convergent. For example, $\int_{-a}^{a} x \, dx = 0$ for all a, which implies that P.V. $\int_{-\infty}^{\infty} x \, dx = 0$, but the integral itself is clearly not convergent since $\int_{0}^{\infty} x \, dx = \infty$. However, whenever $\int_{-\infty}^{\infty} f(x) \, dx$ is convergent, then

$$\text{P.V.} \int_{-\infty}^{\infty} f(x) \, dx$$

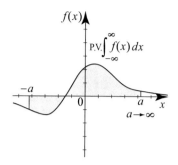

Fig. 5.14 The Cauchy principal value of the integral.

exists, and the two integrals will be the same. This is because $\lim_{a \to \infty} \int_{-a}^{0} f(x) \, dx$ and $\lim_{a \to \infty} \int_{0}^{a} f(x) \, dx$ both exist, and so

$$
\begin{aligned}
\text{P.V.} \int_{-\infty}^{\infty} f(x) \, dx &= \lim_{a \to \infty} \int_{-a}^{a} f(x) \, dx \\
&= \lim_{a \to \infty} \left(\int_{-a}^{0} f(x) \, dx + \int_{0}^{a} f(x) \, dx \right) \\
&= \lim_{a \to \infty} \int_{-a}^{0} f(x) \, dx + \lim_{a \to \infty} \int_{0}^{a} f(x) \, dx \\
&= \int_{-\infty}^{\infty} f(x) \, dx.
\end{aligned}
$$

In view of this fact, we can evaluate a convergent integral over the real line by computing its principal value, which can often be obtained via complex methods and the residue theorem.

The following test of convergence for improper integrals is similar the analogous one for infinite series. We omit the proof.

Proposition 5.3.2. *Let* $-\infty \leq A < B \leq \infty$ *and let* f *be a continuous function on the closure of* (A, B).
(i) If $\int_{A}^{B} |f(x)| \, dx$ *is convergent, then* $\int_{A}^{B} f(x) \, dx$ *is convergent and we have*

$$\left| \int_{A}^{B} f(x) \, dx \right| \leq \int_{A}^{B} |f(x)| \, dx.$$

(ii) If $|f(x)| \leq g(x)$ *for all* $A < x < B$ *and* $\int_{A}^{B} g(x) \, dx$ *is convergent, then* $\int_{A}^{B} f(x) \, dx$ *is convergent and we have*

$$\left| \int_{A}^{B} f(x) \, dx \right| \leq \int_{A}^{B} g(x) \, dx.$$

Example 5.3.3. (Improper integrals and residues: the main ideas) Evaluate

$$I = \int_{-\infty}^{\infty} \frac{x^2}{x^4 + 1} \, dx.$$

Solution. To highlight the main ideas, we present the solution in basic steps.

Step 1: Show that the improper integral is convergent. Because the integrand is continuous on the real line, it is enough to show that the integral outside a finite interval, say $[-1, 1]$, is convergent. For $|x| \geq 1$, we have $\frac{x^2}{x^4+1} \leq \frac{x^2}{x^4} = \frac{1}{x^2}$, and since $\int_1^{\infty} \frac{1}{x^2} \, dx$ is convergent, it follows by Proposition 5.3.2(ii) that $\int_1^{\infty} \frac{x^2}{x^4+1} \, dx$ and $\int_{-\infty}^{-1} \frac{x^2}{x^4+1} \, dx$ are convergent. Thus $\int_{-\infty}^{\infty} \frac{x^2}{x^4+1} \, dx$ is convergent, and so

$$\int_{-\infty}^{\infty} \frac{x^2}{x^4 + 1} \, dx = \lim_{R \to \infty} \int_{-R}^{R} \frac{x^2}{x^4 + 1} \, dx. \tag{5.3.4}$$

Step 2: Set up the contour integral. We replace x by z and consider the function $f(z) = \frac{z^2}{z^4+1}$. The general guideline is to integrate this function over a contour that consists partly of the interval $[-R, R]$, so as to recapture the integral $\int_{-R}^{R} \frac{x^2}{x^4+1} \, dx$ as part of the contour integral on γ_R.

Choosing the appropriate contour is not obvious in general. For a rational function, such as $\frac{x^2}{x^4+1}$, where the denominator does not vanish on the x-axis and the degree of the denominator is larger than the degree of the numerator by 2, a closed semicircle γ_R as in Figure 5.15 will work. Since γ_R consists of the interval $[-R, R]$ followed by the semicircle σ_R, using the additive property of path integrals [Proposition 3.2.12(i)], we write

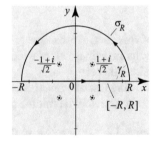

Fig. 5.15 The path and poles for Example 5.3.3.

$$I_{\gamma_R} = \int_{\gamma_R} f(z) \, dz = \int_{[-R,R]} f(z) \, dz + \int_{\sigma_R} f(z) \, dz = I_R + I_{\sigma_R}. \tag{5.3.5}$$

For $z = x$ in $[-R, R]$, we have $f(z) = f(x) = \frac{x^2}{x^4+1}$ and $dz = dx$, and so $I_R = \int_{-R}^{R} \frac{x^2}{x^4+1} \, dx$, which according to (5.3.4) converges to the desired integral as $R \to \infty$. So, in order to compute the desired integral, we must get a handle on the other quantities, I_{γ_R} and I_{σ_R}, in (5.3.5). Our strategy is as follows: In Step 3, we compute I_{γ_R} by the residue theorem; and in Step 4, we show that $\lim_{R \to \infty} I_{\sigma_R} = 0$. These provide the necessary ingredients to complete the solution in Step 5.

Step 3: Compute I_{γ_R} by the residue theorem. For $R > 1$, the function $f(z) = \frac{z^2}{z^4+1}$ has two poles inside γ_R. These are the roots of $z^4 + 1 = 0$ in the upper half-plane. To solve $z^4 = -1$, we write $-1 = e^{i\pi}$; then using the formula for the nth roots (Proposition 1.3.10), we find the four roots

$$z_1 = \frac{1+i}{\sqrt{2}}, \quad z_2 = \frac{-1+i}{\sqrt{2}}, \quad z_3 = \frac{-1-i}{\sqrt{2}}, \quad z_4 = \frac{1-i}{\sqrt{2}}$$

(see Figure 5.15). Since these are simple roots, we have simple poles at z_1 and z_2 and, in view of Proposition 5.1.3(ii), the residues there are

$$\operatorname{Res}(z_1) = \frac{z^2}{\frac{d}{dz}(z^4+1)}\bigg|_{z=z_1} = \frac{z^2}{4z^3}\bigg|_{z=z_1} = \frac{1}{4z}\bigg|_{z=z_1} = \frac{\sqrt{2}}{4(1+i)} = \frac{1-i}{4\sqrt{2}},$$

and similarly

$$\operatorname{Res}(z_2) = \frac{1}{4z}\bigg|_{z=z_2} = \frac{-1-i}{4\sqrt{2}}.$$

So, by the residue theorem, for all $R > 1$

$$I_{\gamma_R} = \int_{\gamma_R} \frac{z^2}{z^4+1}\, dz = 2\pi i \big(\operatorname{Res}(z_1) + \operatorname{Res}(z_2)\big) = 2\pi i \frac{-2i}{4\sqrt{2}} = \frac{\pi}{\sqrt{2}}. \tag{5.3.6}$$

Step 4: Show that $\lim_{R\to\infty} I_{\sigma_R} = 0$. For z on σ_R, we have $|z| = R$, and so

$$\left|\frac{z^2}{z^4+1}\right| \le \frac{R^2}{R^4-1} = M_R.$$

Using the inequality of Theorem 3.2.19 we obtain

$$|I_{\sigma_R}| = \left|\int_{\sigma_R} \frac{z^2}{z^4+1}\, dz\right| \le \ell(\sigma_R) M_R = \pi R \frac{R^2}{R^4-1} = \frac{\pi}{R - 1/R^3} \to 0, \quad \text{as } R \to \infty.$$

Step 5: Compute the improper integral. Using (5.3.5) and (5.3.6), we obtain

$$\frac{\pi}{\sqrt{2}} = I_R + I_{\sigma_R}.$$

Let $R \to \infty$, then $I_R \to \int_{-\infty}^{\infty} \frac{x^2}{x^4+1}\, dx$ and $I_{\sigma_R} \to 0$, and so

$$\int_{-\infty}^{\infty} \frac{x^2}{x^4+1}\, dx = \frac{\pi}{\sqrt{2}}. \qquad \square$$

In Example 5.3.3, we could have used the contour γ_R' in the lower half-plane as shown in Figure 5.16. In this case, it is easier to take the orientation of γ_R' to be negative in order to coincide with the orientation of the interval $[-R, R]$. A calculation similar to that in Example 5.3.3 using the path γ_R' instead of γ_R yields the same answer.

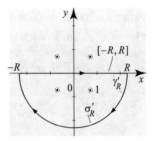

Fig. 5.16 An alternative path.

The idea used in the solution of Example 5.3.3 can be used to prove the following general result.

Proposition 5.3.4. *Let p, q be polynomials of a real variable and assume that q has no real roots; assume furthermore that*

$$degree\, q \geq 2 + degree\, p.$$

Let σ_R denote the arc $\{Re^{i\theta} : 0 \leq \theta \leq \pi\}$. Then we have

$$\lim_{R \to \infty} \left| \int_{\sigma_R} \frac{p(z)}{q(z)}\, dz \right| = 0. \tag{5.3.7}$$

Moreover, if $z_1, z_2, \ldots, z_N$ denote all the poles of p/q in the upper half-plane, then

$$\int_{-\infty}^{\infty} \frac{p(x)}{q(x)}\, dx = 2\pi i \sum_{j=1}^{N} \operatorname{Res}\left(\frac{p}{q}, z_j\right). \tag{5.3.8}$$

Proof. We use the same idea as in Example 5.3.3. If $p(x) = a_n x^n + \cdots + a_1 x + a_0$ and $q(x) = b_m x^m + \cdots + b_1 x + b_0$, where $a_n \neq 0 \neq b_m$ and $m \geq n + 2$, then when $z = Re^{i\theta}$, for R sufficiently large, we have

$$\left| \int_{\sigma_R} \frac{p(z)}{q(z)}\, dz \right| \leq \frac{|a_n|R^n + \cdots + |a_1|R + |a_0|}{|b_m|R^m - \cdots - |b_1|R - |b_0|} 2\pi R \to 0$$

as $R \to \infty$, since $m > n + 1$. This proves (5.3.7). All the poles of $p(x)/q(x)$ are inside the interior of the contour $\sigma_R \cup [-R, R]$ for large R and consequently, identity (5.3.8) follows from Theorem 5.1.2. ∎

Example 5.3.5. (Improper integral of a rational function) Evaluate

$$\int_{-\infty}^{\infty} \frac{1}{(x^2 + 1)(x^2 + 4)}\, dx.$$

Solution. The integrand satisfies the two conditions of Proposition 5.3.4:

$$\text{degree } q = 4 \geq 2 + \text{degree } p = 2$$

and the denominator $q(x) = (x^2 + 1)(x^2 + 4)$ has no real roots. So we may apply (5.3.8). We have

$$(z^2 + 1)(z^2 + 4) = (z + i)(z - i)(z + 2i)(z - 2i),$$

and hence the function

$$\frac{1}{(z^2 + 1)(z^2 + 4)}$$

has simple poles at i and $2i$ in the upper half-plane (Figure 5.17). The residues there are

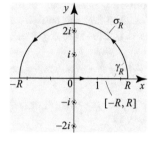

$$\text{Res}\,(i) = \lim_{z \to i}(z - i)\,\frac{1}{(z - i)(z + i)(z^2 + 4)}$$

$$= \frac{1}{(2i)(-1 + 4)}$$

$$= -\frac{i}{6}$$

Fig. 5.17 The path and poles for the contour integral in Example 5.3.5.

and

$$\text{Res}\,(2i) = \lim_{z \to 2i}(z - 2i)\,\frac{1}{(z^2 + 1)(z - 2i)(z + 2i)} = \frac{i}{12}.$$

According to (5.3.8), we have

$$\int_{-\infty}^{\infty} \frac{1}{(x^2 + 1)(x^2 + 4)}\,dx = 2\pi i\left(-\frac{i}{6} + \frac{i}{12}\right) = \frac{\pi}{6}. \qquad \square$$

Integrals Involving Exponential Functions

The success of the method of contour integration crucially depends on the choice of contours. In the previous examples, we used expanding semicircular contours. To evaluate integrals involving exponential functions, we use rectangular contours. We consider integrals of the form

$$\int_{-\infty}^{\infty} \frac{e^{ax}}{e^{bx} + c}\,dx, \quad \text{where } 0 < a < b,\ c > 0.$$

Since $c > 0$ and $e^{bx} > 0$, the denominator does not vanish. Also the condition $a < b$ guarantees that the integral is convergent near $+\infty$, while the integrand is bounded by e^{ax}/c, hence it is integrable near $-\infty$ as well.

Example 5.3.6. (Integral involving exponential functions) Let $\alpha > 1$. Establish the identity

$$\int_{-\infty}^{+\infty} \frac{e^x}{e^{\alpha x} + 1}\, dx = \frac{\pi}{\alpha \sin \frac{\pi}{\alpha}}. \tag{5.3.9}$$

Solution Step 1: As noted before, since $\alpha > 1$, the integral converges. The integrand leads us to define the analytic function $f(z) = \frac{e^z}{e^{\alpha z}+1}$, whose poles are the roots of $e^{\alpha z} + 1 = 0$. Since the exponential function is $2\pi i$-periodic, then

$$e^{\alpha z} = -1 = e^{i\pi} \quad \Leftrightarrow \quad \alpha z = i\pi + 2k\pi i \quad \Leftrightarrow \quad z_k = (2k+1)\frac{\pi}{\alpha}i,\ k = 0, \pm1, \pm2, \ldots.$$

Thus $f(z) = \frac{e^z}{e^{\alpha z}+1}$ has infinitely many poles at z_k, all lying on the imaginary axis (Figure 5.18).

Step 2: Select the contour of integration. Our contour should expand in the x-direction in order to cover the entire x-axis. To avoid including infinitely many poles on the y-axis, we do not expand the contour in the upper half-plane, as we did with the semicircles in the previous examples. Instead, we use a rectangular contour γ_R consisting of the paths γ_1, γ_2, γ_3, and γ_4, as in Figure 5.18, and let I_j denote the path integral of f over γ_j ($j = 1, \ldots, 4$). As $R \to \infty$, γ_R will expand in the horizontal direction, but the length of the vertical sides remains fixed at $\frac{2\pi}{\alpha}$.

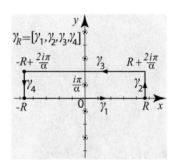

Fig. 5.18 The path and poles for the contour integral in Example 5.3.6.

To understand the reason for our choice of the vertical length, let us consider I_1 and I_3. On γ_1, we have $z = x$, $dz = dx$,

$$I_1 = \int_{-R}^{R} \frac{e^x}{e^{\alpha x} + 1}\, dx, \tag{5.3.10}$$

and so I_1 converges to the desired integral I as $R \to \infty$. On γ_3, we have $z = x + i\frac{2\pi}{\alpha}$, $dz = dx$, and using the $2\pi i$-periodicity of the exponential function, we get

$$I_3 = \int_{R}^{-R} \frac{e^{x+i\frac{2\pi}{\alpha}}}{e^{\alpha(x+i\frac{2\pi}{\alpha})} + 1}\, dx = -e^{\frac{2\pi}{\alpha}i}\int_{-R}^{R} \frac{e^x}{e^{\alpha x} + 1}\, dx = -e^{\frac{2\pi}{\alpha}i}I_1. \tag{5.3.11}$$

This last equality explains the choice of the vertical sides: They are chosen so that the integral on the returning horizontal side γ_3 is equal to a constant multiple of the integral on γ_1. From here the solution is straightforward.

Step 3: Apply the residue theorem. From Step 2, we have only one pole of f inside γ_R at $z_0 = \frac{\pi}{\alpha}i$. Since $e^{\alpha z} + 1$ has a simple root, this is a simple pole. Using Proposition 5.1.3(ii), we find

$$\mathrm{Res}\left(\frac{e^z}{e^{\alpha z}+1}, \frac{\pi}{\alpha}i\right) = \frac{e^{\frac{\pi}{\alpha}i}}{\alpha e^{\alpha\frac{\pi}{\alpha}i}} = \frac{e^{\frac{\pi}{\alpha}i}}{\alpha e^{i\pi}} = -\frac{e^{\frac{\pi}{\alpha}i}}{\alpha},$$

and so by the residue theorem

$$I_1 + I_2 + I_3 + I_4 = \int_{\gamma_R} \frac{e^z}{e^{\alpha z}+1}\,dz = 2\pi i\,\mathrm{Res}\left(\frac{e^z}{e^{\alpha z}+1}, \frac{\pi}{\alpha}i\right) = -2\pi i\frac{e^{\frac{\pi}{\alpha}i}}{\alpha}. \quad (5.3.12)$$

Step 4: Show that the integrals on the vertical sides tend to 0 as $R \to \infty$. For I_2, $z = R + iy$ $(0 \le y \le \frac{2\pi}{\alpha})$ on γ_2, hence $|e^{\alpha z}| = |e^{\alpha R}e^{i\alpha y}| = e^{\alpha R}$,

$$|e^{\alpha z}+1| \ge |e^{\alpha z}| - 1 = e^{\alpha R} - 1 \quad \Rightarrow \quad \frac{1}{|e^{\alpha z}+1|} \le \frac{1}{e^{\alpha R}-1},$$

and so for z on γ_2

$$|f(z)| = \left|\frac{e^z}{e^{\alpha z}+1}\right| \le \frac{|e^{R+iy}|}{e^{\alpha R}-1} = \frac{e^R}{e^{\alpha R}-1} = \frac{1}{e^{(\alpha-1)R}-e^{-R}}.$$

Consequently, by the *ML*-inequality for path integrals,

$$|I_2| = \left|\int_{\gamma_2}\frac{e^z}{e^{\alpha z}+1}\,dz\right| \le \frac{\ell(\gamma_2)}{e^{(\alpha-1)R}-e^{-R}} = \frac{\frac{2\pi}{\alpha}}{e^{(\alpha-1)R}-e^{-R}} \to 0, \text{ as } R \to \infty.$$

The proof that $I_4 \to 0$ as $R \to \infty$ is done similarly; we omit the details.
Step 5: Compute the desired integral (5.3.9). Using (5.3.10), (5.3.11), and (5.3.12), we find that

$$I_1\left(1 - e^{\frac{2\pi}{\alpha}i}\right) + I_2 + I_4 = -2\pi i\frac{e^{\frac{\pi}{\alpha}i}}{\alpha}.$$

Letting $R \to \infty$ and using the result of Step 4, we get

$$\left(1 - e^{\frac{2\pi}{\alpha}i}\right)\int_{-\infty}^{\infty}\frac{e^x}{e^{\alpha x}+1}\,dx = -2\pi i\frac{e^{\frac{\pi}{\alpha}i}}{\alpha},$$

and after simplifying

$$\int_{-\infty}^{\infty}\frac{e^x}{e^{\alpha x}+1}\,dx = \frac{2\pi i}{\alpha\left(e^{\frac{\pi}{\alpha}i} - e^{-\frac{\pi}{\alpha}i}\right)} = \frac{\pi}{\alpha\sin\frac{\pi}{\alpha}},$$

which implies (5.3.9). □

There are interesting integrals of rational functions that are not computable using semicircular contours as in Example 5.3.3. One such integral is

$$\int_0^\infty \frac{dx}{x^3 + 1}. \tag{5.3.13}$$

This integral can be reduced to an integral involving exponential functions, via the substitution $x = e^t$. We outline this useful technique in the following example.

Example 5.3.7. (The substitution $x = e^t$) For $\alpha > 1$ establish the identity

$$\int_0^\infty \frac{1}{x^\alpha + 1} \, dx = \frac{\pi}{\alpha \sin \frac{\pi}{\alpha}}. \tag{5.3.14}$$

Solution. Step 1: Show that the integral converges. The integrand is continuous, so it is enough to show that the integral converges on $[1, \infty)$. We have $\frac{1}{x^\alpha + 1} \leq \frac{1}{x^\alpha}$, and the integral is convergent since $\int_1^\infty \frac{1}{x^\alpha} \, dx < \infty$.

Step 2: Apply the substitution $x = e^t$. Let $x = e^t$, $dx = e^t \, dt$, and note that as x varies from 0 to ∞, t varies from $-\infty$ to ∞, and so

$$I = \int_0^\infty \frac{1}{x^\alpha + 1} \, dx = \int_{-\infty}^\infty \frac{e^t}{e^{\alpha t} + 1} \, dt = \int_{-\infty}^\infty \frac{e^x}{e^{\alpha x} + 1} \, dx,$$

where, for convenience, in the last integral we have used x as a variable of integration instead of t. Identity (5.3.14) follows now from Example 5.3.6. □

The tricky part in Example 5.3.6 is choosing the contour. Let us clarify this part with one more example. For instance, we compute the integral

$$\int_0^\infty \frac{\ln x}{x^4 + 1} \, dx.$$

This integral is improper as the interval of integration is infinite and the integrand tends to $-\infty$ as $x \downarrow 0$. To define the convergence of such integrals, we follow the general procedure of taking all one-sided limits one at a time; see Figure 5.19. Thus

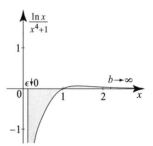

Fig. 5.19 Splitting an improper integral.

$$\int_0^\infty \frac{\ln x}{x^4 + 1} \, dx = \lim_{\varepsilon \downarrow 0} \int_\varepsilon^1 \frac{\ln x}{x^4 + 1} \, dx + \lim_{b \to \infty} \int_1^b \frac{\ln x}{x^4 + 1} \, dx.$$

It is not difficult to show that both limits exist and thus the integral converges. We look at this integral in the next example.

Example 5.3.8. (An integral involving $\ln x$) Derive the identity

$$\int_0^\infty \frac{\ln x}{x^4+1}\,dx = -\frac{\pi^2}{8\sqrt{2}}. \tag{5.3.15}$$

Solution. Let $x = e^t$, $\ln x = t$, $dx = e^t\,dt$. This transforms the integral into

$$\int_{-\infty}^\infty \frac{t}{e^{4t}+1}e^t\,dt = \int_{-\infty}^\infty \frac{xe^x}{e^{4x}+1}\,dx.$$

To evaluate this integral, we integrate the function

$$f(z) = \frac{ze^z}{e^{4z}+1}$$

over the rectangular contour γ_R in Figure 5.20. Let I_j denote the integral of f over γ_j. Here again, we chose the vertical sides of γ_R so that on the returning path γ_3 the denominator equals to $e^{4x}+1$. As we will see momentarily, this enables us to relate I_3 to I_1. Let us now compute $I_{\gamma_R} = \int_{\gamma_R} f(z)\,dz$.

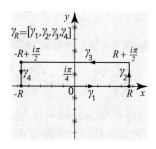

Fig. 5.20 The path and poles for the contour integral in Example 5.3.8.

The function f has one (simple) pole at $z = i\frac{\pi}{4}$ inside γ_R. By Proposition 5.1.3(ii) the residue there is

$$\text{Res}\left(\frac{ze^z}{e^{4z}+1}, i\frac{\pi}{4}\right) = \frac{i\frac{\pi}{4}e^{i\frac{\pi}{4}}}{4e^{i\pi}} = -\frac{i\pi(1+i)}{16\sqrt{2}}.$$

So by the residue theorem

$$I_{\gamma_R} = 2\pi i\left(-\frac{i\pi(1+i)}{16\sqrt{2}}\right) = \frac{\pi^2(1+i)}{8\sqrt{2}} = I_1 + I_2 + I_3 + I_4. \tag{5.3.16}$$

Examining each I_j ($j = 1, \ldots, 4$), we have

$$I_1 = \int_{\gamma_1} \frac{ze^z}{e^{4z}+1}\,dz = \int_{-R}^R \frac{xe^x}{e^{4x}+1}\,dx.$$

On γ_3, $z = x + i\frac{\pi}{2}$, $dz = dx$, so using $e^{\frac{i\pi}{2}} = i$, we get

$$I_3 = \int_{\gamma_3} \frac{ze^z}{e^{4z}+1}\,dz = \int_R^{-R} \frac{(x+i\frac{\pi}{2})e^x e^{i\frac{\pi}{2}}}{e^{4x}+1}\,dx$$

$$= -i\int_{-R}^R \frac{xe^x}{e^{4x}+1}\,dx + \frac{\pi}{2}\int_{-R}^R \frac{e^x}{e^{4x}+1}\,dx$$

$$= -iI_1 + \frac{\pi}{2}\int_{-R}^R \frac{e^x}{e^{4x}+1}\,dx.$$

To show that I_2 and I_4 tend to 0 as $R \to \infty$, we proceed as in Step 5 of Example 5.3.6. For $z = R + iy$ ($0 \le y \le \frac{\pi}{2}$) on γ_2, we have $|z| \le R + y \le R + \frac{\pi}{2}$, and so, as in Example 5.3.6,

$$|f(z)| = \left|\frac{ze^z}{e^{4z}+1}\right| = |z|\left|\frac{e^z}{e^{4z}+1}\right| \le (R+\frac{\pi}{2})\frac{e^R}{e^{4R}-1} = \frac{R+\frac{\pi}{2}}{e^{3R}-e^{-R}}.$$

Consequently, by the *ML*-inequality for path integrals

$$|I_2| = \left|\int_{\gamma_2} \frac{ze^z}{e^{4z}+1}\,dz\right| \le \frac{\ell(\gamma_2)(R+\frac{\pi}{2})}{e^{3R}-e^{-R}} = \frac{\frac{\pi}{2}(R+\frac{\pi}{2})}{e^{3R}-e^{-R}} \to 0, \text{ as } R \to \infty.$$

The proof that $I_4 \to 0$ as $R \to \infty$ is done similarly; we omit the details. Substituting our finding into (5.3.16) and taking the limit as $R \to \infty$, we get

$$\frac{\pi^2(1+i)}{8\sqrt{2}} = \lim_{R\to\infty}\left(I_1 + I_2 - iI_1 + \frac{\pi}{2}\int_{-R}^R \frac{e^x}{e^{4x}+1}\,dx + I_4\right)$$

$$= (1-i)\int_{-\infty}^\infty \frac{xe^x}{e^{4x}+1}\,dx + \frac{\pi}{2}\int_{-\infty}^\infty \frac{e^x}{e^{4x}+1}\,dx.$$

Equating imaginary parts of both sides, we obtain identity (5.3.15). If we equate real parts of both sides we obtain the value of the integral in (5.3.14) that corresponds to $\alpha = 4$.   □

The substitution $x = e^t$ is also useful even when we do not use complex methods to evaluate the integral in t. For example,

$$\int_0^\infty \frac{\ln x}{x^2+1}\,dx = \int_{-\infty}^\infty \frac{te^t}{e^{2t}+1}\,dt = 0$$

since the integral is convergent and the integrand $\frac{te^t}{e^{2t}+1} = \frac{t}{e^t+e^{-t}}$ is an odd function.

Exercises 5.3

In Exercises 1–9, evaluate the improper integrals by the method of Example 5.3.3.

1. $\int_{-\infty}^\infty \frac{dx}{x^4+1} = \frac{\pi}{\sqrt{2}}$

2. $\int_{-\infty}^\infty \frac{dx}{x^4+x^2+1} = \frac{\pi}{\sqrt{3}}$

3. $\int_{-\infty}^\infty \frac{dx}{(x^2+1)(x^4+1)} = \frac{\pi}{2}$

4. $\int_{-\infty}^\infty \frac{dx}{(x-i)(x+3i)} = \frac{\pi}{2}$

5. $\displaystyle\int_{-\infty}^{\infty} \frac{dx}{(x^2+1)^3} = \frac{3\pi}{8}$

6. $\displaystyle\int_{-\infty}^{\infty} \frac{dx}{(x^4+1)^2} = \frac{3\pi}{4\sqrt{2}}$

7. $\displaystyle\int_{-\infty}^{\infty} \frac{dx}{(4x^2+1)(x-i)} = \frac{\pi}{3}i$

8. $\displaystyle\int_{-\infty}^{\infty} \frac{dx}{(x+i)(x-3i)} = \frac{\pi}{2}$

9. $\displaystyle\int_{-\infty}^{\infty} \frac{dx}{(1+x^2)^{n+1}} = \frac{(2n)!}{2^{2n}(n!)^2}\pi$

In Exercises 10–13, evaluate the improper integrals by the method of Example 5.3.6.

10. $\displaystyle\int_0^{\infty} \frac{x^2 e^x}{e^{2x}+1}dx = \frac{\pi^3}{16}$

11. $\displaystyle\int_{-\infty}^{\infty} \frac{1}{3e^x+e^{-x}}dx = \frac{\pi}{2\sqrt{3}}$

12. $\displaystyle\int_{-\infty}^{\infty} \frac{xe^x}{e^{2x}+1}dx = 0$

13. $\displaystyle\int_{-\infty}^{\infty} \frac{e^{ax}}{e^{bx}+1}dx = \frac{\pi}{b\sin\frac{\pi a}{b}} \quad (0 < a < b)$

In Exercises 14–23, evaluate the improper integrals by the method of Example 5.3.7 or Example 5.3.8.

14. $\displaystyle\int_0^{\infty} \frac{x}{x^5+1}dx = \frac{\pi}{5\sin\left(\frac{2\pi}{5}\right)}$

15. $\displaystyle\int_0^{\infty} \frac{x}{x^\alpha+1}dx = \frac{\pi}{\alpha\sin\left(\frac{2\pi}{\alpha}\right)} \quad (\alpha > 2)$

16. $\displaystyle\int_0^{\infty} \frac{\sqrt{x}}{x^3+1}dx = \frac{\pi}{3}$

17. $\displaystyle\int_0^{\infty} \frac{x^\alpha}{(x+1)^2}dx = \frac{\pi\alpha}{\sin\pi\alpha} \quad (-1 < \alpha < 1)$

18. $\displaystyle\int_0^{\infty} \frac{x\ln x}{x^3+1}dx = \frac{2\pi^2}{27}$

19. $\displaystyle\int_0^{\infty} \frac{x\ln(2x)}{x^3+1}dx = \frac{2\pi^2}{27} + \frac{2\pi\ln 2}{3\sqrt{3}}$

20. $\displaystyle\int_0^{\infty} \frac{\ln(2x)}{x^2+4}dx = \frac{\pi}{2}\ln 2$

21. $\displaystyle\int_0^{\infty} \frac{\ln(ax)}{x^2+b^2}dx = \frac{\pi}{2b}\ln(ab) \quad (a, b > 0)$

22. $\displaystyle\int_0^{\infty} \frac{(\ln x)^2}{x^2+1}dx = \frac{\pi^3}{8}$

23. $\displaystyle\int_0^{\infty} \frac{(\ln x)^2}{x^3+1}dx = \frac{10\pi^3}{81\sqrt{3}}$

24. Use the contour γ_R in the adjacent figure to evaluate

$$\int_0^{\infty} \frac{1}{x^3+1}dx.$$

[Hint: $I_{\gamma_R} = I_1 + I_2 + I_3$; $I_3 = -e^{i\frac{2\pi}{3}}I_1$; and $I_2 \to 0$ as $R \to \infty$.]

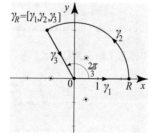

25. A property of the gamma function. (a) Show that for $0 < \alpha < 1$,

$$\int_0^{\infty} \frac{x^{\alpha-1}}{1+x}dx = \frac{\pi}{\sin\pi\alpha}.$$

(b) Use the definition of the gamma function (Exercise 24, Section 4.2) to derive the identity

$$\Gamma(\alpha)\Gamma(1-\alpha) = \int_0^{\infty}\int_0^{\infty} e^{-(s+t)}s^{-\alpha}t^{\alpha-1}\,ds\,dt, \quad 0 < \alpha < 1.$$

(c) Make the change of variables $x = s+t$, $y = \frac{t}{s}$ in (b), then use (a) to derive the following **formula of the complements** of the gamma function:

$$\Gamma(\alpha)\Gamma(1-\alpha) = \frac{\pi}{\sin\pi\alpha}, \quad 0 < \alpha < 1.$$

(d) Use the identity principle (Theorem 4.5.5) to extend the formula of the complements to all z satisfying $0 < \mathrm{Re}\, z < 1$.

(e) Derive the integral $\displaystyle\int_0^{\frac{\pi}{2}} \sqrt{\cot x}\, dx = \frac{\pi}{\sqrt{2}}$. [*Hint*: Exercise 25(d), Section 4.2.]

5.4 Products of Rational and Trigonometric Functions

In this section, we use residues to evaluate improper integrals of the form

$$\int_{-\infty}^{\infty} \frac{p(x)}{q(x)} \cos(ax)\, dx \quad \text{and} \quad \int_{-\infty}^{\infty} \frac{p(x)}{q(x)} \sin(bx)\, dx, \qquad (5.4.1)$$

where a and b are real and $\frac{p}{q}$ is a rational function. The method is similar to that in the previous section. Observe that

$$\cos(ax) = \mathrm{Re}\left(e^{iax}\right) \quad \text{and} \quad \sin(bx) = \mathrm{Im}\left(e^{ibx}\right),$$

for all real numbers a, b, x. So

$$\int_{-\infty}^{\infty} \frac{p(x)}{q(x)} \cos(ax)\, dx = \int_{-\infty}^{\infty} \frac{p(x)}{q(x)} \mathrm{Re}\left(e^{iax}\right) dx = \int_{-\infty}^{\infty} \mathrm{Re}\left(\frac{p(x)}{q(x)} e^{iax}\right) dx$$

$$= \mathrm{Re}\left(\int_{-\infty}^{\infty} \frac{p(x)}{q(x)} e^{iax}\, dx\right), \qquad (5.4.2)$$

and similarly

$$\int_{-\infty}^{\infty} \frac{p(x)}{q(x)} \sin(bx)\, dx = \mathrm{Im}\left(\int_{-\infty}^{\infty} \frac{p(x)}{q(x)} e^{ibx}\, dx\right). \qquad (5.4.3)$$

Example 5.4.1. Let $a > 0$ and $s \geq 0$ be real numbers. Derive the identity

$$\int_{-\infty}^{\infty} \frac{\cos(sx)}{x^2 + a^2}\, dx = \frac{\pi}{a} e^{-sa}. \qquad (5.4.4)$$

Solution. Step 1: Show that the improper integral is convergent. We have

$$\left|\frac{\cos(sx)}{x^2 + a^2}\right| \leq \frac{1}{x^2 + a^2},$$

and since

$$\int_{-\infty}^{\infty} \frac{dx}{x^2 + a^2}$$

is convergent, the integral in (5.4.4) converges by Proposition 5.3.2(*ii*).

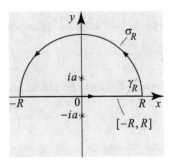

Fig. 5.21 The contour is the union of γ_R and σ_R.

Step 2: Set up and evaluate the contour integral. Since

$$\int_{-\infty}^{\infty} \frac{\cos(sx)}{x^2 + a^2} \, dx = \int_{-\infty}^{\infty} \text{Re}\left(\frac{e^{isx}}{x^2 + a^2}\right) dx = \text{Re}\left(\int_{-\infty}^{\infty} \frac{e^{isx}}{x^2 + a^2} \, dx\right), \qquad (5.4.5)$$

we will consider the contour integral

$$I_{\gamma_R} = \int_{\gamma_R} \frac{e^{isz}}{z^2 + a^2} \, dz = \int_{\sigma_R} \frac{e^{isz}}{z^2 + a^2} \, dz + \int_{-R}^{R} \frac{e^{isx}}{x^2 + a^2} \, dx = I_{\sigma_R} + I_R, \qquad (5.4.6)$$

where γ_R and σ_R are as in Figure 5.21. For $R > a$, $\frac{e^{isz}}{z^2 + a^2}$ has one simple pole inside γ_R at $z = ia$. In view of Proposition 5.1.3(ii), the residue there is

$$\text{Res}\left(\frac{e^{isz}}{z^2 + a^2}, ia\right) = \frac{e^{is(ia)}}{2ia} = \frac{e^{-sa}}{2ia}.$$

By the residue theorem, for all $R > a$, we have

$$I_{\gamma_R} = I_{\sigma_R} + I_R = 2\pi i \frac{e^{-sa}}{2ia} = \frac{\pi}{a} e^{-sa}. \qquad (5.4.7)$$

Step 3: Show that $\lim_{R \to \infty} I_{\sigma_R} = 0$. For $s \geq 0$ and $0 \leq \theta \leq \pi$, we have $\sin \theta \geq 0$, hence $-sR \sin \theta \leq 0$, and so $e^{-sR \sin \theta} \leq 1$. Write z on σ_R, as $z = Re^{i\theta} = R(\cos \theta + i \sin \theta)$, where $0 \leq \theta \leq \pi$. Then

$$\left| e^{isz} \right| = \left| e^{isR(\cos \theta + i \sin \theta)} \right| = \left| \overbrace{e^{isR \cos \theta}}^{1} \right| \left| e^{-sR \sin \theta} \right| = e^{-sR \sin \theta} \leq 1. \qquad (5.4.8)$$

Hence, for $R > a$ and z on the semicircle σ_R, we have

$$\left| \frac{e^{isz}}{z^2 + a^2} \right| \leq \frac{1}{|z^2 + a^2|} \leq \frac{1}{|z|^2 - a^2} = \frac{1}{R^2 - a^2},$$

and so the ML-inequality for path integrals yields

$$\left| \int_{\sigma_R} \frac{e^{isz}}{z^2 + a^2} \, dz \right| \leq \ell(\sigma_R) \frac{1}{R^2 - a^2} = \frac{\pi R}{R^2 - a^2} \to 0, \text{ as } R \to \infty.$$

This estimate works because the degree of the polynomial in the denominator is greater than the degree of the polynomial in the numerator by 2.

Step 4: Compute the desired improper integral. Let $R \to \infty$ in (5.4.7), use Step 3, and get

$$\lim_{R \to \infty} I_{\sigma_R} + \lim_{R \to \infty} I_R = 0 + \int_{-\infty}^{\infty} \frac{e^{isx}}{x^2 + a^2} \, dx = \frac{\pi}{a} e^{-sa}.$$

Taking real parts on both sides and using (5.4.5), we obtain the desired integral

$$\int_{-\infty}^{\infty} \frac{\cos(sx)}{x^2 + a^2}\, dx = \frac{\pi}{a} e^{-sa}. \qquad \square$$

By observing that $\sin\theta \geq 0$ for $0 \leq \theta \leq \pi$, we were able in (5.4.8) to obtain the inequality $\left| e^{isz} \right| \leq 1$ for all $s \geq 0$ and $z = Re^{i\theta}$, $0 \leq \theta \leq \pi$. A more precise analysis of $\sin\theta$ yields a better estimate on $\left| e^{isz} \right|$, which in turn makes possible to compute integrals of the form (5.4.1), where $\mathrm{degree}\, q \geq 1 + \mathrm{degree}\, p$.

Lemma 5.4.2. (Jordan's Lemma) *The following inequality is valid for $R > 0$:*

$$\int_0^\pi e^{-R\sin\theta}\, d\theta \leq \frac{\pi}{R}. \tag{5.4.9}$$

Proof. We begin by writing

$$\int_0^\pi e^{-R\sin\theta}\, d\theta = \int_0^{\frac{\pi}{2}} e^{-R\sin\theta}\, d\theta + \int_{\frac{\pi}{2}}^\pi e^{-R\sin\theta}\, d\theta.$$

Then we change variables $t = \pi - \theta$ in the second integral above and, noting that $\sin(\theta) = \sin(\pi - \theta) = \sin t$ and $d\theta = -dt$, we obtain

$$\int_0^\pi e^{-R\sin\theta}\, d\theta = \int_0^{\frac{\pi}{2}} e^{-R\sin\theta}\, d\theta - \int_{\frac{\pi}{2}}^0 e^{-R\sin t}\, dt = 2\int_0^{\frac{\pi}{2}} e^{-R\sin\theta}\, d\theta. \tag{5.4.10}$$

At this point, we need an estimate on $\sin\theta$. On the interval $[0, \frac{\pi}{2}]$, the graph of $\sin\theta$ is concave down, because the second derivative is negative. Hence the graph of $\sin\theta$ for $0 \leq \theta \leq \frac{\pi}{2}$ is above the chordal line that joins two points on the graph. In particular, it is above the chord that joins the origin to the point $(\frac{\pi}{2}, 1)$, whose equation is $y = \frac{2}{\pi}\theta$.

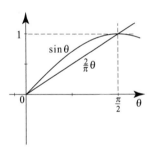

Fig. 5.22 Proof of (5.4.11).

This fact is expressed analytically by the inequality

$$\sin\theta \geq \frac{2}{\pi}\theta, \tag{5.4.11}$$

valid for $0 \leq \theta \leq \frac{\pi}{2}$, whose geometric proof is illustrated in Figure 5.22.

Inequality (5.4.11) implies that $-\sin\theta \leq -\frac{2}{\pi}\theta$ and combining this with (5.4.10) we deduce

$$\int_0^\pi e^{-R\sin\theta}\, d\theta \leq 2\int_0^{\frac{\pi}{2}} e^{-\frac{2}{\pi}R\theta}\, d\theta. \tag{5.4.12}$$

Changing variables $t = \frac{2}{\pi}R\theta$, we write

$$2\int_0^{\frac{\pi}{2}} e^{-\frac{2}{\pi}R\theta}\,d\theta = 2\int_0^R e^{-t}\,dt\,\frac{\pi}{2R} = 2(1-e^{-R})\frac{\pi}{2R} \le \frac{\pi}{R}\,,$$

and combing this inequality with (5.4.12) concludes the proof of the lemma. ∎

Lemma 5.4.3. (General Version of Jordan's Lemma) *Let $R_0 > 0$ and $0 \le \theta_1 < \theta_2 \le \pi$. For $R \ge R_0$, let σ_R be the circular arc of all $z = Re^{i\theta}$ with $0 \le \theta_1 \le \theta \le \theta_2 \le \pi$ as shown in Figure 5.23. Let f be a continuous complex-valued function defined on all arcs σ_R and let $M(R)$ denote the maximum value of $|f|$ on σ_R. If $\lim_{R\to\infty} M(R) = 0$, then for all $s > 0$*

$$\lim_{R\to\infty} \int_{\sigma_R} e^{isz} f(z)\,dz = 0.$$

Proof. For $s > 0$, we have from (5.4.8), $\left|e^{isz}\right| = e^{-sR\sin\theta}$.

Note that since $e^{-sR\sin\theta} > 0$, its integral increases if we increase the size of the interval of integration. Thus

$$\int_{\theta_1}^{\theta_2} e^{-sr\sin\theta}\,d\theta \le \int_0^{\pi} e^{-sr\sin\theta}\,d\theta,$$

if $0 \le \theta_1 \le \theta_2 \le \pi$. Parametrize σ_R by $\gamma(\theta) = Re^{i\theta}$, where $\theta_1 \le \theta \le \theta_2$. Then

$$\gamma'(\theta) = Rie^{i\theta}, \qquad |\gamma'(\theta)|d\theta = R\,d\theta$$

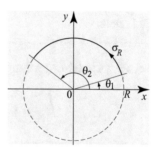

Fig. 5.23 The circular arcs σ_R.

and hence

$$\left|\int_{\sigma_R} e^{isz} f(z)\,dz\right| \le \int_{\theta_1}^{\theta_2} \left|e^{isz} f(z)\right| R\,d\theta$$

$$\le RM(R) \int_{\theta_1}^{\theta_2} e^{-sR\sin\theta}\,d\theta$$

$$\le RM(R) \int_0^{\pi} e^{-sR\sin\theta}\,d\theta. \qquad (5.4.13)$$

From inequality (5.4.9) we obtain that the integral in (5.4.13) is bounded by $\pi/(sR)$. Thus

$$\left|\int_{\sigma_R} e^{isz} f(z)\,dz\right| \le RM(R)\frac{\pi}{sR} = \frac{\pi}{s}M(R) \to 0$$

as $R \to \infty$. This concludes the proof of the lemma. ∎

An analog of Jordan's lemma holds for $s < 0$ if the circular arc σ_R is in the lower half-plane. Applying Jordan's lemma in the special case when $f(z)$ is a rational function, we obtain the following useful result.

Corollary 5.4.4. *Let p,q are complex polynomials with degree $q \geq 1 + $ degree p. Let σ_R denote the semicircular arc consisting of all $z = Re^{i\theta}$, where $0 \leq \theta \leq \pi$. Then $\lim_{R \to \infty} \int_{\sigma_R} e^{isz} \frac{p(z)}{q(z)} dz = 0$ for all $s > 0$.*

Proof. Let $M(R)$ denote the maximum of $|p(z)/q(z)|$ for z on σ_R. Since degree $q \geq 1 + $ degree p, $M(R) \to 0$ as $R \to \infty$. Applying Lemma 5.4.3 we obtain the claimed assertion. ∎

Next we evaluate the improper integral

$$\int_{-\infty}^{\infty} \frac{x \sin x}{x^2 + a^2} dx.$$

It is not difficult to show that this integral is convergent using integration by parts (Exercise 19). However, because the degree of $x^2 + a^2$ is only one more than the degree of x, the estimate in Step 3 of Example 5.4.1 will not be sufficient to show that the integral on the expanding semicircle tends to 0. For this purpose we appeal to Jordan's lemma.

Example 5.4.5. (Applying Jordan's lemma) Derive the identity

$$\int_{-\infty}^{\infty} \frac{x \sin x}{x^2 + a^2} dx = \frac{\pi}{e^a}, \quad a > 0. \tag{5.4.14}$$

Solution. Consider the contour integral

$$I_{\gamma_R} = \int_{\sigma_R} \frac{z}{z^2 + a^2} e^{iz} dz + \int_{-R}^{R} \frac{x}{x^2 + a^2} e^{ix} dx = I_{\sigma_R} + I_R, \tag{5.4.15}$$

where σ_R is the circular arc shown in Figure 5.24 and $\gamma_R = \sigma_R \cup [-R, R]$. By Jordan's lemma (precisely by Corollary 5.4.4), $\lim_{R \to \infty} I_{\sigma_R} = 0$. For $R > a$, $\frac{z}{z^2 + a^2} e^{iz}$ has a simple pole inside γ_R at ia. By the residue theorem, for all $R > a$, we obtain

$$I_{\gamma_R} = 2\pi i \operatorname{Res}\left(\frac{z e^{iz}}{z^2 + a^2}, ia\right)$$

$$= 2\pi i \frac{(ia) e^{i(ia)}}{2(ia)}$$

$$= \frac{\pi}{e^a} i.$$

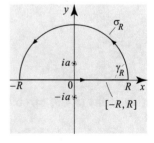

Fig. 5.24 The path and poles for the contour integral in Example 5.4.5.

Taking the limit as $R \to \infty$ in (5.4.15) and using the fact that $I_{\sigma_R} \to 0$, we deduce

$$\frac{\pi}{e^a}i = \int_{-\infty}^{\infty} \frac{x}{x^2+a^2} e^{ix}\,dx = \int_{-\infty}^{\infty} \frac{x\cos x}{x^2+a^2}\,dx + i\int_{-\infty}^{\infty} \frac{x\sin x}{x^2+a^2}\,dx,$$

and the desired identity follows upon taking imaginary parts on both sides. Notice that the convergence of the improper integral in (5.4.14) follows by letting $R \to \infty$ in (5.4.15). □

Indenting Contours

Taking $a = 0$ in Example 5.4.5, we obtain the formula

$$\int_{-\infty}^{\infty} \frac{\sin x}{x}\,dx = \pi.$$

This identity is in fact needed in many applications. In the remainder of this section, we develop a method to calculate similar and other interesting integrals.

In Section 5.3 we defined the Cauchy principal value of an improper integral over the real line. However, an integral can also be improper if the integrand becomes unbounded at a point inside the interval of integration. To make our discussion concrete, consider

$$\int_{-1}^{1} f(x)\,dx,$$

where f is a continuous function on $[-1,0)$ and $(0,1]$ but might have infinite limits as x approaches 0 from the left or right. Such an integral is said to be **convergent** if both $\lim_{b\to 0^-} \int_{-1}^{b} f(x)\,dx$ and $\lim_{a\to 0^+} \int_{a}^{1} f(x)\,dx$ are convergent. In this case, we set (Figure 5.25)

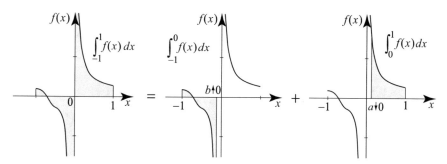

Fig. 5.25 Splitting an improper integral.

$$\int_{-1}^{1} f(x)\,dx = \lim_{b\to 0^-} \int_{-1}^{b} f(x)\,dx + \lim_{a\to 0^+} \int_{a}^{1} f(x)\,dx. \qquad (5.4.16)$$

This expression should be contrasted with the one in which a function is integrated on intervals that approach the singular point $x = 0$ in a symmetric fashion. We define the Cauchy principal value of the integral $\int_{-1}^{1} f(x)\,dx$, with a singularity at $x = 0$, to be (Figure 5.26)

$$\text{P.V.} \int_{-1}^{1} f(x)\, dx = \lim_{r \to 0^+} \left(\int_{-1}^{-r} f(x)\, dx + \int_{r}^{-1} f(x)\, dx \right). \tag{5.4.17}$$

The Cauchy principal value of an integral may exist even though the integral itself is not convergent. For example,

$$\int_{-1}^{-r} \frac{dx}{x} + \int_{r}^{1} \frac{dx}{x} = 0$$

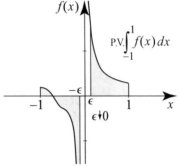

for all $r > 0$, so P.V. $\int_{-1}^{1} \frac{dx}{x} = 0$, but the integral itself is clearly not convergent since $\int_{0}^{1} \frac{dx}{x} = \infty$. However, whenever $\int_{-1}^{1} f(x)\, dx$ is convergent the P.V. $\int_{-1}^{1} f(x)\, dx$ exists and the two integrals are the same, since we can split the limit of a sum in (5.4.17) into a sum of limits and recover (5.4.16).

Fig. 5.26 The Cauchy principal value.

This fact allows us to compute convergent integrals by computing their principal values. We illustrate these ideas in an example.

Example 5.4.6. (Cauchy principal values and singular points) Show that

$$\text{P.V.} \int_{-\infty}^{\infty} \frac{1}{x-1}\, dx = 0 \tag{5.4.18}$$

by writing the integral in terms of limits of integrals over finite intervals.

Solution. The integral is improper as it extends over the infinite real line and the integrand is singular at $x = 1$. Accordingly, the principal value involves two limits

$$\text{P.V.} \int_{-\infty}^{\infty} \frac{1}{x-1}\, dx = \lim_{R \to \infty} \left(\int_{-R}^{0} \frac{1}{x-1}\, dx + \int_{2}^{R} \frac{1}{x-1}\, dx \right) \tag{5.4.19}$$

$$+ \lim_{r \to 0^+} \left(\int_{0}^{1-r} \frac{1}{x-1}\, dx + \int_{1+r}^{2} \frac{1}{x-1}\, dx \right),$$

where the choices $x = 0$ and $x = 2$ were arbitrary; in fact, any pair of numbers with the first being in $(-\infty, 1)$ and the second in $(1, \infty)$ will work equally well in the splitting of the integrals. See Figure 5.27. Using elementary methods, we can see that both limits on the right of (5.4.19) exist, so the principal value exists, and we can write

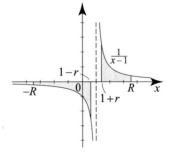

Fig. 5.27 The integral in (5.4.18).

$$\text{P.V.} \int_{-\infty}^{\infty} \frac{1}{x-1}\,dx = \lim_{R\to\infty, r\to 0^+} \left(\int_{-R}^{1-r} \frac{1}{x-1}\,dx + \int_{1+r}^{R} \frac{1}{x-1}\,dx \right),$$

where the limits may be taken in any order. Using basic calculus, we find that for large $R > 1$

$$\int_{-R}^{1-r} \frac{1}{x-1}\,dx = \ln|x-1|\Big|_{-R}^{1-r} = \ln(r) - \ln(R+1);$$

and

$$\int_{1+r}^{R} \frac{1}{x-1}\,dx = \ln|x-1|\Big|_{1+r}^{R} = \ln(R-1) - \ln(r).$$

So

$$\text{P.V.} \int_{-\infty}^{\infty} \frac{1}{x-1}\,dx = \lim_{R\to\infty} \left[\ln(R-1) - \ln(R+1) \right] = \lim_{R\to\infty} \ln\left(\frac{R-1}{R+1} \right) = 0. \quad \square$$

Now, we return to the integral $\int_{-\infty}^{\infty} \frac{\sin x}{x}\,dx$. Since the integrand is well-behaved at $x = 0$ and the integral over the infinite interval converges, we can compute it by taking the imaginary part of $\text{P.V.} \int_{-\infty}^{\infty} \frac{e^{ix}}{x}\,dx$.

This principal value integral involves the limit as $r \to 0^+$ and $R \to \infty$ of an integral over $[-R, -r]$ and $[r, R]$. We envision this as a contour integral of the complex function $\frac{e^{iz}}{z}$ and close the contour with a large positively oriented semicircle of radius R and a small negatively oriented semicircle of radius r, which bypasses the problem at 0. The resulting path $\gamma_{r,R}$ is shown in Figure 5.28. The limit of the path integral over the small circle will be computed using the following result.

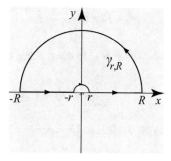

Fig. 5.28 Bypassing the point 0 with the contour $\gamma_{r,R}$.

Lemma 5.4.7. (Shrinking Path Lemma) *Suppose that f is a continuous complex-valued function on a closed disk $\overline{B_{r_0}}(z_0)$ with center at z_0 and radius r_0. For $0 < r \le r_0$, let σ_r denote the positively oriented circular arc at angle α (Figure 5.29), consisting of all $z = z_0 + re^{i\theta}$, where $\theta_0 \le \theta \le \theta_0 + \alpha$, θ_0 and α are fixed, and $\alpha \ne 0$. Then*

$$\lim_{r\to 0^+} \frac{1}{i\alpha} \int_{\sigma_r} \frac{f(z)}{z - z_0}\,dz = f(z_0). \tag{5.4.20}$$

Proof. Parametrize the integral in (5.4.20) by $z = z_0 + re^{i\theta}$, where $\theta_0 \leq \theta \leq \theta_0 + \alpha$, $dz = rie^{i\theta} \, d\theta$, $z - z_0 = re^{i\theta}$. Then

$$\frac{1}{i\alpha} \int_{\sigma_r} \frac{f(z)}{z - z_0} \, dz$$

$$= \frac{1}{i\alpha} \int_{\theta_0}^{\theta_0 + \alpha} \frac{f(z_0 + re^{i\theta})}{re^{i\theta}} \, ire^{i\theta} \, d\theta$$

$$= \frac{1}{\alpha} \int_{\theta_0}^{\theta_0 + \alpha} f(z_0 + re^{i\theta}) \, d\theta.$$

Define the function

$$F(r) = \frac{1}{\alpha} \int_{\theta_0}^{\theta_0 + \alpha} f(z_0 + re^{i\theta}) \, d\theta.$$

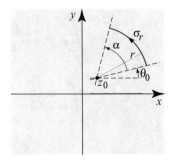

Fig. 5.29 Circular arc of angle α.

Since f is continuous on $\overline{B_{r_0}(z_0)}$, it follows that it is uniformly continuous on this set. Thus given $\varepsilon > 0$ there is $\delta > 0$ such that $|w - w'| < \delta$ implies $|f(w) - f(w')| < \varepsilon$.

Then $|r - r'| < \delta$ implies $|(z_0 + re^{i\theta}) - (z_0 + r'e^{i\theta})| < \delta$ which in turn implies $|f(z_0 + re^{i\theta}) - f(z_0 + r'e^{i\theta})| < \varepsilon$. The ML-inequality yields

$$|F(r) - F(r')| = \left| \frac{1}{\alpha} \int_{\theta_0}^{\theta_0 + \alpha} \left(f(z_0 + re^{i\theta}) - f(z_0 + r'e^{i\theta}) \right) d\theta \right| \leq \varepsilon.$$

It follows that F is also continuous on $[0, r_0]$. Consequently,

$$\lim_{r \to 0^+} F(r) = F(0) = \frac{1}{\alpha} \int_{\theta_0}^{\theta_0 + \alpha} f(z_0) d\theta = f(z_0),$$

which is equivalent to (5.4.20). ∎

The following simple consequence of Lemma 5.4.7 is useful.

Corollary 5.4.8. *Suppose that g is an analytic function in a deleted neighborhood of z_0 with a simple pole at z_0. For $0 < r \leq r_0$, let σ_r be the circular arc at angle α (see Figure 5.29). Then we have*

$$\lim_{r \to 0^+} \int_{\sigma_r} g(z) \, dz = i\alpha \operatorname{Res}(g, z_0). \qquad (5.4.21)$$

Proof. Let $f(z) = (z - z_0)g(z)$ for $z \neq z_0$ and define $f(z_0) = \lim_{z \to z_0}(z - z_0)g(z) = \operatorname{Res}(g, z_0)$. Since g has a simple pole at z_0, it follows from Theorem 4.5.15 that f is analytic at z_0. Now apply Lemma 5.4.7 to f and (5.4.21) follows from (5.4.20), since for $z \neq z_0$, $f(z)/(z - z_0) = g(z)$. ∎

The improper integrals that we consider next differ from previous ones in that they are not always convergent; however, their Cauchy principal values do exist.

Example 5.4.9. (Cauchy principal value: Use of indented contours) Derive the integral identities

$$\text{P.V.} \int_{-\infty}^{\infty} \frac{\sin x}{x-a}\, dx = \pi \cos a$$

$$\text{P.V.} \int_{-\infty}^{\infty} \frac{\cos x}{x-a}\, dx = -\pi \sin a,$$

for all $-\infty < a < \infty$. In particular, when $a = 0$, the first integral yields the interesting identity

$$\int_{-\infty}^{\infty} \frac{\sin x}{x}\, dx = \pi.$$

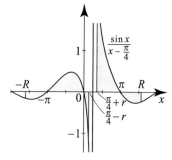

Fig. 5.30 Principal value integral.

Solution. The integrals are improper as the interval of integration is infinite and they are singular at $x = a$. The first integral's principal value is

$$\text{P.V.} \int_{-\infty}^{\infty} \frac{\sin x}{x-a}\, dx = \lim_{\substack{r\to 0^+ \\ R\to\infty}} \left(\int_{-R}^{a-r} \frac{\sin x}{x-a}\, dx + \int_{a+r}^{R} \frac{\sin x}{x-a}\, dx \right)$$

(Figure 5.30 represents the case $a > 0$). Consider the integral of $f(z) = \frac{e^{iz}}{z-a}$ over the closed contour $\gamma_{r,R}$ in Figure 5.31, where we have indented the contour around $x = a$. The larger semicircle σ_R has radius R and positive orientation. The smaller semicircle has radius r; it is negatively oriented and we denote it by σ_r^*. Since f is analytic on and inside $\gamma_{r,R}$, Cauchy's theorem implies that

$$0 = \int_{-R}^{a-r} \frac{e^{ix}}{x-a}\, dx + \int_{\sigma_r^*} \frac{e^{iz}}{z-a}\, dz + \int_{a+r}^{R} \frac{e^{ix}}{x-a}\, dz + \int_{\sigma_R} \frac{e^{iz}}{z-a}\, dz$$
$$= I_1 + I_2 + I_3 + I_4.$$

By Jordan's lemma, $I_4 \to 0$ as $R \to \infty$. By the shrinking path lemma,

$$I_2 \to -i\pi e^{ia} \quad \text{as} \quad r \to 0$$

(note the negative sign due to the fact that the path of integration is negatively oriented). Thus

$$\lim_{\substack{r\to 0 \\ R\to\infty}} \left(\int_{-R}^{a-r} \frac{e^{ix}}{x-a}\, dx + \int_{a+r}^{R} \frac{e^{ix}}{x-a}\, dx \right)$$
$$= i\pi e^{ia}$$
$$= \pi(-\sin a + i\cos a),$$

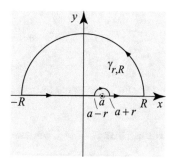

Fig. 5.31 Indenting the contour around a.

and the desired integral identities follow upon taking real and imaginary parts. □

Exercises 5.4

In Exercises 1–10, evaluate the improper integral. Make sure to outline the steps in your solutions. Carry out the details even if the integral follows from previously derived formulas. Here a, b are real and $c > 0$.

1. $\displaystyle\int_{-\infty}^{\infty} \frac{\cos(4x)}{x^2+1} \, dx = \pi e^{-4}$

2. $\displaystyle\int_{-\infty}^{\infty} \frac{\sin(\frac{\pi x}{2})}{x^2+2x+4} \, dx = -\frac{\pi}{\sqrt{3}} e^{-\frac{\sqrt{3}}{2}\pi}$

3. $\displaystyle\int_{-\infty}^{\infty} \frac{x\sin(3x)}{x^2+2} \, dx = e^{-3\sqrt{2}}\pi$

4. $\displaystyle\int_{-\infty}^{\infty} \frac{\sin(\pi x)}{x^2+2x+4} \, dx = 0$

5. $\displaystyle\int_{-\infty}^{\infty} \frac{x^2\cos(2x)}{(x^2+1)^2} \, dx = -\frac{\pi}{2} e^{-2}$

6. $\displaystyle\int_{-\infty}^{\infty} \frac{\cos x}{(x^2+1)^2} \, dx = \frac{\pi}{e}$

7. $\displaystyle\int_{-\infty}^{\infty} \frac{\cos(a(x-b))}{x^2+c^2} \, dx = \frac{\pi}{c} e^{-|a|c} \cos(ab)$

8. $\displaystyle\int_{-\infty}^{\infty} \frac{\cos(4\pi x)}{2x^2+x+1} \, dx = -2e^{-\sqrt{7}\pi}\frac{\pi}{\sqrt{7}}$

9. $\displaystyle\int_{-\infty}^{\infty} \frac{x\cos(\pi x)}{x^2+x+9} \, dx = \pi e^{-\frac{\sqrt{35}}{2}\pi}$

10. $\displaystyle\int_{-\infty}^{\infty} \frac{\sin(\pi x)}{x^2+x+1} \, dx = -\frac{2\pi}{\sqrt{3}} e^{-\frac{\sqrt{3}}{2}\pi}$

In Exercises 11–18, use an indented contour to evaluate the Cauchy principal value of the improper integrals.

11. P.V. $\displaystyle\int_{-\infty}^{\infty} \frac{\sin x \cos x}{x} \, dx = \frac{\pi}{2}$

12. P.V. $\displaystyle\int_{0}^{\infty} \frac{\sin x \cos(2x)}{x} \, dx = 0$

13. P.V. $\displaystyle\int_{-\infty}^{\infty} \frac{1-\cos x}{x^2} \, dx = \pi$

14. P.V. $\displaystyle\int_{-\infty}^{\infty} \frac{2x\sin x}{x^2-a^2} \, dx = 2\pi\cos a$

15. P.V. $\displaystyle\int_{-\infty}^{\infty} \frac{\sin^2 x}{x^2} \, dx = \pi$

16. P.V. $\displaystyle\int_{-\infty}^{\infty} \frac{\sin(ax)}{x-b} \, dx = \pi\cos(ab)$ and P.V. $\displaystyle\int_{-\infty}^{\infty} \frac{\cos(ax)}{x-b} \, dx = -\pi\sin(ab)$

17. P.V. $\displaystyle\int_{-\infty}^{\infty} \frac{\sin x}{x(x^2+1)} \, dx = \pi\left(1-\frac{1}{e}\right)$

18. P.V. $\displaystyle\int_{-\infty}^{\infty} \frac{\cos x}{x^2-a^2} \, dx = -\pi\frac{\sin a}{a}$ $(a \neq 0)$

19. Show that the improper integral
$$\int_{0}^{\infty} \frac{x\sin x}{x^2+a^2} \, dx$$
is convergent. [*Hint:* For $A > 0$, integrate $\int_0^A \frac{x\sin x}{x^2+a^2} \, dx$ by parts by letting $u = \frac{x}{x^2+a^2}$, $dv = \sin x \, dx$.]

20. Show that the improper integral $\int_0^\infty \frac{\sin x}{x} \, dx$ is convergent. [*Hint:* For $A > 0$, integrate $\int_0^A \frac{\sin x}{x} \, dx$ by parts by letting $u = \frac{1}{x}$, $dv = \sin x$, $du = -\frac{dx}{x^2}$ and $v = 1 - \cos x$.]

21. (a) Use Example 5.4.9 and a suitable change of variables to establish the formula

$$\frac{2}{\pi} \int_0^\infty \frac{\sin(ax)}{x}\, dx = \operatorname{sgn} a,$$

where $\operatorname{sgn} a = -1$ if $a < 0$, 0 if $a = 0$, and 1 if $a > 0$.

(b) Use part (a) and a suitable trigonometric identity to prove that

$$\int_0^\infty \frac{\sin(ax)\cos(bx)}{x}\, dx = \begin{cases} 0 & \text{if } 0 < a < b, \\ \frac{\pi}{4} & \text{if } a = b > 0, \\ \frac{\pi}{2} & \text{if } 0 < b < a. \end{cases}$$

22. Use Example 5.4.1 to establish the identity

$$\int_{-\infty}^\infty \frac{\cos(sx)}{x^2 + a^2}\, dx = \frac{\pi}{|a|} e^{-|sa|}, \quad -\infty < s < \infty, \; a \neq 0.$$

23. Use a suitable change of variables in Example 5.4.5 to establish the identity

$$\int_{-\infty}^\infty \frac{x\sin(sx)}{x^2 + a^2}\, dx = \pi \operatorname{sgn}(s) e^{-|as|}.$$

24. Project Problem: Fourier transforms of the hyperbolic secant and cosecant. In this exercise, we derive the identities

$$\int_0^\infty \frac{\cos(wx)}{\cosh(\pi x)}\, dx = \frac{1}{2}\operatorname{sech}\frac{w}{2} \quad \text{and} \quad \int_0^\infty \frac{\sin(wx)}{\sinh(\pi x)}\, dx = \frac{1}{2}\tanh\frac{w}{2}, \tag{5.4.22}$$

where w is a real number. Up to a constant multiple, these integrals give the Fourier cosine transform of the hyperbolic secant $1/\cosh(\pi x)$ and the Fourier sine transform of the hyperbolic cosecant $1/\sinh(\pi x)$.

(a) Let $\gamma_{\varepsilon,R}$ denote the indented contour shown in the adjacent figure. Show that

$$I_{\gamma_{\varepsilon,R}} = \int_{\gamma_{\varepsilon,R}} \frac{e^{(\pi + iw)z}}{e^{2\pi z} + 1}\, dz = 0.$$

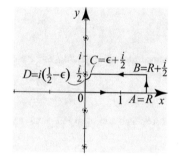

(b) Show that $\lim_{R \to \infty} I_{AB} = 0$. [*Hint:* Use the idea of Step 5, Example 5.3.6.]

(c) Show that

$$I_{OA} = \frac{1}{2}\int_0^R \frac{e^{iwx}}{\cosh(\pi x)}\, dx, \quad I_{DO} = -\frac{i}{2}\int_0^{\frac{1}{2}-\varepsilon} \frac{e^{-wy}}{\cos(\pi y)}\, dy, \quad I_{BC} = i\frac{e^{-\frac{w}{2}}}{2}\int_\varepsilon^R \frac{e^{iwx}}{\sinh(\pi x)}\, dx.$$

(d) Use Corollary 5.4.8 to show that $\lim_{\varepsilon \to 0} I_{CD} = -\frac{1}{4}e^{-\frac{w}{2}}$.

(e) Letting $R \to \infty$ then $\varepsilon \to 0$ and using (a), conclude that

$$\frac{1}{2}\int_0^\infty \frac{e^{iwx}}{\cosh(\pi x)}\, dx - \frac{1}{4}e^{-\frac{w}{2}} + \lim_{\varepsilon \to 0}\left[i\frac{e^{-\frac{w}{2}}}{2}\int_\varepsilon^\infty \frac{e^{iwx}}{\sinh(\pi x)}\, dx - \frac{i}{2}\int_0^{\frac{1}{2}-\varepsilon} \frac{e^{-wy}}{\cos(\pi y)}\, dy \right] = 0.$$

(f) Let

$$A = \int_0^\infty \frac{\cos(wx)}{\cosh(\pi x)}\, dx \quad \text{and} \quad B = \int_0^\infty \frac{\sin(wx)}{\sinh(\pi x)}\, dx.$$

Equating real parts in (e) conclude that $A - e^{-\frac{w}{2}}B = \frac{1}{2}e^{-\frac{w}{2}}$. Use the fact that A is even in w and B is odd in w, replace w by $-w$, and conclude $A + e^{\frac{w}{2}}B = \frac{1}{2}e^{\frac{w}{2}}$. Then solve for A and B.

25. Project Problem: Bernoulli numbers and residues.   In this exercise, we evaluate a sine integral related to the sine Fourier transform of a hyperbolic function and then use our answer to give an integral representation of the Bernoulli numbers, which are defined in Example 4.3.12.

(a) Follow the steps outlined in the previous exercise as you integrate the function $z \mapsto \frac{e^{iwz}}{e^{2\pi z}-1}$ on the indented contour in Figure 5.32, and obtain the identity

$$\int_0^\infty \frac{\sin(wx)}{e^{2\pi x}-1}\,dx = \frac{1}{4}\frac{1+e^{-w}}{1-e^{-w}} - \frac{1}{2w}, \qquad w > 0.$$

(b) With the help of the Maclaurin series of $\frac{z}{2}\coth\frac{z}{2}$ (see the details following Example 4.3.12), obtain the Laurent series

$$\frac{1}{2}\frac{1+e^{-w}}{1-e^{-w}} - \frac{1}{w} = \sum_{k=1}^\infty \frac{B_{2k}}{(2k)!}w^{2k-1}$$

for all $|w| < 2\pi$.

(c) Replace $\sin(wx)$ in the integral in part (a) by its Taylor series

$$\sin(wx) = \sum_{k=1}^\infty (-1)^{k-1}\frac{w^{2k-1}x^{2k-1}}{(2k-1)!},$$

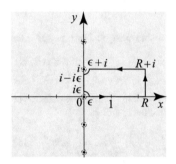

Fig. 5.32 Indented contour for Exercise 25.

then interchange order of integration, and conclude that

$$\int_0^\infty \frac{x^{2k-1}}{e^{2\pi x}-1}\,dx = \frac{(-1)^{k-1}}{4k}B_{2k}.$$

(The diligent reader should justify the interchange of the sum and the integral.)

5.5 Advanced Integrals by Residues

In this section, we evaluate some classical integrals that arise in applied mathematics. The examples are independent of each other. We will need the formula

$$I = \int_{-\infty}^\infty e^{-ax^2}\,dx = \sqrt{\frac{\pi}{a}}, \qquad a > 0. \tag{5.5.1}$$

To evaluate the integral, consider

$$I^2 = \int_{-\infty}^\infty e^{-ax^2}\,dx \int_{-\infty}^\infty e^{-ay^2}\,dy = \int_{-\infty}^\infty \int_{-\infty}^\infty e^{-a(x^2+y^2)}\,dx\,dy.$$

Now use polar coordinates, $x = r\cos\theta$, $y = r\sin\theta$, $dx\,dy = r\,dr\,d\theta$. Then

$$I^2 = \int_0^{2\pi} d\theta \int_0^\infty e^{-ar^2} r\,dr = 2\pi\left[-\frac{1}{2a}e^{-ar^2}\right]_{r=0}^\infty = \frac{\pi}{a},$$

which is equivalent to (5.5.1).

Example 5.5.1. Derive the **Poisson integral**

$$\int_{-\infty}^\infty e^{-ax^2}\cos(wx)\,dx = \sqrt{\frac{\pi}{a}}\,e^{-\frac{w^2}{4a}}, \qquad a>0,\ -\infty < w < \infty. \tag{5.5.2}$$

Solution   The case $w = 0$ follows from (5.5.1). Also, it is enough to deal with the case $w > 0$. Since $\sin(wx)\,e^{-ax^2}$ is an odd function of x, its integral over symmetric intervals is 0. So

$$\int_{-\infty}^\infty \cos(wx)\,e^{-ax^2}\,dx = \int_{-\infty}^\infty e^{iwx}e^{-ax^2}\,dx = \int_{-\infty}^\infty e^{-ax^2+iwx}\,dx.$$

Completing the square in the exponent of the integrand, we get

$$e^{-ax^2+iwx} = e^{-a(x^2-i\frac{w}{a}x)} = e^{-a(x-i\frac{w}{2a})^2 + a(i\frac{w}{2a})^2} = e^{-\frac{w^2}{4a}}e^{-a(x-i\frac{w}{2a})^2},$$

and since $e^{-\frac{w^2}{4a}}$ is independent of x,

$$\int_{-\infty}^\infty \cos(wx)\,e^{-ax^2}\,dx = e^{-\frac{w^2}{4a}}\int_{-\infty}^\infty e^{-a(x-i\frac{w}{2a})^2}\,dx. \tag{5.5.3}$$

To evaluate the integral on the right, we integrate $e^{-a(z-i\frac{w}{2a})^2}$ over the rectangular contour in Figure 5.33. Let I_j ($j = 1, 2, 3, 4$) denote the integral of this function over the path γ_j as indicated in Figure 5.33. The choice of a rectangular contour is not surprising, as similar contours were used on path integrals involving exponential functions in Section 5.3. The choice of the y-intercept is related to the shift in the exponent of e^{-az^2}. It will be justified as we compute I_3.

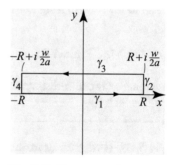

Fig. 5.33 The rectangular contour for Example 5.5.1.

Since e^{-az^2} is entire, Cauchy's theorem implies that

$$I_1 + I_2 + I_3 + I_4 = 0. \tag{5.5.4}$$

We have

$$I_1 = \int_{-R}^R e^{-a(x-i\frac{w}{2a})^2}\,dx \rightarrow \int_{-\infty}^\infty e^{-a(x-i\frac{w}{2a})^2}\,dx, \text{ as } R \rightarrow \infty.$$

For I_3, we have $z = x + i\frac{w}{2a}$, $dz = dx$, and so

$$I_3 = \int_R^{-R} e^{-ax^2} dx = -\int_{-R}^{R} e^{-ax^2} dx \to -\sqrt{\frac{\pi}{a}}, \text{ as } R \to \infty,$$

by (5.5.1). For I_2, $z = R + iy$, $0 \le y \le \frac{w}{2a}$, $dz = i\,dy$, and

$$e^{-a(z-i\frac{w}{2a})^2} = e^{-a(R+i(y-\frac{w}{2a}))^2} = e^{-a\left(R^2-(y-\frac{w}{2a})^2\right)}e^{-2aRi(y-\frac{w}{2a})}.$$

Since $\left|e^{-2aRi(y-\frac{w}{2a})}\right| = 1$, for $0 \le y \le \frac{w}{2a}$,

$$\left|e^{-a(z-\frac{iw}{2a})^2}\right| = e^{-a\left(R^2-(y-\frac{w}{2a})^2\right)} = e^{-aR^2}e^{a(\frac{w}{2a}-y)^2} \le e^{-aR^2}e^{a(\frac{w}{2a})^2},$$

using the usual inequality for path integrals, we obtain

$$|I_2| = \left|\int_0^{\frac{w}{2a}} e^{-a(R+i(y-\frac{w}{2a}))^2} i\,dy\right| \le \int_0^{\frac{w}{2a}} \left|e^{-a(R+i(y-\frac{w}{2a}))^2}\right| dy$$

$$\le e^{-aR^2}e^{a(\frac{w}{2a})^2} \frac{w}{2a} \to 0 \text{ as } R \to \infty.$$

Similarly, $I_4 \to 0$ as $R \to \infty$. Taking the limit as $R \to \infty$ in (5.5.4) and using what we know about the limits of the I_j, we get

$$0 = \int_{-\infty}^{\infty} e^{-a(x-i\frac{w}{2a})^2} dx - \sqrt{\frac{\pi}{a}} \quad \text{or} \quad \int_{-\infty}^{\infty} e^{-a(x-i\frac{w}{2a})^2} dx = \sqrt{\frac{\pi}{a}},$$

and (5.5.2) follows upon using (5.5.3). $\square$

Integrals Involving Branch Cuts

The remaining examples illustrate the useful technique of integration around branch points and branch cuts.

Example 5.5.2. (Integrating around a branch point) For $0 < \alpha < 1$, derive the integral identities

$$\int_0^\infty \frac{\cos x}{x^\alpha} dx = \Gamma(1-\alpha)\sin\frac{\alpha\pi}{2} \quad \text{and} \quad \int_0^\infty \frac{\sin x}{x^\alpha} dx = \Gamma(1-\alpha)\cos\frac{\alpha\pi}{2}, \quad (5.5.5)$$

where $\Gamma(z) = \int_0^\infty e^{-t}t^{z-1}\,dt$ is the gamma function (Exercise 24, Section 4.2).

Solution. For these integrands, it is natural to consider the function $f(z) = \frac{e^{iz}}{z^\alpha}$. The only trouble is that z^α is multiply-valued and so we need to specify a single-valued branch of z^α.

The choice of this branch is usually affected by the choice of the contour of integration. In this case, we use the contour γ in Figure 5.34, and choose a branch of $z^\alpha = e^{\alpha \log z}$ that equals the real-numbered power x^α on the x-axis and is analytic on the contour. There are several possibilities, but clearly the easiest one is $z^\alpha = e^{\alpha \operatorname{Log} z}$ where $\operatorname{Log} z$ denotes the principal value branch of the logarithm, with a branch cut on the negative x-axis. Recall that for $z \neq 0$, $\operatorname{Log} z = \ln|z| + i \operatorname{Arg} z$, where $-\pi < \operatorname{Arg} z \leq \pi$.

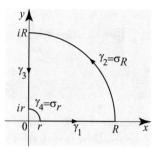

Fig. 5.34 The contour for Example 5.5.1.

Since the function

$$\frac{e^{iz}}{z^\alpha} = \frac{e^{iz}}{e^{\alpha \operatorname{Log} z}}$$

is analytic inside and on the contour γ, Cauchy's theorem implies that

$$0 = \int_\gamma \frac{e^{iz}}{e^{\alpha \operatorname{Log} z}}\, dz = I_1 + I_2 + I_3 + I_4, \qquad (5.5.6)$$

where I_j is the integral over γ_j, as indicated in Figure 5.34. For I_1, $z = x$, $r < x < R$, $dz = dx$, $\frac{e^{ix}}{e^{\alpha \operatorname{Log} x}} = \frac{e^{ix}}{x^\alpha}$, and so

$$\lim_{\substack{r \to 0 \\ R \to \infty}} I_1 = \lim_{\substack{r \to 0 \\ R \to \infty}} \int_r^R \frac{e^{ix}}{x^\alpha}\, dx = \int_0^\infty \frac{e^{ix}}{x^\alpha}\, dx = \int_0^\infty \frac{\cos x}{x^\alpha}\, dx + i \int_0^\infty \frac{\sin x}{x^\alpha}\, dx.$$

For I_3, $z = iy$, $r < y < R$, $dz = i\, dy$,

$$\frac{e^{i(iy)}}{e^{\alpha \operatorname{Log}(iy)}} = \frac{e^{-y}}{e^{\alpha(\ln y + i \operatorname{Arg}(iy))}} = \frac{e^{-y}}{y^\alpha e^{i\frac{\alpha\pi}{2}}} = e^{-i\frac{\alpha\pi}{2}} e^{-y} y^{-\alpha},$$

and so

$$\lim_{\substack{r \to 0 \\ R \to \infty}} I_3 = \lim_{\substack{r \to 0 \\ R \to \infty}} i e^{-i\frac{\alpha\pi}{2}} \int_R^r e^{-y} y^{-\alpha}\, dy = -i e^{-i\frac{\alpha\pi}{2}} \int_0^\infty e^{-y} y^{-\alpha}\, dy$$

$$= -i e^{-i\frac{\alpha\pi}{2}} \Gamma(1-\alpha) = -i\left(\cos\frac{\alpha\pi}{2} - i \sin\frac{\alpha\pi}{2}\right) \Gamma(1-\alpha),$$

where the second-to-last equality follows from the definition of the gamma function.

We now deal with the integrals I_2 and I_4. For z on the circular arc σ_R, write $z = R e^{i\theta}$, where $0 \leq \theta \leq \frac{\pi}{2}$. Then

$$|z^\alpha| = \left|e^{\alpha(\ln|z| + i\,\mathrm{Arg}\,z)}\right| = e^{\alpha\ln|z|} = |z|^\alpha = R^\alpha;$$

also, $|e^{iz}| = \left|e^{R(-\sin\theta + i\cos\theta)}\right| = e^{-R\sin\theta}$, and so

$$\left|\frac{e^{iz}}{z^\alpha}\right| = \frac{e^{-R\sin\theta}}{R^\alpha}. \tag{5.5.7}$$

Let $M(R)$ denote the maximum of (5.5.7) for z in σ_R. Then $M(R) = \frac{1}{R^\alpha}$, which tends to 0 as $R \to \infty$. Thus $I_2 \to 0$ as $R \to \infty$, by Lemma 5.4.2 (Jordan's lemma). On I_4, we have $\left|\frac{e^{iz}}{z^\alpha}\right| \le \frac{1}{r^\alpha}$, and using the ML-inequality we find

$$|I_4| \le \ell(\sigma_r)\frac{1}{r^\alpha} = \frac{2\pi}{4}r^{1-\alpha}.$$

Since $1 - \alpha > 0$, we see that $|I_4| \to 0$ as $r \to 0$. Going back to (5.5.6) and taking the limits as $r \to 0$ and $R \to \infty$, we get

$$0 = \int_0^\infty \frac{\cos x}{x^\alpha}\,dx + i\int_0^\infty \frac{\sin x}{x^\alpha}\,dx - i\left(\cos\frac{\alpha\pi}{2} - i\sin\frac{\alpha\pi}{2}\right)\Gamma(1-\alpha),$$

which is equivalent to (5.5.5). $\qquad\square$

Taking $\alpha = \frac{1}{2}$ in (5.5.5) and using $\Gamma\left(\frac{1}{2}\right) = \sqrt{\pi}$ (see Exercise 25(a), Section 4.2), we obtain

$$\int_0^\infty \frac{\cos x}{\sqrt{x}}\,dx = \Gamma\left(\frac{1}{2}\right)\sin\frac{\pi}{4} = \sqrt{\frac{\pi}{2}} \quad \text{and} \quad \int_0^\infty \frac{\sin x}{\sqrt{x}}\,dx = \Gamma\left(\frac{1}{2}\right)\cos\frac{\pi}{4} = \sqrt{\frac{\pi}{2}}.$$

Letting $x = u^2$, $dx = 2u\,du$, we derive the famous **Fresnel integrals**,

$$\int_0^\infty \cos u^2\,du = \frac{1}{2}\sqrt{\frac{\pi}{2}}$$

$$\int_0^\infty \sin u^2\,du = \frac{1}{2}\sqrt{\frac{\pi}{2}},$$

named after the French mathematician and physicist Augustin Fresnel (1788–1827), one of the founders of the wave theory of light. The cosine integral and its convergence are illustrated in Figure 5.35. The integrals can also be obtained using a different contour integral (see Exercise 5).

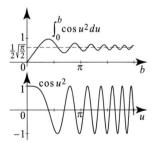

Fig. 5.35 An illustration of the Fresnel cosine integral.

In our next example, we illustrate an important method that is based on the fact that different branches of the logarithm differ by an integer multiple of $2\pi i$. We note that the example can also be evaluated using the substitution $x = e^t$, as in Section 5.3.

Example 5.5.3. (Integrating around a branch cut) For $0 < \alpha < 1$, derive the integral identity

$$\int_0^\infty \frac{dx}{x^\alpha(x+1)} = \frac{\pi}{\sin(\pi\alpha)}. \tag{5.5.8}$$

Solution. We integrate branches of $f_1(z) = \frac{1}{z^\alpha(z+1)}$ on the contours $\Gamma_1 = [\gamma_1, \gamma_3, \gamma_5, \gamma_7]$ and $\Gamma_2 = [\gamma_2, \gamma_4, \gamma_6, \gamma_8]$, shown in Figures 5.36, 5.37, and 5.38.

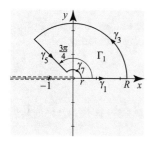

Fig. 5.36 The contour $\Gamma_1 = [\gamma_1, \gamma_3, \gamma_5, \gamma_7]$ and the branch cut of $f_1(z)$.

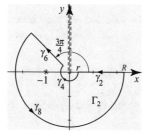

Fig. 5.37 The contour $\Gamma_2 = [\gamma_2, \gamma_4, \gamma_6, \gamma_8]$ and the branch cut of $f_2(z)$.

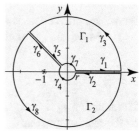

Fig. 5.38 The contours Γ_1 and Γ_2.

On Γ_1, we can take the principal branch of z^α, which coincides with the real power on the real axis and allows us to recover the integral (5.5.8). We cannot use the residue theorem to help integrate this one branch all the way around the origin, because of its branch cut on the negative real axis. Thus we have closed Γ_1, bringing it back at the ray $\theta = \frac{3\pi}{4}$, and use a different branch of z^α to integrate on Γ_2. For the integral on Γ_2, we choose the branch $z^\alpha = e^{\alpha \log_{\frac{\pi}{2}} z}$. In the second quadrant, this branch coincides with the principal branch (and so integrals over γ_5 and γ_6 cancel), but the new branch continues to be analytic as we wind around the origin into the third and fourth quadrants. It is important to note that integrals over γ_1 and γ_2 will not cancel because the two branches of z^α are not the same here; the logarithm branches differ by $2\pi i$. With this in mind, let

$$f_1(z) = \frac{1}{e^{\alpha \operatorname{Log} z}(z+1)} \quad \text{and} \quad f_2(z) = \frac{1}{e^{\alpha \log_{\frac{\pi}{2}} z}(z+1)},$$

where $\operatorname{Log} z = \ln|z| + i \operatorname{Arg} z$, $-\pi < \operatorname{Arg} z \le \pi$, and $\log_{\frac{\pi}{2}} z = \ln|z| + i \arg_{\frac{\pi}{2}} z$, $\frac{\pi}{2} < \arg_{\frac{\pi}{2}} \le \frac{5\pi}{2}$. We integrate f_1 on Γ_1 and f_2 on Γ_2. Let I_j denote the integral of the appropriate branch f_1 or f_2 on γ_j. On γ_1, $z = x > 0$, $\mathrm{Log}\, z = \mathrm{Log}\, x = \ln x$, $e^{\alpha \ln x} = x^\alpha$, and so

$$I_1 = \int_r^R \frac{dx}{x^\alpha(x+1)};$$

also for $z = x > 0$, $\log_{\frac{\pi}{2}} z = \log_{\frac{\pi}{2}} x = \ln x + 2\pi i$, $e^{\alpha \log_{\frac{\pi}{2}} x} = x^\alpha e^{2\pi i \alpha}$, thus

$$I_2 = \int_R^r \frac{dx}{e^{2\pi i \alpha} x^\alpha(x+1)} = -e^{-2\pi i \alpha} \int_r^R \frac{dx}{x^\alpha(x+1)} = -e^{-2\pi i \alpha} I_1. \qquad (5.5.9)$$

For z on γ_5 (and hence γ_6) we have $\mathrm{Log}\, z = \log_{\frac{\pi}{2}} z$, hence $f_1(z) = f_2(z)$, and consequently $I_5 = -I_6$; thus it follows

$$I_5 + I_6 = 0. \qquad (5.5.10)$$

From here on the details of the solution are very much like the previous example.
    The function f_1 is analytic on and inside Γ_1, so by Cauchy's theorem

$$\int_{\Gamma_1} f_1(z)\, dz = I_1 + I_3 + I_5 + I_7 = 0. \qquad (5.5.11)$$

The function f_2 is analytic on and inside Γ_2 except for a simple pole at $z = -1$, so by the residue theorem

$$\int_{\Gamma_2} f_2(z)\, dz = I_2 + I_4 + I_6 + I_8 = 2\pi i \, \mathrm{Res}\, (f_2, -1) = \frac{2\pi i}{e^{\alpha \log_{\frac{\pi}{2}}(-1)}} = \frac{2\pi i}{e^{\pi i \alpha}}. \qquad (5.5.12)$$

Using (5.5.9)–(5.5.12), we obtain

$$\frac{2\pi i}{e^{\pi i \alpha}} = \int_{\Gamma_1} f_1(z)\, dz + \int_{\Gamma_2} f_2(z)\, dz = (1 - e^{-2\pi i \alpha}) I_1 + I_3 + I_4 + I_7 + I_8. \qquad (5.5.13)$$

Letting $r \to 0$ and $R \to \infty$ we obtain $I_1 \to \int_0^\infty \frac{dx}{x^\alpha(x+1)}$. Additionally, if we show that I_3, I_4, I_7, I_8 tend to 0, it will follow from (5.5.13) that

$$\frac{2\pi i}{e^{\pi i \alpha}} = (1 - e^{-2\pi i \alpha}) \int_0^\infty \frac{dx}{x^\alpha(x+1)}.$$

Solving for the integral, we find

$$\int_0^\infty \frac{dx}{x^\alpha(x+1)} = \frac{2\pi i}{e^{\pi i \alpha}} \frac{1}{1 - e^{-2\pi i \alpha}} = \pi \frac{2i}{e^{\pi i \alpha} - e^{-\pi i \alpha}} = \frac{\pi}{\sin \pi \alpha},$$

as desired. So let us show that I_3, I_4, I_7, I_8 tend to 0. This part is similar to Example 5.5.2; we just sketch the details. We have $|z^\alpha| = |z|^\alpha$, so $|f_j(z)| \le \frac{1}{|z|^\alpha|1-|z||}$ $(j = 1, 2)$. For z on γ_7 or γ_8, $|z| = r$ (we may take $r < 1$), and so

$$|I_7| \le \ell(\gamma_7)\frac{1}{r^\alpha(1-r)} \le 2\pi r\frac{1}{r^\alpha(1-r)} = 2\pi\frac{r^{1-\alpha}}{1-r} \to 0, \quad \text{as } r \to 0,$$

because $1 - \alpha > 0$. Similarly, $|I_8| \to 0$ as $r \to 0$. For z on γ_3 or γ_4, $|z| = R$ (we may take $R > 1$), and so

$$|I_3| \le \ell(\gamma_3)\frac{1}{R^\alpha(R-1)} \le 2\pi R\frac{1}{R^\alpha(R-1)} = 2\pi\frac{R^{1-\alpha}}{R-1} \to 0, \quad \text{as } R \to \infty,$$

because the degree of the numerator, $1 - \alpha$, is smaller than the degree of the denominator, which is 1. Similarly, $|I_4| \to 0$ as $R \to \infty$. □

In Example 5.5.3, the point of introducing different branches of the multiple-valued function $\frac{1}{z^\alpha(z+1)}$ was to explicitly indicate how the function can continuously change as we wind around the origin, and how its value on the real axis is affected by the winding. Now we illustrate a useful idea that can be used as an alternative to multiple branches. The idea is to think of a function as taking different values at a point on the branch cut depending on how we approach this point. Consider, for example,

$$h(z) = \sqrt{z-a} \quad (a \text{ real}), \tag{5.5.14}$$

where we use the $\log_0$ branch of the square root, so that the branch cut of $h(z)$ is on the part of the positive x-axis $x > a$. Explicitly, $h(z) = |z-a|^{\frac{1}{2}}e^{\frac{i}{2}\arg_0(z-a)}$, where $0 < \arg_0(z-a) \le 2\pi$. We will use this function and allow it to take different values on the branch cut, depending on how we approach the branch cut. For example, if z approaches the part $x > a$ of the x-axis from the upper half-plane (Figure 5.39), then $\arg_0(z-a)$ approaches 0 and so the values of the function h to the right of a and above the x-axis are $\sqrt{z-a}$ which tend to $\sqrt{|x-a|} = \sqrt{x-a}$. If z approaches the part $x > a$ of the x-axis from the lower half-plane (Figure 5.40), then $\arg_0(z-a)$

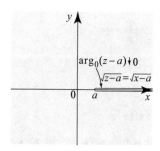

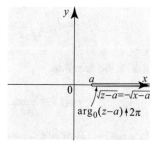

Fig. 5.39 Defining $\sqrt{z-a}$ as z approaches the branch cut from above.

Fig. 5.40 Defining $\sqrt{z-a}$ as z approaches the branch cut from below.

approaches 2π and so the values of the function h to the right of a and below the x-axis tend to $e^{i\frac{1}{2}(2\pi)}\sqrt{|x-a|} = -\sqrt{x-a}$. The mapping $w = h(z)$ is illustrated in Figure 5.41. Even though we omit the justification of the use of functions with multiple values, this can be achieved by the use of different branches of the square root, as in Example 5.5.3.

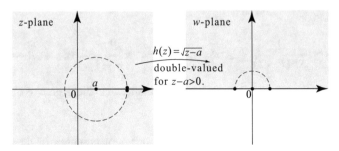

Fig. 5.41 The mapping $w = \sqrt{z-a}$ maps the upper half of the branch cut to the right half of the real axis and the lower half of the branch cut to the left half of the real axis.

In our final example we prove a useful property of **elliptic integrals**, which are integrals of the form $\int \frac{dx}{\sqrt{p(x)}}$, where p is a polynomial of degree ≥ 2. These integrals arise when computing arc length of ellipses, and thus their name. They also arise when computing Schwarz-Christoffel transformations.

Example 5.5.4. (A property of elliptic integrals) Let $a < b < c$ be real numbers. Show that

$$\int_a^b \frac{dx}{\sqrt{(x-a)(b-x)(c-x)}}$$
$$= \int_c^\infty \frac{dx}{\sqrt{(x-a)(x-b)(x-c)}}. \tag{5.5.15}$$

Solution. Consider the function

$$f(z) = \frac{1}{\sqrt{z-a}\sqrt{z-b}\sqrt{z-c}},$$

where we choose the branches of the square roots with branch cuts along the positive real axis. The function f is analytic at all z except possibly at $z = x \geq a$ (in fact, the singularities are removable along $b < z < c$, see Exercise 23). Since f is analytic inside and on the contour in Figure 5.42, Cauchy's theorem implies that its integral on this contour is 0. Letting I_j denote the integral of the limiting values of f on γ_j, we obtain

$$I_1 + I_2 + I_3 + I_4 + I_5 + I_6 + I_7 + I_8 + I_9 + I_{10} + I_{11} + I_{12} = 0. \tag{5.5.16}$$

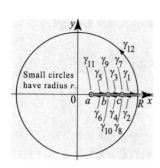

Fig. 5.42 The contour in Example 5.5.4 and its twelve components.

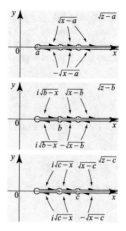

Fig. 5.43 The limiting values of $\sqrt{z-a}$, $\sqrt{z-b}$, and $\sqrt{z-c}$.

It is straightforward to show that I_7, I_8, I_9, I_{10}, I_{11}, I_{12} tend to 0 as $r \to 0$ and $R \to \infty$. The details can be found in Examples 5.5.2 and 5.5.3 and are omitted here. To handle the remaining integrals, we compute the limiting values of f (still denoted by f) in each case with the aid of the values in Figure 5.43:

$$I_1: \quad f(z) = \frac{1}{\sqrt{x-a}\sqrt{x-b}\sqrt{x-c}} \quad \Rightarrow \quad I_1 = \int_{c+r}^{R} \frac{dx}{\sqrt{x-a}\sqrt{x-b}\sqrt{x-c}};$$

$$I_2: \quad f(z) = \frac{-1}{\sqrt{x-a}\sqrt{x-b}\sqrt{x-c}} \quad \Rightarrow \quad I_2 = I_1;$$

$$I_3: \quad f(z) = \frac{-i}{\sqrt{x-a}\sqrt{x-b}\sqrt{c-x}} \quad \Rightarrow \quad I_3 = \int_{b+r}^{c-r} \frac{(-i)\,dx}{\sqrt{x-a}\sqrt{x-b}\sqrt{c-x}};$$

$$I_4: \quad f(z) = \frac{-i}{\sqrt{x-a}\sqrt{x-b}\sqrt{c-x}} \quad \Rightarrow \quad I_4 = -I_3;$$

$$I_5: \quad f(z) = \frac{-1}{\sqrt{x-a}\sqrt{b-x}\sqrt{c-x}} \quad \Rightarrow \quad I_5 = \int_{a+r}^{b-r} \frac{(-1)\,dx}{\sqrt{x-a}\sqrt{b-x}\sqrt{c-x}};$$

$$I_6: \quad f(z) = \frac{-1}{\sqrt{x-a}\sqrt{b-x}\sqrt{c-x}} \quad \Rightarrow \quad I_6 = I_5.$$

When we add these integrals I_3 and I_4 cancel. Taking the limit as $R \to \infty$ and $r \to 0$ and using (5.5.16), we obtain (5.5.15). $\qquad\qquad\square$

Exercises 5.5

In Exercises 1–4, evaluate the integrals using an appropriate contour. Carry out the details of the solution even if the integral follows from previously derived material.

1. $\dfrac{1}{\sqrt{2\pi}} \displaystyle\int_{-\infty}^{\infty} e^{-\frac{x^2}{2}} \cos(wx)\,dx = e^{-\frac{w^2}{2}}$

2. $\displaystyle\int_{-\infty}^{\infty} e^{-6x^2 + iwx}\,dx = \sqrt{\dfrac{\pi}{6}}\, e^{-\frac{w^2}{24}}$

3. $\displaystyle\int_{0}^{\infty} \dfrac{\sin 2x}{\sqrt{x}}\,dx = \dfrac{\sqrt{\pi}}{2}$

4. $\displaystyle\int_{0}^{\infty} \dfrac{dx}{\sqrt{x}(x^2 + 1)} = \dfrac{\pi}{\sqrt{2}}$

5. The Fresnel integrals. We present an alternative way to derive the Fresnel integrals. This is based on an inequality similar to the one used in the proof of Jordan's lemma.

(a) Consider the graph of $\cos 2x$ on the interval $[0, \frac{\pi}{4}]$ (Figure 5.44). Since the graph concaves down, it is above the chord that joins two points on the graph. Explain how this implies the inequality

$$\cos 2x \geq 1 - \frac{4}{\pi}x, \quad \text{for } 0 \leq x \leq \frac{\pi}{4}. \tag{5.5.17}$$

Let I_j $(j = 1, 2, 3)$ denote the integral of e^{-z^2} on the closed contour in Figure 5.45. Prove the following.

(b) $I_1 + I_2 + I_3 = 0$.

(c) $I_1 \to \int_0^\infty e^{-x^2} dx = \frac{\sqrt{\pi}}{2}$, as $R \to \infty$, and $I_2 \to 0$ as $R \to \infty$. [*Hint*: Use (5.5.17) to show that for $z = Re^{i\theta}$, $\left| e^{-z^2} \right| \leq e^{R^2(\frac{4}{\pi}\theta - 1)} = e^{-R^2} e^{\frac{4R^2}{\pi}\theta}$. So $|I_2| \leq Re^{-R^2} \int_0^{\frac{\pi}{4}} e^{\frac{4R^2}{\pi}\theta} d\theta = \frac{\pi}{4R}\left(1 - e^{-R^2}\right) \to 0$ as $R \to \infty$.]

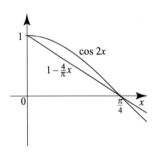

Fig. 5.44 Proof of (5.5.17).

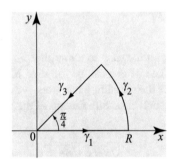

Fig. 5.45 Contour of integration.

(d) $I_3 \to -e^{i\frac{\pi}{4}} \left(\int_0^\infty \cos r^2 dr - i \int_0^\infty \sin r^2 dr \right)$, as $R \to \infty$. Note that this limit incorporates the Fresnel integrals where the variable of integration is r.

(e) Derive the Fresnel integrals using (b)–(d).

6. Using identities (5.5.5) in Example 5.5.2 derive the formulas

$$\int_0^\infty \frac{\cos x}{x^{1-\beta}} dx = \Gamma(\beta) \cos \frac{\beta\pi}{2} \quad \text{and} \quad \int_0^\infty \frac{\sin x}{x^{1-\beta}} dx = \Gamma(\beta) \sin \frac{\beta\pi}{2},$$

where $0 < \beta < 1$.

7. (a) Use the contour in Figure 5.46 to establish the identity

$$\int_0^\infty \frac{(\ln x)^2}{x^2 + 1} dx = \frac{\pi^3}{8}.$$

(b) Prove that

$$\int_0^1 \frac{(\ln x)^2}{x^2 + 1} dx = \int_1^\infty \frac{(\ln x)^2}{x^2 + 1} dx = \frac{\pi^3}{16}.$$

[*Hint*: Change of variables for the first equality, and (a) for the second one.]

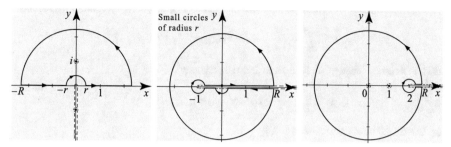

Fig. 5.46 Exercise 7.   **Fig. 5.47** Exercise 8.   **Fig. 5.48** Exercise 9.

8. Use the contour in Figure 5.47 to establish the identity

$$\int_0^\infty \frac{\ln(x+1)}{x^{1+\alpha}}\,dx = \frac{\pi}{\alpha \sin(\pi\alpha)}, \qquad (0 < \alpha < 1).$$

9. Use the contour in Figure 5.48 to establish the identity

$$\int_2^\infty \frac{dx}{x(x-1)\sqrt{x-2}} = \pi\left(1 - \frac{1}{\sqrt{2}}\right).$$

10. Establish the identity $\displaystyle \int_3^\infty \frac{dx}{\sqrt{x-3}\,(x-2)(x-1)} = \pi - \frac{\pi}{\sqrt{2}}.$

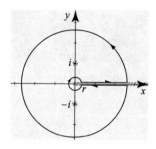

Fig. 5.49 Exercise 11.

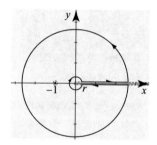

Fig. 5.50 Exercise 12.

11. Use the contour in Figure 5.49 to establish the identity

$$\int_0^\infty \frac{x^{\alpha-1}}{x^2+1}\,dx = \frac{\pi}{2\sin\left(\frac{\alpha\pi}{2}\right)} \qquad (0 < \alpha < 2).$$

12. Use the contour in Figure 5.50 to establish the identity

$$\int_0^\infty \frac{x^{\alpha-1}}{(x+1)^2}\,dx = \frac{(1-\alpha)\pi}{\sin(\alpha\pi)} \qquad (0 < \alpha < 2).$$

13. (a) Use the method of Example 5.5.4 and the contour in Figure 5.51 to establish the identity

$$\text{P.V.} \int_0^\infty \frac{x^p}{x(1-x)} \, dx = \pi \cot(p\pi) \qquad (0 < p < 1).$$

(b) Use a suitable change of variables to derive the identity

$$\text{P.V.} \int_{-\infty}^\infty \frac{e^{px}}{1-e^x} \, dx = \pi \cot(p\pi) \qquad (0 < p < 1).$$

(c) Use (b) to show that for $-1 < w < 1$

$$\text{P.V.} \int_{-\infty}^\infty \frac{e^{wx}}{\sinh x} \, dx = \pi \tan\left(\frac{\pi w}{2}\right).$$

(d) Use a suitable change of variables to show that

$$\text{P.V.} \int_{-\infty}^\infty \frac{e^{ax}}{\sinh(bx)} \, dx = \frac{\pi}{b} \tan\left(\frac{\pi a}{2b}\right) \qquad (b > |a|).$$

(e) Conclude that

$$\int_{-\infty}^\infty \frac{\sinh(ax)}{\sinh(bx)} \, dx = \frac{\pi}{b} \tan\left(\frac{\pi a}{2b}\right) \qquad (b > |a|).$$

Note that the integral is convergent so there is no need to use the principal value.

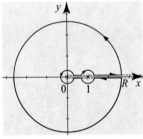

Fig. 5.51 Exercise 13.

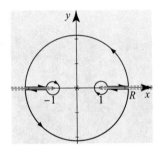

Fig. 5.52 Exercise 14.

14. Use the contour in Figure 5.52 to establish the identity

$$\int_1^\infty \frac{dx}{x\sqrt{x^2-1}} = \frac{\pi}{2}.$$

In Exercises 15–20, derive the identities.

15. $\displaystyle\int_0^\infty \frac{dx}{(x+2)\sqrt{x+1}} = \frac{\pi}{2}$

16. $\displaystyle\int_0^\infty \frac{dx}{(x+2)^2\sqrt{x+1}} = \frac{\pi}{4} - \frac{1}{2}$

17. $\displaystyle\int_0^\infty \frac{\sqrt{x}}{x^2+x+1} \, dx = \frac{\pi}{\sqrt{3}}$

18. $\displaystyle\int_0^\infty \frac{x^a}{x^2+1} \, dx = \frac{\pi}{2} \sec\left(\frac{a\pi}{2}\right) \quad (-1 < a < 1)$

19. $\displaystyle\int_0^\infty \frac{x}{\sinh x} \, dx = \frac{\pi^2}{4}$

20. $\displaystyle\int_0^\infty \frac{dx}{\cosh x} = \frac{\pi}{2}$

21. Integral of the Gaussian with complex parameters. We show that, for α and β complex with $\text{Re}\,\alpha > 0$,

$$\int_{-\infty}^{\infty} e^{-\alpha(t-\beta)^2} \, dt = \sqrt{\frac{\pi}{\alpha}}, \tag{5.5.18}$$

where on the right side we use the principal branch of the square root.

(a) Recognize the left side of (5.5.18) as the parametrized form of $\frac{1}{\sqrt{\alpha}} \int_{\gamma} e^{-z^2} \, dz$, where $\gamma(t) = \sqrt{\alpha}(t - \beta)$, the line at angle $\operatorname{Arg} \sqrt{\alpha}$ through $-\beta\sqrt{\alpha}$. Show $|\operatorname{Arg}\sqrt{\alpha}| < \frac{\pi}{4}$.

(b) Consider the contour in Figure 5.53. Argue that for fixed α and β, there exists $\varepsilon > 0$ such that for large enough R, γ_2 lies below the ray $\theta = \frac{\pi}{4} - \varepsilon$ in the right half-plane, and γ_4 lies above the ray $\theta = -\frac{3\pi}{4} - \varepsilon$ in the left half-plane.

(c) If I_j denotes the integral of e^{-z^2} on γ_j, show that $I_2 \to 0$ and $I_4 \to 0$ as $R \to \infty$.

(d) Use $\int_{-\infty}^{\infty} e^{-x^2} \, dx = \sqrt{\pi}$ to derive (5.5.18).

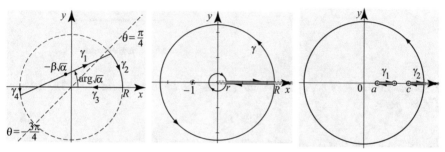

Fig. 5.53 Exercise 21.   **Fig. 5.54** Exercise 22.   **Fig. 5.55** Exercise 23.

22. Another look at Example 5.5.3. Derive (5.5.8) by considering the $\log_0$ branch of z^α for $\frac{1}{z^\alpha(z+1)}$. Use the contour in Figure 5.54, and treat this function to have different values on the upper and lower sides of the real axis, as in the discussion following Example 5.5.3.

23. Another look at Example 5.5.4. It turns out that the branch-cut singularity in $f(z) = \frac{1}{\sqrt{z-a}\sqrt{z-b}\sqrt{z-c}}$ is removable on the real interval (b, c). Here we prove this and rederive (5.5.15).

(a) Use Figure 5.43 to show that the limits of $f(z)$ as z approaches the interval (b, c) from the upper half-plane and from the lower half-plane are the same.

(b) Define $f(z)$ on (b, c) to be this common value. To show that f is analytic here, use Morera's theorem. [*Hint*: If a triangle crosses the real axis, subdivide it into a triangle and a quadrilateral, each with one side on the real axis. Approximate these by triangles and quadrilaterals that lie entirely in the upper or the lower half-plane over which the corresponding integrals vanish.]

(c) Rederive (5.5.15) using Figure 5.55 and the fact that the paths γ_1 and γ_2 are homotopic.

5.6 Summing Series by Residues

In this section, we use residue theory to sum infinite series of the form $\sum_k f(k)$, where $f = \frac{p}{q}$ is a rational function with degree $q \geq 2 + \text{degree } p$. For example,

$$\sum_{k=-\infty}^{\infty} \frac{1}{k^2+1}; \qquad \sum_{k=1}^{\infty} \frac{1}{k^6}; \qquad \sum_{k=-\infty}^{\infty} \frac{k^2}{k^4+1}.$$

Our starting point is the result of Example 5.1.8: if k is an integer and f is analytic at k, then

$$\mathrm{Res}\left(f(z)\cot(\pi z), k\right) = \frac{1}{\pi}f(k). \tag{5.6.1}$$

We integrate $f(z)\cot(\pi z)$ on squares centered at the origin. For a positive integer N we define Γ_N to be a positively oriented square with corners at $(N+\frac{1}{2})(1+i)$, $(N+\frac{1}{2})(-1+i)$, $(N+\frac{1}{2})(-1-i)$, $(N+\frac{1}{2})(1-i)$, as shown in Figure 5.56.

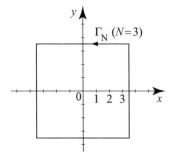

Fig. 5.56 Square contour Γ_N.

Lemma 5.6.1. *Let N be a positive integer and Γ_N be as in Figure 5.56. For all z on Γ_N, we have*

$$|\cot(\pi z)| \leq 2. \tag{5.6.2}$$

Moreover, if $f = \frac{p}{q}$ is a rational function with degree $q \geq 2 + $ degree p, then

$$\lim_{N\to\infty}\left|\int_{\Gamma_N}\frac{p(z)}{q(z)}\cot(\pi z)\,dz\right| = 0. \tag{5.6.3}$$

Proof. To prove (5.6.2), we deal with each side of the square Γ_N separately. On the right vertical side, $z = N + \frac{1}{2} + iy$, where $-N - \frac{1}{2} \leq y \leq N + \frac{1}{2}$. Using formulas (1.7.18) and (1.7.19) we obtain

$$|\cot(\pi z)| = \left|\frac{\cos(\pi z)}{\sin(\pi z)}\right| = \frac{\sqrt{\cos^2\left(\pi(N+\frac{1}{2})\right) + \sinh^2(\pi y)}}{\sqrt{\sin^2\left(\pi(N+\frac{1}{2})\right) + \sinh^2(\pi y)}}$$

$$= \frac{|\sinh(\pi y)|}{\sqrt{1 + \sinh^2(\pi y)}} \leq \frac{|\sinh(\pi y)|}{|\sinh(\pi y)|} = 1 \leq 2.$$

This establishes (5.6.2) for z on the right vertical side of Γ_N. For z on left vertical side the estimate follows immediately, since $\cot(\pi z)$ is an odd function. The horizontal sides are handled similarly (Exercise 19). To prove (5.6.3), we use the inequality for path integrals, Theorem 3.2.19, the fact that the perimeter of the square Γ_N is $4(2N+1)$, and (5.6.2) and obtain

$$\left|\int_{\Gamma_N}\frac{p(z)}{q(z)}\cot(\pi z)\,dz\right| \leq \ell(\Gamma_N)M_N = 4(2N+1)\,2M_N,$$

where M_N is the maximum value of $\left|\frac{p(z)}{q(z)}\right|$ for z on Γ_N. We have $4(2N+1)(2M_N) \to 0$ as $N \to \infty$, since degree $q \geq 2 + $ degree p. Thus (5.6.3) holds. ∎

The following result illustrates how identity (5.6.1) and Lemma 5.6.1 can be used to sum infinite series. Variations on this result are presented in the exercises. In what follows, all doubly infinite series are to be interpreted as the limit of symmetric partial sums; that is,

$$\sum_{k=-\infty}^{\infty} a_k = \lim_{N \to \infty} \sum_{k=-N}^{N} a_k,$$

whenever the limit exists.

Proposition 5.6.2. (Summing Infinite Series via Residues) *Suppose that $f = \frac{p}{q}$ is a rational function with degree $q \geq 2 + $ degree p. Suppose further that f has no poles at the integers. Then*

$$\sum_{k=-\infty}^{\infty} f(k) = -\pi \sum_{j} \operatorname{Res}\left(f(z)\cot(\pi z), z_j\right), \tag{5.6.4}$$

where the (finite) sum on the right runs over all the poles z_j of f.

Proof. The poles of $f(z)\cot(\pi z)$ occur at the integers where $\cot(\pi z)$ has poles and at the points z_j where f has poles. For large enough N, all z_j are on the inside of Γ_N. Applying Theorem 5.1.2 and (5.6.1), we obtain

$$\int_{\Gamma_N} f(z)\cot(\pi z)\, dz = 2\pi i \sum_{k=-N}^{N} \frac{1}{\pi} f(k) + 2\pi i \sum_{j} \operatorname{Res}\left(f(z)\cot(\pi z), z_j\right). \tag{5.6.5}$$

Letting $N \to \infty$, the left side of (5.6.5) goes to zero and we get (5.6.4). $\blacksquare$

Example 5.6.3. (Summing series by residues) Evaluate

$$\sum_{k=-\infty}^{\infty} \frac{1}{k^2 + 1}.$$

Solution. We apply (5.6.4) with $f(z) = \frac{1}{z^2+1}$ and find

$$\sum_{k=-\infty}^{\infty} \frac{1}{k^2 + 1} = -\pi \sum_{j} \operatorname{Res}\left(\frac{1}{z^2 + 1}\cot(\pi z), z_j\right),$$

where the sum on the right runs over the poles of $\frac{1}{z^2+1}$. The function $\frac{1}{z^2+1}$ has simple poles at $\pm i$ and

$$\operatorname{Res}\left(\frac{1}{z^2 + 1}\cot(\pi z), i\right) = \lim_{z \to i}(z - i)\frac{1}{(z-i)(z+i)}\cot(\pi z) = \frac{1}{2i}\cot(\pi i)$$

$$= \frac{1}{2i}\frac{\cos(\pi i)}{\sin(\pi i)} = \frac{1}{2i}\frac{\cosh(\pi)}{i\sinh(\pi)} = -\frac{1}{2}\coth\pi,$$

using (1.7.14) and (1.7.15), and similarly

$$\text{Res}\left(\frac{1}{z^2+1}\cot(\pi z), -i\right) = -\frac{1}{2i}\cot(-\pi i) = \frac{1}{2i}\cot(\pi i) = -\frac{1}{2}\coth \pi.$$

Thus,

$$\sum_{k=-\infty}^{\infty}\frac{1}{k^2+1} = -\pi\left(-\frac{1}{2}\coth \pi - \frac{1}{2}\coth \pi\right) = \pi\coth \pi. \qquad \square$$

Exercises 5.6

In Exercises 1–12, derive the identity using Proposition 5.6.2.

1. $\displaystyle\sum_{k=-\infty}^{\infty}\frac{1}{k^2+9} = \frac{\pi}{3}\coth(3\pi)$

2. $\displaystyle\sum_{k=-\infty}^{\infty}\frac{1}{(k^2+1)^2} = \frac{\pi}{2}\coth \pi + \frac{\pi^2}{2}\operatorname{csch}^2\pi$

3. $\displaystyle\sum_{k=-\infty}^{\infty}\frac{1}{k^2+a^2} = \frac{\pi}{a}\coth(a\pi)$ $(a \neq ik,\ k\text{ is an integer})$

4. $\displaystyle\sum_{k=-\infty}^{\infty}\frac{1}{(k^2+a^2)^2} = \frac{\pi}{2a^3}\coth(a\pi) + \frac{\pi^2}{2a^2}\operatorname{csch}^2(a\pi)$ $(a \neq ik,\ k\text{ is an integer})$

5. $\displaystyle\sum_{k=-\infty}^{\infty}\frac{1}{4k^2-1} = 0$

6. $\displaystyle\sum_{k=-\infty}^{\infty}\frac{k^2}{(k^2-\frac{1}{4})^2} = \frac{\pi^2}{2}$

7. $\displaystyle\sum_{k=-\infty}^{\infty}\frac{1}{(4k^2-1)^2} = \frac{\pi^2}{8}$

8. $\displaystyle\sum_{k=-\infty}^{\infty}\frac{1}{(k-\frac{1}{2})^2+1} = \pi\tanh \pi$

9. $\displaystyle\sum_{k=-\infty}^{\infty}\frac{1}{(k-2)(k-1)+1} = \frac{2\pi\tanh\left(\frac{\sqrt{3}\pi}{2}\right)}{\sqrt{3}}$

10. $\displaystyle\sum_{k=-\infty}^{\infty}\frac{k^2}{(k^2+1)^2} = \frac{\pi\operatorname{csch}^2\pi}{4}(\sinh(2\pi)-2\pi)$

11. $\displaystyle\sum_{k=-\infty}^{\infty}\frac{1}{k^4+4} = \frac{\pi\sinh(2\pi)}{4(\cosh(2\pi)-1)}$

12. $\displaystyle\sum_{k=-\infty}^{\infty}\frac{1}{k^4+4a^4} = \frac{\pi}{4a^3}\frac{\sinh(2a\pi)+\sin(2a\pi)}{\cosh(2a\pi)-\cos(2a\pi)}$ $(a>0)$.

13. Project Problem: Summation of series with a pole at 0. In Proposition 5.6.2 if the function f has a pole at 0, then (5.6.4) has to be modified to account for the residue at 0. In this case, we have the following useful result: Suppose that $f = \frac{p}{q}$ is a rational function with degree $q \geq 2 +$ degree p. Suppose further that f has no poles at the integers, except possibly at 0. Then

$$\sum_{\substack{k=-\infty \\ k\neq 0}}^{\infty} f(k) = -\pi \sum_j \operatorname{Res}\left(f(z)\cot(\pi z), z_j\right), \tag{5.6.6}$$

where the (finite) sum on the right runs over all the poles z_j of $f(z)$, including 0. Prove (5.6.6) by modifying the proof of Proposition 5.6.2; more specifically, explain what happens to (5.6.4) under the current conditions.

In Exercises 14–16, use (5.6.6) to derive the identity.

14. $\displaystyle\sum_{k=1}^{\infty} \frac{1}{k^2} = \frac{\pi^2}{6}$

15. $\displaystyle\sum_{k=1}^{\infty} \frac{1}{k^4} = \frac{\pi^4}{90}$

16. $\displaystyle\sum_{k=1}^{\infty} \frac{1}{k^2(k^2+4)} = \frac{3+4\pi^2 - 6\pi\coth(2\pi)}{96}$

17. Project Problem: Sums of the reciprocals of even powers of integers. In this exercise, we use (5.6.6) to derive

$$\sum_{k=1}^{\infty} \frac{1}{k^{2n}} = (-1)^{n-1}\frac{2^{2n-1}B_{2n}\pi^{2n}}{(2n)!}, \tag{5.6.7}$$

where n is a positive integer, B_{2n} is the Bernoulli number (Example 4.3.12). This remarkable identity sums the reciprocals of the even powers of the integers. There is no known finite expression corresponding to any powers.
(a) Show that if $f(z) = \frac{1}{z^{2n}}$ then (5.6.6) becomes

$$\sum_{\substack{k=-\infty \\ k\neq 0}}^{\infty} \frac{1}{k^{2n}} = -\pi\operatorname{Res}\left(\frac{\cot(\pi z)}{z^{2n}}, 0\right),$$

and so

$$\sum_{k=1}^{\infty} \frac{1}{k^{2n}} = -\frac{\pi}{2}\operatorname{Res}\left(\frac{\cot(\pi z)}{z^{2n}}, 0\right). \tag{5.6.8}$$

(b) Using the Taylor series expansion of $z\cot z$ from Exercise 31, Section 4.3, obtain

$$\operatorname{Res}\left(\frac{\cot(\pi z)}{z^{2n}}, 0\right) = (-1)^n\frac{2^{2n}B_{2n}\pi^{2n-1}}{(2n)!};$$

then derive (5.6.7).

18. Project Problem: Sums with alternating signs. (a) Modify the proof of Proposition 5.6.2 to prove the following summation result. Suppose that $f = \frac{p}{q}$ is a rational function with degree $q \geq 2 + $ degree p. Suppose further that f has no poles at the integers, except possibly at 0. Then,

$$\sum_{\substack{k=-\infty \\ k\neq 0}}^{\infty} (-1)^k f(k) = -\pi \sum_j \operatorname{Res}\left(f(z)\csc(\pi z), z_j\right), \tag{5.6.9}$$

where the (finite) sum on the right is taken over all the poles z_j of f, including 0. [*Hint:* You need a version of Lemma 5.6.1 for the cosecant.]
(b) Show that if $f(z) = \frac{1}{z^{2n}}$ then (5.6.9) becomes

$$\sum_{\substack{k=-\infty \\ k\neq 0}}^{\infty} \frac{(-1)^k}{k^{2n}} = -\pi\operatorname{Res}\left(\frac{\csc(\pi z)}{z^{2n}}, 0\right),$$

and so

$$\sum_{k=1}^{\infty} \frac{(-1)^k}{k^{2n}} = -\frac{\pi}{2} \operatorname{Res}\left(\frac{\csc(\pi z)}{z^{2n}}, 0\right). \tag{5.6.10}$$

(c) Using the Taylor series expansion of $z \csc z$ from Exercise 31, Section 4.3, obtain

$$\operatorname{Res}\left(\frac{\csc(\pi z)}{z^{2n}}, 0\right) = (-1)^{n-1} \frac{(2^{2n} - 2)B_{2n}\pi^{2n-1}}{(2n)!}.$$

(d) Show that for an integer $n \geq 1$ we have

$$\sum_{k=1}^{\infty} \frac{(-1)^k}{k^{2n}} = (-1)^n \frac{(2^{2n-1} - 1)B_{2n}\pi^{2n}}{(2n)!}.$$

19. The horizontal sides of Γ_N. Here we prove inequality (5.6.2) for the upper and lower horizontal sides of the square Γ_N.
(a) Argue that we need only consider the upper side, since $\cot(\pi z)$ is an odd function of z.
(b) For $z = x + i(N + \frac{1}{2})$, $-N - \frac{1}{2} \leq x \leq N + \frac{1}{2}$, justify

$$|\cot(\pi z)| = \frac{\sqrt{\cos^2(\pi x)\cosh^2\left(\pi(N+\frac{1}{2})\right) + \sin^2(\pi x)\sinh^2\left(\pi(N+\frac{1}{2})\right)}}{\sqrt{\sin^2(\pi x)\cosh^2\left(\pi(N+\frac{1}{2})\right) + \cos^2(\pi x)\sinh^2\left(\pi(N+\frac{1}{2})\right)}} \leq \coth\left[\pi(N+\frac{1}{2})\right].$$

Prove that this is less than or equal to 2, for all $N \geq 1$. This is not a best possible estimate, but it is sufficient to prove (5.6.3).

5.7 The Counting Theorem and Rouché's Theorem

In this section, we derive several properties of analytic functions. We start with a simple but useful lemma.

Lemma 5.7.1. *Suppose that f is analytic and not identically zero in a region Ω.*
(i) If z_0 is a zero of f of order $m \geq 1$, then $\frac{f'(z)}{f(z)}$ has a simple pole at z_0 and the residue there is m.
(ii) If z_0 is a pole of f of order $m \geq 1$, then $\frac{f'(z)}{f(z)}$ has a simple pole at z_0 and the residue there is $-m$.

Proof. (i) Write $f(z) = (z - z_0)^m \lambda(z)$, where λ is analytic and nonvanishing in a neighborhood $B_r(z_0)$, as in Theorem 4.5.2. For z in $B_r(z_0)$,

$$\frac{f'(z)}{f(z)} = \frac{m(z-z_0)^{m-1}\lambda(z) + (z-z_0)^m \lambda'(z)}{(z-z_0)^m \lambda(z)} = \frac{m}{z-z_0} + \frac{\lambda'(z)}{\lambda(z)}. \tag{5.7.1}$$

Then $\frac{\lambda'}{\lambda}$ is analytic in $B_r(z_0)$ since λ is nonvanishing in $B_r(z_0)$. From this it follows that z_0 is a simple pole of $\frac{f'}{f}$ and the residue there is m.
(ii) Write $f(z) = (z - z_0)^{-m} \lambda(z)$, where λ is analytic and nonvanishing in a neighborhood $B_r(z_0)$ (Theorem 4.5.15(ii)). Then for z in $B_r(z_0)$,

$$\frac{f'(z)}{f(z)} = \frac{-m(z-z_0)^{-m-1}\lambda(z) + (z-z_0)^{-m}\lambda'(z)}{(z-z_0)^{-m}\lambda(z)} = -\frac{m}{z-z_0} + \frac{\lambda'(z)}{\lambda(z)}, \quad (5.7.2)$$

where $\frac{\lambda'}{\lambda}$ is analytic in $B_r(z_0)$ since λ is nonvanishing in $B_r(z_0)$. Hence z_0 is a simple pole of $\frac{f'}{f}$ and the residue there is $-m$. ∎

The next result bears the name **the counting theorem** because it counts the number of zeros (according to multiplicity) of an analytic function inside a simple path. Counting zeros according to multiplicity means that each zero is counted as many times as its order.

For example, the polynomial

$$p(z) = (z-1)(z-i)^2(z+2i)$$

has zeros at $z_1 = 1$ (order 1), $z_2 = i$ (order 2), and $z_3 = -2i$ (order 1). The number of zeros, counting multiplicity, inside the circle $C_{\frac{3}{2}}(0)$ is 3 (Figure 5.57).

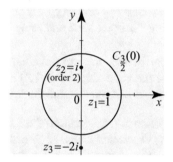

Fig. 5.57 The roots of p.

Theorem 5.7.2. (Counting Theorem) *Suppose that C is a simple closed positively oriented path, Ω is the region inside C, and f is analytic inside and on C and nonvanishing on C. Let $N(f)$ denote the number of zeros of f inside Ω, counted according to multiplicity. Then $N(f)$ is finite and*

$$N(f) = \frac{1}{2\pi i} \int_C \frac{f'(z)}{f(z)}\, dz. \quad (5.7.3)$$

Proof. If f has no zeros in Ω (hence $N(f) = 0$), then $\frac{f'}{f}$ is analytic on Ω and (5.7.3) follows from Cauchy's theorem. Note that f cannot have infinitely many zeros in Ω, by Corollary 4.5.7. Let $z_1, z_2, \ldots, z_n$ denote the zeros of f inside Ω. By Lemma 5.7.1, $\frac{f'}{f}$ has simple poles in Ω located at z_j and the residue at z_j is the order of the zero at z_j. Thus (5.7.3) follows at once from the residue theorem. ∎

A function is called **meromorphic** on a region if it is analytic in this region except at its poles. For example, $\frac{\sin z}{z^2+1}$ is meromorphic in the complex plane. Like zeros, we will count poles according to multiplicity. Next we generalize Theorem 5.7.2 to meromorphic functions.

Theorem 5.7.3. (Meromorphic Counting Theorem) *Suppose that C is a simple closed positively oriented path, Ω is the region inside C, and f is meromorphic on Ω and analytic and nonvanishing on C. Let $N(f)$ denote the number of zeros of f inside Ω and $P(f)$ denote the number of poles of f inside Ω, counted according to multiplicity. Then $N(f)$ and $P(f)$ are finite and*

$$N(f) - P(f) = \frac{1}{2\pi i} \int_C \frac{f'(z)}{f(z)} \, dz. \tag{5.7.4}$$

Proof. That $N(f)$ and $P(f)$ are finite follows as in the proof of Theorem 5.7.2. Applying Theorem 5.1.2 and using Lemma 5.7.1 we see that (5.7.4) holds.   ∎

Either of the preceding theorems is also known as the **argument principle** because the right sides of (5.7.3) and (5.7.4) can be interpreted as the change in argument as one runs around the image path $f[C]$. We now investigate this.

Proposition 5.7.4. (Branch of the Logarithm) *If f is analytic and nonvanishing on a simply connected region Ω, then there exists an analytic branch of the logarithm, $\log f = \ln|f| + i \arg f$, such that for all z in Ω we have*

$$\frac{d}{dz} \log f(z) = \frac{f'(z)}{f(z)}. \tag{5.7.5}$$

Proof. We know from Corollary 3.6.9(*ii*) that every analytic function on a simply connected region Ω has an antiderivative. In particular, if f is analytic and nonvanishing on Ω, then f'/f is analytic on Ω and thus has an antiderivative on Ω. We call this antiderivative a branch of the logarithm of f and write it as $\log f$. Indeed, the branch of the logarithm is of the form

$$\log f = \ln|f| + i \arg f,$$

where $\arg f$ is a continuous branch of the argument. Then

$$e^{\log f} = f \qquad \text{hence} \qquad (\log f)' e^{\log f} = f',$$

which implies (5.7.5).   ∎

The integrand in (5.7.4) suggests a connection with the logarithm of f, which we now explore. Let f and C be as in Theorem 5.7.3. Since $f[C]$ is a closed and bounded set that does not contain the origin, we can partition C into small subarcs γ_j $(j = 1, \ldots, n)$ such that each image $f[\gamma_j]$ is contained in a simply connected region that does not contain the origin, as shown in Figure 5.58.

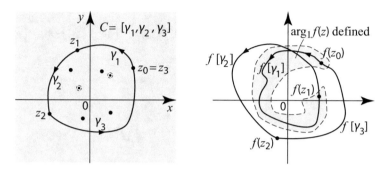

Fig. 5.58 The image $f[\gamma_1]$ is contained in a simply connected region not containing the origin. We can define the branch $\arg_1 z$ on this region.

Denote the initial and terminal points of γ_j by z_{j-1} and z_j, with $z_n = z_0$. For each $j = 1, \ldots, n$, we have a branch of the logarithm of f, $\log_j f$, which is an antiderivative of $\frac{f'}{f}$ on γ_j. For each j, we have $\log_j f = \ln|f| + i \arg_j f$, where $\arg_j$ is a branch of the argument, determined up to an additive constant multiple of 2π. These constants are determined in the following way. Starting with $j = 1$, pick and fix $\arg_1 f$. Then choose $\arg_2 f$ in such a way that $\arg_1 f(z_1) = \arg_2 f(z_1)$. Continue in this manner to determine the remaining branches of the argument. To simplify the notation, denote the resulting function by $\arg f$. Note that $\arg f$ is continuous on C, except at $z_0 = z_n$, and we will make the convention that $\arg f$ takes different values at this point depending on whether we approach the point from γ_1 or from γ_n. Thus $\arg f(z_n) = \arg_n f(z_n)$ and $\arg f(z_0) = \arg_1 f(z_0)$, and the difference is always an integer multiple of 2π that we denote by $\Delta_C \arg f = \arg_n f(z_n) - \arg_1 f(z_0)$. The quantity $\Delta_C \arg f$ measures the net change in the argument of $f(z)$ as z travels once around the curve C (Figure 5.59).

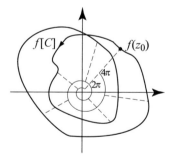

Fig. 5.59 The point $f(z)$ travels twice around the origin and $\Delta_C \arg f = 4\pi$.

We evaluate the integral on the right side of (5.7.3) or (5.7.4) with a telescoping sum:

$$\frac{1}{2\pi i}\int_C \frac{f'(z)}{f(z)}\,dz = \frac{1}{2\pi i}\sum_{j=1}^{n}\int_{\gamma_j}\frac{f'(z)}{f(z)}\,dz$$

$$= \frac{1}{2\pi i}\sum_{j=1}^{n}\ln|f(z)|+i\arg f(z)\Big|_{z_{j-1}}^{z_j}$$

$$= \frac{1}{2\pi i}\Big(\ln|f(z_n)|-\ln|f(z_0)|+i\big(\arg f(z_n)-\arg f(z_0)\big)\Big)$$

$$= \frac{1}{2\pi}\Delta_C\arg f.$$

Comparing with (5.7.4) we find that

$$\frac{1}{2\pi}\Delta_C\arg f = N(f)-P(f). \tag{5.7.6}$$

This formula links two somewhat unrelated quantities: the net change in the argument of f as we travel once around C, and the number of zeros and poles of f inside C. The formula has many interesting applications, as we now illustrate.

Example 5.7.5. (The argument principle) Find the number of zeros of the polynomial $z^6+6z+10$ in the first quadrant.

Solution. We apply the argument principle, which in this case is expressed by $\frac{1}{2\pi}\Delta_C\arg f = N(f)$ since polynomials have no poles. The contour C is a closed quarter circle of radius $R>0$, as shown in Figure 5.60.

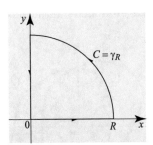

Fig. 5.60 The closed quarter circle C of radius $R>0$.

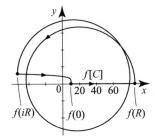

Fig. 5.61 The path $f[C]$ (pictured for $R=2$) goes around the origin once.

Since the given polynomial has finitely many zeros (at most six), if R is large enough, the contour C will contain all the zeros in the first quadrant. Our goal is to determine the change of the argument of $z^6+6z+10$ as z runs through C. We proceed in steps.

Step 1: Show that there are no roots on the contour C. We can always choose $R>0$ so that no roots lie on the circular arc γ_R. We need to prove that there are no roots on the nonnegative x-axis and y-axis. If $z=x\geq 0$, then $f(z)=x^6+6x+10$, and this is clearly positive. If $z = iy$, with $y \geq 0$, then $f(iy) = (iy)^6 + 6iy + 10 = (-y^6 + 10) + 6iy$ and because $\mathrm{Im}(f(iy)) = 0 \Leftrightarrow y = 0 \Rightarrow \mathrm{Re}\,(f(iy)) \neq 0$, we see that $f(iy) \neq 0$.

Step 2: Compute the change of the argument of $f(z)$ as z varies from the initial to the terminal point of $I_1 = [0, R]$. For $z = 0$, $f(z) = 10$, and for $0 \leq z = x \leq R$, $f(z) = x^6 + 6x + 10 > 0$. So the image of the interval $[0, R]$ is the interval $[10, R^6 + 6R + 10]$ and the argument of $f(z)$ does not change on I_1.

Step 3: Compute the change of the argument of $f(z)$ as z varies from the initial to the terminal point of the arc γ_R. Here we are not looking for the exact image of γ_R by f, but only a rough picture that gives us the change in the argument of f. For very large R and z on γ_R, write $z = Re^{i\theta}$, where $0 \leq \theta \leq \frac{\pi}{2}$. Then

$$f(z) = R^6 e^{6i\theta} \left(1 + \frac{6}{R^5} e^{-i5\theta} + \frac{10}{R^6} e^{-6i\theta}\right) \approx R^6 e^{6i\theta},$$

because $\frac{6}{R^5} e^{-i5\theta} + \frac{10}{R^6} e^{-6i\theta} \approx 0$. So as θ varies from 0 to $\frac{\pi}{2}$, the argument of $f(Re^{i\theta})$ varies from 0 to $6 \cdot \frac{\pi}{2} = 3\pi$. In fact, the point $f(iR) = -R^6 + 10 + 6iR$ lies in the second quadrant and has argument very close to 3π. See Figure 5.61.

Step 4: Compute the change of the argument of $f(z)$ as z varies from iR to 0. As z varies from iR to 0, $f(z)$ varies from $w_3 = -R^6 + 10 + 6iR$ to $w_0 = 10$. Since $\mathrm{Im} f(z) \geq 0$, this tells us that the point $f(z)$ remains in the upper half-plane as $f(z)$ moves from w_3 to w_0. Hence the change in the argument of $f(z)$ is $-\pi$.

Step 5: Apply the argument principle. The net change of the argument of $f(z)$ as we travel once around C is $3\pi - \pi = 2\pi$. According to (5.7.6), the number of zeros of f inside C, and hence in the first quadrant, is $\frac{1}{2\pi} 2\pi = 1$. $\qquad\square$

We give one more version of the counting theorem.

Theorem 5.7.6. (Variant of the Counting Theorem) *Let C, Ω, and f be as in Theorem 5.7.3, let g be analytic on an open set that contains C and its interior. Let $z_1, z_2, \ldots, z_{n_1}$ denote the zeros of f in Ω and $p_1, p_2, \ldots, p_{n_2}$ denote the poles of f in Ω. Let $m(z_j)$ be the order of the root z_j of f and $m(p_j)$ denote the order of the pole p_j of f. Then*

$$\frac{1}{2\pi i} \int_C g(z) \frac{f'(z)}{f(z)} \, dz = \sum_{j=1}^{n_1} m(z_j) g(z_j) - \sum_{j=1}^{n_2} m(p_j) g(p_j). \qquad (5.7.7)$$

Proof. We modify the proof of the previous theorem as follows. If z_j is a zero of f, then, since g is analytic and $\frac{f'}{f}$ has a simple pole at z_j, using Lemma 5.7.1 and Proposition 5.1.3 we obtain

$$\mathrm{Res}\left(g\frac{f'}{f}, z_j\right) = \lim_{z \to z_j}(z - z_j)g(z)\frac{f'(z)}{f(z)} = g(z_j) \lim_{z \to z_j}(z - z_j)\frac{f'(z)}{f(z)}$$

$$= g(z_j)\,\mathrm{Res}\left(\frac{f'}{f}, z_j\right) = g(z_j)\,m(z_j).$$

Similarly for the poles p_j,

$$\mathrm{Res}\left(g\,\frac{f'}{f},p_j\right)=g(p_j)\,\mathrm{Res}\left(\frac{f'(z)}{f(z)},p_j\right)=-m(p_j)\,g(p_j).$$

Now (5.7.7) follows from the residue theorem. $\blacksquare$

Example 5.7.7. Evaluate

$$\int_{C_1(0)}\frac{e^z\cos z}{e^z-1}\,dz,$$

where $C_1(0)$ is the positively oriented unit circle.

Solution. The function $f(z)=e^z-1$ has a zero at $z=0$ and, because $f'(0)=1\neq0$, this zero is simple. Also, using the $2\pi i$-periodicity of e^z, it is easy to see that $e^z=1$ inside the unit disk $|z|<1$ only at $z=0$. Applying (5.7.7) with $f(z)=e^z-1$ and $g(z)=\cos z$ it follows immediately that

$$\int_{C_1(0)}\frac{e^z\cos z}{e^z-1}\,dz=2\pi i\cos0=2\pi i.\qquad\square$$

As an application of the counting principle we derive a famous result known as Rouché's theorem, named after the French mathematician and educator Eugène Rouché (1832–1910). We need the following lemma.

Lemma 5.7.8. *Let ϕ be a continuous function on a region Ω that takes only integer values. Then ϕ is constant in Ω.*

Proof. Assume that ϕ is not constant, and let z_1 and z_2 in Ω be such that $\phi(z_1)=n_1<\phi(z_2)=n_2$. Let r be a real number such that $n_1<r<n_2$. Since ϕ is continuous, the sets

$$A=\{z\in\Omega:\ \phi(z)<r\}$$
$$B=\{z\in\Omega:\ \phi(z)>r\}$$

are open. Also, $z_1\in A$ and $z_2\in B$, hence A and B are nonemtpy. They are also disjoint and satisfy $A\cup B=\Omega$. This contradicts the fact that Ω is connected. Thus ϕ is constant. $\blacksquare$

In what follows, we need a version of Lemma 5.7.8, where ϕ is a continuous, integer-valued function on an interval $[a,b]$. The preceding proof can be easily modified to cover this case.

Theorem 5.7.9. (Rouché's Theorem) *Suppose that C is a simple closed path, Ω is the region inside C, f and g are analytic inside and on C. If $|g(z)|<|f(z)|$ for all z on C, then*

$$N(f+g)=N(f)$$

in other words, $f+g$ and f have the same number of zeros on Ω.

Proof. For z on C, the inequality $|g(z)| < |f(z)|$ implies that $f(z) \neq 0$, and so $|g|/|f|$ is continuous and strictly less than 1 on C. Since C is a closed and bounded set, $|g|/|f|$ attains its maximum $\delta < 1$ on C. Pick $\delta < \delta_1 < 1$. For z on C and $0 < |\lambda| \leq 1/\delta_1$, if $f(z) + \lambda g(z) = 0$, then $|g(z)|/|f(z)| = 1/|\lambda| \geq \delta_1$, which contradicts that $|g|/|f| \leq \delta$ on C. So $f(z) + \lambda g(z)$ is not equal to zero for all z on C and $|\lambda| \leq 1/\delta_1$. We claim that the function

$$\phi(\lambda) = \frac{1}{2\pi i} \int_C \frac{f'(z) + \lambda g'(z)}{f(z) + \lambda g(z)} \, dz \tag{5.7.8}$$

is continuous in λ on $[-1/\delta_1, 1/\delta_1]$. To verify this, notice that the integrand $G(\lambda, z)$ in (5.7.8) is continuous and thus uniformly continuous on $[0, 1/\delta_1] \times C$. Consequently, $G(\lambda, z)$ is close to $G(\lambda', z)$ uniformly in z and thus the integral of $G(\lambda, z)$ in z is close to that of $G(\lambda', z)$ in z. By Theorem 5.7.2, $\phi(\lambda) = N(f + \lambda g)$, and so ϕ is integer-valued. By Lemma 5.7.8, ϕ is constant for all $|\lambda| < 1/\delta$. In particular, $\phi(0) = N(f) = \phi(1) = N(f+g)$. ∎

We can give a geometric interpretation of Rouché's theorem, based on the argument principle. According to (5.7.6), we are merely claiming that

$$\Delta_C \arg(f+g) = \Delta_C \arg f. \tag{5.7.9}$$

As z traces C, $f(z)$ winds around the origin a specific number of times. Since $|g(z)| < |f(z)|$, the point $f(z) + g(z)$ must lie in the disk of radius $|f(z)|$ centered at $f(z)$ (Figure 5.62), and so $f(z) + g(z)$ must wind around the origin the same number of times as $f(z)$ (see Exercise 25).

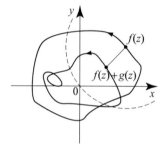

Fig. 5.62 Suppose that for each z on C there is a rope, shorter than the distance from $f(z)$ to the origin, joining $f(z)$ to $f(z) + g(z)$. Then $f(z) + g(z)$ goes around the origin the same number of times as $f(z)$ goes around the origin, when z traces the curve C.

The following are typical applications of Rouché's theorem.

Example 5.7.10. (Counting zeros with Rouché's theorem)
(a) Show that all zeros of $p(z) = z^4 + 6z + 3$ lie inside the circle $|z| = 2$.
(b) Show that if a is a real number with $a > e$, then in $|z| < 1$ the equation $e^z = az^n$ has n roots (counting orders).

Solution. (a) Since p is a polynomial of degree 4, it is enough to show that $N(p) = 4$ inside the circle $|z| = 2$. Take $f(z) = z^4$ and $g(z) = 6z + 3$ and note that $f(z) + g(z) = p(z)$. For $|z| = 2$, we have $|f(z)| = 2^4 = 32$ and $|g(z)| \leq |6z| + 3 = 15$. Hence $|g(z)| < |f(z)|$ for all z on the circle $|z| = 2$, and so by Rouché's theorem $N(f) = N(f+g) = N(p)$. Clearly f has one zero with multiplicity 4 at $z = 0$. Thus $N(f) = 4$ and so $N(p) = 4$, as desired.

(b) Take $f(z) = az^n$ and $g(z) = -e^z$. Counted according to multiplicity, f has n zeros in $|z| < 1$ and so $N(f) = n$. For $|z| = 1$, $|f(z)| = |az^n| = a$ and

$$|g(z)| = |e^z| = \left|e^{\cos\theta + i\sin\theta}\right| = e^{\cos\theta} \leq e < a.$$

Thus $|g(z)| < |f(z)|$ for all $|z| = 1$ and so by Rouché's theorem $n = N(az^n) = N(az^n - e^z)$, which implies that $az^n - e^z = 0$ has n roots in $|z| < 1$.   □

As a further application we give a very simple proof of the fundamental theorem of algebra.

Example 5.7.11. (The fundamental theorem of algebra) Let

$$p(z) = a_n z^n + a_{n-1} z^{n-1} + \cdots + a_1 z + a_0$$

with $a_n \neq 0$ be a polynomial of degree $n \geq 1$. Show that p has exactly n roots, counting multiplicity.

Solution. Take $f(z) = a_n z^n$. Then f has a zero of multiplicity n at $z = 0$. Also, for $|z| = R$, we have $|f(z)| = |a_n| R^n$, which is a polynomial of degree n in R. Now let $g(z) = a_{n-1} z^{n-1} + \cdots + a_1 z + a_0$, then $|g(z)| \leq |a_{n-1}| |z^{n-1}| + \cdots + |a_1| |z| + |a_0|$, and so for $|z| = R$, $|g(z)| \leq |a_{n-1}| R^{n-1} + \cdots + |a_1| R + |a_0|$. Since the modulus of f grows at a faster rate than the modulus of g, in the sense that

$$\lim_{R \to \infty} \frac{1}{|a_n| R^n} \left(|a_{n-1}| R^{n-1} + \cdots + |a_1| R + |a_0|\right) = 0,$$

we can find R_0 large enough so that for $R \geq R_0$, we have

$$|a_{n-1}| R^{n-1} + \cdots + |a_1| R + |a_0| < |a_n| R^n.$$

This implies that $|g(z)| < |f(z)|$ for all $|z| = R$ with $R \geq R_0$. By Rouché's theorem, $N(f)$, the number of zeros of f in the region $|z| < R$, is the same as $N(f + g)$, the number of zeros of $f + g$. But $N(f) = n$ and $f + g = p$, so $N(p) = n$ showing that p has exactly n zeros.   □

The Local Mapping Theorem

In this part, we investigate fundamental properties of the inverse function of analytic functions When do they exist? Are they analytic? Are there explicit formulas for them? All these questions can be answered with the help of Rouché's theorem and the counting theorem. Our investigation leads to interesting new properties of analytic functions and shed new light on some classical results studied earlier. In particular, we give a simple proof of the maximum modulus principle. Interesting applications of these topics are presented in the exercises, including a formula due to Lagrange for the inverse of analytic functions with some of its applications to the solution to transcendental equations.

Suppose that f is analytic at z_0 and let $w_0 = f(z_0)$. We say that f has **order** $m \geq 1$ at z_0 if the zero of $f(z) - w_0$ has order m at z_0. The order of f at z_0 is the order of the first nonvanishing term in the Taylor series expansion, past the constant term. Thus f has order m at z_0 if and only if $f^{(j)}(z_0) = 0$ for $j = 1, \ldots, m-1$ and $f^{(m)}(z_0) \neq 0$. In particular, f has order 1 at z_0 if and only if $f'(z_0) \neq 0$.

Theorem 5.7.12. (Local Mapping Theorem) *Let f be a nonconstant analytic function in a neighborhood of z_0. Let $w_0 = f(z_0)$ and n be the order of f at z_0. Then there exist $R > 0$ and $\rho > 0$ such that for every $w \in B_\rho(w_0) \setminus \{w_0\}$ there are precisely n distinct points in $B_R(z_0)$ whose image under f is w.*

Proof. Since $f - w_0$ has a zero of order $n \geq 1$, write

$$f(z) - w_0 = a_n(z - z_0)^n + a_{n+1}(z - z_0)^{n+1} + \cdots,$$

where $a_n \neq 0$. Then $\lim_{z \to z_0} \frac{f(z) - w_0}{(z - z_0)^n} = a_n \neq 0$. So we can find $R > 0$ such that

$$\left| \frac{f(z) - w_0}{(z - z_0)^n} \right| > \frac{|a_n|}{2}, \quad \text{for all } 0 < |z - z_0| \leq R. \tag{5.7.10}$$

In particular, for $z \neq z_0$ we have $f(z) \neq w_0$, and for $|z - z_0| = R$ we have

$$\frac{|f(z) - w_0|}{R^n} > \frac{|a_n|}{2} \quad \Rightarrow \quad |f(z) - w_0| > \frac{|a_n| R^n}{2}. \tag{5.7.11}$$

By taking a smaller value of R if necessary, we can also assume that $f'(z) \neq 0$ for all $z \neq z_0$ with $|z - z_0| < R$. Now take $\rho = \frac{|a_n| R^n}{2}$, then for $|w_0 - w| < \rho$, set $g(z)$ to be the constant $w_0 - w$. For $|z - z_0| = R$, using (5.7.11), we find

$$|g(z)| = |w_0 - w| \leq \frac{|a_n| R^n}{2} < |f(z) - w_0|,$$

and so by Rouché's theorem, $N(f - w) = N(f - w_0 + g) = N(f - w_0)$, and since $N(f - w_0) = n$ it follows that $N(f - w) = n$ for all $|w - w_0| < \rho$. So w has n antecedents in $|z - z_0| < R$. As $f'(z) \neq 0$ for all z satisfying $0 < |z - z_0| < R$, we do not have repeated roots, and so the antecedents are all distinct. ∎

The case $n = 1$ of the theorem deserves a separate statement.

Theorem 5.7.13. (Inverse Function Theorem) *Suppose that f is analytic on a region Ω and z_0 is in Ω. Then f is one-to-one on some neighborhood $B_r(z_0)$ if and only if $f'(z_0) \neq 0$.*

Proof. If $f'(z_0) \neq 0$, f has order 1 and by Theorem 5.7.12 we can find R and ρ such that for $0 < |w - w_0| < \rho$, f takes on the value w exactly once in $B_R(z_0)$. Also, f takes on the value w_0 only at z_0. Let $U = f^{-1}[B_\rho(w_0)]$ be the pre-image of

$B_\rho(w_0)$ under f. Since $B_\rho(w_0)$ is open and f is continuous, U is open (Exercise 41, Section 2.2) and it clearly contains z_0. So we can find an open disk $B_r(z_0)$ that is contained in U and $B_R(z_0)$. Then f is one-to-one on $B_r(z_0)$, for if $f(z_1) = f(z_2)$ is a point in $B_\rho(w_0)$, Theorem 5.7.12 guarantees $z_1 = z_2$. Conversely, if $f'(z_0) = 0$, then the order of f at z_0 is at least 2. Applying Theorem 5.7.12 we find a neighborhood of z_0 on which f is at least two-to-one; so f is not one-to-one in this case. ∎

A few comments are in order regarding the inverse function theorem. The theorem can be obtained from the classical inverse function theorem in two variables. The latter states that the mapping $(x, y) \mapsto (u(x, y), v(x, y))$ is one-to-one in a neighborhood of (x_0, y_0) if its Jacobian is nonzero at (x_0, y_0); recall that the Jacobian of this mapping at (x, y) is

$$J(x, y) = \det \begin{vmatrix} u_x(x, y) & u_y(x, y) \\ v_x(x, y) & v_y(x, y) \end{vmatrix} = u_x(x, y)v_y(x, y) - u_y(x, y)v_x(x, y).$$

Hence, using the Cauchy-Riemann equations, we find that

$$J(x, y) = u_x^2(x, y) + u_y^2(x, y) = |f'(x+iy)|^2.$$

So if $f'(x_0 + iy_0) \neq 0$, then $J(x_0, y_0) \neq 0$ and Theorem 5.7.13 follows from the inverse function theorem for functions of two variables as claimed.

It is important to keep in mind that the condition $f'(z) \neq 0$ for all z in Ω does not imply that f is one-to-one on Ω. Consider $f(z) = e^z$; then $f'(z) = e^z \neq 0$ for all z, yet f is not one-to-one on the whole complex plane. Theorem 5.7.13 only guarantees that f is one-to-one in *some* neighborhood of a point z_0 where $f'(z_0) \neq 0$. Because this neighborhood depends on z_0 in general, an obvious question is whether we can estimate its size in terms of the sizes of f and f'. An answer to this question was provided by the German mathematician Edmund Landau (1877–1938) and is known as **Landau's estimate**. See Exercise 36.

The following two corollaries describe important properties of analytic functions, which are direct applications of the local mapping theorem.

Corollary 5.7.14. (Open Mapping Property) *Let Ω be a region. Then for a nonconstant analytic function f on Ω and an open subset U of Ω the set $f[U]$ is open. A function with this property is said to be* **open**.

Proof. Let U be an open subset of Ω. Given w_0 in $f[U]$, let z_0 be some point in U where $f(z_0) = w_0$. Since U is open, we can find a neighborhood $B_R(z_0)$ contained in Ω; by applying Theorem 5.7.12 to $B_R(z_0)$, we find that each point in the associated $B_\rho(w_0)$ is assumed by f. Thus $B_\rho(w_0)$ is contained in $f[U]$, and $f[U]$ is open. ∎

Corollary 5.7.15. (Mapping of Regions) *If Ω is a nonempty region (open and connected set) and f is a nonconstant analytic function on Ω, then $f[\Omega]$ is a region.*

Proof. By Corollary 5.7.14, $f[\Omega]$ is open. To show that $f[\Omega]$ is connected, suppose that $f[\Omega] = U \cup V$, where U and V are open and disjoint. Since f is continuous,

$f^{-1}[U]$ and $f^{-1}[V]$ are open (note that since f is defined on Ω, when taking a pre-image, we only consider those points in Ω). Clearly, $f^{-1}[U]$ and $f^{-1}[V]$ are disjoint and their union is Ω. Since Ω is connected, either $\Omega = f^{-1}[U]$ or $\Omega = f^{-1}[V]$, and hence $f[\Omega] = U$ or $f[\Omega] = V$, implying that $f[\Omega]$ is connected. ∎

Another interesting application of the open mapping property of analytic functions is a simple proof of the maximum principle.

Corollary 5.7.16. (Maximum Modulus Principle) *If f is analytic on a region Ω, such that $|f|$ attains a maximum at some point in Ω, then f is constant in Ω.*

Proof. Suppose f is nonconstant. Let z_0 be an arbitrary point in Ω; we will show that $|f(z_0)|$ is not a maximum. By Corollary 5.7.14, f is open. Then for a small open ball $B_\varepsilon(z_0)$ the set $f[B_\varepsilon(z_0)]$ is open and thus it contains a neighborhood $B_\rho(w_0)$ of $w_0 = f(z_0)$. Clearly there exists w in $B_\rho(w_0)$ such that $|w| > |w_0|$. See Figure 5.63. Then $w = f(z)$ for some $z \in B_\varepsilon(z_0)$ and $|f(z)| > |f(z_0)|$. Consequently $f(z_0)$ does not have the largest modulus among all the points in $B_\rho(w_0)$ and thus $|f(z_0)|$ cannot be a maximum of $|f|$. ∎

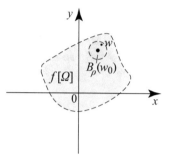

Fig. 5.63 $B_\rho(w_0) \subset f[\Omega]$.

We next use the variant of the counting theorem to give a formula for the inverse function of f in terms of an integral involving f.

Theorem 5.7.17. (Inverse Function Formula) *Suppose that f is analytic in a neighborhood of a point z_0 and $f'(z_0) \neq 0$. Let $w_0 = f(z_0)$. Then there are $R, \rho > 0$ such that $f^{-1}[B_\rho(w_0)] \subset B_R(z_0)$, f is one-to-one on $f^{-1}[B_\rho(w_0)]$ and has an inverse function $g : B_\rho(w_0) \to f^{-1}[B_\rho(w_0)]$ which is is analytic on $B_\rho(w_0)$ and satisfies*

$$g(w) = \frac{1}{2\pi i} \int_{C_R(z_0)} z \frac{f'(z)}{f(z) - w} dz, \qquad w \in B_\rho(w_0). \qquad (5.7.12)$$

Proof. Let R and ρ be as in Theorem 5.7.12. For w in $B_\rho(w_0)$, the function $f - w$ has exactly one zero in $B_R(z_0)$, which we call $g(w)$. Applying identity (5.7.7) (with $n_1 = 1$, $z_1 = g(w)$, $m(z_1) = 1$, $n_2 = 0$) we see that identity (5.7.12) holds. Then g is continuous on $B_\rho(w_0)$, as for $w \to w_1$ we can pass the limit inside the integral in (5.7.12) to obtain $g(w) \to g(w_1)$. Applying Theorem 2.3.12 [with $h = f \circ g$ being the identity map on $B_\rho(w_0)$] we obtain that g is analytic on $B_\rho(w_0)$. ∎

Theorem 5.7.17 can be viewed as a statement about the local existence of inverse functions. If f is one-to-one on a region Ω, there is no ambiguity in defining the inverse function f^{-1} on the whole region $f[\Omega]$. Indeed, for w in $f[\Omega]$, we define $f^{-1}(w)$ to be the unique z in Ω with $f(z) = w$. For f analytic and one-to-one, we have the following result, which should be compared to Theorem 2.3.12. Theorem 5.7.17 can be used to derive a useful formula due to Lagrange for the inversion of power series. See Exercise 33.

A function f that is analytic and one-to-one is called a **univalent** function. The next corollary says that univalent functions have global inverses.

Corollary 5.7.18. (A Global Inverse Function) *Suppose that f is univalent function on a region Ω. Then its inverse function f^{-1} exists and is analytic on the region $f[\Omega]$. Moreover,*

$$\frac{d}{dw} f^{-1}(w) = \frac{1}{f'(z)}, \quad where \; w = f(z). \tag{5.7.13}$$

Proof. By Theorem 5.7.17, f^{-1} is analytic in a neighborhood of each point in Ω, and hence it is analytic on Ω. Also, since f is one-to-one, Theorem 5.7.13 implies that $f'(z) \neq 0$ for all z in Ω. Differentiating both sides of the identity $z = f^{-1}(f(z))$, we obtain $1 = \frac{d}{dw} f^{-1}(w) f'(z)$, which is equivalent to (5.7.13). ∎

Exercises 5.7

In Exercises 1–6, use the method of Example 5.7.5 to find the number of zeros in the first quadrant of the following polynomials.

1. $z^2 + 2z + 2$ **2.** $z^2 - 2z + 2$

3. $z^3 - 2z + 4$ **4.** $z^3 + 5z^2 + 8z + 6$

5. $z^4 + 8z^2 + 16z + 20$ **6.** $z^5 + z^4 + 13z^3 + 10$

In Exercises 7–14, use Rouché's theorem to determine the number of zeros of the functions in the indicated region.

7. $z^3 + 3z + 1$, $|z| < 1$ **8.** $z^4 + 4z^3 + 2z^2 - 7$, $|z| < 2$

9. $7z^3 + 3z^2 + 11$, $|z| < 1$ **10.** $7z^3 + z^2 + 11z + 1$, $1 < |z|$

11. $4z^6 + 41z^4 + 46z^2 + 9$, $2 < |z| < 4$ **12.** $z^4 + 50z^2 + 49$, $3 < |z| < 4$

13. $e^z - 3z$, $|z| < 1$ **14.** $e^{z^2} - 4z^2$, $|z| < 1$

15. Show that the equation

$$3 - z + 2e^{-z} = 0$$

has exactly one root in the right half-plane $\operatorname{Re} z > 0$. [*Hint:* Use Rouché's theorem and contours such as the one in the adjacent figure.]

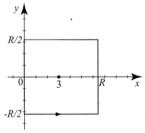

16. Suppose that $\operatorname{Re} w > 0$ and let a be a complex number. Show that $w - z + ae^{-z} = 0$ has exactly one root in the right half-plane $\operatorname{Re} z > 0$.

In Exercises 17–20, evaluate the path integrals. As usual, $C_R(z_0)$ stands for the positively oriented circle with radius $R > 0$ centered at z_0.

17. $\displaystyle \int_{C_1(0)} \frac{dz}{z^5 + 3z + 5}$ **18.** $\displaystyle \int_{C_1(0)} \frac{e^z - 12z^3}{e^z - 3z^4} \, dz$

19. $\displaystyle\int_{C_2(0)} \frac{ze^{iz}}{z^2+1}\, dz$

20. $\displaystyle\int_{C_1(0)} \frac{z^3 e^{z^2}}{e^{z^2}-1}\, dz$

21. Summing roots of unity. We use the variant of the counting theorem to show that for $n \geq 2$, the sum of the n nth roots of unity is 0 (see Exercise 63, Section 1.3). Let S denote this sum.
(a) Using Theorem 5.7.6, explain why

$$S = \frac{1}{2\pi i}\int_{C_R(0)} z \frac{nz^{n-1}}{z^n-1}\, dz = \frac{n}{2\pi i}\int_{C_{1/R}(0)} \frac{1}{1-z^n}\frac{dz}{z^2},$$

where $R > 1$. [*Hint*: To prove the second equality make a suitable change of variables.]
(b) Evaluate the second integral in part (a) using Cauchy's generalized integral formula and conclude that $S = 0$.

22. Examples concerning Lemma 5.7.1. Give an example of a function f with an essential singularity at 0 such that $\frac{f'}{f}$ has
(a) an essential singularity at 0;
(b) a pole of order $m \geq 2$ at 0.
[*Hint*: Use suitable compositions of the function $e^{1/z}$ in your examples.]

23. Minimum modulus principle. Show that for a nonzero nonconstant analytic function f on a region Ω, $|f|$ does not attain a minimum in Ω. [*Hint*: Use that the function f is open.]

24. Meromorphic Rouché's theorem. Suppose that C is a simple closed path, Ω is the region inside C, and f and g are meromorphic inside and on C, having no zeros or poles on C. Show that if $|g(z)| < |f(z)|$ for all z on C, then $N(f+g) - P(f+g) = N(f) - P(f)$. [*Hint*: Repeat the proof of Rouché's theorem. What can you say about the values of ϕ in the present case?]

25. Complete argument that provides a geometric proof of Theorem 5.7.9 (Rouché's theorem).
(i) Show that for each $z \in C$, we can find a branch of the argument where

$$\arg f(z) - \frac{\pi}{2} < \arg(f(z)+g(z)) < \arg f(z) + \frac{\pi}{2}.$$

(ii) Using connectedness, we can show that this inequality holds for the specific argument function $\arg f$ used to define $\Delta_C \arg f$, see page 353. Show that

$$\Delta_C \arg f - \pi < \Delta_C \arg(f+g) < \Delta_C \arg f + \pi,$$

and use the fact that $\Delta_C \arg(f+g)$ must be an integer multiple of 2π to prove that $\Delta_C \arg f = \Delta_C \arg(f+g)$ on C.

26. Project Problem: Hurwitz's theorem. We outline a proof of a useful theorem due to the German mathematician Adolf Hurwitz (1859–1919). The theorem states the following: Suppose that $\{f_n\}_{n=1}^{\infty}$ is a sequence of analytic functions on a region Ω converging uniformly on every closed and bounded subset of Ω to a function f. Then either
(i) f is identically 0 on Ω; or
(ii) if $B_r(z_0)$ is an open disk in Ω such that f does not vanish on $C_r(z_0)$, then f_n and f have the same number of zeros in $B_r(z_0)$ for all sufficiently large n. In particular, if f is not identically 0 and f has p distinct zeros in Ω, then so do the functions f_n for all sufficiently large n.
   Observe that f is analytic by Theorem 4.1.10. Also, note that the theorem guarantees that, for large n, f_n and f have the same number of zeros, but these zeros are not necessarily the same for f_n and f. To see this, take $f_n(z) = z - \frac{1}{n}$ and $f(z) = z$ for all $z \in \Omega$. Finally, observe that the possibility that f is identically zero can arise, even if the f_n's are all nonzero. Simply take $f_n = \frac{1}{n}$.
   Fill in the details in the following proof. Suppose that f is not identically 0 in Ω. Let $\overline{B_r(z_0)}$ be a closed disk such that f is nonvanishing on $C_r(z_0)$. Let $m = \min|f|$ on $C_r(z_0)$. Then $m > 0$ (why?). Apply uniform convergence to get an index N such that $n > N$ implies that $|f_n - f| < m \leq |f|$ on $C_r(z_0)$. Complete the proof by applying Rouché's theorem.

Hurwitz's theorem has many interesting applications. We start with some theoretical properties and then give some applications to counting zeros of analytic functions (Exercises 28–30).

27. Suppose that f_n converges to f uniformly on every closed and bounded subset of a region Ω with f_n analytic and vanishing nowhere on Ω. Then either f is identically 0 or f has no zero on Ω.

28. Univalent functions. (a) Suppose that $\{f_n\}_{n=1}^{\infty}$ is a sequence of univalent functions on a region Ω that converges to f uniformly on every closed and bounded subset of Ω, and f is not identically constant on Ω. Then f is univalent. [*Hint*: Fix z_0 in Ω and apply Exercise 26 to the sequence of functions $\{f_n - f_n(z_0)\}_{n=1}^{\infty}$ defined on $\Omega \setminus \{z_0\}$.]
(b) Give an example of a sequence of univalent functions converging uniformly on the closed unit disk to a constant function.

29. Counting zeros with Hurwitz's theorem. If we want to find the number of zeros inside the unit disk of the polynomial $p(z) = z^5 + z^4 + 6z^2 + 3z + 1$, then we cannot just apply Rouché's theorem since there is not one single coefficient of the polynomial whose absolute value dominates the sum of the absolute values of the other coefficients. Here is how we can handle this problem:
(a) Consider the polynomials $p_n = p - \frac{1}{n}$. Show that p_n converges to p uniformly on the closed unit disk.
(b) Apply Rouché's theorem to show that p_n has two zeros in the unit disk.
(c) Apply Hurwitz's theorem to show that p has two zeros inside the unit disk.

In Exercises 30–31, *modify the steps in Exercise* 29 *to find the zeros of the polynomials in the indicated region.*

30. $z^5 + z^4 + 6z^2 + 3z + 11$, $|z| < 1$.   **31.** $4z^4 + 6z^2 + z + 1$, $|z| < 1$.

32. Project Problem: Lagrange's inversion formula. In this exercise, we outline a proof of the following useful inversion formula for analytic functions due to the French mathematician Joseph-Louis Lagrange (1736–1813).

Suppose that f is analytic at z_0, $w_0 = f(z_0)$, and $f'(z_0) \neq 0$, and let

$$\phi(z) = \frac{z - z_0}{f(z) - w_0}, \quad z \neq z_0. \tag{5.7.14}$$

Show that the inverse function $z = g(w)$ has a power series expansion

$$g(w) = z_0 + \sum_{n=1}^{\infty} b_n(w - w_0)^n, \quad \text{where} \quad b_n = \frac{1}{n!} \frac{d^{n-1}}{dz^{n-1}} [\phi(z)]^n \bigg|_{z=z_0}. \tag{5.7.15}$$

Fill in the details in the following proof. Note that ϕ is analytic at z_0 (why?). To prove (5.7.15), start with the formula for the inverse function (5.7.12) and differentiate with respect to w using Theorem 3.8.5 and then integrate by parts to obtain

$$g'(w) = \frac{1}{2\pi i} \int_{C_R(z_0)} z \frac{f'(z)}{(f(z) - w)^2} \, dz$$

$$= -\frac{1}{2\pi i} \int_{C_R(z_0)} z \, d((f(z) - w)^{-1}) = \frac{1}{2\pi i} \int_{C_R(z_0)} \frac{dz}{f(z) - w}.$$

Differentiate under the integral sign $n - 1$ more times and evaluate at $w = w_0$ to obtain

$$g^{(n)}(w_0) = \frac{(n-1)!}{2\pi i} \int_{C_R(z_0)} \frac{dz}{(f(z) - w_0)^n} = \frac{(n-1)!}{2\pi i} \int_{C_R(z_0)} \phi(z)^n \frac{dz}{(z - z_0)^n}.$$

Hence by Cauchy's generalized integral formula

$$g^{(n)}(w_0) = \frac{d^{n-1}}{dz^{n-1}} \phi(z)^n \bigg|_{z=z_0},$$

and (5.7.15) follows from the formula for the Taylor coefficients.

33. Let a be an arbitrary complex number. Consider the equation $z = a + we^z$. Show that a solution of this equation is

$$z = a + \sum_{n=1}^{\infty} \frac{n^{n-1} e^{na}}{n!} w^n,$$

when $|w| < e^{-1 - \text{Re}\, a}$. [*Hint*: Let $z_0 = a$, $w = (z - a)e^{-z}$, $\phi(z) = \frac{z-a}{(z-a)e^{-z}} = e^z$, and apply Lagrange's inversion formula (5.7.15).]

34. Lambert's w-function. This function has been applied in quantum physics, fluid mechanics, biochemistry, and combinatorics. It is named after the German mathematician Johann Heinrich Lambert (1728–1777). The **Lambert function** or **Lambert w-function** is defined as the inverse function of $f(z) = ze^z$. Using the technique of Exercise 34, based on Lagrange's formula, show that the solution of $w = ze^z$ is

$$z = \sum_{n=1}^{\infty} \frac{(-1)^{n-1} n^{n-2}}{(n-1)!} w^n \quad \text{whenever } |w| < \frac{1}{e}.$$

35. Project Problem: Landau's estimate. In this exercise we present Landau's solution of the following problem: Given an analytic function f in a neighborhood of a closed disk $\overline{B_R(z_0)}$ with $f'(z_0) \neq 0$, find $r > 0$ such that f is one-to-one on the open disk $B_r(z_0)$. Landau's solution: It suffices to take $r = R^2 |f'(z_0)|/(4M)$, where M is the maximum value of $|f|$ on $C_R(z_0)$.

Fill in the details in the following argument. It suffices to choose r so that $f'(z) \neq 0$ for all z in $B_r(z_0)$ (why?). Without loss of generality we can take $z_0 = 0$ and $w_0 = f(z_0) = 0$ (why?). Then for $|z| < R$, $f(z) = a_1 z + a_2 z^2 + \cdots$. Write $r = \lambda R$, where $0 < \lambda < 1$ is to be determined so that $f'(z) \neq 0$ for all z in $B_{\lambda R}(0)$. For z_1 and z_2 in $B_{\lambda R}(0)$, we have

$$\left| \frac{f(z_1) - f(z_2)}{z_1 - z_2} \right| = \left| a_1 + \sum_{n=2}^{\infty} a_n \left(z_1^{n-1} + z_1^{n-2} z_2 + \cdots + z_1 z_2^{n-2} + z_2^{n-1} \right) \right| \geq |a_1| - \sum_{n=2}^{\infty} n |a_n| \lambda^{n-1} R^{n-1}.$$

If we could choose a number λ such that

$$\sum_{n=2}^{\infty} n |a_n| \lambda^{n-1} R^{n-1} < |a_1|, \tag{5.7.16}$$

this would make the absolute value of the difference quotient for the derivative positive, independently of z_1 and z_2 in $B_{\lambda R}(0)$, and this would in turn imply that $f'(z) \neq 0$ for all z in $B_{\lambda R}(0)$ and complete the proof. So let us show that we can choose λ so that (5.7.16) holds. Let $M = \max_{|z|=R} |f(z)|$. Cauchy's estimate yields $|a_n| \leq \frac{M}{R^n}$. Then

$$\sum_{n=2}^{\infty} n |a_n| \lambda^{n-1} R^{n-1} \leq \frac{M}{R} \sum_{n=2}^{\infty} n \lambda^{n-1} = \frac{M}{R} \frac{\lambda(2 - \lambda)}{(1 - \lambda)^2} < \frac{M}{R} \frac{2\lambda}{(1 - \lambda)^2}.$$

(Use $\sum_{n=2}^{\infty} n \lambda^{n-1} = \frac{d}{d\lambda}(\lambda^2 + \lambda^3 + \cdots) = \frac{d}{d\lambda} \frac{\lambda^2}{1 - \lambda} = \frac{2\lambda - \lambda^2}{(1 - \lambda)^2}$.) Consider the choice $\lambda = \frac{R|a_1|}{4M}$. This yields $\lambda \leq \frac{1}{4}$ and $\sum_{n=2}^{\infty} n |a_n| \lambda^{n-1} R^{n-1} < \frac{|a_1|}{4\lambda} \frac{2\lambda}{(1 - \lambda)^2} = \frac{|a_1|}{2} \frac{1}{(1 - \lambda)^2} \leq \frac{8}{9} |a_1|$. (The maximum of $1/(1 - \lambda)^2$ on the interval $[0, 1/4]$ occurs at $\lambda = 1/4$ and is equal to $\frac{16}{9}$.) Hence (5.7.16) holds for this choice of λ, and for this choice, we get $r = \lambda R = \frac{R^2 |a_1|}{4M}$, which is what Landau's estimate says.

Chapter 6
Harmonic Functions and Applications

> *The mathematician's patterns, like the painter's or the poet's must be beautiful; the ideas, like the colors or the words must fit together in a harmonious way.*
>
> -Godfrey Harold Hardy (1877–1947)

There are many important applications of complex analysis to real-world problems. The ones studied in this chapter are related to the fundamental differential equation

$$\Delta u = \frac{\partial^2 u}{\partial x^2} + \frac{\partial^2 u}{\partial y^2} = 0,$$

known as **Laplace's equation**. This partial differential equation models phenomena in engineering and physics, such as steady-state temperature distributions, electrostatic potentials, and fluid flow, just to name a few. A real-valued function that satisfies Laplace's equation is said to be harmonic. There is an intimate relationship between harmonic and analytic functions. This is investigated in Section 6.1 along with other fundamental properties of harmonic functions.

To illustrate an application, consider a two-dimensional plate of homogeneous material, with insulated lateral surfaces. We represent this plate by a region Ω in the complex plane (see Figure 6.1). Suppose that the temperature of the points on the boundary of the plate is described by the function $b(x, y)$ that does not change with time. It is a fact of thermodynamics that the temperature inside the plate will eventually reach and remain at an equilibrium distribution $u(x, y)$, known as the **steady-state temperature distribution**, which satisfies the equation $\Delta u = 0$.

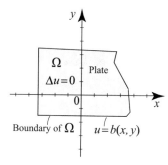

Fig. 6.1 The steady-state temperature distribution of a plate satisfies Laplace's equation.

The Laplacian is named after the great French mathematician and physicist Pierre-Simon de Laplace (1749–1827). This operator appeared for the first time in a memoir of Laplace in 1784, in which he completely determined the attraction of a spheroid on the points outside it. The Laplacian of a function measures the difference between the value of the function at a point and the average value of the function in a neighborhood of that point. Thus a function that does not vary abruptly has a very small Laplacian. Harmonic functions have a zero Laplacian; they vary in a very regular way. Examples of such functions include the temperature distribution

© Springer International Publishing AG, part of Springer Nature 2018
N. H. Asmar and L. Grafakos, *Complex Analysis with Applications*,
Undergraduate Texts in Mathematics, https://doi.org/10.1007/978-3-319-94063-2_6

in a plate, the potential of the attractive force due to a sphere, or the function that gives the brightness of colors in an image.

6.1 Harmonic Functions

Definition 6.1.1. A real-valued function u defined on an open subset Ω of the complex plane is called **harmonic** if it has continuous partial derivatives of first and second order in Ω and satisfies **Laplace's equation**

$$\Delta u = \frac{\partial^2 u}{\partial x^2} + \frac{\partial^2 u}{\partial y^2} = 0 \qquad \text{for all } (x,y) \text{ in } \Omega. \tag{6.1.1}$$

Examples of harmonic functions exist in abundance. In fact the real and imaginary parts of analytic functions are harmonic. Let us verify this assertion. Consider an analytic function $f = u + iv$ on an open set Ω. We know that u and v have derivatives of all orders (Corollary 3.8.9) and satisfy the Cauchy-Riemann equations (Theorem 2.5.1):

$$\frac{\partial u}{\partial x} = \frac{\partial v}{\partial y} \qquad \text{and} \qquad \frac{\partial u}{\partial y} = -\frac{\partial v}{\partial x}. \tag{6.1.2}$$

Since u and v have partial derivatives of all orders, we can interchange the order of partial derivatives. Using the Cauchy-Riemann equations (6.1.2) we can write

$$\frac{\partial^2 u}{\partial x^2} = \frac{\partial}{\partial x}\left(\frac{\partial u}{\partial x}\right) = \frac{\partial}{\partial x}\left(\frac{\partial v}{\partial y}\right) = \frac{\partial}{\partial y}\left(\frac{\partial v}{\partial x}\right) = \frac{\partial}{\partial y}\left(-\frac{\partial u}{\partial y}\right) = -\frac{\partial^2 u}{\partial y^2},$$

$$\frac{\partial^2 v}{\partial x^2} = \frac{\partial}{\partial x}\left(\frac{\partial v}{\partial x}\right) = \frac{\partial}{\partial x}\left(-\frac{\partial u}{\partial y}\right) = -\frac{\partial}{\partial y}\left(\frac{\partial u}{\partial x}\right) = -\frac{\partial}{\partial y}\left(\frac{\partial v}{\partial y}\right) = -\frac{\partial^2 v}{\partial y^2},$$

and hence

$$\frac{\partial^2 u}{\partial x^2} + \frac{\partial^2 u}{\partial y^2} = \frac{\partial^2 v}{\partial x^2} + \frac{\partial^2 v}{\partial y^2} = 0$$

for all (x,y) in Ω. In other words, u and v satisfy Laplace's equation. We have therefore proved the following important result.

Theorem 6.1.2. *The real and imaginary parts of analytic functions defined on open sets are harmonic.*

This result effectively provides us with many examples of harmonic functions; simply consider the real or imaginary parts of analytic functions.

Example 6.1.3. (Harmonic functions) Show that the following functions are harmonic in the stated regions.
(a) $u(x, y) = x^2 - y^2$ on $\mathbb{C}$.
(b) $u(x, y) = e^x \sin y$ on $\mathbb{C}$.

(c) $u(x, y) = \text{Arg}\, z$ $(z = x + iy)$ on the region $\Omega = \mathbb{C} \setminus (-\infty, 0]$.

(d) $u(x, y) = \ln |z| = \ln \sqrt{x^2 + y^2}$ on the region $\mathbb{C} \setminus \{0\}$.

Solution. (a) It is not hard to see that u is harmonic here by direct verification of Laplace's equation, but we recognize that $u(x, y) = x^2 - y^2$ is the real part of the entire function $f(z) = z^2 = (x^2 - y^2) + 2ixy$. Thus from Theorem 6.1.2 we conclude that u is harmonic on $\mathbb{C}$.

(b) Recognizing $u(x, y) = e^x \sin y$ as the imaginary part of the entire function e^z (Corollary 1.6.4), we conclude from Theorem 6.1.2 that u is harmonic on $\mathbb{C}$.

(c) We have $\text{Arg}\, z = \text{Im}\,(\text{Log}\, z)$, where $\text{Log}\, z = \ln |z| + i\, \text{Arg}\, z$ is the principal branch of the logarithm defined in (1.8.4). Since $\text{Log}\, z$ is analytic on Ω, we conclude that $\text{Arg}\, z$ is harmonic on Ω, by Theorem 6.1.2.

(d) We need two different analytic functions to establish the claim. Arguing as in (c), it follows that $\ln |z| = \ln \sqrt{x^2 + y^2}$ is harmonic on $\Omega = \mathbb{C} \setminus (-\infty, 0]$, being the real part of $\text{Log}\, z$, which is analytic on $\mathbb{C} \setminus (-\infty, 0]$. This establishes that $\ln |z|$ is harmonic on the desired region except on the negative x-axis, $(-\infty, 0)$. To show that $\ln |z|$ is harmonic on $(-\infty, 0)$, we use $\log_\alpha z$, which is a branch of $\log z$ with branch cut at angle α, with $-\pi < \alpha < \pi$.

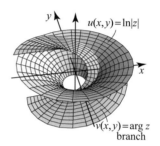

Fig. 6.2 The function $\ln |z|$.

It follows from (1.8.6) that $\log_\alpha z = \ln |z| + \arg_\alpha z$ and $\log_\alpha z$ is analytic except at the branch cut and, in particular, is analytic on the negative x-axis. Hence, its real part, $\ln |z|$, is harmonic on the negative x-axis. See Figure 6.2.

We may also answer part (d) directly by verifying that $\ln \sqrt{x^2 + y^2} = \frac{1}{2} \ln(x^2 + y^2)$ satisfies Laplace's equation and the first and second partial derivatives are continuous everywhere except at the origin. □

Proposition 6.1.4. *Suppose that u and v are harmonic on an open set S, and a, b are real constants. Then the function $au + bv$ is harmonic on S.*

Proof. The proof is left to the reader. ■

Example 6.1.5. (a) The function $\phi(x, y) = 2(x^2 - y^2) + 7$ is harmonic on $\mathbb{C}$, being the linear combination of the harmonic functions $x^2 - y^2$ and 7.

(b) The function $\phi(x, y) = (ax + b)(cy + d)$, where a, b, c, d are real constants, is harmonic, being a linear combination of the harmonic functions x, y, xy, and constants.

(c) The product of two harmonic functions need not be harmonic. For instance, the function $u(x, y) = x$ is harmonic but $(u(x, y))^2 = x^2$ is not. □

We have seen that analytic functions have derivatives of all orders (Corollary 3.8.9). The same is true for harmonic functions. The proof of this result is facilitated via a definition and a lemma.

Definition 6.1.6. For a harmonic function u on Ω, the function $\phi = u_x - i u_y$ defined on Ω is called the **conjugate gradient** of u.

Lemma 6.1.7. (Analyticity of the Conjugate Gradient) *Suppose that u is harmonic on a region Ω. Then its conjugate gradient is analytic on Ω.*

Proof. Write $\phi = \text{Re}(\phi) + i\,\text{Im}(\phi) = u_x - i u_y$. Since u has continuous second partial derivatives, it follows that $\text{Re}\,\phi$ and $\text{Im}\,\phi$ have continuous first partial derivatives. To show that ϕ is analytic, in view of Corollary 2.5.2, it suffices to show that $\text{Re}\,\phi$ and $\text{Im}\,\phi$ satisfy the Cauchy-Riemann equations. Indeed, we have

$$\frac{\partial}{\partial x}\text{Re}\,\phi = \frac{\partial}{\partial x}u_x = u_{xx} \quad \text{and} \quad \frac{\partial}{\partial y}\text{Im}\,\phi = \frac{\partial}{\partial y}(-u_y) = -u_{yy}.$$

But as u is harmonic, $u_{xx} = -u_{yy}$, and so $\frac{\partial}{\partial x}\text{Re}\,\phi = \frac{\partial}{\partial y}\text{Im}\,\phi$. Thus, the first of the Cauchy-Riemann equations is satisfied. Now, since u has continuous second partial derivatives, we have $u_{xy} = u_{yx}$. Thus

$$\frac{\partial}{\partial y}\text{Re}\,\phi = u_{xy} \quad \text{and} \quad \frac{\partial}{\partial x}\text{Im}\,\phi = -u_{yx} = -u_{xy}.$$

So $\frac{\partial}{\partial y}\text{Re}\,\phi = -\frac{\partial}{\partial x}\text{Im}\,\phi$ and the second of the Cauchy-Riemann equations holds. Hence ϕ is analytic on Ω.   ∎

Example 6.1.8. (Conjugate gradient)
Consider the function

$$u(x,y) = \frac{x}{x^2 + y^2}$$

defined on the upper half-plane $y > 0$.
(a) Show that u is harmonic.
(b) Find the conjugate gradient of u.

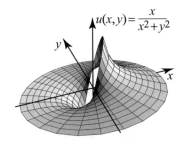

Fig. 6.3 The plot of $u(x,y)$.

Solution. (a) The function u has a simpler expression in polar coordinates:

$$u(r,\theta) = \frac{r\cos\theta}{r^2} = \frac{\cos\theta}{r} = r^{-1}\cos\theta.$$

From this, it is easy to see that u is the real part of the function

$$f(z) = \frac{1}{z} = z^{-1} = r^{-1}e^{-i\theta} = r^{-1}(\cos\theta - i\sin\theta).$$

Since f is analytic for all $z \neq 0$, it follows that its real part $u = r^{-1}\cos\theta$ is harmonic for all $z \neq 0$ by Theorem 6.1.2. In particular, u is harmonic in the upper half-plane.

(b) We have

$$u_x = \frac{y^2 - x^2}{(x^2 + y^2)^2} \quad \text{and} \quad u_y = \frac{-2xy}{(x^2 + y^2)^2}.$$

Thus the conjugate gradient in the upper half-plane is

$$\phi = u_x - i u_y = \frac{(y^2 - x^2) + 2ixy}{(x^2 + y^2)^2} = -\frac{(x - iy)^2}{(x^2 + y^2)^2} = -\frac{(\bar{z})^2}{(z\bar{z})^2} = -\frac{1}{z^2}. \qquad \square$$

Suppose that $f = u + iv$ is analytic on a region Ω. From the Cauchy-Riemann equations, $f'(z) = u_x - i u_y$, and thus f' is the conjugate gradient of u; equivalently, f is an antiderivative of the conjugate gradient of u. This fact allows us to use the conjugate gradient to construct f when only $\operatorname{Re} f = u$ is known. In Example 6.1.8, $u(x, y) = \frac{x}{x^2 + y^2} = \operatorname{Re}(f(z))$ where $f(z) = \frac{1}{z} = \frac{1}{x + iy}$ and the conjugate gradient of u is $\phi(z) = -\frac{1}{z^2} = f'(z)$.

Using the conjugate gradient $\phi = u_x - i u_y$, we can express partial derivatives of u as the real or imaginary part of an analytic function, namely a derivative of ϕ. For example, to obtain u_{xxx}, differentiate ϕ with respect to x twice and get $\phi'' = u_{xxx} - i u_{yxx}$. (Note that since the derivatives of ϕ exist, we can obtain them by differentiating with respect to any one of the variables x, y, or z.) Since ϕ is analytic, all its higher-order derivatives are analytic (Corollary 3.8.9), and so u_{xxx} and u_{yxx} are both harmonic by Theorem 6.1.2. We can carry these ideas further and arrive at the following useful result.

Corollary 6.1.9. *A harmonic function defined on a region has continuous partial derivatives of all orders.*

We now discuss a result concerning the composition of a harmonic function with an analytic one.

Theorem 6.1.10. *Let f be analytic on a region Ω, and let w be a function with two continuous partial derivatives defined on a region that contains $f[\Omega]$. Then the identity holds:*

$$\Delta(w \circ f) = |f'|^2 \Delta w. \tag{6.1.3}$$

Moreover, if w is harmonic, then $w \circ f$ is harmonic on Ω.

Proof. We write $f(x + iy) = u(x + iy) + iv(x + iy)$. Since f is analytic, Theorem 6.1.2 gives that u and v are harmonic, and thus they satisfy $u_{xx} + u_{yy} = 0$, $v_{xx} + v_{yy} = 0$ on their domains. We have

$$(w \circ f)_x = w_u u_x + w_v v_x$$
$$(w \circ f)_y = w_u u_y + w_v v_y$$

and another differentiation using Leibniz's rule yields:

$$(w \circ f)_{xx} = w_{uu}(u_x)^2 + w_{uv} u_x v_x + w_u u_{xx} + w_{vv}(v_x)^2 + w_{vu} v_x u_x + w_v v_{xx}$$
$$(w \circ f)_{yy} = w_{uu}(u_y)^2 + w_{uv} u_y v_y + w_u u_{yy} + w_{vv}(v_y)^2 + w_{vu} v_y u_y + w_v v_{yy}.$$

Adding these equations and using that $u_{xx} + v_{yy} = 0$, $v_{xx} + v_{yy} = 0$ and that $u_x v_x + u_y v_y = 0$ (which is a consequence of the Cauchy-Riemann identity (2.5.7) in Theorem 2.5.1), we deduce that

$$
\begin{aligned}
(w \circ f)_{xx} + (w \circ f)_{yy} &= w_{uu}(u_x)^2 + w_{uu}(u_y)^2 + w_{vv}(v_x)^2 + w_{vv}(v_y)^2 \\
&= w_{uu}(u_x)^2 + w_{uu}(v_x)^2 + w_{vv}(v_x)^2 + w_{vv}(u_x)^2 \\
&= \left(w_{uu} + w_{vv} \right) \left((u_x)^2 + (v_x)^2 \right)
\end{aligned}
$$

and so (6.1.3) follows. If w is harmonic, then $\Delta w = 0$ and it follows from (6.1.3) that $\Delta(w \circ f) = 0$, and thus $w \circ f$ is harmonic if w is harmonic. ∎

Harmonic Conjugates

Definition 6.1.11. Suppose that u and v are harmonic functions that satisfy the Cauchy-Riemann equations on some open set Ω; in other words, the function $f = u + iv$ is analytic in Ω. Then v is called the **harmonic conjugate** of u.

Can we always find a harmonic conjugate of a harmonic function u? As it turns out, the answer depends on the function u and its domain of definition. For example, the function $\ln|z|$ is harmonic in $\Omega = \mathbb{C} \setminus \{0\}$ (Example 6.1.3(d)), but $\ln|z|$ has no harmonic conjugate in that region (Exercise 34). It does, however, have a harmonic conjugate in $\mathbb{C} \setminus (-\infty, 0]$, namely $\operatorname{Arg} z$. Our next example shows one way of using the Cauchy-Riemann equations to find the harmonic conjugate in a region such as the entire complex plane, a disk, or a rectangle.

Example 6.1.12. (Finding harmonic conjugates) Show that $u(x, y) = x^2 - y^2 + x$ is harmonic in the entire plane and find a harmonic conjugate for it.

Solution. That u is harmonic follows from $u_{xx} = 2$ and $u_{yy} = -2$. To find a harmonic conjugate v, we use the Cauchy-Riemann equations as follows. We want $u + iv$ to be analytic. Hence v must satisfy the Cauchy-Riemann equations

$$
\frac{\partial u}{\partial x} = \frac{\partial v}{\partial y}, \quad \text{and} \quad \frac{\partial u}{\partial y} = -\frac{\partial v}{\partial x}. \tag{6.1.4}
$$

Since $\frac{\partial u}{\partial x} = 2x + 1$, the first equation implies that

$$
2x + 1 = \frac{\partial v}{\partial y}.
$$

To get v we integrate both sides of this equation with respect to y. However, since v is a function of x and y, the constant of integration may be a function of x. Thus integrating with respect to y yields

$$
v(x, y) = (2x + 1)y + c(x),
$$

where $c(x)$ is a function of x alone. Plugging this into the second equation in (6.1.4), we get

$$-2y = -\left(2y + \frac{d}{dx}c(x)\right),$$

and hence $c(x)$ has zero derivative and must be a constant. We pick such constant and write $c(x) = C$. Substituting into the expression for v we get

$$v(x, y) = (2x+1)y + c(x) = 2xy + y + C.$$

The pair of functions u and v satisfies the Cauchy-Riemann equations, and thus the function $(x^2 - y^2 + x) + i(2xy + y + C)$ is entire. □

So, from Example 6.1.12, a harmonic conjugate of $u(x, y) = x^2 - y^2 + x$ is $v(x, y) = 2xy + y$. What is a harmonic conjugate of $v(x, y) = 2xy + y$? Surely it is related to u. Indeed, you can check that a conjugate of v is $-u(x, y) = -x^2 + y^2 - x$. More generally, we have the following useful result.

Proposition 6.1.13. *Suppose that u is harmonic on an open set Ω and that v is a harmonic conjugate of u on Ω. Then $-u$ is a harmonic conjugate of v on Ω.*

Proof. We know that $f = u + iv$ is analytic on Ω. It follows that the function $(-i)f = v - iu$ is analytic on Ω, and hence $-u$ is a harmonic conjugate of v on Ω. ■

In Example 6.1.12, we found the harmonic conjugate of u up to an arbitrary additive real constant. In fact, the following properties of the harmonic conjugate are not hard to prove and are left to the reader.

Proposition 6.1.14. *Suppose that u is a harmonic function in a region Ω. Then*
(i) if v_1 and v_2 are harmonic conjugates of u in Ω, then v_1 and v_2 must differ by a real constant.
(ii) If v is a harmonic conjugate of u, then v is also a harmonic conjugate of $u + c$ where c is a real constant.

Remark 6.1.15. Notice that in the hypothesis of Proposition 6.1.14 we are assuming that Ω is a connected set. Otherwise the constant c in assertion (ii) may vary on each different connected component of Ω.

Our next result guarantees the existence of harmonic conjugates on simply connected regions.

Theorem 6.1.16. (Existence of Harmonic Conjugates) *Suppose that u is a harmonic function on a simply connected region Ω. Let z_0 be a fixed point in Ω and $\gamma(z_0, z)$ be a path that joins z_0 to an arbitrary point z in Ω. Then the function*

$$v(z) = \int_{\gamma(z_0,z)} -u_y\,dx + u_x\,dy \tag{6.1.5}$$

is a harmonic conjugate of u in Ω; the integral is independent of the path.

Proof. Consider the analytic conjugate gradient $\phi = u_x - iu_y$. The integral of ϕ is independent of path in Ω, in view of Corollary 3.6.9. Define

$$f(z) = \int_{\gamma(z_0,z)} \phi(\zeta)\,d\zeta.$$

Then f is analytic and $f' = \phi$. Write $\zeta = x + iy$, $d\zeta = dx + i\,dy$. Then

$$f(z) = \int_{\gamma(z_0,z)} u_x\,dx + u_y\,dy + i\overbrace{\int_{\gamma(z_0,z)} -u_y\,dx + u_x\,dy}^{v(z)}.$$

We claim that

$$\int_{\gamma(z_0,z)} u_x\,dx + u_y\,dy = u(z) - u(z_0).$$

From this it will follow that $v(z) = \int_{\gamma(z_0,z)} -u_y\,dx + u_x\,dy$ is a harmonic conjugate of $u(z)$, since $(u - u(z_0)) + iv = f$ is analytic as the additive constant $u(z_0)$ does not affect analyticity. To prove the claim, parametrize the path from z_0 to z by $\zeta(t) = x(t) + iy(t)$, $a \le t \le b$. Then

$$u_x\,dx + u_y\,dy = \left(u_x\frac{dx}{dt} + u_y\frac{dy}{dt}\right)dt = \frac{d}{dt}u(\zeta(t))\,dt,$$

by the chain rule in two dimensions. Hence

$$\int_{\gamma(z_0,z)} u_x\,dx + u_y\,dy = \int_a^b \frac{d}{dt}u(\zeta(t))\,dt = u(\zeta(t))\Big|_a^b = u(z) - u(z_0),$$

as claimed. ∎

Suppose that u is harmonic on a region Ω, and let z_0 be in Ω. Since Ω is open, we can find an open disk $B_R(z_0)$ in Ω. Since $B_R(z_0)$ is simply connected, u has a harmonic conjugate in $B_R(z_0)$. This means that Theorem 6.1.16 holds *locally* in Ω (Figure 6.4). This is a useful fact that we record in the following corollary.

Corollary 6.1.17. *A harmonic function defined on a region has a harmonic conjugate locally.*

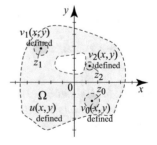

Fig. 6.4 Local existence of the harmonic conjugate.

The function u and its conjugate v have a very interesting geometric relationship based on the notion of orthogonal curves.

Suppose that two curves C_1 and C_2 meet at a point A. The curves are said to be **orthogonal** if their respective tangent lines L_1 and L_2 (at A) are orthogonal (Figure 6.5). We also say that C_1 and C_2 intersect at a right angle at A. Recall that if m_1 and m_2 denote the respective slopes of the tangent lines, and if neither is zero, then L_1 and L_2 are orthogonal if and only if

$$m_1 m_2 = -1. \tag{6.1.6}$$

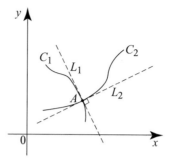

Fig. 6.5 The curves C_1 and C_2 are orthogonal at A if the tangent lines L_1 and L_2 (at A) are orthogonal.

Two families of curves are said to be **orthogonal** if each curve from one family intersects the curves from the other family at right angles.

Consider the level curves of a harmonic function u in a region Ω. These are the curves determined by the implicit relation

$$u(x, y) = C_1, \tag{6.1.7}$$

where C_1 is a constant (in the range of u). Since u is harmonic, it has continuous partial derivatives, and hence differentiating both sides of (6.1.7) with respect to x and using the chain rule we obtain

$$\frac{\partial u}{\partial x}\frac{dx}{dx} + \frac{\partial u}{\partial y}\frac{dy}{dx} = 0.$$

But $\frac{dx}{dx} = 1$, so if $\frac{\partial u}{\partial y} \neq 0$, we can solve for $\frac{dy}{dx}$ and get

$$\frac{dy}{dx} = -\frac{\frac{\partial u}{\partial x}}{\frac{\partial u}{\partial y}}. \tag{6.1.8}$$

This gives the slope of the tangent line at a point on a level curve. Now suppose that we can find a harmonic conjugate v of u in Ω, and let us consider the level curves

$$v(x, y) = C_2, \tag{6.1.9}$$

where C_2 is a constant (in the range of v). Since v is harmonic, arguing as for u, we find that the slope of the tangent line at a point on a level curve is

$$\frac{dy}{dx} = -\frac{\frac{\partial v}{\partial x}}{\frac{\partial v}{\partial y}} = \frac{\frac{\partial u}{\partial y}}{\frac{\partial u}{\partial x}}, \tag{6.1.10}$$

since by the Cauchy-Riemann equations, $\partial v/\partial x = -\partial u/\partial y$ and $\partial v/\partial y = \partial u/\partial x$. Comparing (6.1.9) and (6.1.10), we see that the slopes of the tangent lines satisfy the orthogonality relation (6.1.6), and hence the level curves of u are orthogonal to the level curves of v. This orthogonality relation also holds when the tangents are horizontal and vertical. We thus have the following result.

Theorem 6.1.18. (Orthogonality of Level Curves) *Suppose that u is a harmonic function in a region Ω, and let v be a harmonic conjugate of u in Ω, so that $f = u + iv$ is analytic in Ω. Then, the two families of level curves, $u(x, y) = C_1$ and $v(x, y) = C_2$, are orthogonal at every point $z = x + iy$ for which $f'(z) \neq 0$.*

As an illustration, we show in Figure 6.6 the level curves of the harmonic function $u = x^2 - y^2 + x$ and in Figure 6.7 the level curves of its conjugate $v(x, y) = 2xy + y$ (see Example 6.1.12). The graphs of the two families are superposed in Figure 6.8 to illustrate their orthogonality.

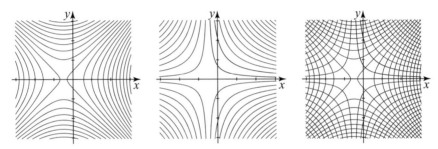

Fig. 6.6 Level curves of $u(x, y) = x^2 - y^2 + x$.

Fig. 6.7 Level curves of $v(x, y) = 2xy + y$.

Fig. 6.8 The level curves of u and v are orthogonal.

Mean Value and Maximum/Minimum Modulus Principle

Using the local existence of a harmonic conjugate, we obtain the mean value property of harmonic functions from the corresponding property of analytic functions.

Corollary 6.1.19. (Gauss' Mean Value Property) *Let u be a harmonic function on a region Ω. Then u satisfies the **mean value property** in the following sense: If z is in Ω, and the closed disk $\overline{B_r(z)}$ $(r > 0)$ is contained in Ω, then*

$$u(z) = \frac{1}{2\pi} \int_0^{2\pi} u(z + re^{it}) \, dt. \tag{6.1.11}$$

Proof. Let B be an open disk in Ω containing the closed disk $\overline{B_r(z_0)}$. Since B is simply connected, u has a harmonic conjugate v in B. So $f = u + iv$ is analytic in B.

By the mean value property of analytic functions (3.9.4), we have

$$f(z) = \frac{1}{2\pi} \int_0^{2\pi} f(z+re^{it})\,dt = \frac{1}{2\pi} \int_0^{2\pi} u(z+re^{it})\,dt + \frac{i}{2\pi} \int_0^{2\pi} v(z+re^{it})\,dt.$$

Now take real parts on both sides to obtain (6.1.11). ∎

We now prove the maximum-minimum modulus principle for harmonic functions. Note the role of connectedness in the proof.

Theorem 6.1.20. (Maximum and Minimum Modulus Principle) *Suppose that u is a harmonic function on a region Ω. If u attains a maximum or a minimum in Ω, then u is constant in Ω.*

Proof. By considering $-u$, we need only to prove the statement for maxima. We first prove the result under the assumption that Ω is simply connected. Applying Theorem 6.1.16 we find an analytic function $f = u + iv$ on Ω. Consider the function

$$g = e^f = e^u e^{iv}.$$

Then g is analytic in Ω and $|g| = e^u$. Since the real exponential function is strictly increasing, a maximum of e^u corresponds to a maximum of u. By Theorem 3.9.6, if $|g|$ attains a maximum or a minimum in Ω, then g is constant, implying that u is constant in Ω.

We now deal with an arbitrary region Ω. Suppose that u attains a maximum M at a point in Ω. Let

$$\Omega_0 = \{z \in \Omega : u(z) < M\}$$
$$\Omega_1 = \{z \in \Omega : u(z) = M\}.$$

We have $\Omega = \Omega_0 \cup \Omega_1$, Ω_0 is open, and Ω_1 is nonempty by assumption. It is enough to show that Ω_1 is open. By connectedness this will imply that $\Omega = \Omega_1$.

Suppose that z_0 is in Ω_1, and let $B_r(z_0)$ be an open disk in Ω centered at z_0 (Figure 6.9). Since $B_r(z_0)$ is simply connected and the restriction of u to $B_r(z_0)$ is a harmonic function that attains its maximum at z_0 inside $B_r(z_0)$, it follows from the previous case that u is constant in $B_r(z_0)$. Thus $u(z) = M$ for all z in $B_r(z_0)$, implying that $B_r(z_0)$ is contained in Ω_1. Hence Ω_1 is open. ∎

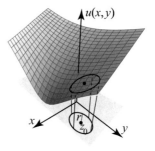

Fig. 6.9 Local existence of the harmonic conjugate.

Note that in Theorem 6.1.20 the minimum principle holds without the further assumption that $u \neq 0$, which was required for the minimum principle for analytic functions. The following corollaries of Theorem 6.1.20 are similar to already proved results regarding the modulus of analytic functions. We relegate their proofs to the exercises.

Corollary 6.1.21. *Suppose Ω is a bounded region, and u is harmonic on Ω and continuous on the closure of Ω. Then*
(i) u attains its maximum M and minimum m on the boundary of Ω;
(ii) either u is constant or $m < u < M$ for all points in Ω.

Corollary 6.1.22. *Suppose Ω is a bounded region, and u is harmonic on Ω and continuous on the boundary of Ω. If u is constant on the boundary of Ω, then u is constant in Ω.*

Corollary 6.1.23. *Suppose Ω is a bounded region, and u_1 and u_2 are harmonic on Ω and continuous on the boundary of Ω. If $u_1 = u_2$ on the boundary of Ω, then $u_1 = u_2$ in Ω.*

Exercises 6.1

In Exercises 1–4, (a) verify that the function u is harmonic on the shown region Ω. (b) Find the conjugate gradient of u and verify that it is analytic on Ω.

1. $u(x, y) = xy$, $\Omega = \mathbb{C}$
2. $u(x, y) = e^y \cos x$, $\Omega = \mathbb{C}$
3. $u(x, y) = \frac{y}{x^2 + y^2}$, $\Omega = \{z : \text{Im} z > 0\}$
4. $u(x, y) = \ln(x^2 + y^2)$, $\Omega = \mathbb{C} \setminus (-\infty, 0]$

In Exercises 5–16, determine the set on which the functions are harmonic. You may verify Laplace's equation (6.1.1) directly or use Theorem 6.1.2.

5. $x^2 - y^2 + 2x - y$
6. $\frac{1}{x^2 - y^2}$.
7. $e^x \cos y$

8. $\frac{y}{x^2 + y^2}$
9. $\frac{1}{x + y}$
10. $\sinh x \cos y$

11. $\cos x \cosh y$
12. $e^{2x} \cos(2y)$
13. $e^{x^2 - y^2} \cos(2xy)$

14. $\ln(x^2 + y^2)$
15. $\ln((x-1)^2 + y^2)$
16. $\arg_{\frac{\pi}{2}} z$

In Exercises 17–20, find a harmonic conjugate of the indicated harmonic function via the method of Example 6.1.12. Check your answer by verifying the Cauchy-Riemann equations.

17. $x + 2y$
18. $x^2 - y^2 - xy$
19. $e^y \cos x$
20. $\cos x \sinh ys$

21. Suppose that $f = u + iv$ is analytic on a region Ω and that u, v are real-valued. Show that the conjugate gradient of u is f'. [*Hint*: Use the Cauchy-Riemann equations.]

22. (a) Suppose that u is harmonic and bounded on $\mathbb{C}$. Show that u is constant.
(b) Suppose that u is harmonic on $\mathbb{C}$ and bounded above by M. Show that u is constant. [*Hint*: Let v be a harmonic conjugate of $M + 1 - u$. Apply Liouville's theorem to $g(x, y) = \frac{1}{M+1-u(x,y)+iv(x, y)}$.]
(c) Derive the same conclusion if u is bounded below.

23. Suppose that u is harmonic on $\mathbb{C}$ and there is an interval (a, b) with $a < b$ such that $u(x, y)$ does not lie in (a, b) for all $(x, y) \in \mathbb{C}$. Show that u is constant. [*Hint*: Let v be a harmonic conjugate of u. Apply Liouville's theorem to $g(x, y) = \frac{1}{u(x,y) - \frac{a+b}{2} + iv(x, y)}$.]

24. Find a harmonic conjugate of the function $u(y) = y$ using Theorem 6.1.16.

25. Consider $u(x, y) = e^x \cos y$, where (x, y) lies in the square with vertices at $\pm \pi \pm i\pi$. Find the maximum and minimum values of u and determine where these values occur.

26. Show that $u(x, y) = xy$ is harmonic in the upper half-plane. Does u attain a maximum or a minimum on the boundary of the upper half-plane? Does this contradict Corollary 6.1.21?

27. Failure of the identity principle for harmonic functions. Give an example of a harmonic function u in a region Ω that vanishes identically on a line segment in Ω, but such that u is not identically zero in Ω. [*Hint*: Think of a linear function.]

28. Let n be an integer.
(a) Show that $u(r, \theta) = r^n \cos(n\theta)$ and $v(r, \theta) = r^n \sin(n\theta)$ are harmonic on $\mathbb{C}$ if $n \geq 0$ and on $\mathbb{C} \setminus \{0\}$ if $n < 0$.
(b) Find their respective harmonic conjugates. [*Hint*: Consider $f(z) = z^n$ in polar coordinates.]

29. Show that if u and u^2 are both harmonic in a region Ω, then u must be constant. [*Hint*: Plug u^2 into Laplace's equation and show that $u_x = u_y = 0$.]

30. Show that if u, v, and $u^2 + v^2$ are harmonic on a region Ω, then u and v must be constant. [*Hint*: Plug $u^2 + v^2$ into Laplace's equation and show that $u_x = u_y = 0$ and $v_x = v_y = 0$.]

31. Suppose that u is harmonic and v is a harmonic conjugate of u. Show that $u^2 - v^2$ and uv are both harmonic. [*Hint*: Consider $(u + iv)^2$.]

32. Translating and dilating a harmonic function. Suppose that u is harmonic on $\mathbb{C}$. Show that the following functions of (x, y) are also harmonic:
(a) $u(x - \alpha, y - \beta)$, where α and β are real numbers;
(b) $u(\alpha x, \alpha y)$, where $\alpha \neq 0$ is a real number.

33. Suppose that u is harmonic on $\mathbb{C}$. Show that $u(x, -y)$ is also harmonic on $\mathbb{C}$.

34. Nonexistence of a harmonic conjugate. The goal here is to show that $\ln|z|$ does not have a harmonic conjugate in $\mathbb{C} \setminus \{0\}$.
(a) Suppose that $\phi(z)$ is a harmonic conjugate of $\ln|z|$ in $\mathbb{C} \setminus \{0\}$. Show that $\phi(z) = \text{Arg}(z) + c$ for all z in $\mathbb{C} \setminus (-\infty, 0]$. [*Hint*: The functions $\ln|z| + i\phi(z)$ and $\text{Log}\, z$ are analytic in the region $\mathbb{C} \setminus (-\infty, 0]$ and have the same real parts. Use Corollary 2.5.8.]
(b) Argue that, since ϕ is harmonic in $\mathbb{C} \setminus \{0\}$, ϕ is continuous on $(-\infty, 0)$. Obtain a contradiction using (a) and the fact that the discontinuities of $\text{Arg}\, z$ are not removable on the negative x-axis (Theorem 2.2.21).

35. Prove Corollaries 6.1.21–6.1.23 using Theorem 6.1.20.

36. Let f be analytic on a region Ω with nonvanishing derivative, and let w be a twice continuously differentiable function on the region $f[\Omega]$. Suppose that $w \circ f$ is harmonic on Ω. Show that w is harmonic. [*Hint*: Use (6.1.3).]

37. Suppose that both u and e^u are harmonic functions on a region Ω. What conclusion can you draw about u?

38. Suppose that a polynomial u of two real variables is a harmonic function on $\mathbb{C}$. Show that the harmonic conjugate of u is also a polynomial of two real variables. [*Hint*: Define $v(x, y) = \int_0^y u_x(x, s)\, ds - \int_0^x u_y(t, 0)\, dt$.]

6.2 Dirichlet Problems

To determine the steady-state temperature distribution inside a plate Ω, we must solve Laplace's equation inside Ω subject to the condition that $u(x, y)$ is a given function $b(x, y)$ on the boundary of Ω; such a condition is known as a **boundary condition**. A problem consisting of a partial differential equation along with specified boundary conditions is known as a **boundary value problem**. The special case involving Laplace's equation with specified boundary values is known as a **Dirichlet problem**. Solving Dirichlet problems is of paramount importance in applied mathematics, engineering, and physics. Many methods have been developed for this purpose. The ones that we present in this section provide beautiful applications of complex analysis.

Solving and Interpreting Dirichlet Problems

An interesting example of a harmonic function is $u(x, y) = a \operatorname{Arg} z + b$, which is harmonic on $\mathbb{C} \setminus (-\infty, 0]$ by Example 6.1.3(c) and Proposition 6.1.4. Because $\operatorname{Arg} z$ is constant on rays (independent of r), it follows that $u = a \operatorname{Arg} z + b$ is also constant on rays. (In fact, this is the only harmonic function with such a property. See Exercise 7.) Thus u is a good candidate for a solution of Dirichlet problems in which the boundary data is constant on rays or independent of r. We illustrate these ideas with an example.

Example 6.2.1. (Dirichlet problem in a quadrant)
The boundary of a very large sheet of metal (thought of as the quarter plane Ω) is kept at the constant temperatures $100°$ on the bottom and $50°$ on the left, as illustrated in Figure 6.10. After a long enough period of time, the temperature inside the plate reaches an equilibrium distribution. Find this steady-state temperature $u(x, y)$.

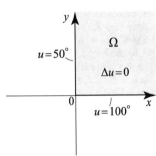

Fig. 6.10 Dirichlet problem on Ω.

Solution. The steady-state temperature is a solution of the Dirichlet problem, which consists of Laplace's equation

$$\Delta u = 0, \qquad \text{inside } \Omega;$$

along with the boundary conditions

$$u(x, 0) = 100, \, x > 0, \qquad u(0, y) = 50, \, y > 0.$$

Based on our discussion preceding the example, because the boundary data is independent of $r = |x + iy|$, we try for a solution of the harmonic function

$$u(x, y) = a \operatorname{Arg}(x + iy) + b,$$

where a and b are real constants to be determined so as to satisfy the boundary conditions. From the first condition we obtain

$$u(x, 0) = 100 \Rightarrow a\, \text{Arg}\, x + b = 100 \Rightarrow b = 100,$$

as $\text{Arg}\, x = 0$ for $x > 0$. From the second condition

$$u(0, y) = 50 \Rightarrow a\, \text{Arg}\, (iy) + b = 50$$

$$\Rightarrow a\frac{\pi}{2} + 100 = 50$$

$$\Rightarrow a = -\frac{100}{\pi},$$

since $\text{Arg}\, (iy) = \frac{\pi}{2}$ for $y > 0$, and $b = 100$. Thus the steady-state temperature inside the plate is

$$u(x, y) = -\frac{100}{\pi} \text{Arg}\, (z) + 100.$$

Now for $z = x + iy$ with $x > 0$, we have

$$\text{Arg}\, z = \tan^{-1}\left(\frac{y}{x}\right),$$

and so another way of expressing the solution is

$$u(x, y) = -\frac{100}{\pi} \tan^{-1}\left(\frac{y}{x}\right) + 100.$$

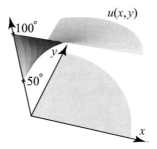

Fig. 6.11 A three-dimensional picture representing the temperature distribution of the plate. Note the boundary values on the graph.

The graph of u is shown in Figure 6.11. Note the temperature on the boundary; it matches the boundary conditions. ☐

In contrast to $\text{Arg}\, z$ we can find harmonic functions which are independent of the argument and depend only on $r = |z|$. An example of such a function is

$$u(z) = a \ln |z| + b,$$

where a and b are real constants. By Example 6.1.3(d), this function is harmonic in $\mathbb{C} \setminus \{0\}$. It is a good candidate for a solution of Dirichlet problems in which the boundary data is constant on circles. See Exercises 29–32 for illustrations.

Example 6.2.2. (Dirichlet problem in an infinite strip) Solve the Dirichlet problem shown in Figure 6.12.

Solution. Since the boundary data does not depend on x, it is plausible to guess that the solution will also not depend on x. In this case, the function $u(x, y)$ depends on y alone, hence $u_x = 0$ and $u_{xx} = 0$, and Laplace's equation becomes $u_{yy} = 0$, implying that u is a linear function of y. Hence $u(x, y) = ay + b$. Using the boundary conditions, we have

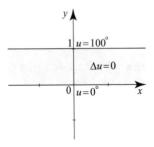

Fig. 6.12 Dirichlet problem in Example 6.2.2

$$u(x, 0) = 0 \Rightarrow b = 0;$$
$$u(x, 1) = 100 \Rightarrow a = 100.$$

Hence the solution of the problem is $u(x, y) = 100y$, which is clearly harmonic and satisfies the boundary conditions. $\qquad\square$

In Example 6.2.2 we used a harmonic function that was independent of x. Similarly, we can find harmonic functions that are independent of y (hence $u_y = 0$ and $u_{yy} = 0$). In this case we have that $u = ax + b$, where a and b are real constants. This function is a good candidate for solving Dirichlet problems in infinite vertical strips with constant boundary data. On this, see Exercises 1 and 2.

Harmonic Conjugates, Isotherms, and Heat Flow

In Example 6.2.1, the temperature of the boundary is kept at two constant values, $100°$ and $50°$. Our physical intuition tells us that, as the plate is insulated, the temperature values of the points inside the plate vary between these two values and equal those values only at the boundary. It is natural to ask for those points (x, y) inside the plate with the same temperature $u(x, y) = T$, where $50 < T < 100$. These points lie on curves inside the plate, called curves of constant temperature or **isotherms**. Isotherms have many practical applications. Computing them leads to interesting properties of harmonic functions.

Example 6.2.3. (Isotherms) Find the isotherms in Example 6.2.1.

Solution. Since the temperature inside the plate will vary between $50°$ and $100°$, to find the isotherms, we must solve

$$u(x, y) = T, \quad \text{where} \quad 50 < T < 100, \tag{6.2.1}$$

where (x, y) is a point inside the first quadrant, not on the boundary. Appealing to the solution of Example 6.2.1, (6.2.1) becomes

$$-\frac{100}{\pi} \tan^{-1}\left(\frac{y}{x}\right) + 100 = T, \quad \text{where} \quad 50 < T < 100, \quad x > 0, \, y > 0. \tag{6.2.2}$$

Thus,

$$\tan^{-1}\left(\frac{y}{x}\right) = \pi - \frac{\pi T}{100} \Rightarrow \frac{y}{x} = \tan\left(\pi - \frac{\pi T}{100}\right)$$

$$\Rightarrow \frac{y}{x} = -\tan\frac{\pi T}{100}$$

$$\Rightarrow y = -\overbrace{\left(\tan\frac{\pi T}{100}\right)}^{\text{a positive constant}} x,$$

where we have used the identity $\tan(\pi - \alpha) = -\tan\alpha$. Since $50 < T < 100$, $\frac{\pi}{2} < \frac{\pi T}{100} < \pi$ and so $-\tan\frac{\pi T}{100} > 0$.

Thus the equation of the isotherms $y = -\left(\tan\frac{\pi T}{100}\right)x$ corresponds to rays in the first quadrant emanating from the origin. As $T \to 100$, the slope of the ray $y = -\left(\tan\frac{\pi T}{100}\right)x$ tends to 0, indicating that the ray tends to the positive x-axis. As $T \to 50$, the slope of the ray $-\tan\frac{\pi T}{100}$ tends to ∞, showing that the ray tends to the positive y-axis. This agrees with our intuition, since points near the x-axis have temperature close to $100°$, and points near the y-axis have temperature close to $50°$. The isotherms corresponding to $T = 90°, 80°, 70°$, and $60°$ are shown in Figure 6.13. □

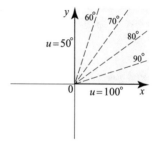

Fig. 6.13 The isotherms in Example 6.2.1.

Related to the topic of isotherms is the topic of **curves of heat flow**. These are the curves along which the heat is flowing inside the plate. To determine these curves, we use **Fourier's law of heat conduction**, which states that heat flows from hot to cold in the direction in which the temperature difference is the greatest. Clearly the change in temperature is greatest along directions moving away from the isotherms, i.e., along curves perpendicular to them. Hence the curves of heat flow are orthogonal to the isotherms.

Recall from vector calculus that the **gradient vector** $\nabla u = (u_x, u_y)$ points in the direction of greatest change in a function. The gradient is perpendicular to the level curves of $u(x, y)$. Fourier's law states that heat flows along $-\nabla u$, and thus curves of heat flow are orthogonal to level curves of u, and hence coincident with level curves of v. So to find the curves of heat flow in a plate, it is enough to find a harmonic conjugate $v(x, y)$ of the steady-state temperature distribution $u(x, y)$, since by Theorem 6.1.18 the level curves of v are orthogonal to the level curves of u.

Example 6.2.4. (Curves of heat flow) Find the curves of heat flow in Example 6.2.1.

Solution. The isotherms found in Example 6.2.3 are $u(x, y) = C_1$, where $u(x, y) = -\frac{100}{\pi} \text{Arg}(z) + 100$, where $z = x + iy$.

To determine the curves of heat flow, we must find a harmonic conjugate of u. By Proposition 6.1.14(ii), it is enough to find a harmonic conjugate of $-\frac{100}{\pi} \text{Arg}(z)$. From our knowledge of analytic functions, we see that

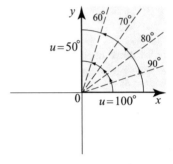

$$-\frac{100}{\pi}(\ln|z| + i\,\text{Arg}(z)) = -\frac{100}{\pi}\text{Log}\,z$$

is analytic in $\mathbb{C} \setminus (-\infty, 0]$. Hence a harmonic conjugate of $-\frac{100}{\pi}\ln|z|$ is $-\frac{100}{\pi}\text{Arg}\,z$.

Fig. 6.14 Curves of heat flow and isotherms in Example 6.2.4.

Consequently, in view of Proposition 6.1.13, a harmonic conjugate of $-\frac{100}{\pi}\text{Arg}\,z$ is $\frac{100}{\pi}\ln|z|$. Hence the curves of heat flow are

$$\frac{100}{\pi}\ln|z| = \text{constant} \quad \Leftrightarrow \quad |z| = \text{constant} \quad \Leftrightarrow \quad x^2 + y^2 = c^2.$$

Thus the curves of heat flow are arcs of circles centered at the origin. In Figure 6.14, we show the isotherms along with the curves of heat flow to illustrate their orthogonality. $\qquad \square$

So far we have used our knowledge of analytic functions to solve Dirichlet problems by guessing the solution. Guessing is certainly a legitimate method that can be used in solving differential equations and computing indefinite integrals. Further development of the theory of analytic and harmonic functions will be necessary to tackle more general Dirichlet problems. Topics such as the Poisson integral formula, Fourier series, and conformal mappings are examples of theories and tools that provide us with systematic ways for solving Dirichlet problems.

Other issues that could be addressed concern the uniqueness of the solution of a Dirichlet problem. Consider the problem in Example 6.2.1, whose solution is

$$u(x, y) = -\frac{100}{\pi}\tan^{-1}\left(\frac{y}{x}\right) + 100.$$

If we add to this solution the harmonic function $\phi(x, y) = xy$, we obtain the harmonic function $\psi(x, y) = u(x, y) + xy$, which solves Laplace's equation. Moreover, because xy vanishes on the boundary of the first quadrant, ψ and u have the same boundary values. Thus, ψ is another solution of the Dirichlet problem in Example 6.2.1. This situation is counterintuitive since we would like to think of the solution as representing a temperature distribution and as such it must be unique. As it turns out, on unbounded regions, such as the one in Example 6.2.1, if we require the (natural) condition that the temperature distribution be bounded, then the solution will be unique. For example, with this additional boundedness assumption, we can no longer add the function xy to the solution in Example 6.2.1 to obtain another solution.

Exercises 6.2

1. Harmonic functions independent of y.
(a) Suppose that $u(x, y)$ is a harmonic function whose values depend only on x and not on y. Using Laplace's equation, show that $u(x, y) = ax + b$, where a and b are real constants.
(b) Consider the Dirichlet problem in the infinite vertical strip shown in the adjacent figure. Because the boundary values do not depend on y, it is plausible to try for a solution a harmonic function whose values do not depend on y. Solve the problem, using (a).
(c) Determine and plot the isotherms and curves of heat flow.

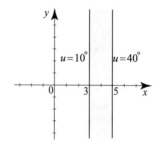

2. A Dirichlet problem in an infinite vertical strip.
(a) Solve the Dirichlet problem in Figure 6.15.
(b) Determine and plot the isotherms and curves of heat flow.

3. A Dirichlet problem in a wedge, I. Harmonic functions independent of r.
(a) Solve the Dirichlet problem in Figure 6.16. Since the boundary values do not depend on $r = |z|$, look for a harmonic function solution whose values do not depend on r.
(b) Determine and plot the isotherms and curves of heat flow.

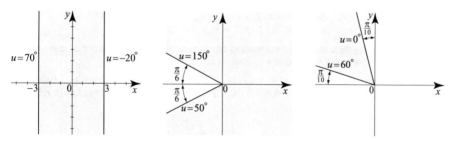

Fig. 6.15 Exercise 2.   **Fig. 6.16** Exercise 3.   **Fig. 6.17** Exercise 4.

4. A Dirichlet problem in a wedge, II. Harmonic functions independent of r.
(a) Solve the Dirichlet problem in Figure 6.17.
(b) Determine and plot the isotherms and curves of heat flow.

5. Project Problem: Laplacian in polar coordinates. We derive the polar form of the Laplacian

$$\Delta u = \frac{\partial^2 u}{\partial r^2} + \frac{1}{r}\frac{\partial u}{\partial r} + \frac{1}{r^2}\frac{\partial^2 u}{\partial \theta^2}. \tag{6.2.3}$$

(a) Recall the relationship between rectangular and polar coordinates

$$x = r\cos\theta, \qquad y = r\sin\theta, \qquad r^2 = x^2 + y^2, \qquad \tan\theta = \frac{y}{x}.$$

Differentiating $r^2 = x^2 + y^2$ with respect to x once, then a second time, obtain

$$\frac{\partial r}{\partial x} = \frac{x}{r}, \qquad \frac{\partial^2 r}{\partial x^2} = \frac{y^2}{r^3}.$$

(b) Differentiate $\tan\theta = \frac{y}{x}$ with respect to x once, and then a second time, obtain

$$\frac{\partial \theta}{\partial x} = -\frac{y}{r^2}, \qquad \frac{\partial^2 \theta}{\partial x^2} = \frac{2xy}{r^4}.$$

(c) In a similar way, differentiate with respect to y and obtain

$$\frac{\partial r}{\partial y} = \frac{y}{r}, \qquad \frac{\partial^2 r}{\partial y^2} = \frac{x^2}{r^3}, \qquad \frac{\partial \theta}{\partial y} = \frac{x}{r^2}, \qquad \frac{\partial^2 \theta}{\partial y^2} = -\frac{2xy}{r^4}.$$

(d) From the previous identities, derive

$$\frac{\partial^2 \theta}{\partial x^2} + \frac{\partial^2 \theta}{\partial y^2} = 0 \quad \text{and} \quad \frac{\partial \theta}{\partial x}\frac{\partial r}{\partial x} + \frac{\partial \theta}{\partial y}\frac{\partial r}{\partial y} = 0.$$

(What does the first equation say about the function θ?)

(e) Use the chain rule in two dimensions to obtain

$$\frac{\partial^2 u}{\partial x^2} = \frac{\partial^2 u}{\partial r^2}\left(\frac{\partial r}{\partial x}\right)^2 + 2\frac{\partial^2 u}{\partial \theta \partial r}\frac{\partial r}{\partial x}\frac{\partial \theta}{\partial x} + \frac{\partial u}{\partial r}\frac{\partial^2 r}{\partial x^2} + \frac{\partial^2 u}{\partial \theta^2}\left(\frac{\partial \theta}{\partial x}\right)^2 + \frac{\partial u}{\partial \theta}\frac{\partial^2 \theta}{\partial x^2}.$$

Interchanging x with y obtain

$$\frac{\partial^2 u}{\partial y^2} = \frac{\partial^2 u}{\partial r^2}\left(\frac{\partial r}{\partial y}\right)^2 + 2\frac{\partial^2 u}{\partial \theta \partial r}\frac{\partial r}{\partial y}\frac{\partial \theta}{\partial y} + \frac{\partial u}{\partial r}\frac{\partial^2 r}{\partial y^2} + \frac{\partial^2 u}{\partial \theta^2}\left(\frac{\partial \theta}{\partial y}\right)^2 + \frac{\partial u}{\partial \theta}\frac{\partial^2 \theta}{\partial y^2}.$$

(f) Add $\frac{\partial^2 u}{\partial x^2}$ and $\frac{\partial^2 u}{\partial y^2}$ and simplify with the help of the identity in part (d) to derive (6.2.3).

6. (a) Use (6.2.3) to give a direct proof of the result of Exercise 28(a), Section 6.1.
(b) Use (6.2.3) to give a direct proof that $\mathrm{Log}\,|z|$ is harmonic for all $z \neq 0$.

7. Show that if u is harmonic and independent of r, then $u_{\theta\theta} = 0$. Conclude that $u = a\theta + b$; equivalently, $u = a\arg_\alpha z + b$, where $\arg_\alpha z$ is a branch of the argument.

8. Harmonic functions independent of θ. Let $u(re^{i\theta}) = v(r)$ be a harmonic function that depends only on r and not θ. Show that v satisfies the second-order linear differential equation in r, known as an **Euler equation**,

$$v_{rr} + \frac{1}{r}v_r = 0.$$

Verify that the general solution of this equation is $v(r) = c_1 \ln r + c_2$, where c_1, c_2 are constants.

9. Dirichlet problems in annular regions. The annular region A_{R_1,R_2} in Figure 6.18 is centered at the origin with inner radius R_1 and outer radius R_2. Consider the Dirichlet problem in A_{R_1,R_2} with constant boundary conditions $u(R_1 e^{i\theta}) = T_1$ and $u(R_2 e^{i\theta}) = T_2$ for all θ. Find a solution u of this problem. [*Hint:* Since the boundary conditions are independent of θ, try a harmonic function of the form $c_1 \ln r + c_2$ as in Exercise 8. The answer is $u(re^{i\theta}) = u(r) = T_1 + (T_2 - T_1)\frac{\ln(r/R_1)}{\ln(R_2/R_1)}$.]

Exercises 10–12 are Dirichlet problems described by the Figures 6.18–6.20, respectively. In each case, (a) solve the Dirichlet problem. (b) Determine the isotherms. (c) Determine the curves of heat flow. (d) Plot the isotherms and curves of heat flow.

10.   **11.**   **12.**

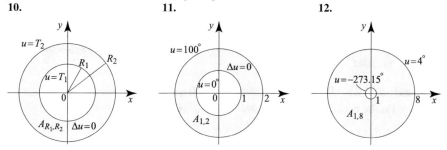

Fig. 6.18 Exercise 10.   **Fig. 6.19** Exercise 11.   **Fig. 6.20** Exercise 12.

6.3 Dirichlet Problem and the Poisson Integral on a Disk

In this section we consider the Dirichlet problem on a disk and derive an interesting formula for its solution. For convenience, we use polar coordinates. Suppose that a bounded piecewise continuous function $f(\theta)$ represents boundary data of the points $Re^{i\theta}, 0 \leq R < 1, \theta \in \mathbb{R}$ (Figure 6.21). Since θ and $\theta + 2\pi$ represent the same polar angle, we think of f as being 2π-periodic; that is $f(\theta + 2\pi) = f(\theta)$ for all θ. Our problem is to find a function $u(re^{i\theta})$ such that

$$\Delta u(re^{i\theta}) = 0, \qquad 0 \leq r < R, \ 0 < \theta \leq 2\pi; \qquad (6.3.1)$$
$$u(Re^{i\theta}) = f(\theta), \qquad 0 < \theta \leq 2\pi, \qquad (6.3.2)$$

where the equality holds at all points of continuity of f (Figures 6.21, 6.22).

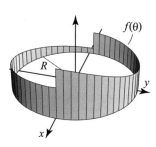

Fig. 6.21 The boundary values of u are $f(\theta) = u(Re^{i\theta})$.

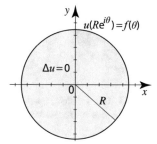

Fig. 6.22 Dirichlet problem on a disk with radius R.

A fundamental Dirichlet problem is the one that has boundary data:

$$f(\theta) = a_0 + \sum_{n=1}^{N} \big(a_n \cos(n\theta) + b_n \sin(n\theta)\big), \tag{6.3.3}$$

i.e., a finite linear combination of functions from the 2π-periodic trigonometric system: $1, \cos x, \cos 2x, \ldots, \sin x, \sin 2x, \ldots.$

Proposition 6.3.1. (Dirichlet Problem on a Disk) *The solution of the Dirichlet problem on the disk $|z| \le R$ with boundary condition (6.3.3) is*

$$u(re^{i\theta}) = a_0 + \sum_{n=1}^{N} \left(\frac{r}{R}\right)^n \big(a_n \cos(n\theta) + b_n \sin(n\theta)\big), \qquad r < R. \tag{6.3.4}$$

Proof. For $|z| < R$, write $z = re^{i\theta} = r(\cos\theta + i\sin\theta)$. The function

$$f(z) = z^n = r^n \big(\cos(n\theta) + i\sin(n\theta)\big)$$

is analytic on the disk $|z| < R$. Hence its real and imaginary parts are harmonic (Theorem 6.1.2). This shows that the functions $1, r\cos\theta, r^2\cos(2\theta), \ldots, r\sin\theta, r^2\sin(2\theta), \ldots$ are harmonic (in the variable $re^{i\theta}$) on the disk $B_R(0)$. By Proposition 6.1.4, linear combinations of such functions are also harmonic on the unit disk, and so (6.3.4) is harmonic. Setting $r = R$ in (6.3.4), we see that $u(Re^{i\theta}) = f(\theta)$, where f is as in (6.3.3). Hence (6.3.4) is the solution of the Dirichlet problem with boundary data (6.3.3). ∎

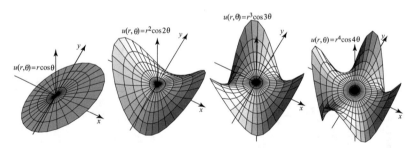

Fig. 6.23 The saddle-shaped graphs of the harmonic functions $r\cos\theta$, $r^2\cos 2\theta$, $r^3\cos 3\theta$, $r^4\cos 4\theta$, respectively.

Each term in the finite sum (6.3.4) is a constant multiple of the harmonic functions $1, r\cos\theta, r^2\cos(2\theta), \ldots, r\sin\theta, r^2\sin(2\theta), \ldots.$ With the exception of the constant function 1, the graphs of these functions over the disk $|z| < R$ are saddle-shaped; see Figure 6.23.

Example 6.3.2. (A steady-state problem on a disk)

The temperature on the boundary of a circular plate of radius 1 with insulated lateral surface and center placed at the origin is a function of the radial angle θ which varies between $0°$ and $100°$ according to the formula $f(\theta) = 50 - 50\cos\theta$ (Figure 6.24).

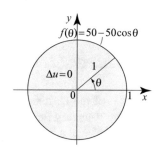

(a) Find the steady-state temperature inside the plate.

(b) Describe the isotherms and lines of heat flow.

Fig. 6.24 The boundary function, as a function of θ.

Solution. (a) To find the steady-state temperature, we must solve the Dirichlet problem with boundary data f. According to (6.3.4), the solution is $u(r, \theta) = 50 - 50r\cos\theta$, $0 \le r < 1$. This function is harmonic and equals $f(\theta)$ when $r = 1$ (Figure 6.25).

(b) Since the temperature on the boundary varies between 0 and 100, by the maximum-minimum principle for harmonic functions, the temperature inside the plate will vary between these two limits. To find the isotherms, let $0 < T < 100$ and solve the equation $u(r, \theta) = T$ or $50 - 50r\cos\theta = T$. Using $x = r\cos\theta$, the equation becomes $50 - 50x = T$ or $x = \frac{50-T}{50}$. Thus the isotherms are the intersections with the unit disk of the vertical lines $x = \frac{50-T}{50}$. The isotherms vary between $x = 1$ and $x = -1$ as T varies between 0 and 100. In view of Theorem 6.1.18 the heat flows along the curves that are orthogonal to the isotherms. In this case, it is not difficult to verify that heat flows inside the unit disk in the direction of the horizontal lines $y = b$ (Figure 6.26). □

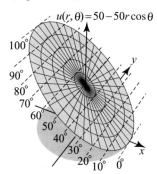

Fig. 6.25 The solution of the Dirichlet problem. Note how on the boundary the values of this solution coincide with the values of the boundary function.

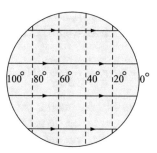

Fig. 6.26 The isotherms and lines of heat flow. The orthogonality of these two families of curves is obvious in this case.

In the next section, we show that the solution of the Dirichlet problem with arbitrary piecewise continuous boundary data f can be expressed in the form (6.3.4) if we allow the series to have infinitely many terms. The series we thus obtain is called **Fourier series**.

Poisson's Integral Formula

The solution of the Dirichlet problem on the disk can be expressed as an integral involving the boundary function, known as the **Poisson integral formula**.

Suppose that u is a harmonic function on the open unit disk, continuous on the closed disk $\overline{B_1(0)}$, and equal to a given function f on the boundary of the disk. Then the mean value property of u at 0 (Corollary 6.1.19) implies that

$$u(0) = \frac{1}{2\pi} \int_0^{2\pi} u(e^{i\theta}) \, d\theta = \frac{1}{2\pi} \int_0^{2\pi} f(\theta) \, d\theta,$$

where f is the boundary function. In this equation, the value of u at the center of the disk is expressed as an integral of the boundary function f. Our goal is to derive an integral formula for all other values of $u(z_0)$, where z_0 is a point inside the unit disk.

For fixed $|z_0| < 1$, consider the linear fractional transformation

$$\phi_{-z_0}(z) = \frac{z + z_0}{1 + \overline{z_0}z}, \quad |z| < 1.$$

We know from Proposition 4.6.2 that $\phi_{-z_0}(z)$ is analytic and one-to-one and maps the unit disk onto itself and the unit circle, $C_1(0)$, onto itself. Let

$$U(z) = u \circ \phi_{-z_0}(z), \quad |z| < 1.$$

In view of Theorem 6.1.10, U is harmonic on the open unit disk. Applying the mean value property of U at 0, we find that

$$u(z_0) = U(0) = \frac{1}{2\pi} \int_0^{2\pi} U(e^{it}) \, dt = \frac{1}{2\pi} \int_0^{2\pi} u \circ \phi_{-z_0}(e^{it}) \, dt. \qquad (6.3.5)$$

Thus the value of u at the interior point z_0 is expressed as an integral involving the boundary values of $u \circ \phi_{-z_0}$. To get the desired formula that involves the boundary values of u, we will perform the change of variables

$$e^{is} = \phi_{-z_0}(e^{it}).$$

Recall that ϕ_{-z_0} maps the unit circle into itself and the inverse of ϕ_{-z_0} is ϕ_{z_0}. So

$$\phi_{z_0}(e^{is}) = e^{it} \Rightarrow \phi'_{z_0}(e^{is})ie^{is}\,ds = ie^{it}\,dt$$

$$\Rightarrow \frac{\phi'_{z_0}(e^{is})}{e^{it}}e^{is}\,ds = dt$$

$$\Rightarrow dt = \frac{\phi'_{z_0}(e^{is})}{\phi_{z_0}(e^{is})}e^{is}\,ds.$$

In view of identity (v) in Proposition 4.6.2 we have $\phi'_{z_0}(z) = \dfrac{1-|z_0|^2}{(1-\overline{z}_0 z)^2}$. Hence,

$$\frac{\phi'_{z_0}(e^{is})}{\phi_{z_0}(e^{is})}e^{is} = e^{is}\frac{1-|z_0|^2}{(1-\overline{z}_0 e^{is})(e^{is}-z_0)} = \frac{1-|z_0|^2}{(e^{-is}-\overline{z}_0)(e^{is}-z_0)} = \frac{1-|z_0|^2}{|e^{is}-z_0|^2}.$$

Substituting into (6.3.5), we obtain

$$u(z_0) = \frac{1-|z_0|^2}{2\pi}\int_0^{2\pi}\frac{u(e^{is})}{|e^{is}-z_0|^2}\,ds, \qquad |z_0| < 1. \tag{6.3.6}$$

This is the **Poisson integral formula** on the unit disk. If u is harmonic in a disk of radius $R > 0$, centered at the origin, we consider the function $u(Rz)$, which is harmonic in $|z| < 1$, and so according to (6.3.6),

$$u(Rz_0) = \frac{1-|z_0|^2}{2\pi}\int_0^{2\pi}\frac{u(Re^{is})}{|e^{is}-z_0|^2}\,ds \qquad (|z_0| < 1).$$

Let $z = Rz_0$, $z_0 = z/R$, $|z| = r < R$, then

$$u(z) = \frac{R^2-r^2}{2\pi}\int_0^{2\pi}\frac{u(Re^{is})}{|Re^{is}-z|^2}\,ds \qquad (|z| < R). \tag{6.3.7}$$

This is the **Poisson integral formula** on the disk of radius $R > 0$, centered at the origin. Another common way of expressing the Poisson integral formula is obtained by realizing that for $z = re^{i\theta}$, $0 < r < R$,

$$|Re^{i\phi}-z|^2 = (Re^{i\phi}-re^{i\theta})(Re^{-i\phi}-re^{-i\theta}) = R^2 - 2rR\cos(\theta-\phi)+r^2;$$

and so from (6.3.7) (with the variable s replaced by ϕ) we obtain the alternative Poisson integral formula

$$u(re^{i\theta}) = \frac{R^2-r^2}{2\pi}\int_0^{2\pi}\frac{u(Re^{i\phi})}{R^2 - 2rR\cos(\theta-\phi)+r^2}\,d\phi. \tag{6.3.8}$$

The importance of (6.3.8) lies in the fact that it recovers a harmonic function u in the disk $B_R(0)$ from only its (piecewise continuous) values on the circle $C_R(0)$. Precisely, we have the following result.

Definition 6.3.3. The **Poisson kernel** on the disk $B_R(0)$ is the function

$$P(r,\theta) = \frac{R^2 - r^2}{R^2 - 2rR\cos\theta + r^2} \qquad (6.3.9)$$

for $z = re^{i\theta}$ with $0 \le r < R$.

Proposition 6.3.4. *The Poisson kernel $P(r,\theta)$, viewed as a function of the variable $z = re^{i\theta}$, can be written as*

$$P(r,\theta) = \operatorname{Re}\frac{R+z}{R-z} = \frac{R^2 - |z|^2}{|R-z|^2}, \qquad |z| < R. \qquad (6.3.10)$$

Thus $P(r,\theta)$ is a harmonic function of $re^{i\theta}$ on the disk $B_R(0)$.

Proof. For $z = re^{i\theta}$ and $0 \le r < R$ define

$$V(z) = \frac{R+z}{R-z}, \qquad |z| < R.$$

Then V is analytic on the disk $|z| < R$. Moreover, we have

$$\operatorname{Re}V(z) = \operatorname{Re}\frac{(R+re^{i\theta})(R-re^{-i\theta})}{|Re^{i\phi} - re^{i\theta}|^2} = \frac{R^2 - r^2}{|R-re^{i\theta}|^2} = \frac{R^2 - |z|^2}{|R-z|^2}.$$

Note that this is a positive function that coincides with that in (6.3.9). Applying Theorem 6.1.2, we find that $\operatorname{Re}V(z)$ is a harmonic function. ∎

The harmonicity of the Poisson kernel combined with property (6.1.11) in Corollary 6.1.19 yields that

$$\frac{1}{2\pi}\int_0^{2\pi} P(r,\theta)\,d\theta = P(0,0) = 1, \qquad 0 \le r < R. \qquad (6.3.11)$$

We record another important property of the Poisson kernel.

Proposition 6.3.5. *For $0 < \delta < \pi$ we have*

$$\frac{1}{2\pi}\int_{-\pi}^{-\delta} P(r,\theta)\,d\theta + \frac{1}{2\pi}\int_{\delta}^{\pi} P(r,\theta)\,d\theta \to 0 \qquad (6.3.12)$$

as $r \uparrow R$.

Proof. Notice that if $\delta \le |\theta| \le \pi$, then $\cos\theta \le \cos\delta$ and this implies

$$\frac{R^2 - r^2}{|R-re^{i\theta}|^2} = \frac{R^2 - r^2}{R^2 - 2rR\cos\theta + r^2} \le \frac{R^2 - r^2}{R^2 - 2rR\cos\delta + r^2} = \frac{R^2 - r^2}{|R-re^{i\delta}|^2}$$

and this converges to zero as $r \uparrow R$, since $e^{i\delta} \ne 1$. By the ML-inequality it follows that the expression in (6.3.12) is bounded by $\frac{2\pi - 2\delta}{2\pi}\frac{R^2 - r^2}{|R-re^{i\delta}|^2}$ and thus it tends to zero as $r \uparrow R$. ∎

The following result guarantees the existence of a unique solution to the Dirichlet problem on a disk.

Theorem 6.3.6. (Poisson's Integral Formula) *Consider the Dirichlet problem (6.3.1) on the disk $|z| \leq R$, with boundary conditions $u(Re^{i\theta}) = f(\theta)$, where f is a 2π-periodic and piecewise continuous function on the line. Let $0 \leq r < R$. Then the function*

$$u(re^{i\theta}) = \frac{R^2 - r^2}{2\pi} \int_0^{2\pi} \frac{f(\phi)}{R^2 - 2rR\cos(\theta - \phi) + r^2} \, d\phi, \tag{6.3.13}$$

is a unique solution to this Dirichlet problem. Precisely, u is harmonic in the open disk $B_R(0)$ and $u(re^{i\theta}) \to f(\theta)$ as $r \uparrow R$ at all points of continuity of f.

Proof. The function u in (6.3.13) can be written as

$$u(re^{i\theta}) = \frac{1}{2\pi} \int_0^{2\pi} f(\phi) \operatorname{Re} \frac{R + re^{i(\theta-\phi)}}{R - re^{i(\theta-\phi)}} \, d\phi = \frac{1}{2\pi} \int_0^{2\pi} f(\phi) \operatorname{Re} \frac{Re^{i\phi} + z}{Re^{i\phi} - z} \, d\phi$$

where $z = re^{i\theta}$. If $f = f_1 + if_2$, where f_1, f_2 are real-valued, then we write

$$u(z) = \frac{1}{2\pi} \operatorname{Re}\left[\int_0^{2\pi} f_1(\phi) \frac{Re^{i\phi} + z}{Re^{i\phi} - z} \, d\phi\right] + \frac{i}{2\pi} \operatorname{Re}\left[\int_0^{2\pi} f_2(\phi) \frac{Re^{i\phi} + z}{Re^{i\phi} - z} \, d\phi\right] \tag{6.3.14}$$

and if we know that the expressions in the square brackets in (6.3.14) are analytic functions of z, then u will be harmonic in view of Theorem 6.1.2. To show this, let $t_1 < t_2 < \cdots < t_M$ be the points of discontinuity of f in the interval $[0, 2\pi]$. Then we write the first term in (6.3.14) as

$$\sum_{j=1}^{M} \frac{1}{2\pi} \operatorname{Re}\left[\int_{t_j}^{t_{j+1}} f_1(\phi) \frac{Re^{i\phi} + z}{Re^{i\phi} - z} \, d\phi\right]. \tag{6.3.15}$$

Note that since f is piecewise continuous (see Definition 3.2.1), it has left and right limits at the t_j and t_{j+1} and can be redefined to be continuous from the right at t_j and from the left at t_{j+1}; the same is true for f_1 and f_2. Then the integrand in (6.3.15) is a continuous function in ϕ and analytic in z. Using Theorem 3.8.5 we deduce that the expression in the square brackets in (6.3.15) is an analytic function of z. We conclude that u is harmonic on $B_R(0)$ by Theorem 6.1.2.

It remains to show that the limit of $u(re^{i\theta_0})$ as $r \uparrow R$ equals $f(\theta_0)$ for every point θ_0 at which f is continuous. First note that f is a bounded function, and let M be a bound for $|f|$. Let θ_0 be a point of continuity of f. Given $\varepsilon > 0$ there is a $\delta > 0$ (which can be taken to be smaller than π) such that

$$|\phi| < \delta \implies |f(\phi - \theta_0) - f(\theta_0)| < \frac{\varepsilon}{2}. \tag{6.3.16}$$

Using property (6.3.11) and by changing variables we write

$$u(re^{i\theta_0}) - f(\theta_0) = \frac{1}{2\pi} \int_0^{2\pi} P(r,\phi) f(\theta_0 - \phi) \, d\phi - f(\theta_0) \left(\frac{1}{2\pi} \int_0^{2\pi} P(r,\phi) \, d\phi \right)$$

$$= \frac{1}{2\pi} \int_0^{2\pi} P(r,\phi) \left(f(\theta_0 - \phi) - f(\theta_0) \right) d\phi$$

$$= \frac{1}{2\pi} \int_{-\pi}^{\pi} P(r,\phi) \left(f(\theta_0 - \phi) - f(\theta_0) \right) d\phi \qquad \text{(see Exercise 8)}$$

$$= \frac{1}{2\pi} \int_{I_\delta} P(r,\phi) \left(f(\theta_0 - \phi) - f(\theta_0) \right) d\phi$$

$$+ \frac{1}{2\pi} \int_{J_\delta} P(r,\phi) \left(f(\theta_0 - \phi) - f(\theta_0) \right) d\phi,$$

where $I_\delta = (-\delta, \delta)$ and $J_\delta = [-\pi, \pi] \setminus (-\delta, \delta)$. Now we have

$$\left| \frac{1}{2\pi} \int_{I_\delta} P(r,\phi) \left(f(\theta_0 - \phi) - f(\theta_0) \right) d\phi \right| \leq \left(\frac{1}{2\pi} \int_{I_\delta} P(r,\phi) d\phi \right) \frac{\varepsilon}{2} \leq \frac{\varepsilon}{2}$$

using (6.3.16) and (6.3.11). Additionally, the ML-inequality gives

$$\left| \frac{1}{2\pi} \int_{J_\delta} P(r,\phi) \left(f(\theta_0 - \phi) - f(\theta_0) \right) d\phi \right| \leq \left(\frac{1}{2\pi} \int_{J_\delta} P(r,\phi) d\phi \right) 2M$$

and this can be made less than $\varepsilon/2$ provided $r \in (r_0, R)$ for some choice of r_0, in view of Proposition 6.3.5. Combining these facts we deduce that $|u(re^{i\theta_0}) - f(\theta_0)| < \varepsilon$ for r sufficiently close to R, that is, $\lim_{r \uparrow R} u(re^{i\theta_0}) = f(\theta_0)$ if f is continuous at θ_0.

If u_1 and u_2 are solutions of the Dirichlet problem (6.3.1) and (6.3.2), then $u_1 - u_2$ will be a harmonic function on the disk $B_R(0)$ which extends continuously to the zero function on the boundary. Then Corollary 6.1.21 implies that $u_1 - u_2 = 0$ on $B_R(0)$. The same conclusion also follows directly from Corollary 6.1.23.   ∎

The following corollary asserts that we can express the mean value of a harmonic function on the unit disk in terms of its values on the boundary of the disk, on which it may not even be continuous.

Corollary 6.3.7. (Mean Value Property) *Suppose that u is harmonic on $B_R(0)$ and $u(Re^{i\phi}) = f(\phi)$, where f is a 2π-periodic and piecewise continuous function on the line. Then*

$$u(0) = \frac{1}{2\pi} \int_0^{2\pi} f(\phi) \, d\phi.$$

Proof. The result is obtained by evaluating (6.3.13) at $r = 0$.   ∎

The Poisson formula is difficult to evaluate, even for simple boundary data. For this reason, we later develop alternative forms of the solution, including some that are based on Fourier series.

Exercises 6.3

In Exercises 1–4, solve the Dirichlet problem (6.3.1)–(6.3.2) *for the indicated boundary function on the disk with center at the origin and radius* $R > 0$.

1. $f(\theta) = 1 - \cos\theta + \sin(2\theta),\ R = 1.$

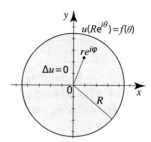

2. $f(\theta) = \cos\theta - \frac{1}{2}\sin(2\theta),\ R = 1.$

3. $f(\theta) = 100\cos^2\theta,\ R = 2.$

4. $f(\theta) = \sum_{n=1}^{10}\frac{\sin(n\theta)}{n},\ R = 1.$

5. Find the isotherms in Exercise 1.

6. For $n = 1, 2, \ldots, 0 \le r < 1$, show that

$$\frac{1-r^2}{2\pi}\int_0^{2\pi}\frac{\cos(n\phi)}{1-2r\cos(\theta-\phi)+r^2}\,d\phi = r^n\cos(n\theta),$$

and

$$\frac{1-r^2}{2\pi}\int_0^{2\pi}\frac{\sin(n\phi)}{1-2r\cos(\theta-\phi)+r^2}\,d\phi = r^n\sin(n\theta).$$

[*Hint*: Identify the integrals as solutions of Dirichlet problems on the unit disk and use Proposition 6.3.1.]

7. For $|z| < 1$, let

$$u(z) = \text{Arg}(z+i) - \text{Arg}(z-i).$$

(a) Write $z = re^{i\theta}$, $-\pi < \theta \le \pi$. Using basic facts from plane geometry, show that

$$\lim_{r\uparrow 1}u(z) = \begin{cases}\frac{\pi}{2} & \text{if } -\frac{\pi}{2} < \theta < \frac{\pi}{2} \\[2mm] \frac{3\pi}{2} & \text{if } -\pi < \theta < -\frac{\pi}{2} \quad\text{or}\quad \frac{\pi}{2} < \theta < \pi.\end{cases}$$

(b) Argue that u is harmonic in the disk $|z| < 1$ and describe the Dirichlet problem it satisfies.

8. **Different ways to express the Poisson integral formula.** Show that the Poisson integral formula (6.3.13) of a 2π-periodic function f can be expressed in the following equivalent forms:

$$u(re^{i\theta}) = \frac{R^2 - r^2}{2\pi}\int_{-\pi}^{\pi}\frac{f(\phi)}{R^2 - 2rR\cos(\theta-\phi)+r^2}\,d\phi; \tag{6.3.17}$$

$$u(re^{i\theta}) = \frac{R^2 - r^2}{2\pi}\int_{a}^{2\pi+a}\frac{f(\phi)}{R^2 - 2rR\cos(\theta-\phi)+r^2}\,d\phi; \tag{6.3.18}$$

$$u(re^{i\theta}) = \frac{R^2 - r^2}{2\pi}\int_{a}^{2\pi+a}\frac{f(\theta-\phi)}{R^2 - 2rR\cos\phi+r^2}\,d\phi, \tag{6.3.19}$$

where a is a real number. [*Hint*: The integrand is 2π-periodic so the integral does not change as long as we integrate over an interval of length 2π.]

6.4 Harmonic Functions and Fourier Series

In previous sections we saw that the Dirichlet problem on the disk $B_R(0)$

$$\Delta u(re^{i\theta}) = 0, \qquad 0 \leq r < R, \text{ all } \theta; \tag{6.4.1}$$

$$u(Re^{i\theta}) = f(\theta) \tag{6.4.2}$$

is solved by the Poisson integral

$$u(re^{i\theta}) = \frac{R^2 - r^2}{2\pi} \int_0^{2\pi} \frac{f(\phi)}{R^2 - 2rR\cos(\theta - \phi) + r^2} d\phi \qquad (0 \leq r < R) \tag{6.4.3}$$

with the understanding that (6.4.2) is interpreted as $\lim_{r \uparrow R} u(Re^{i\theta}) = f(\theta)$ at all points θ of continuity of f. In order to rewrite (6.4.3) in a form that is more suitable for numerical computations, we begin by deriving a series form of the Poisson kernel

$$P(r, \theta) = \frac{R^2 - r^2}{R^2 - 2rR\cos\theta + r^2}, \qquad (0 \leq r < R). \tag{6.4.4}$$

Lemma 6.4.1. *For $0 \leq r < R$ and all θ, we have*

$$P(r, \theta) = \text{Re}\left(\frac{R + re^{i\theta}}{R - re^{i\theta}}\right) = 1 + 2\sum_{n=1}^{\infty} \left(\frac{r}{R}\right)^n \cos(n\theta). \tag{6.4.5}$$

Proof. To prove the second equality in (6.4.5), let $z = re^{i\theta}$. Using a geometric series expansion, we have for $|z| = r < R$,

$$\frac{R + z}{R - z} = (R + z)\frac{1}{R\left(1 - \frac{z}{R}\right)} = \frac{R + z}{R}\frac{1}{1 - \frac{z}{R}}$$

$$= \left(1 + \frac{z}{R}\right)\sum_{n=0}^{\infty}\left(\frac{z}{R}\right)^n = \sum_{n=0}^{\infty}\left(\frac{z}{R}\right)^n + \sum_{n=0}^{\infty}\left(\frac{z}{R}\right)^{n+1}$$

$$= 1 + 2\sum_{n=1}^{\infty}\left(\frac{z}{R}\right)^n = 1 + 2\sum_{n=1}^{\infty}\left(\frac{r}{R}\right)^n\left(\cos(n\theta) + i\sin(n\theta)\right),$$

where in the last equality we used $z^n = r^n e^{in\theta} = r^n(\cos(n\theta) + i\sin(n\theta))$, by Euler's identity. Comparing real parts of both sides to obtain (6.4.5). ∎

If f is piecewise continuous on $[0, 2\pi]$, let

$$a_0 = \frac{1}{2\pi}\int_0^{2\pi} f(\theta)\,d\theta; \tag{6.4.6}$$

$$a_n = \frac{1}{\pi}\int_0^{2\pi} f(\theta)\cos(n\theta)\,d\theta \qquad (n = 1, 2, \ldots); \tag{6.4.7}$$

$$b_n = \frac{1}{\pi}\int_0^{2\pi} f(\theta)\sin(n\theta)\,d\theta \qquad (n = 1, 2, \ldots). \tag{6.4.8}$$

The coefficients a_n are called the **cosine Fourier coefficients** of f and b_n the **sine Fourier coefficients** of f.

Theorem 6.4.2. *Consider the Dirichlet problem (6.4.1)–(6.4.2) with piecewise continuous boundary data f. Then the solution is*

$$u(r\,e^{i\theta}) = a_0 + \sum_{n=1}^{\infty} \left(\frac{r}{R}\right)^n \left(a_n \cos(n\theta) + b_n \sin(n\theta)\right), \quad 0 \le r < R, \qquad (6.4.9)$$

where a_0, a_n, and b_n are the Fourier coefficients of f, defined in (6.4.6)–(6.4.8).

Proof. Starting with the solution (6.4.3), we expand the Poisson integral in a series by using (6.4.5) (replace θ by $\theta - \phi$ in (6.4.5)) and get

$$u(r, \theta) = \frac{1}{2\pi} \int_0^{2\pi} f(\phi) \left(1 + 2 \sum_{n=1}^{\infty} \left(\frac{r}{R}\right)^n \cos\left(n(\theta - \phi)\right)\right) d\phi$$

$$= \frac{1}{2\pi} \int_0^{2\pi} f(\phi)\, d\phi + \frac{1}{\pi} \int_0^{2\pi} \sum_{n=1}^{\infty} \left\{\left(\frac{r}{R}\right)^n f(\phi) \cos\left(n(\theta - \phi)\right)\right\} d\phi.$$

Since f is piecewise continuous, it is bounded on $[0, 2\pi]$. Let $A \ge 0$ be such that $|f(\phi)| \le A$ for all ϕ. For fixed $0 \le r < R$, we have

$$\left|\left(\frac{r}{R}\right)^n f(\phi) \cos\left(n(\theta - \phi)\right)\right| \le A \left(\frac{r}{R}\right)^n,$$

and so the series

$$\sum_{n=1}^{\infty} \left\{\left(\frac{r}{R}\right)^n f(\phi) \cos n(\theta - \phi)\right\}$$

converges uniformly in ϕ on the interval $[0, 2\pi]$, by the Weierstrass M-test, because

$$\sum_{n=1}^{\infty} A \left(\frac{r}{R}\right)^n < \infty.$$

Hence, in view of Corollary 4.1.6 we can integrate term by term. Appealing to (6.4.6)–(6.4.8), we get

$$u(r\,e^{i\theta}) = \frac{1}{2\pi} \int_0^{2\pi} f(\phi)\, d\phi + \sum_{n=1}^{\infty} \left\{\left(\frac{r}{R}\right)^n \frac{1}{\pi} \int_0^{2\pi} f(\phi) \cos\left(n(\theta - \phi)\right) d\phi\right\}$$

$$= a_0 + \sum_{n=1}^{\infty} \left\{\left(\frac{r}{R}\right)^n \frac{1}{\pi} \int_0^{2\pi} f(\phi) \Big(\cos(n\theta)\cos(n\phi) + \sin(n\theta)\sin(n\phi)\Big) d\phi\right\}$$

$$= a_0 + \sum_{n=1}^{\infty} \left(\frac{r}{R}\right)^n \left(a_n \cos(n\theta) + b_n \sin(n\theta)\right),$$

which proves (6.4.9). ∎

Example 6.4.3. (A steady-state problem in a disk)

The temperature on the boundary of a circular plate with radius $R = 2$, center at the origin, and insulated lateral surface is the function

$$f(\theta) = \begin{cases} 100 & \text{if } 0 \leq \theta \leq \pi, \\ 0 & \text{if } \pi < \theta < 2\pi. \end{cases}$$

Express the steady-state temperature inside the plate in terms of Fourier series.

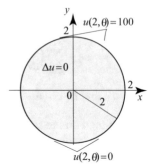

Fig. 6.27 Dirichlet problem in Example 6.4.3.

Solution. According to (6.4.9), the solution inside the disk is

$$u(re^{i\theta}) = a_0 + \sum_{n=1}^{\infty} \left(\frac{r}{2}\right)^n \left(a_n \cos(n\theta) + b_n \sin(n\theta)\right), \qquad 0 \leq r < 2, \quad (6.4.10)$$

where a_0, a_n, and b_n are the Fourier coefficients of f. Using the identities in (6.4.6)–(6.4.8), we obtain

$$a_0 = \frac{1}{2\pi} \int_0^{\pi} 100 \, d\theta = 50, \quad a_n = \frac{1}{\pi} \int_0^{\pi} 100 \cos(n\theta) \, d\theta = 0,$$

and

$$b_n = \frac{1}{\pi} \int_0^{\pi} 100 \sin(n\theta) \, d\theta = \frac{100}{n\pi} \left(1 - \cos(n\pi)\right).$$

Substituting into (6.4.10), we find the solution

$$u(re^{i\theta}) = 50 + \frac{100}{\pi} \sum_{n=1}^{\infty} \frac{1}{n} \left(1 - \cos(n\pi)\right) \left(\frac{r}{2}\right)^n \sin(n\theta), \qquad 0 \leq r < 2.$$

Notice that $1 - \cos(n\pi)$ is either 0 or 2 depending on whether n is even or odd. Thus, only odd terms survive, so we put $n = 2k + 1$ for $k = 0, 1, \ldots$, and get

$$u(re^{i\theta}) = 50 + \frac{200}{\pi} \sum_{k=0}^{\infty} \frac{1}{2k+1} \left(\frac{r}{2}\right)^k \sin\left((2k+1)\theta\right), \qquad 0 \leq r < 2. \quad (6.4.11)$$

This is the Fourier series form of the steady-state temperature inside the plate. $\square$

Fourier Series

One may guess that letting $r \uparrow R$ in Theorem 6.4.2 yields the representation

$$f(\theta) = a_0 + \sum_{n=1}^{\infty} \left[a_n \cos(n\theta) + b_n \sin(n\theta) \right], \qquad (6.4.12)$$

where a_0, a_n, b_n are defined in (6.4.6)–(6.4.8). This representation is called the **Fourier series** of f and is valid for piecewise differentiable functions as well.

Example 6.4.4. (Fourier series of a square wave) Consider the boundary function f in Example 6.4.3. Plot several partial sums of the Fourier series of f and compare the outcome with the function itself.

Solution. The boundary function in Example 6.4.3, plotted in Figure 6.28, is

$$f(\theta) = \begin{cases} 100 & \text{if } 0 \leq \theta \leq \pi, \\ 0 & \text{if } \pi < \theta < 2\pi. \end{cases} \qquad (6.4.13)$$

The graph looks like a square wave that repeats every 2π units.

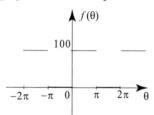

Fig. 6.28 The function f in 6.4.13.

Fig. 6.29 The Fourier series representation of f.

Setting $r = 2$ in the solution (6.4.11) and using the fact that $\lim_{r \uparrow 2} u(r e^{i\theta}) = f(\theta)$, we would expect the following

$$f(\theta) = 50 + \frac{200}{\pi} \sum_{k=0}^{\infty} \frac{\sin((2k+1)\theta)}{2k+1} \qquad (6.4.14)$$

Fourier series representation of f. This is plotted in Figure 6.29

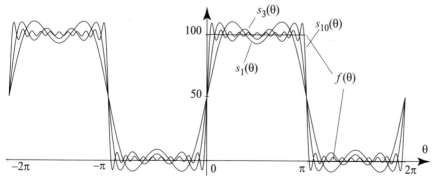

Fig. 6.30 Partial sums of the Fourier series: $s_n(\theta) = 50 + \frac{200}{\pi} \sum_{k=0}^{n} \frac{\sin((2k+1)\theta)}{2k+1}$, for $n = 1, 3, 10$. The frequencies of the sine terms increase with n, causing more oscillation in the partial sums.

To justify the representation in (6.4.14) we consider the partial sums of the Fourier series of f. These are given by

$$s_n(\theta) = 50 + \frac{200}{\pi} \sum_{k=0}^{n} \frac{\sin((2k+1)\theta)}{2k+1} \qquad (6.4.15)$$

for $n = 1, 2, 3, \ldots$ and are plotted in Figure 6.30 for certain values of n.

The Fourier series of f converges pointwise to $f(\theta)$ at each point θ where f is continuous. In particular, we have

$$100 = 50 + \frac{200}{\pi} \sum_{k=0}^{\infty} \frac{\sin((2k+1)\theta)}{2k+1} \qquad \text{for } 0 < \theta < \pi$$

and

$$0 = 50 + \frac{200}{\pi} \sum_{k=0}^{\infty} \frac{\sin((2k+1)\theta)}{2k+1} \qquad \text{for } \pi < \theta < 2\pi.$$

At the points of discontinuity ($\theta = m\pi, m = 0, \pm 1, \pm 2, \ldots$), all terms $\sin((2k+1)\theta)$ are zero and so we know that the series converges to 50. The graph of the infinite Fourier series $50 + \frac{200}{\pi} \sum_{k=0}^{\infty} \frac{\sin((2k+1)\theta)}{2k+1}$ is shown in Figure 6.29. It agrees with the graph of the function, except at the points of discontinuity. □

The Fourier coefficients of piecewise continuous functions are always unique. This requires some work to prove in general, but the case below is straightforward.

Proposition 6.4.5. *Suppose that $a_0 + \sum_{n=1}^{\infty} a_n \cos(n\theta) + b_n \sin(n\theta) = 0$ for all θ in $[0, 2\pi]$, where $\sum_{n=1}^{\infty}(|a_n| + |b_n|) < \infty$. Then $a_0 = a_n = b_n = 0$ for all $n \geq 1$.*

Proof. Let

$$f(\theta) = a_0 + a_1 \cos(\theta) + b_1 \sin(\theta) + a_2 \cos(2\theta) + b_2 \sin(2\theta) + \cdots \qquad (6.4.16)$$

In view of the hypothesis $\sum_{n=1}^{\infty}(|a_n| + |b_n|) < \infty$, Theorem 4.1.7 implies that the series defining f converges uniformly (and in particular f is continuous). Integrating f from 0 to 2π and using Corollary 4.1.6 we obtain that

$$0 = \int_0^{2\pi} a_0 \, d\theta + \sum_{n=1}^{\infty} \int_0^{2\pi} [a_n \cos(n\theta) + b_n \sin(n\theta)] \, d\theta = 2\pi a_0,$$

and from this we conclude that $a_0 = 0$. We now multiply both sides of (6.4.16) by $\cos(m\theta)$ for some $m \geq 1$, and we notice that the new series converges uniformly in θ in view of Theorem 4.1.7. Integrating this series and using Corollary 4.1.6 we find

$$0 = \sum_{n=1}^{\infty} \int_0^{2\pi} [a_n \cos(m\theta) \cos(n\theta) + b_n \cos(m\theta) \sin(n\theta)] \, d\theta. \qquad (6.4.17)$$

Since

$$\int_0^{2\pi} \cos(m\theta)\cos(n\theta)\,d\theta = \frac{1}{2}\int_0^{2\pi} \left[\cos\left((m+n)\theta\right) + \cos\left((m-n)\theta\right)\right] d\theta$$

is equal to 0 when $n \neq m$ and equals π when $n = m$, and

$$\int_0^{2\pi} \cos(m\theta)\sin(n\theta)\,d\theta = \frac{1}{2}\int_0^{2\pi} \left[\sin\left((m+n)\theta\right) + \sin\left((n-m)\theta\right)\right] d\theta = 0,$$

it follows that the expression on the right in (6.4.17) is equal to πa_m. Hence $a_m = 0$ for all $m \geq 1$. Similarly, multiplying both sides of (6.4.16) by $\sin(m\theta)$, integrating in θ, and using that $\int_0^{2\pi} \sin(m\theta)\cos(n\theta)\,d\theta = 0$ (shown above) and that

$$\int_0^{2\pi} \sin(m\theta)\sin(n\theta)\,d\theta = \frac{1}{2}\int_0^{2\pi} \left[\cos\left((m-n)\theta\right) - \cos\left((m+n)\theta\right)\right] d\theta$$

equals 0 when $n \neq m$ and equals π when $n = m$, we conclude that $\pi b_m = 0$; hence $b_m = 0$ for all $m \geq 1$. ∎

As an example, the unique Fourier coefficients of $7\cos\theta - 30\sin(50\theta)$ are $a_1 = 7$, $b_1 = 0$, $a_{50} = 0$, $b_{50} = -30$, and $a_n = b_n = 0$ for $n \neq 1, 50$.

Exercises 6.4

In Exercises 1–6, compute the Fourier coefficients of the functions on $[0, 2\pi)$.

1. $1 - \cos\theta + \sin(2\theta)$.

2. $\cos\theta - \frac{1}{2}\sin(2\theta)$.

3. $\cos^2\theta$.

4. $h(\theta) = \sin\theta$ if $\theta \in [0, \pi)$ and $h(\theta) = \cos\theta - 1$ if $\theta \in [\pi, 2\pi)$.

5. The 2π-periodic extension of $f(\theta) = \theta$ defined on $[0, 2\pi)$.

6. The 2π-periodic extension of $g(\theta) = \theta^2$ defined on $[0, 2\pi)$.

7. Project Problem: A steady-state problem with continuous boundary data, Fourier series of a triangular wave. We apply the results of this section to the Dirichlet problem in the unit disk with boundary data

$$f(\theta) = \begin{cases} \pi + \theta & \text{if } -\pi \leq \theta \leq 0, \\ \pi - \theta & \text{if } 0 < \theta < \pi. \end{cases}$$

(a) Think of the boundary function as a 2π-periodic function of θ. Plot its graph over the interval $[-4\pi, 4\pi]$. (Remember that the graph of a 2π-periodic function repeats every 2π units.)

(b) Using (6.4.9), show that the solution of the Dirichlet problem is

$$u(r, \theta) = a_0 + \sum_{n=1}^{\infty} r^n \left(a_n \cos(n\theta) + b_n \sin(n\theta)\right) \qquad (0 \leq r < 1), \tag{6.4.18}$$

where a_0, a_n, and b_n are the Fourier coefficients of f.

(c) Show that $a_0 = \frac{\pi}{2}$, $a_n = \frac{2}{\pi}\left\{\frac{1}{n^2} - \frac{\cos n\pi}{n^2}\right\}$, and $b_n = 0$.

(d) Plugging the coefficients into (6.4.18), we obtain the solution for $0 < r < 1$,

$$u(re^{i\theta}) = \frac{\pi}{2} + \sum_{n \text{ odd}} \frac{4}{\pi n^2} r^n \cos(n\theta) = \frac{\pi}{2} + \frac{4}{\pi} \sum_{k=0}^{\infty} \frac{r^{2k+1}}{(2k+1)^2} \cos\left((2k+1)\theta\right). \tag{6.4.19}$$

(e) Derive the Fourier series expansion of the triangular wave: For all θ,

$$f(\theta) = \frac{\pi}{2} + \sum_{n \text{ odd}} \frac{4}{\pi n^2} \cos(n\theta) = \frac{\pi}{2} + \frac{4}{\pi} \sum_{k=0}^{\infty} \frac{1}{(2k+1)^2} \cos\left((2k+1)\theta\right). \qquad (6.4.20)$$

(f) Illustrate the convergence of the Fourier series to $f(\theta)$ by plotting several partial sums.

8. Project Problem: Solve the Dirichlet problem on the unit disk, where the boundary values are $f(\theta) = \theta$, $0 < \theta < 2\pi$. In your solution, follow parts (a)-(d) of the previous exercise.

9. (a) Plot the graph of the 2π-periodic extension of the function $f(\theta) = \frac{1}{2}(\pi - \theta)$ defined on $(0, 2\pi]$.

$$f(\theta) = \begin{cases} \frac{1}{2}(\pi - \theta) & \text{if } 0 < \theta \le 2\pi, \\ [6pt] f(\theta + 2\pi) & \text{otherwise.} \end{cases}$$

(b) Derive the Fourier series

$$f(\theta) = \sum_{n=1}^{\infty} \frac{\sin(n\theta)}{n}.$$

10. Let the 2π-periodic function f be defined on the interval $[-\pi, \pi)$ by $f(\theta) = |\theta|$ if $-\pi \le \theta < \pi$.
Derive the Fourier series $f(\theta) = \frac{\pi}{2} - \frac{4}{\pi} \sum_{k=0}^{\infty} \frac{1}{(2k+1)^2} \cos\left((2k+1)\theta\right)$.

11. Let the 2π-periodic function f be defined on the interval $[-\pi, \pi)$ by

$$f(\theta) = \begin{cases} 1 & \text{if } 0 < \theta < \pi/2, \\ -1 & \text{if } -\pi/2 < \theta < 0, \\ 0 & \text{if } \pi/2 < |\theta| < \pi. \end{cases}$$

Prove that the Fourier series of this function is

$$f(\theta) = \frac{2}{\pi} \sum_{n=1}^{\infty} \frac{1}{n} \left(1 - \cos\frac{n\pi}{2}\right) \sin(n\theta).$$

12. Show that the Fourier series of the 2π-periodic function $|\sin\theta|$, $-\pi \le \theta \le \pi$, is

$$\frac{2}{\pi} - \frac{4}{\pi} \sum_{k=1}^{\infty} \frac{1}{(2k)^2 - 1} \cos(2k\theta).$$

13. Show that the Fourier series of the 2π-periodic function $|\cos\theta|$, $-\pi \le \theta \le \pi$, is

$$\frac{2}{\pi} - \frac{4}{\pi} \sum_{k=1}^{\infty} \frac{(-1)^k}{(2k)^2 - 1} \cos(2k\theta).$$

14. Reflecting and translating a Fourier series. Suppose that f is 2π-periodic, and let $g(\theta) = f(-\theta)$ and $h(\theta) = f(\theta - \alpha)$, where α is a fixed real number. To avoid confusion we use $a(\phi, n)$ and $b(\phi, n)$ instead of a_n and b_n to denote the Fourier coefficients of a function ϕ.
(a) Show that $a(f, 0) = a(g, 0)$, $a(f, n) = a(g, n)$, and $b(f, n) = -b(g, n)$ for all $n \ge 1$.
(b) Show that $a(f, 0) = a(h, 0)$ and that for $n \ge 1$ we have

$$a(h, n) = a(f, n)\cos(n\alpha) - b(f, n)\sin(n\alpha)$$
$$b(h, n) = a(f, n)\sin(n\alpha) + b(f, n)\cos(n\alpha).$$

Chapter 7
Conformal Mappings

First, it is necessary to study the facts, to multiply the number of observations, and then later to search for formulas that connect them so as thus to discern the particular laws governing a certain class of phenomena. In general, it is not until after these particular laws have been established that one can expect to discover and articulate the more general laws that complete theories by bringing a multitude of apparently very diverse phenomena together under a single governing principle
-Augustin-Louis Cauchy (1789–1857)

This chapter presents a sampling of successful applications of complex analysis in applied mathematics, engineering, and physics. After laying down the theory and methods of conformal mappings we discuss Dirichlet problems; in particular, we derive Poisson's integral formula in the upper half-plane and other regions, by performing a suitable change of variables. In Section 7.4, we broaden the scope of our applications with the Schwarz-Christoffel transformation, which is a method for finding conformal mappings of regions bounded by polygonal paths. The section contains interesting applications from fluid flow.

In Section 7.5, we derive a Poisson integral formula for simply connected regions in terms of the so-called Green's function. The solution is based on a simple change of variables and the mean value property of harmonic functions. This approach illustrates in a natural way the applicability of Green's functions to the solution of boundary value problems.

In Section 7.6, we tackle other famous boundary value problems, including Laplace's equation and Poisson's equation. We follow the approach of changing variables using conformal mappings, and we derive general formulas for the solutions in terms of Green's functions and the Neumann function. Both functions are computed explicitly in important special cases.

7.1 Basic Properties

In this section we present some basic mapping properties of analytic functions. These properties will be useful in the solutions of partial differential equations involving the **Laplacian** $\Delta u = \frac{\partial^2 u}{\partial x^2} + \frac{\partial^2 u}{\partial y^2}$. We start with a review from calculus of the notion of tangent lines to curves in parametric form. We state these results using the convenient complex notation.

© Springer International Publishing AG, part of Springer Nature 2018
N. H. Asmar and L. Grafakos, *Complex Analysis with Applications*,
Undergraduate Texts in Mathematics, https://doi.org/10.1007/978-3-319-94063-2_7

Suppose that γ is a smooth path parametrized by $z(t) = x(t) + iy(t)$, $a \leq t \leq b$. Assume that $z'(t) = x'(t) + iy'(t) \neq 0$ for all t, which guarantees the existence of a tangent. The tangent line to the curve at $z_0 = z(t)$ has direction of $z'(t)$; we can characterize this direction by specifying $\arg z'(t)$ (see Figure 7.1). Thus the direction of the tangent line at a point on a path $z(t)$ is determined by the argument of $z'(t)$.

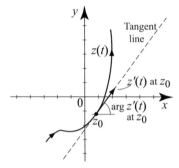
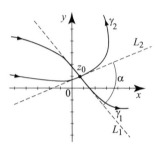

Fig. 7.1 The direction of the tangent line at $z(t)$ is determined by $\arg z'(t)$.

Fig. 7.2 The curves γ_1 and γ_2 intersect at angle α.

Let z_0 be a point in the z-plane, let γ_1 and γ_2 be two smooth paths that intersect at z_0, and let L_1 and L_2 denote the tangent lines to γ_1 and γ_2 at z_0. We say that γ_1 and γ_2 intersect at angle α at z_0 if the lines L_1 and L_2 intersect at angle α at z_0 (Figure 7.2).

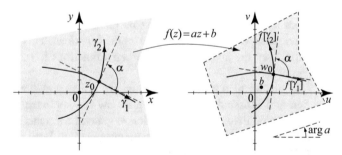

Fig. 7.3 A linear mapping $f(z) = az + b$ $(a \neq 0)$ rotates by an angle $\arg a$, dilates by a factor $|a|$, and translates by b. In particular, it preserves angles between curves.

To explain the geometric meaning of the mapping properties discussed in this section, let us consider the simple example of a linear mapping $f(z) = az + b$, where $a \neq 0$ and b are complex numbers. As usual, we consider a mapping as taking points in the z-plane to points in the w-plane. Using our geometric interpretation of addition and multiplication of complex numbers, we see that the effect of the linear mapping $f(z) = az + b$ is to rotate by the fixed angle $\arg a$, dilate by the factor $|a|$, and then translate by b. (Note that the rotation and dilation commute, so it does not matter which one we apply first. But we cannot change the order of the translation;

it comes last.) In particular, if γ_1 and γ_2 are two smooth paths that intersect at angle α at z_0, then their images under f are two paths in the w-plane that intersect at $w_0 = f(z_0)$; since $f(z)$ has rotated each curve by the same angle $\arg a$, the angle of their intersection in the w-plane is still α. Furthermore, this mapping preserves the orientation of γ_1 (as being either clockwise or counterclockwise) with respect to γ_2 (Figure 7.3). With this example in mind, we introduce the following definition.

Definition 7.1.1. A mapping defined in a neighborhood of a point z_0 is said to be **conformal** at z_0 if it preserves the magnitude and the orientation of the angles between curves intersecting at z_0.

Now suppose that f is analytic at z_0 and $f'(z_0) \neq 0$. Then f is defined and is analytic in a neighborhood of z_0, and from the study of the derivative in Section 2.3 we know that f is approximately linear in a neighborhood of z_0. More precisely, for z near z_0 we have

$$f(z) = f(z_0) + f'(z_0)(z - z_0) + \varepsilon(z)(z - z_0),$$

where $\varepsilon(z) \to 0$ as $z \to z_0$. So, near z_0, we can write $f(z) \approx f'(z_0)z + b$, where $b = -f'(z_0)z_0 + f(z_0)$ is a constant. From what we just discovered about linear mappings, this suggests that f is conformal at z_0; it rotates by an angle $\theta = \arg f'(z_0)$ and scales by a factor $|f'(z_0)|$. Indeed, we have the following important result.

Theorem 7.1.2. (Conformal Property) *Suppose that f is analytic in a neighborhood of a point z_0 in $\mathbb{C}$ and suppose that $f'(z_0) \neq 0$. Then f is conformal at z_0.*

Proof. Let γ be a smooth path through z_0, parametrized by $\gamma(t) = x(t) + iy(t)$, with $\gamma(t_0) = z_0$ and $\gamma'(t_0) \neq 0$. Then the image of γ by f is a path parametrized by $f(\gamma(t))$ that passes through $w_0 = f(z_0)$ in the w-plane. Since $\gamma'(t_0) \neq 0$, the direction of the tangent line to γ at z_0 is $\arg \gamma'(t_0)$. Also, from the chain rule, $\frac{d}{dt} f(\gamma(t))\big|_{t_0} = f'(\gamma(t_0))\gamma'(t_0) \neq 0$ and the direction of the tangent line to $f \circ \gamma$ at $w_0 = f(z_0)$ is

$$\arg \left(\frac{d}{dt} f(\gamma(t))\Big|_{t_0} \right) = \arg \left(f'(\gamma(t_0))\gamma'(t_0) \right) = \arg f'(\gamma(t_0)) + \arg \gamma'(t_0).$$

It follows that
$$\arg(f \circ \gamma)'(t_0) - \arg \gamma'(t_0) = \arg f'(z_0),$$

and consequently

$$\arg(f \circ \gamma_1)'(t_0) - \arg(f \circ \gamma_2)'(t_0) = \arg \gamma_1'(t_0) - \arg \gamma_2'(t_0)$$

for any two curves with $\gamma_1(t_0) = z_0 = \gamma_2(t_0)$ and $\gamma_1'(t_0) \neq 0 \neq \gamma_2'(t_0)$. Thus the oriented angle of intersection between two tangent lines of curves intersecting at z_0 is the same as that for the images of these curves under the mapping f. This proves that f is a conformal mapping. ∎

Boundary Behavior

We use conformal mappings to transform boundary value problems consisting of Laplace's equation along with boundary conditions. To handle the effect of the mapping on the boundary conditions, it would be nice to know that the boundary is mapped to the boundary. But as the following simple example shows, this may fail in general.

Example 7.1.3. (Boundary points mapped to interior points) Consider $f(z) = z^2$ and $\Omega = \{z = re^{i\theta} : \frac{1}{2} < r < 1, \ -\pi < \theta < \pi\}$. Then f is analytic and $f'(z) = 2z \neq 0$ for all z in Ω. Hence f is conformal at each z in Ω. Since $z^2 = r^2 e^{2i\theta}$, it is easy to see that $f[\Omega]$ is the annulus

$$f[\Omega] = \{w = re^{i\theta} : \tfrac{1}{4} < r < 1, \ -2\pi < \theta < 2\pi\} = \{w = re^{i\theta} : \tfrac{1}{4} < r < 1, \ 0 \leq \theta \leq 2\pi\}.$$

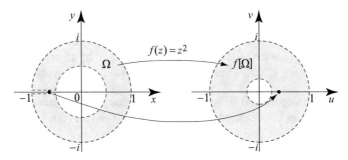

Fig. 7.4 The function $f(z) = z^2$ is conformal in Ω. It takes the interval $(-1, -\frac{1}{2})$ on the boundary of Ω to the interval $(\frac{1}{4}, 1)$ in the interior of $f[\Omega]$. Thus f does not map boundary points to boundary points.

It is also easy to see that the boundary points $z = x$, $-1 \leq x \leq -\frac{1}{2}$ are mapped to the *interior* points $w = u$, $\frac{1}{4} \leq u \leq 1$ (see Figure 7.4). Thus f does not map the boundary of Ω to the boundary of $f[\Omega]$. □

The condition $f'(z) \neq 0$ for all z in a region Ω ensures only that f is one-to-one locally, and not necessarily one-to-one on the entire region; see Theorem 5.7.13 (Inverse Function Theorem) and Example 7.1.3. We show below that if f is analytic and one-to-one, then it maps the boundary of its domain to the boundary of the range. Before we do so, let us clarify certain issues. We know from Corollary 5.7.13 (Mapping of Regions) that if f is analytic and nonconstant on a region Ω, then $f[\Omega]$ is a region. So all the points in Ω are mapped to the open connected set $f[\Omega]$. Now f might not be defined on the boundary of Ω, so we need a special definition to describe how f maps the boundary of Ω.

Definition 7.1.4. Let f be a mapping from a region Ω to the complex plane.
(i) We say that f **maps the boundary of Ω to the boundary of** $f[\Omega]$ if the following

condition holds: If $\{z_n\}_{n=1}^\infty$ is any sequence in Ω converging to α_0 on the boundary of Ω and β is any point in $f[\Omega]$, then $\{f(z_n)\}_{n=1}^\infty$ does not converge to β. So if $\{f(z_n)\}_{n=1}^\infty$ converges, it must converge to a boundary point of $f[\Omega]$ or to infinity. If the region Ω is unbounded, we allow α_0 to be ∞.

(ii) We say that f **maps the boundary of Ω onto the boundary of $f[\Omega]$** if f maps the boundary of Ω to the boundary of $f[\Omega]$ and, additionally, for every point w_0 on the boundary of $f[\Omega]$ there is a sequence z_n in Ω with $z_n \to \alpha_0$, α_0 being on the boundary of Ω such that $f(z_n) \to w_0$.

There are examples where $\{f(z_n)\}_{n=1}^\infty$ does not converge where $\{z_n\}_{n=1}^\infty$ converges to a point on the boundary of a region; see Exercise 19.

Theorem 7.1.5. (Boundary Behavior) *Suppose that f is analytic and one-to-one on a region Ω. Then f maps the boundary of Ω onto the boundary of $f[\Omega]$.*

Proof. We first prove that f maps the boundary of Ω to the boundary of $f[\Omega]$. Since f is one-to-one, it is nonconstant and, in view of Corollary 5.7.15, $f[\Omega]$ is a region. Let β be a point of $f[\Omega]$ and let $\{z_n\}_{n=1}^\infty$ be a sequence in Ω such that $f(z_n) \to \beta$. Since f is analytic and one-to-one, f^{-1} exists and is analytic, and hence continuous. Thus

$$\lim_{n \to \infty} z_n = \lim_{n \to \infty} \left(f^{-1}(f(z_n)) \right) = f^{-1}\left(\lim_{n \to \infty} f(z_n) \right) = f^{-1}(\beta),$$

so $\{z_n\}_{n=1}^\infty$ converges to an interior point of Ω. Thus if $\{z_n\}_{n=1}^\infty$ converges to a point α_0 on the boundary of Ω, then $\{f(z_n)\}_{n=1}^\infty$ cannot converge to β.

We now turn to the "onto" assertion of the theorem. Let w_0 be on the boundary of $f[\Omega]$; then we can find points $w_n = f(z_n)$ such that z_n is in Ω and $w_n \to w_0$. We distinguish two cases. If the sequence $\{z_n\}_{n=1}^\infty$ is unbounded, then a subsequence $\{z_{n_j}\}_{j=1}^\infty$ tends to ∞, which is a point on the boundary of Ω, and $f(z_{n_j}) \to w_0$. If $\{z_n\}_{n=1}^\infty$ is bounded, then by the Bolzano-Weierstrass property, we can find a subsequence $\{z_{n_j}\}_{j=1}^\infty$ that converges to a point α_0 in the closure $\overline{\Omega}$ of Ω. Notice that if α_0 were in Ω, then by the continuity of f we would have $f(z_{n_j}) \to f(\alpha_0)$. But since $f(z_{n_j}) \to w_0$, it would follow that $f(\alpha_0) = w_0$, which is a contradiction since $f(\alpha_0)$ lies in the region $f[\Omega]$ while w_0 lies on the boundary of $f[\Omega]$. Thus α_0 lies on the boundary of Ω and the claim is proved. ∎

If the conformal mapping can be extended to a continuous function on the boundary, the following version of Theorem 7.1.5 will be useful.

Corollary 7.1.6. *Suppose that f is analytic and one-to-one in a region Ω. Suppose that f extends continuously to the closure $\overline{\Omega}$. Then for any point α of the boundary of Ω the point $f(\alpha)$ lies on the boundary of $f[\Omega]$.*

Proof. Let $\{z_n\}_{n=1}^\infty$ be a sequence in Ω converging to α on the boundary of Ω. Since f is continuous at α, $f(z_n)$ converges to $f(\alpha)$. By Theorem 7.1.5, f maps the boundary of Ω to the boundary of Ω, which means that $f(z_n)$ does not converge to an interior point of Ω. Thus $f(\alpha)$ is a boundary point of $f[\Omega]$. ∎

The fact that the boundary is mapped to the boundary helps determine the image of a region from our knowledge of the image of the boundary and one interior point. We illustrate this useful technique by revisiting Example 1.7.8.

Example 7.1.7. (Mapping of regions) Let $\Omega = \{z = x + iy : -\frac{\pi}{2} < x < \frac{\pi}{2}, y > 0\}$ be the semi-infinite vertical strip shown in Figure 7.5. Let $f(z) = \sin z$ defined on Ω. Since $\sin z_1 = \sin z_2$ if and only if $z_1 = z_2 + 2k\pi$ (k an integer), we see that f is one-to-one on Ω. Also, f is continuous on the boundary of Ω.

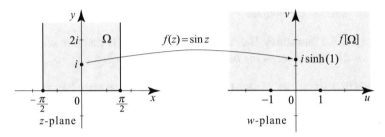

Fig. 7.5 The fact that the boundary of $f[\Omega]$ is the u-axis implies that $f[\Omega]$ is either the upper or lower half-plane. We decide which half it is by checking the image of one point. Note how the right angles at $z = \pm\frac{\pi}{2}$ got flattened by f in the w-plane. The function f is still conformal in Ω, even though f is not conformal at two points on the boundary.

Solution. We start by determining the image of the boundary. For $z = x$ real, we have $f(z) = \sin x$, and so f maps the interval $-\frac{\pi}{2} \leq x \leq \frac{\pi}{2}$ onto the interval $[-1, 1]$. For $z = \frac{\pi}{2} + iy$, we have $f(z) = \sin(\frac{\pi}{2} + iy) = \cosh y$, which is a real number; see (1.7.17). So f maps the vertical semi-infinite line $z = \frac{\pi}{2} + iy$ ($y \geq 0$) onto the semi-infinite interval $[1, \infty)$. For $z = -\frac{\pi}{2} + iy$, we have $f(z) = \sin(-\frac{\pi}{2} + iy) = -\cosh y$; see again (1.7.17). So f maps the vertical semi-infinite line $z = -\frac{\pi}{2} + iy$ ($y \geq 0$) onto the semi-infinite interval $(-\infty, -1]$. Thus, f maps the boundary of Ω to the real axis. According to Theorem 7.1.5, the image of the vertical strip has boundary the u-axis, so it is either the upper or the lower half-plane. Checking the value of f at one point in Ω, say $z = i$, we find $f(i) = \sin(i) = i\sinh(1)$, which is a point in the upper half. Thus the image of Ω is the upper half-plane. $\qquad\square$

As a further application, we consider the **Joukowski function**

$$J(z) = \frac{1}{2}\left(z + \frac{1}{z}\right) \quad (z \neq 0). \tag{7.1.1}$$

This function has applications in aerospace engineering.

Example 7.1.8. (The Joukowski mapping)
(a) Show that the Joukowski function J maps $\sigma = \{z : z = e^{i\theta}, 0 \leq \theta \leq \pi\}$ (the

upper unit semicircle) onto the real interval $J[\sigma] = [-1, 1]$ and also maps the semi-infinite intervals $[1, \infty)$ and $(-\infty, -1]$ onto themselves.

(b) Show that the Joukowski function maps the set $\Omega = \{z : |z| \geq 1, 0 \leq \operatorname{Arg} z \leq \pi\}$ onto the upper half-plane $\{w = u + iv : v \geq 0\}$ (see Figure 7.6). (A more precise description of the Joukowski mapping is outlined in Exercise 17.)

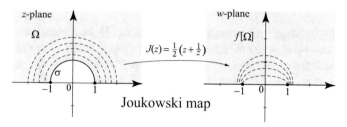

Fig. 7.6 The Joukowski function maps the region Ω one-to-one onto the upper half-plane. It also maps the upper semi-circle of radius $R > 1$, $x^2 + y^2 = R^2$, $y \geq 0$, to the upper semi-ellipse $\dfrac{(\operatorname{Re} w)^2}{\left[\frac{1}{2}\left(R + \frac{1}{R}\right)\right]^2} + \dfrac{(\operatorname{Im} w)^2}{\left[\frac{1}{2}\left(R - \frac{1}{R}\right)\right]^2} = 1$.

Solution. (a) For $z = e^{i\theta}$, $0 \leq \theta \leq \pi$, we have

$$w = J(z) = \frac{1}{2}\left(e^{i\theta} + \frac{1}{e^{i\theta}}\right) = \frac{1}{2}(e^{i\theta} + e^{-i\theta}) = \cos\theta.$$

As θ varies from 0 to π, w varies from 1 to -1, showing that the image of the upper semicircle is the interval $[-1, 1]$. To determine the images of the semi-infinite intervals $[1, \infty)$ and $(-\infty, -1]$, let $z = x$; then $J(z) = \frac{1}{2}(x + \frac{1}{x})$. As x varies through $[1, \infty)$ or $(-\infty, -1]$, $J(x)$ varies through the same interval (in the w-plane).

(b) We showed in (a) that J maps the boundary of Ω onto the real axis. If we can show that J is conformal and one-to-one, it will follow from Corollary 7.1.6 that $J[\Omega]$ is either the upper or the lower half of the w-plane. We can then determine which half it is by checking the image of one point in Ω. That J is conformal at all points in Ω is clear from $J'(z) = \frac{1}{2}\left(1 - \frac{1}{z^2}\right) \neq 0$ for all $|z| > 1$. To show that J is one-to-one, let z_1, z_2 be in Ω, and suppose that $J(z_1) = J(z_2)$. Then

$$z_1 + \frac{1}{z_1} = z_2 + \frac{1}{z_2} \Rightarrow \frac{z_1^2 + 1}{z_1} = \frac{z_2^2 + 1}{z_2}$$

$$\Rightarrow z_2 z_1^2 + z_2 - z_1 z_2^2 - z_1 = 0$$

$$\Rightarrow (z_1 - z_2)(z_1 z_2 - 1) = 0.$$

So either $z_1 = z_2$ or $z_1 z_2 = 1$. Since $1 < |z_1|$ and $1 < |z_2|$, we cannot have $z_1 z_2 = 1$. So $z_1 = z_2$, implying that J is one-to-one. We have $J(2i) = \frac{1}{2}(2i + \frac{1}{2i}) = \frac{3i}{4}$. Since $J(2i)$ is in the upper half-plane, we conclude that $J[\Omega]$ is the upper half-plane. $\square$

As we observed after Example 7.1.3, the condition $f'(z) \neq 0$ for all z in a region Ω is not enough to ensure that f is one-to-one on Ω. However, if f is analytic and one-to-one on the whole region Ω, then $f'(z) \neq 0$ for all z in Ω, so f is a conformal mapping. Moreover, from Section 5.7, we know that the inverse function exists on $\Omega' = f[\Omega]$ and is analytic and one-to-one. In this situation, we say that Ω and Ω' are **conformally equivalent** regions.

The following famous theorem of Riemann states that any simply connected region of the complex plane other than the plane itself is conformally equivalent to the open unit disk.

Theorem 7.1.9. (Riemann Mapping Theorem) *Let Ω be a simply connected region in the complex plane other than the complex plane itself. Then there is a one-to-one analytic function f that maps Ω onto the unit disk $|w| < 1$. The mapping f is unique if we specify that $f(z_0) = 0$ and $f'(z_0) > 0$, for some z_0 in Ω.*

The proof of the Riemann mapping theorem extends beyond the scope of this book. A proof of this theorem can be found in L. Ahlfors: *Complex Analysis, 3rd Ed., Inter. Ser. in Pure & Applied Math.* McGraw-Hill Ed, 1979 or in J. Conway: *Functions of One Complex Variable I, GTM 11, 2nd Ed.*, Springer 1978.

In the next section we pursue the construction of important conformal mappings. Some of these are then used to solve Dirichlet problems in general simply connected domains by transforming the Dirichlet problem to the unit disk. This is achieved in Section 7.3.

Exercises 7.1

In Exercises 1–6, determine where the mappings are conformal.

1. $\dfrac{z^2+1}{e^z}$ **2.** $\dfrac{z+1}{z^2+2z+1}$ **3.** $\dfrac{\sin z}{e^z}$

4. $\dfrac{e^z+1}{e^z-1}$ **5.** $z+\dfrac{1}{z}$ **6.** $\dfrac{z+1}{z+i}+\dfrac{2}{z}$

In Exercises 7–12, determine the angle of rotation $\alpha = \arg f'(z)$ and the dilation factor $|f'(z)|$ of the mappings at the indicated points.

7. $\dfrac{1}{z},\ z=1,i,1+i$ **8.** $\mathrm{Log}\,z,\ z=1,i,-i$ **9.** $\sin z,\ z=0,\ \pi+ia,\ i\pi$

10. $z^2,\ z=1,2i,-1-i$ **11.** $e^{iz},\ z=\pi,i\pi,\dfrac{\pi}{2}$

12. $\dfrac{1+e^z}{1-e^z},\ z=1,i\pi,\ln(3)+2i$

In Exercises 13–16 determine the images of the orthogonal lines $x=a$ and $y=b$ (a and b are real numbers) under the following mappings f. For which values of a and b are the images orthogonal at the point of intersection $f(a+ib)$? Verify your answer by computing the angle between the image curves at the point $f(a+ib)$.

13. e^z **14.** $\sin z$ **15.** $(1+i)z$ **16.** $\dfrac{z+1}{z-1}$

17. The Joukowski function. Refer to Example 7.1.8.
(a) Show that $J(\frac{1}{z}) = J(z)$ for all $z \neq 0$.
(b) Fix $R > 1$. Show that the upper semicircle of radius R, $S_R = \{z : |z| = R,\ 0 \le \mathrm{Arg}\,z\}$, is mapped onto the upper semi-ellipse

$$J[S_R] = \left\{ w = u+iv : \frac{u^2}{\left[\frac{1}{2}\left(R+\frac{1}{R}\right)\right]^2} + \frac{v^2}{\left[\frac{1}{2}\left(R-\frac{1}{R}\right)\right]^2} = 1,\ v \ge 0 \right\}.$$

18. Let $f(z) = \frac{z}{(1-z)^2}$. (a) Show that f is analytic on $\mathbb{C} \setminus \{1\}$ and $f'(z) \neq 0$ for $z \neq -1$. (b) Show that f is one-to-one in the open unit disk but is not one-to-one in any larger disk centered at 0.

19. Let Ω be the slit plane, $\mathbb{C} \setminus \{z : z \leq 0\}$, and $f(z) = \operatorname{Log} z$. Prove that $z_n = -1 + i\frac{(-1)^n}{n}$ $(n = 1, 2, \ldots)$ converge to -1 on the boundary of Ω and yet $\{f(z_n)\}_{n=1}^{\infty}$ does not converge.

7.2 Linear Fractional Transformations

Our success in solving boundary value problems involving Laplace's equation is closely tied to our ability to construct conformal mappings between regions in the plane. A good place to start our study of special conformal mappings is on the unit disk, since there we have a general formula for the solution of the Dirichlet problem, namely the Poisson integral formula. As we will soon see, the most suitable mappings for regions involving disks and lines are the linear fractional transformations:

$$\phi(z) = \frac{az+b}{cz+d} \qquad (ad \neq bc). \tag{7.2.1}$$

Since $\phi'(z) = \frac{ad-bc}{(cz+d)^2}$, the condition $ad \neq bc$ ensures that ϕ does not degenerate into a constant. If $c = 0$, the linear fractional transformation reduces to a linear function, which is analytic everywhere or entire. If $c \neq 0$, then ϕ is analytic for all $z \neq -\frac{d}{c}$ and has a simple pole at $z = -\frac{d}{c}$.

Proposition 7.2.1. *Let ϕ be the linear fractional transformation defined in (7.2.1), where a, b, c, d are complex numbers with $c \neq 0$ and $ad \neq bc$.*
(i) Then ϕ has a pole $z = -\frac{d}{c}$ and is one-to-one and conformal on $\mathbb{C} \setminus \{-\frac{d}{c}\}$; its range is $\mathbb{C} \setminus \{\frac{a}{c}\}$.
(ii) The inverse of ϕ is the linear fractional transformation ψ from $\mathbb{C} \setminus \{\frac{a}{c}\}$ onto $\mathbb{C} \setminus \{-\frac{d}{c}\}$ defined by

$$\psi(w) = \frac{dw-b}{-cw+a}. \tag{7.2.2}$$

Proof. We have

$$\phi'(z) = \frac{ad-bc}{(cz+d)^2} \neq 0 \quad \text{for all} \quad z \neq -\frac{d}{c}.$$

Hence, by Theorem 7.1.2, the mapping is conformal at all $z \neq -\frac{d}{c}$ and obviously it has a pole of order 1 at $z = -\frac{d}{c}$. To find the inverse function, we solve $w = \frac{az+b}{cz+d}$ for $z \neq -\frac{d}{c}$ and get $z = \frac{dw-b}{-cw+a}$ for $w \neq \frac{a}{c}$. Since ϕ has an inverse ψ, it must be one-to-one, since $\phi(z_1) = \phi(z_2)$, implies $(\psi \circ \phi)(z_1) = (\psi \circ \phi)(z_2)$ i.e., $z_1 = z_2$. ∎

It is not hard to see that every linear fractional transformation (7.2.1) with $c \neq 0$ is a composition of a linear mapping $w_1 = cz + d$, of an inversion $w_2 = \frac{1}{w_1}$, and of a linear mapping $w = \frac{a}{c} + (b - \frac{ad}{c})w_2$ (see Exercise 39). Now linear mappings have a very useful property that is easy to verify: They map a line to a line and a circle to a circle. The inversion $w_2 = \frac{1}{w_1}$ has a somewhat similar property: It maps the collection of lines and circles in the z-plane to the collection of lines and circles in the w-plane. (Unlike linear mappings, the inversion may map a line to a circle or a circle to a line; see Figure 7.7.)

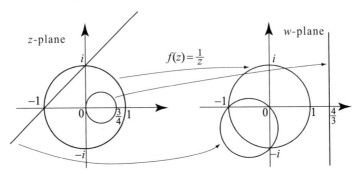

Fig. 7.7 The inversion $f(z) = \frac{1}{z}$ preserves the collection of lines and circles. To verify the images of the given lines and circles, use the fact that f is conformal (preserves angles) and the values $f(0) = \infty$; $f(\infty) = 0$; $f(1) = 1$; $f(-1) = -1$; $f(i) = -i$; $f(-i) = i$; $f(\frac{3}{4}) = \frac{4}{3}$.

This property of the inversion is a bit cumbersome to verify and is sketched in Exercises 33–39. Since a linear fractional transformation is a composition of linear mappings and an inversion, it inherits this property too. Thus we obtain the following very useful result.

Proposition 7.2.2. (Images of Lines and Circles) *A linear fractional transformation, as in (7.2.1), maps lines and circles to lines and circles.*

It follows immediately from Proposition 7.2.1 and Theorem 7.1.5 that a linear fractional transformation maps the boundary of its domain to boundary of its range. As we now illustrate, this property is very useful in determining the image of a region.

Example 7.2.3. (Mappings between the unit disk and the upper half-plane)
(a) Show that the linear fractional transformation

$$\phi(z) = i\frac{1-z}{1+z}$$

maps the unit disk onto the upper half-plane.
(b) Show that the linear fractional transformation

$$\psi(z) = \frac{i-z}{i+z}$$

maps the upper half-plane onto the unit disk.

Solution. (a) By Proposition 7.2.2, the image of the circle $C_1(0)$ is either a line or a circle in the w-plane. Since three points will determine either a line or circle, it suffices to check the images of three points on $C_1(0)$. We have

$$\phi(1) = 0; \quad \phi(i) = i\frac{1-i}{1+i} = 1; \quad \phi(-i) = i\frac{1+i}{1-i} = -1.$$

Thus $\phi(1)$, $\phi(i)$, and $\phi(-i)$ lie on the u-axis (the real axis in the w-plane), and so the image of $C_1(0)$ is the u-axis. As ϕ is one-to-one, it maps the boundary $C_1(0)$ onto the boundary of the image of the unit disk. Thus the image of the unit disk is either the upper half-plane or the lower half-plane. Checking $\phi(0) = i$ (a point in the upper half-plane), we conclude that ϕ maps the unit disk one-to-one onto the upper half-plane (see Figure 7.8). Note also that since ϕ maps the closed unit disk to an unbounded region (the upper half-plane), it has to be discontinuous somewhere in the closed unit disk. Indeed it is singular at $z = -1$.

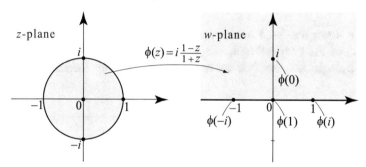

Fig. 7.8 The linear fractional transformation $\phi(z) = i\frac{1-z}{1+z}$ maps the unit circle to the real line and the unit open unit disk to the open upper half plane.

(b) We can do this part in two ways. One way is to use Proposition 7.2.1(ii) and notice that ψ is the inverse of ϕ. Another way is to check the image by ψ of the boundary and one interior point. We leave it as an exercise to verify that $\psi(0) = 1$, $\psi(1) = i$, and $\psi(-1) = -i$. Since the images of the three points are not collinear, we conclude that the real axis is mapped onto the circle that goes through the points 1, i, and $-i$, which is clearly the unit circle. (Here again, we are using the fact that three points determine a circle.) Also, $\psi(i) = 0$; hence ψ maps the upper half-plane onto the unit disk.   □

Another way to realize that the image of the unit circle is a line in Example 7.2.3(a) is to consider the point -1 on $C_1(0)$ and note that $\lim_{z \to -1} \phi(z) = \infty$. So the image of $C_1(0)$ is not bounded and since it is either a line or a circle, it has to be a line (which tends to infinity). Sometimes it is convenient to express the fact that the limit at the point $z_0 = -\frac{d}{c}$ is infinity by writing $\phi(z_0) = \infty$. Likewise, it will be convenient to express that $\lim_{z \to \infty} \frac{az+b}{cz+d} = \frac{a}{c}$ by simply writing $\phi(\infty) = \frac{a}{c}$.

Before the next example we discuss another useful property of linear fractional transformations.

Proposition 7.2.4. (Composition of Mappings) *The composition of any two linear fractional transformations is another linear fractional transformation.*

Proof. Let

$$\phi_1(z) = \frac{a_1 z + b_1}{c_1 z + d_1} \quad \text{and} \quad \phi_2(z) = \frac{a_2 z + b_2}{c_2 z + d_2}.$$

Then

$$\phi(z) = (\phi_2 \circ \phi_1)(z) = \frac{a_2 \frac{a_1 z + b_1}{c_1 z + d_1} + b_2}{c_2 \frac{a_1 z + b_1}{c_1 z + d_1} + d_2}.$$

Multiplying numerator and denominator by $c_1 z + d_1$, we get

$$\phi(z) = \frac{(a_2 a_1 + b_2 c_1) z + a_2 b_1 + b_2 d_1}{(c_2 a_1 + d_2 c_1) z + c_2 b_1 + d_2 d_1},$$

which is a linear fractional transformation. Notice that when we multiplied by the quantity $c_1 z + d_1$, we removed the singularity at $z = -\frac{d_1}{c_1}$; the resulting composition $\phi(z)$ has a single pole, and it is not necessarily at the same location as the poles of ϕ_1 or ϕ_2. ∎

Example 7.2.5. (Composition of linear fractional transformations) Find a linear fractional transformation that maps the disk $B_2(-1)$ with radius 2 and center at -1 onto the right half-plane $\text{Re}\, w > 0$.

Solution. We describe two methods to solve this problem. Let us start with the quickest one based on the result of Example 7.2.3 and a simple application of Proposition 7.2.4. We know that

$$\phi(z) = i\frac{1-z}{1+z}$$

maps the unit disk onto the upper half-plane. It is also easy to see that the linear mapping $\tau(z) = \frac{1}{2}(z+1)$ translates the center of $B_2(-1)$ to the origin and then scales the radius by $\frac{1}{2}$. Thus τ maps $B_2(-1)$ onto the unit disk. Consequently, $\phi \circ \tau$ is a linear fractional transformation that maps $B_2(-1)$ onto the upper half-plane. To map onto the right half-plane, it suffices to rotate the upper half-plane by $-\frac{\pi}{2}$. This can be achieved by multiplying by $e^{-i\frac{\pi}{2}} = -i$. So the desired linear fractional transformation (Figure 7.9) is

$$f(z) = -i\phi \circ \tau(z) = (-i)i\frac{1-\frac{1}{2}(z+1)}{1+\frac{1}{2}(z+1)} = \frac{1-z}{3+z}.$$

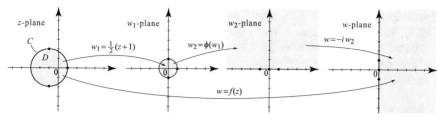

Fig. 7.9 To map a disk to a half-plane, it is always advantageous to map first to the unit disk and then use the transformation ϕ in Example 7.2.3.

Another way to go about this problem is to start from scratch; we want a linear fractional transformation $g(z) = \frac{az+b}{cz+d}$ to map the boundary of the disk onto the boundary of the right half-plane. We can pick any three points on the circle C and map them to any three points on the imaginary axis. Since our image boundary is a line (which extends to infinity), we may map one of our points to ∞. We pick

$$g(1) = 0, \quad g(-3) = \infty, \quad g(i\sqrt{3}) = i. \tag{7.2.3}$$

We use these equations to solve for the coefficients a, b, c, and d, and then we check whether the interior of the disk is mapped to the right half-plane or the left half-plane. Again writing $g(z) = \frac{az+b}{cz+d}$, from $g(1) = 0$ we get $a = -b$. From $g(-3) = \infty$ we get $3c = d$. Thus $g(z) = \frac{az-a}{cz+3c} = \frac{a}{c}\frac{z-1}{z+3}$. From $g(i\sqrt{3}) = i$ we obtain

$$i = \frac{a}{c}\frac{i\sqrt{3}-1}{i\sqrt{3}+3} \Rightarrow \frac{a}{c} = i\frac{3+i\sqrt{3}}{-1+i\sqrt{3}} = \sqrt{3}.$$

Then $g(z) = \sqrt{3}\frac{z-1}{z+3}$ will map the circle C onto the y-axis. Note that any function of the form αg, where $\alpha \neq 0$ is real, will also map the circle C onto the y-axis, since multiplying by a nonzero *real* constant leaves a line through the origin unchanged. So for simplicity we divide by $\sqrt{3}$, still calling the function g, and obtain a mapping $g(z) = \frac{z-1}{z+3}$ of the circle C onto the y-axis. Does $g(z)$ take the region inside C onto the right half-plane? We check the image of one point inside C, say -1, and find $g(-1) = \frac{-2}{2} = -1$, which is a point in the left half-plane. So we modify g by multiplying it by -1 and obtain the desired linear fractional transformation $g(z) = \frac{1-z}{3+z}$. As *positive* multiples of g also work, the solution to this problem is not unique. $\square$

The previous examples illustrate how a linear fractional transformation can be determined from the images of three distinct points. In fact, we have the following useful formula.

Proposition 7.2.6. *There is a unique linear fractional transformation $w = \phi(z)$ that maps three distinct points z_1, z_2, and z_3 onto three distinct points w_1, w_2, and w_3. The mapping w is implicitly given by*

$$\frac{z-z_1}{z-z_3}\frac{z_2-z_3}{z_2-z_1} = \frac{w-w_1}{w-w_3}\frac{w_2-w_3}{w_2-w_1}. \tag{7.2.4}$$

Proof. That w is a linear fractional transformation follows by solving for w in
(7.2.4). To see that w maps z_j to w_j (j=1, 2, 3) it suffices to note that (7.2.4) holds
if we replace z by z_j and w by w_j. (For $j = 3$, you must take reciprocals in (7.2.4)
before replacing z by z_3 and w by w_3.) The uniqueness part is delegated to Exercises 9–10.   ∎

Mapping a circle onto a line by a linear fractional transformation, as we saw in
Example 7.2.5, can be achieved by requiring that $f(z) = \infty$ for some z on the circle.
In this case, the formula in Proposition 7.2.6 can be simplified as follows. Say we
want $f(z_3) = \infty$. As $w_3 \to \infty$, the fraction $\frac{w_2-w_3}{w-w_3} = \frac{1-w_2/w_3}{1-w/w_3} \to 1$. This suggests that
we set $\frac{w_2-w_3}{w-w_3} = 1$ on the right side of (7.2.4), obtaining the formula, stated in the
following proposition, whose verification is left to Exercise 11.

Proposition 7.2.7. (Mapping to a point at infinity) *Let z_1, z_2, and z_3 be three distinct points. There is a unique linear fractional transformation $w = \phi(z)$ that maps
z_1 and z_2 onto two distinct points w_1 and w_2 and maps z_3 to ∞. The mapping w is
implicitly given by*

$$\frac{z-z_1}{z-z_3}\frac{z_2-z_3}{z_2-z_1} = \frac{w-w_1}{w_2-w_1}. \tag{7.2.5}$$

There is also a corresponding identity for a linear fractional transformation mapping ∞ to a point, obtained by reversing the roles of z and w in (7.2.5) (see Exercise 11). Our next example uses the conformal property of linear fractional transformations.

Example 7.2.8. (Mapping of a lens-shaped region) The lens-shaped region Ω in
Figure 7.10 is bounded by the arcs of two circles.
(a) Use a linear fractional transformation ϕ to map Ω onto a sector in the
w-plane in such a way that

$$\phi(-2) = 0$$
$$\phi(-i) = 1$$
$$\phi(2) = \infty$$

(b) Determine the angle between the
circles at the point -2.

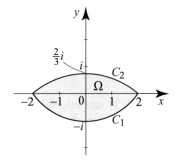

Fig. 7.10 A lens-shaped region.

Solution. (a) We apply (7.2.5) with $z_1 = -2$, $w_1 = 0$, $z_2 = -i$, $w_2 = 1$, $z_3 = 2$, and
get

$$\frac{z+2}{z-2}\frac{-i-2}{-i+2} = \frac{w}{1} \Rightarrow w = \phi(z) = \frac{2+i}{2-i}\frac{2+z}{2-z}.$$

(b) Of course we can determine the angle between C_1 and C_2 by finding their equations, then the slopes of the tangent lines at -2, and then the angle between the tangent lines. A better way is to calculate the angle between the images of C_1 and C_2 and use the conformal property of ϕ. As ϕ maps lines and circles to lines and circles, it is clear from its action on the points -2, $-i$, and 2 that ϕ maps the circle C_1 onto the u-axis. It also maps the circle C_2 onto a line that goes through the point $\phi(-2) = 0$. To determine this line, it suffices to check the image of another point on C_2. We have $\phi(\frac{2}{3}i) = i$. Thus ϕ maps C_2 onto the v-axis. Since ϕ is conformal at $z = -2$, it preserves the angles at this point. Thus, the angle between the circles at $z = -2$ is equal to the angle between their images, the u- and v-axes, which is clearly $\frac{\pi}{2}$.   □

Some applications concerning the electrostatic potential inside a capacitor formed by two cylinders lead to Dirichlet problems in regions bounded by two circles in the plane, which are not necessarily concentric. The problems are easier to solve when the two circles are concentric, giving rise to an annular region. Thus there is a great advantage in using a conformal mapping to center the circles before solving the Dirichlet problem. In what follows, we use a specific example to illustrate this process. More general examples are presented in the exercises.

Example 7.2.9. (Centering disks)

Let C_2 be the boundary of the unit disk, and let C_1 be the circle of radius $\frac{5}{28}$ centered at the point $\frac{9}{28}$ as shown in Figure 7.11. Find a one-to-one analytic mapping that maps the region between the nonconcentric circles C_1 and C_2 to an annular region bounded by two concentric circles.

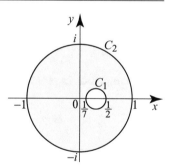

Fig. 7.11 Nonconcentric circles.

Solution. The idea is to use one of the linear fractional transformations

$$\phi_\alpha(z) = \frac{z - \alpha}{1 - \overline{\alpha} z}, \qquad (7.2.6)$$

where α is a complex number such that $|\alpha| < 1$. According to Proposition 4.6.2 we have that ϕ_α maps the unit disk onto itself and the its boundary onto itself. Since ϕ_α is a linear fractional transformation, it will map the circle C_1 onto a circle or a line, but because the image has to be inside the image of the unit disk, it follows that $\phi[C_1]$ is bounded and hence it must be a circle. We now ask the following question: Can we find α so that $\phi_\alpha[C_1]$ is a circle centered at the origin? Suppose for a moment that α were real. Then clearly $\phi_\alpha(x)$ is also real, and so ϕ_α maps the real line to the real line. Note that $\phi_\alpha(1/7)$ and $\phi_\alpha(1/2)$ are the points where $\phi_\alpha[C_1]$ meets the u-axis. Also, the circle $\phi_\alpha[C_1]$ must meet the u-axis in a perpendicular fashion (Figure 7.12), for the following three reasons:

(i) C_1 itself meets the real axis in a perpendicular fashion;
(ii) the x-axis is mapped to the u-axis; and
(iii) the map ϕ_α is conformal.

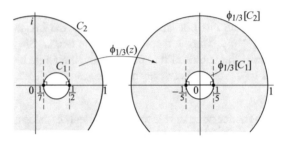

Fig. 7.12 For all $|\alpha| < 1$, $\phi_\alpha(z)$ maps the unit circle C_2 onto itself. But for one special value of α, with $|\alpha| < 1$, ϕ_α will also map the circle C_1 onto a circle centered at the origin, thus centering the images of C_1 and C_2.

So if we want $\phi_\alpha[C_1]$ to be a circle centered at the origin, it is enough to require that $\phi_\alpha(1/7) = -\phi_\alpha(1/2)$. This implies that

$$\frac{\frac{1}{7} - \alpha}{-\frac{\alpha}{7} + 1} = -\frac{\frac{1}{2} - \alpha}{-\frac{\alpha}{2} + 1} \Rightarrow \frac{1 - 7\alpha}{7 - \alpha} = -\frac{1 - 2\alpha}{2 - \alpha} \Rightarrow 9\alpha^2 - 30\alpha + 9 = 0.$$

The last equation in α is equivalent to $3\alpha^2 - 10\alpha + 3 = 0$, with solutions

$$\alpha = \frac{5 \pm \sqrt{16}}{3} = \frac{5 \pm 4}{3} \Rightarrow \alpha = 3 \text{ or } \alpha = \frac{1}{3}.$$

Since we want $|\alpha| < 1$, we take $\alpha = \frac{1}{3}$. Thus

$$\phi(z) = \phi_{\frac{1}{3}}(z) = \frac{z - \frac{1}{3}}{-\frac{1}{3}z + 1} = \frac{3z - 1}{3 - z}$$

will map C_2 onto C_2 and C_1 onto the circle with center at the origin and radius $r = |\phi(1/2)| = \frac{\frac{3}{2} - 1}{3 - \frac{1}{2}} = \frac{1}{5}$.   □

Composing Elementary Mappings

Our next goal is to compose elementary mappings studied thus far to construct non-trivial conformal mappings of regions in the plane. We provide four examples to illustrate this process.

Example 7.2.10. (Mapping a lens onto a disk) Construct a sequence of analytic functions that maps the lens-shaped region of Example 7.2.8 in a one-to-one way onto the unit disk. Write down the mapping that results from your construction.

Solution. The first step is to use the function of Example 7.2.8,

$$w_1 = \phi(z) = \frac{2+i}{2-i}\frac{2+z}{2-z} = \alpha\frac{2+z}{2-z} \qquad \text{where } \alpha = \frac{2+i}{2-i},$$

which maps the lens to the first quadrant. The first quadrant is then mapped onto the upper half-plane by the function $w_2 = w_1^2$. (Note that our mapping is no longer a linear fractional transformation.) Finally, the linear fractional transformation ψ in Example 7.2.3(b) will take the upper half-plane onto the unit disk. Each of these mappings is analytic and one-to-one on the region of interest. So the resulting function is analytic and one-to-one. The mapping is illustrated in Figure 7.13.

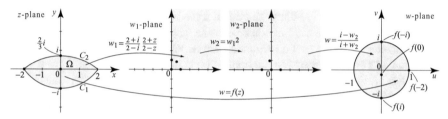

Fig. 7.13 Mapping a lens onto the unit disk. Note the conformal property in the first mapping at the point -2, which preserved the right angle. Note the failure of the conformal property in the second mapping at the point 0, where the angle is doubled.

The explicit formula in terms of z of the conformal mapping of the lens onto the unit disk is

$$w = f(z) = \frac{i-w_2}{i+w_2} = \frac{i-w_1^2}{i+w_1^2} = \frac{i-[\phi(z)]^2}{i+[\phi(z)]^2}, \qquad (7.2.7)$$

where

$$\phi(z) = \frac{2+i}{2-i}\frac{2+z}{2-z}$$

is from Example 7.2.8. Replacing $\phi(z)$ by its value in terms of z and simplifying, we obtain that the expression in (7.2.7) is equal to

$$w = f(z) = -i\frac{4-28iz+z^2}{28+4iz+7z^2}.$$

The following values of w confirm the fact that the mapping takes the lens-shaped region onto the unit disk, also taking boundary points to boundary points:

$$f(0) = -\frac{i}{7}, \quad f(-2) = 1, \quad f\left(\frac{2}{3}i\right) = -i, \quad f(-i) = i. \qquad \square$$

Example 7.2.11. (Mapping a half-disk onto a disk) The sequence of one-to-one analytic mappings in Figure 7.14 takes the upper half of the unit disk onto the unit disk.

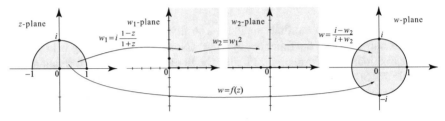

Fig. 7.14 Mapping the upper half of the unit disk onto the unit disk.

The first mapping is the linear fractional transformation $w_1 = \phi(z) = i\frac{1-z}{1+z}$ from Example 7.2.3. It maps the unit disk onto the upper half-plane. It also maps the upper half-disk onto the first quadrant, as can be verified by using the fact that it is conformal at $z = 1$ and checking the image of one interior point, say $\phi(\frac{i}{2}) = \frac{4}{5} + i\frac{3}{5}$, which lies in the first quadrant. The action of the second mapping is clear. The third mapping is the mapping ψ from Example 7.2.3(ii). The explicit formula of the composition of these three functions is the final mapping $w = f(z)$ is

$$w = \psi(w_2) = \frac{i - w_2}{i + w_2} = \frac{i - w_1^2}{i + w_1^2} = \frac{i - \left(i\frac{1-z}{1+z}\right)^2}{i + \left(i\frac{1-z}{1+z}\right)^2} = -i\frac{1 + 2iz + z^2}{1 - 2iz + z^2}.$$

The intermediary mapping $w_2 = w_1^2 = -\left(\frac{1-z}{1+z}\right)^2$ is also of interest. It takes the upper half-disk onto the upper half-plane. $\qquad \square$

Example 7.2.12. (Mapping a crescent-shaped region onto the upper half-plane) The crescent-shaped region in Figure 7.15 is bounded by two circles that intersect at angle 0 at the origin. We describe a sequence of one-to-one analytic mappings that takes this region onto the upper half-plane.

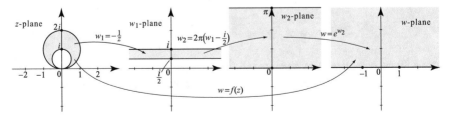

Fig. 7.15 Mapping a crescent onto the upper half-plane.

The first mapping $w_1 = -\frac{1}{z}$, being conformal at $z = i$ and $z = 2i$, preserves the right angles at these points. Since it maps the imaginary axis onto the imaginary axis, and 0 to ∞, consequently it will map the two circles onto two lines that intersect the imaginary axis at right angle. Thus the images of the circles are parallel horizontal lines as shown in Figure 7.15. As we move counterclockwise around the circles in the z-plane, we move rightward on the lines in the w_1-plane. The mapping $w_2 = 2\pi(w_1 - \frac{i}{2})$ translates and then scales the horizontal strip appropriately to set us up for an exponential mapping to the upper half-plane. $\quad\square$

Example 7.2.13. (Mapping the unit disk onto an infinite horizontal strip) We describe a sequence of analytic and one-to-one mappings that take the unit circle onto an infinite horizontal strip. The first linear fractional transformation, $w_1 = -i\phi(z)$, is obtained by multiplying by $-i$ the linear fractional transformation $\phi(z)$ in Example 7.2.3(i). Since $\phi(z)$ maps the unit disk onto the upper half-plane, and multiplication by $-i$ rotates by the angle $-\frac{\pi}{2}$, the effect of $-i\phi(z)$ is to map the unit disk onto the right half-plane.

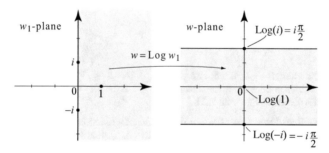

Fig. 7.16 The principal branch of the logarithm, $\mathrm{Log}\, z$, maps the right half-plane onto an infinite horizontal strip.

In Figure 7.16, $\mathrm{Log}\, w_1 = \ln|w_1| + i\,\mathrm{Arg}\, w_1$ is the principal branch of the logarithm. As w_1 varies in the right half-plane, $\mathrm{Arg}\, w_1$ varies between $-\frac{\pi}{2}$ and $\frac{\pi}{2}$, which explains the location of the horizontal boundary of the infinite strip. The desired mapping is

$$w = f(z) = \mathrm{Log}\,(-i\phi(z)) = \mathrm{Log}\,\frac{1-z}{1+z}. \qquad\square$$

Exercises 7.2

In Exercises 1–4, a linear fractional transformation ϕ and three points z_1, z_2, and z_3 are given. Let L_1 denote the line through z_1 and z_2 and L_2 the line through z_2 and z_3. In each case, (a) compute the images w_1, w_2, and w_3 of the points z_1, z_2, and z_3. (b) Describe the images by ϕ of L_1 and L_2. Are they lines or circles? (You need the images of three points on each line.)

1. $\phi(z) = i\dfrac{1-z}{1+z}, z_1 = 1, z_2 = 0, z_3 = i.$

2. $\phi(z) = \dfrac{i+z}{i-z}$, $z_1 = 1+i, z_2 = 0, z_3 = i$.

3. $\phi(z) = \dfrac{1+i-2z}{i-iz}$, $z_1 = i, z_2 = 1, z_3 = -i$.

4. $\phi(z) = \dfrac{1+2z}{i-(1+i)z}$, $z_1 = 1+i, z_2 = 1, z_3 = 1-i$.

5. Find the inverse ψ of the linear fractional transformation in Exercise 1, and verify that ψ maps w_1, w_2, and w_3 to z_1, z_2, and z_3.

6. Repeat Exercise 5 with the linear fractional transformation of Exercise 2.

7. (a) What is the inverse of the function

$$\frac{2z+1}{z-3}?$$

You may use (7.2.2).
(b) Describe the images of the unit circle, the unit disk, and the region outside the unit circle under the map $z \mapsto \frac{2z+1}{z-3}$.

8. Consider the circle passing through the points $-2, ai, 2$ and the one passing through the points $-2, -i, 2$. Let α denote the angle between these two circles at the point -2 as shown in the adjacent figure. Show that

$$\tan \alpha = \frac{4(a+1)(a-4)}{3(a+6)(a-\frac{2}{3})}.$$

Discuss the cases when $a = -1, 4, -6$, and $\frac{2}{3}$.
[*Hint*: Map the circles to lines using the linear fractional transformation in Example 7.2.8.]

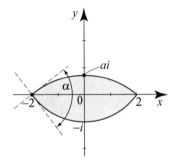

9. Fixed points. Recall that a point z_0 is a fixed point of a function f if $f(z_0) = z_0$. Show that a linear fractional transformation ϕ can have at most two fixed points in the complex plane, unless $\phi(z) = z$, in which case all points are fixed points. [*Hint*: Discuss solutions of $z = \frac{az+b}{cz+d}$.]

10. Uniqueness of a linear fractional transformation. Let z_1, z_2, and z_3 be three distinct points, and let w_1, w_2, and w_3 be three distinct points (we allow ∞). Show that there is a unique linear fractional transformation mapping z_j to w_j. [*Hint*: The existence is guaranteed by Propositions 7.2.6 and 7.2.7. To prove uniqueness, suppose that f and g are two linear fractional transformations mapping z_j to w_j for $j = 1, 2, 3$. Apply the result of Exercise 9 to $g \circ f^{-1}$. How many fixed points does $g \circ f^{-1}$ have?]

11. (a) **Mapping a point to infinity.** Prove Proposition 7.2.7.
(b) **Mapping infinity to a point.** Let z_1 and z_2 be two distinct points, and w_1, w_2, and w_3 be three distinct points. Show that there is a unique linear fractional transformation $w = \phi(z)$ that maps z_1 to w_1, z_2 to w_2 and ∞ to w_3. The mapping w is implicitly given by

$$\frac{z-z_1}{z_2-z_1} = \frac{w-w_1}{w-w_3} \frac{w_2-w_3}{w_2-w_1}. \tag{7.2.8}$$

In Exercises 12–24, (a) supply the formulas of the analytic mappings in each sequence of mappings shown in the accompanying figure (Figures 12-24). (b) Verify that the boundary and the interior of the shaded regions are mapped to the boundary and interior of the shaded regions. (c) Derive the formula for the final composite mapping $w = f(z)$. As usual, we start in the z-plane and end in the w-plane, going through the w_j-planes.

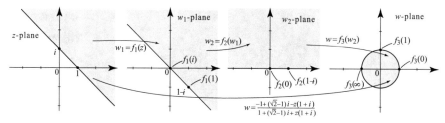

Fig. 7.17 Exercise 12.

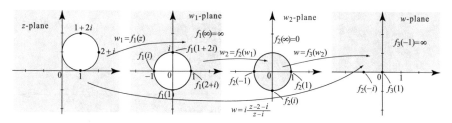

Fig. 7.18 Exercise 13.

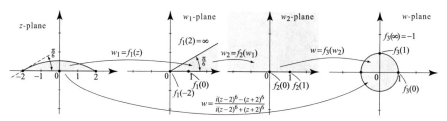

Fig. 7.19 Exercise 14.

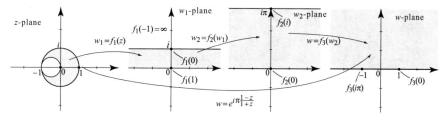

Fig. 7.20 Exercise 15.

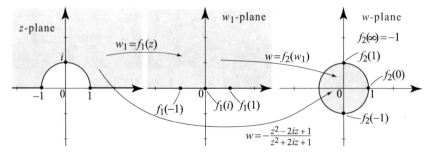

Fig. 7.21 Exercise 16.

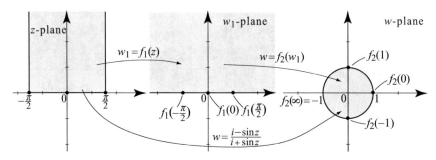

Fig. 7.22 Exercise 17.

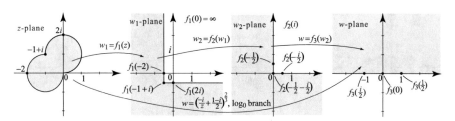

Fig. 7.23 Exercise 18.

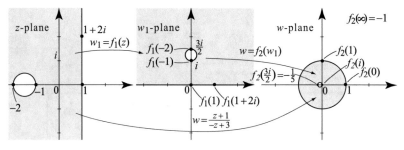

Fig. 7.24 Exercise 19.

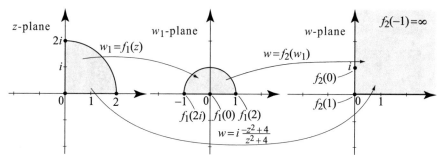

Fig. 7.25 Exercise 20.

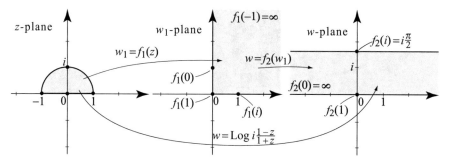

Fig. 7.26 Exercise 21.

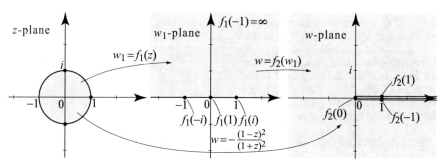

Fig. 7.27 Exercise 22.

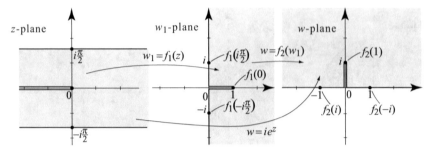

Fig. 7.28 Exercise 23.

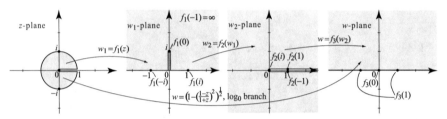

Fig. 7.29 Exercise 24.

25. Project Problem: Centering disks. We generalize the process in Example 7.2.9 to any region bounded by two nonintersecting circles, C_2 and C_1, such that C_1 is in the interior of C_2.

By translating the center of C_2 to the origin, scaling, and rotating, we can always reduce the picture to the one described in Figure 7.30, where $|a| < b < 1$, C_2 is the unit circle, and C_1 is centered on the x-axis with x-intercepts $(a,0)$ and $(b,0)$. We show that we can choose α such that $-1 < \alpha < 1$ and ϕ_α maps C_1 onto a circle centered at the origin. Here ϕ_α is the linear fractional transformation (7.2.6), which maps C_2 onto C_2. As explained in Example 7.2.9, it suffices to choose α so that $\phi_\alpha(a) = -\phi_\alpha(b)$.

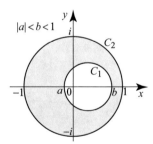

Fig. 7.30 Exercise 25.

(a) Show that the latter condition leads to the equation in α:

$$\alpha^2 - 2\,\frac{1+ab}{a+b}\,\alpha + 1 = 0,$$

with roots

$$\alpha_1 = \frac{1+ab}{a+b} + \sqrt{\left(\frac{1+ab}{a+b}\right)^2 - 1} \quad \text{and} \quad \alpha_2 = \frac{1+ab}{a+b} - \sqrt{\left(\frac{1+ab}{a+b}\right)^2 - 1}.$$

(b) Show that if $|a| < 1$ and $|b| < 1$, then $1 + ab \geq a + b$. [*Hint:* $1 - b \geq a(1 - b)$.]

(c) Show that $\alpha_1 > 1$, while $0 < \alpha_2 < 1$. [*Hint:* The first inequality follows from (b). For the second inequality, use the fact that the product of the roots of $ax^2 + bx + c = 0$ is $\frac{c}{a}$.]

(d) Conclude that $\phi(z) = \frac{z - \alpha}{1 - \alpha z}$ with $\alpha = \frac{1 + ab}{a + b} - \sqrt{\left(\frac{1 + ab}{a + b}\right)^2 - 1}$ maps C_2 onto C_2, C_1 onto a circle centered at the origin with radius $r = \phi(b)$, and the region between C_2 and C_1 onto the annular region bounded by $\phi[C_1]$ and the unit circle.

In Exercises 26–29 (see Figures 7.31–7.34), derive the linear fractional transformation that maps the shaded region between the two circles (or circle and line in Exercise 29) onto an annular region centered at the origin. Refer to Exercise 25 for instructions. In Exercises 28–29, you need to reduce to the situation described in Exercise 25.

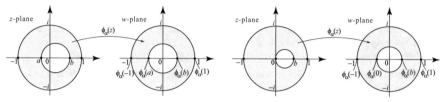

Fig. 7.31 Exercise 26. Here $a = -\frac{1}{4}$, $b = \frac{4}{7}$.   **Fig. 7.32** Exercise 27. Here $a = 0$, $b = \frac{8}{17}$.

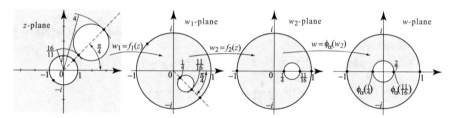

Fig. 7.33 Exercise 28. (The figures are not to scale.)

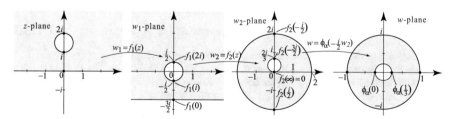

Fig. 7.34 Exercise 29. [*Hint:* In the last step, start by rotating the inner circle to center it on the real axis, then scale the outer radius to 1.]

30. A geometric problem. The following is an interesting illustration of the use of linear fractional transformations to prove geometric facts. Consider a circle C and a point z_0 inside C (Figure 7.35). We show that all the circles C' through z_0 that intersect C at right angle also intersect at a common point z_1 as in Figure 7.35. The point z_1 is called the **reflection** of z_0 in C. We also say that the points z_0 and z_1 are **symmetric** with respect to the circle C.

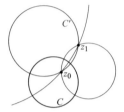

Fig. 7.35 Exercise 30.

(a) Let ϕ be a linear fractional transformation that maps C to the real axis and the interior of C to the upper half-plane. Show that $\phi[C']$ is a circle or a line that intersects the real axis at a right angle.
(b) Observe that $\phi[C']$ passes through the point $-\phi(z_0)$. Setting $z_1 = \phi^{-1}(-\phi(z_0))$ conclude that C' passes through z_1.

31. (a) Characterize all one-to-one analytic mappings of the upper half-plane onto the unit disk. [*Hint*: You may want to use Exercise 6 in Section 4.6.]
(b) Characterize all one-to-one analytic mappings of the upper half-plane onto itself.

32. Matrix correspondence. Define a mapping Φ that associates to each linear fractional transformation (7.2.1) the 2×2 matrix with complex entries

$$S = \begin{pmatrix} a & b \\ c & d \end{pmatrix}.$$

Thus $\Phi(\phi) = S$. Suppose that ϕ and ψ are two linear fractional transformations with matrices $\Phi(\phi) = S$ and $\Phi(\psi) = T$. Show that the $\Phi(\phi \circ \psi) = ST$, where ST denotes the product of the two matrices S and T.

Lines and circles under inversion, part I. *In Exercises 33–35 we show that the function $f(z) = \frac{1}{z}$ maps lines and circles to lines and circles.*

33. (a) Show that with $w = u + iv$ and $z = x + iy$, the mapping $w = 1/z$ can be written as

$$u(x, y) = \frac{x}{x^2 + y^2}, \qquad v(x, y) = -\frac{y}{x^2 + y^2}$$

(b) Deduce that the inverse transformation $z = 1/w$ is

$$x(u, v) = \frac{u}{u^2 + v^2}, \qquad y(u, v) = -\frac{v}{u^2 + v^2}. \tag{7.2.9}$$

34. (a) Show that any circle of the form $(x - x_0)^2 + (y - y_0)^2 = r^2$, $r > 0$ can be written in the form

$$A(x^2 + y^2) + Bx + Cy + D = 0, \text{ where } B^2 + C^2 - 4AD > 0. \tag{7.2.10}$$

(b) Show that any line in the plane can be written in the form (7.2.10).
(c) Show that any set of points satisfying $A(x^2 + y^2) + Bx + Cy + D = 0$ with $B^2 + C^2 - 4AD > 0$ is either a circle or a line depending on whether $A = 0$ or $A \neq 0$.
(d) Show that such circles or lines pass through the origin if and only if $D = 0$.

35. Suppose S is the set of points (x, y) satisfying (7.2.10). Use (7.2.9) and conclude that points (u, v) in $f[S]$ satisfy an equation of the same form as (7.2.10), including the associated constant inequality. Conclude that under the mapping $f(z) = 1/z$ lines and circles are mapped to lines and circles, with the exception of the origin.

Lines and circles under inversion, part II. *In Exercises 36–39, we investigate how specific lines and circles are mapped under the function $f(z) = \frac{1}{z}$ and describe a quick method to obtain the images. These exercises depend on Exercises 33–35, and in particular, (7.2.9) and (7.2.10).*

36. (a) Suppose that S is a circle that does not pass through the origin. Show that $f[S]$ is also a circle that does not pass through the origin.
(b) Let z_1 and z_2 denote the points in S with the smallest and largest moduli, respectively. Show that $f(z_1)$ and $f(z_2)$ have the largest and smallest moduli, respectively, of those points in $f[S]$. Argue that the circle $f[S]$ is uniquely determined by these two points $f(z_1)$ and $f(z_2)$; see Figure 7.36 in which S and $f[S]$ are plotted on the same plane.

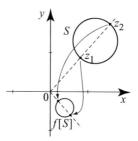

Fig. 7.36 Exercise 36.

37. (a) Suppose S is a line that passes through the origin. Show that $f[S \setminus \{0\}] \cup \{0\}$ is also a line passing through the origin.
(b) Argue that the image $f(z_0)$ of any nonzero point z_0 in S uniquely determines the line $f[S]$ (see Figure 7.37, in which S and $f[S]$ are plotted on the same plane).

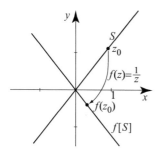

Fig. 7.37 Exercise 37.

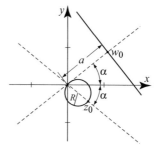

Fig. 7.38 Exercise 38.

38. (a) Suppose S is a circle that passes through the origin. Show that with the exception of mapping from the origin, $f[S]$ is a line that does not pass through the origin.
(b) Suppose that S is a line which does not pass through the origin. Show that with the exception of mapping to the origin, $f[S]$ is a circle that passes through the origin.
(c) Let S be a circle that passes through the origin and $f[S]$ be the associated line that does not. Show that the point z_0 of maximum modulus on the circle maps to the point w_0 of minimum modulus on the line, and vice versa. Argue that each of these points uniquely determines the corresponding circle or line. If we let R denote the radius of the circle and a the perpendicular distance from the origin to the line, show that $2R = \frac{1}{a}$ (see Figure 7.38 in which S and $f[S]$ are plotted on the same plane).

39. Lines and circles under linear fractional transformations.
(a) Verify that every linear fractional transformation is a composition of a linear transformation, followed by an inversion, and then followed by a linear transformation.
(b) Using part (a) and the result of Exercise 35, show that any linear fractional transformation maps lines and circles to lines and circles.

7.3 Solving Dirichlet Problems with Conformal Mappings

In this section we use conformal mappings to solve Dirichlet problems. Very often the difficulty in solving Dirichlet problems is due to the geometry of the region on which the problem is stated. Conformal mappings can be used to transform a region to one on which the ensuing boundary value problem is easier to solve. Roughly speaking, the conformal mapping provides the *change of variables* that leaves Laplace's equation unchanged but transforms the boundary conditions. At the heart of the method lie the Riemann mapping theorem (Theorem 7.1.9) and the invariance of Laplace's equation by conformal mappings [identity 6.1.3].

To make things precise, let Ω be a simply connected region on the plane. Suppose we have the Dirichlet problem

$$\Delta u = 0 \text{ on } \Omega \quad u = b \text{ on } \partial\Omega. \tag{7.3.1}$$

We fix a conformal map f from Ω onto another domain $W = f[\Omega]$. Then we consider the new Dirichlet problem

$$\Delta U = 0 \text{ on } W, \quad U = b \circ f^{-1} \text{ on } \partial W. \tag{7.3.2}$$

Let U be a solution of the Dirichlet problem (7.3.2) on W. We claim that the function $u = U \circ f$ solves the Dirichlet problem (7.3.1) on Ω. Indeed, the conformal mapping transfers one boundary function into the other. Moreover, identity (6.1.3) yields

$$\Delta u = \Delta(U \circ f) = |f'|^2 \Delta U$$

and since f' is never vanishing, it follows that $\Delta u = 0$ on Ω if and only if $\Delta U = 0$ on W. This method is illustrated in Figure 7.39 when W is the unit disk D. In practice W could also be the upper half space or any other domain on which we know how to solve a Dirichlet problem.

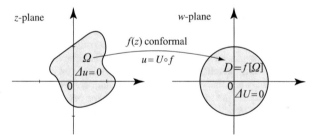

Fig. 7.39 A one-to-one conformal mapping f of Ω onto the unit disk $D = f[\Omega]$ takes boundary to boundary and preserves Laplace's equation. It transforms the conditions $\Delta u = 0$ on Ω and $u = b$ on $\partial\Omega$ to $\Delta U = 0$ on D and $U = b \circ f^{-1}$ on ∂D. The solution of the initial Dirichlet problem in Ω is then $u(z) = U(f(z))$.

This method turns out to be phenomenally powerful. Not only we will be able to solve specific problems, but we can also take general formulas, such as the Poisson integral formula on a disk, and produce similar formulas for the solution of Dirichlet problems on new regions in the plane.

Recall that for Dirichlet problems where the boundary data is constant along rays, we can find a solution using a branch of the argument. We denote by $\arg_\alpha z$ the branch of the argument with a branch cut at angle α, and by $\operatorname{Arg} z$ the principal

branch with a branch cut along the negative real axis. These functions, being the
imaginary parts of the corresponding logarithm branches, are harmonic everywhere
except on their branch cuts.

Recall also that a linear combination of harmonic functions with real scalars is
again a harmonic function (Proposition 6.1.4). For example, $u(z) = \frac{100}{\pi}(\pi - \text{Arg} z)$
is harmonic in the upper half-plane with boundary values $u(x) = 100$ if $x > 0$ and
$u(x) = 0$ if $x < 0$. This solution to a very simple Dirichlet problem in the upper
half-plane helps us solve a somewhat difficult Dirichlet problem on the unit disk.

Example 7.3.1. (Steady-state temperature distribution in a disk) The boundary
of a circular plate of unit radius with insulated lateral surface is kept at a fixed
temperature distribution equal to $100°$ on the upper semicircle and $0°$ on the lower
semicircle (see Figure 7.40). Find the steady-state temperature inside the plate.

Solution. To answer this question, we must solve $\Delta u = 0$ inside the unit disk with
boundary values $u = 100$ on the upper semicircle and $u = 0$ on the lower semicircle.
While the geometry of the circle makes it difficult to understand the effect of the
boundary conditions on the solution inside the unit disk, the corresponding boundary value problem in the upper half-plane has a simple solution. To transform the
original problem into a problem in the upper half-plane, we use the linear fractional
transformation $\phi(z) = i\frac{1-z}{1+z}$ from Example 7.2.3(ii). It is easy to verify that ϕ takes
the upper semicircle onto the positive real axis, and the lower semicircle onto the
negative real axis. Thus ϕ transforms the given Dirichlet problem into a Dirichlet
problem in the upper half-plane with boundary values shown in Figure 7.41.

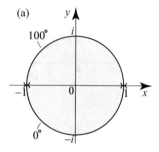

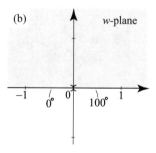

Fig. 7.40 Transforming the
Dirichlet problem on the disk
into one on the upper half space.

Fig. 7.41 The transformation is
via the linear fractional transformation $\phi(z) = i\frac{1-z}{1+z}$.

According to our preceding discussion, the solution in the upper half of the w-plane
is $U(w) = \frac{100}{\pi}(\pi - \text{Arg} w)$. By composing the solution in the w-plane with the con-
formal map, we get a solution of our original problem, $u(z) = U(\phi(z))$, $|z| < 1$.
Hence the solution of the Dirichlet problem in the unit disk is

$$u(z) = \frac{100}{\pi}(\pi - \text{Arg} \phi(z)) = \frac{100}{\pi}\left(\pi - \text{Arg}\left(i\frac{1-z}{1+z}\right)\right).$$

For example, the temperature of the center of the circular plate is

$$u(0) = \frac{100}{\pi}(\pi - \text{Arg}\,(i)) = \frac{100}{\pi}\frac{\pi}{2} = 50,$$

which is, as one might expect, the average value of the temperature on the circumference. With a little extra work, we can express the solution $u(z)$ in terms of x and y instead of $z = x + iy$, x, y real, (see Exercise 18).   □

We next solve a Dirichlet problem in a lens-shaped region.

Example 7.3.2. (Dirichlet problem in a lens-shaped region) Find a harmonic function u in the lens-shaped region Ω in Figure 7.42, with boundary values $u = 100$ on the upper circular arc and $u = 0$ on the lower circular arc.

Solution. The region Ω was discussed in Example 7.2.8 from which we recall the linear fractional transformation $\phi(z) = \frac{2+i}{2-i}\frac{2+z}{2-z}$. It is straightforward to check that ϕ maps the lower boundary of Ω onto the positive real axis, and the upper boundary of Ω onto the positive imaginary axis. By checking the image of one point in Ω, say $z = 0$, we find $\phi(0) = \frac{2+i}{2-i} = \frac{1}{5}(3+4i)$, which is a point in the first quadrant. Thus ϕ maps the region Ω onto the first quadrant and transforms the original problem into a Dirichlet problem in the first quadrant of the w-plane, with boundary conditions as shown in Figure 7.43.

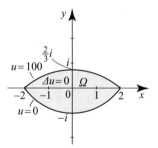

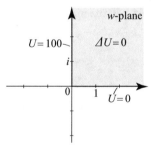

Fig. 7.42 Dirichlet problem in a lens.

Fig. 7.43 Analogous Dirichlet problem in the first quadrant.

It is clear that the solution in the w-plane is $U(w) = \frac{200}{\pi}\,\text{Arg}\,w$. Thus the solution of the Dirichlet problem on Ω is

$$u(z) = \frac{200}{\pi}\,\text{Arg}\,\phi(z) = \frac{200}{\pi}\,\text{Arg}\,\left(\frac{2+i}{2-i}\frac{2+z}{2-z}\right).\qquad\qquad □$$

For our next application, we recall the solution of the Dirichlet problem in an annular region (Figure 7.44), with constant boundary values T_1 and T_2 on each piece of the boundary, respectively:

$$u(z) = T_1 + (T_2 - T_1)\frac{\ln(|z|/R_1)}{\ln(R_2/R_1)}, \quad R_1 < |z| < R_2. \qquad (7.3.3)$$

This function is harmonic in the complex plane minus the origin, and so it is harmonic in the annulus

$$R_1 < |z| < R_2.$$

The fact that it takes the values T_1 and T_2 on the boundary can be verified directly. This solution can be derived from the solution of Exercise 9, Section 6.2. When the outer circle C_2 is the unit circle ($R_2 = 1$), the solution becomes

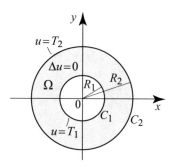

Fig. 7.44 A Dirichlet problem in an annulus.

$$u(z) = T_1 + (T_1 - T_2)\frac{\ln|z| - \ln R_1}{\ln R_1}, \quad R_1 < |z| < 1. \qquad (7.3.4)$$

Example 7.3.3. (A problem on a region between nonconcentric circles) Find a harmonic function u in the nonregular annular region Ω in Figure 7.45, such that $u = 50$ on the inner circle C_1 and $u = 100$ on the outer unit circle C_2.

Solution. We transform the problem into a Dirichlet problem on an annulus using the linear fractional transformation of Example 7.2.9

$$w = \phi(z) = \frac{3z - 1}{3 - z}.$$

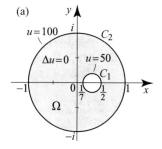

Fig. 7.45 A Dirichlet problem in a nonregular annulus.

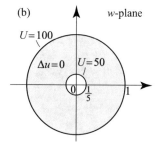

Fig. 7.46 The Dirichlet problem is transformed to one for a regular annulus via a linear fractional transformation.

As shown in Example 7.2.9, ϕ maps the unit circle C_2 onto the unit circle, the inner circle C_1 onto a circle centered at the origin with radius $\frac{1}{5}$, and the region between the unit circle and C_2 onto the annular region $\frac{1}{5} < |w| < 1$. The boundary values in the transformed problem are shown in Figure 7.46, and so according to (7.3.4) the solution of the Dirichlet problem in the w-plane is

$$U(w) = 50 + 50\frac{\ln|w| + \ln 5}{\ln 5}, \qquad \frac{1}{5} < |w| < 1.$$

Substituting $w = \phi(z) = \frac{3z-1}{3-z}$, we obtain the solution in the z-plane

$$u(z) = 50 + 50\frac{\ln\left|\frac{3z-1}{3-z}\right| + \ln 5}{\ln 5}.$$

With the help of a computer, we have evaluated the solution at various points inside the nonregular annular region in Figure 7.45. The values are shown in Table 3.

(x,y)	$(0,0)$	$(\frac{1}{7}-0.001,0)$	$(\frac{1}{2}+0.001,0)$	$(0.99,0.01)$	$(0.99,-0.01)$	$(\frac{1}{2},\frac{1}{3})$	$(\frac{1}{2},-\frac{1}{3})$
$u(x+iy)$	65.87	50.15	50.20	99.38	99.38	74.73	74.73

Table 3. Temperature of various points inside the nonregular annular region in Figure 7.45.

The table seems to confirm the solution that we found. The values of u are between 50 and 100. They are closer to the boundary values as z approaches the boundary (inner and outer). Note also the symmetric property of u, due to the symmetries in the problem: We have $u(x+iy) = u(x-iy)$. You should expand the table of values and make your own conclusions about the solution. $\square$

We next derive the Poisson formula in the upper half-plane using the corresponding formula in the unit disk.

Theorem 7.3.4. (Poisson Integral Formula in the Upper Half-plane) *Let f be a bounded piecewise smooth function on the real line. A solution of the Dirichlet problem $\Delta u(x+iy) = 0$ for $x+iy$ in the upper half-plane $(y > 0)$ with boundary condition $u(x) = f(x)$ for all $-\infty < x < \infty$ is given by the Poisson integral formula*

$$u(x+iy) = \frac{y}{\pi}\int_{-\infty}^{\infty}\frac{f(s)}{(x-s)^2 + y^2}\,ds = \frac{y}{\pi}\int_{-\infty}^{\infty}\frac{f(x-s)}{s^2 + y^2}\,ds. \qquad (7.3.5)$$

Proof. We prove the first equality in (7.3.5); the second one follows by making the change of variables $s' = x - s$. Transform the problem into a Dirichlet problem in the unit disk using the linear fractional transformation $w = \psi(z) = \frac{i-z}{i+z}$, $z = x+iy$, $y > 0$ [Example 7.2.3(b)]. This mapping takes the real line onto the unit circle, and the upper half-plane onto the unit disk. Denote the image of a point s on the real line by $e^{i\phi}$ on the unit circle, so $e^{i\phi} = \psi(s)$.

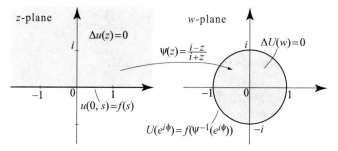

Fig. 7.47 How the boundary values in a Dirichlet problem are transformed, after using a linear fractional transformation that takes boundary to boundary.

As illustrated in Figure 7.47, the boundary data that we get for the problem on the unit disk is $U(e^{i\phi}) = f(\psi^{-1}(e^{i\phi}))$, for all $e^{i\phi}$ on the unit circle. The solution of the Dirichlet problem in the unit disk ($|w| < 1$) with this boundary data is obtained from the Poisson integral formula [see (6.3.7)]

$$U(w) = \frac{1 - |w|^2}{2\pi} \int_0^{2\pi} \frac{f(\psi^{-1}(e^{i\phi}))}{|e^{i\phi} - w|^2} \, d\phi. \tag{7.3.6}$$

Setting $w = \psi(z)$ in (7.3.6), we obtain the solution in the upper half-plane:

$$u(x + iy) = U(\psi(x + iy)) = U(\psi(z)) = \frac{1 - |\psi(z)|^2}{2\pi} \int_0^{2\pi} \frac{f(\psi^{-1}(e^{i\phi}))}{|e^{i\phi} - \psi(z)|^2} \, d\phi. \tag{7.3.7}$$

Our goal is to show that (7.3.7) is precisely (7.3.5). The details are straightforward but a little tedious. Make the change of variables $s = \psi^{-1}(e^{i\phi})$. Since the integrand in (7.3.7) is 2π-periodic, we get the same result if we integrate from $-\pi$ to π instead of 0 to 2π, and as ϕ runs from $-\pi$ to π, s runs from $-\infty$ to ∞. We have

$$s = \psi^{-1}(e^{i\phi}) \;\Rightarrow\; \psi(s) = \frac{i - s}{i + s} = e^{i\phi}.$$

Taking differentials and using $e^{i\phi} = \frac{i-s}{i+s}$, we get

$$\frac{-2i}{(i+s)^2} \, ds = ie^{i\phi} \, d\phi = i\frac{i-s}{i+s} d\phi \;\Rightarrow\; d\phi = \frac{2}{1+s^2} \, ds.$$

Substituting what we have so far into (7.3.7), we obtain

$$u(x + iy) = \frac{1 - |\psi(z)|^2}{\pi} \int_{-\infty}^{\infty} \frac{f(s)}{|e^{i\phi} - \psi(z)|^2} \frac{ds}{1+s^2}. \tag{7.3.8}$$

Comparing (7.3.8) and (7.3.5), it suffices to show that for $z = x + iy$ and $e^{i\phi} = s$,

$$\frac{1-|\psi(z)|^2}{|e^{i\phi}-\psi(z)|^2} = (1+s^2)\,\frac{y}{(x-s)^2+y^2}. \tag{7.3.9}$$

This part is straightforward and is left to Exercise 21.   ■

Definition 7.3.5. For $y>0$, the function $P_y(x) = \sqrt{\frac{2}{\pi}}\frac{y}{x^2+y^2}$ $(-\infty < x < \infty)$ is called the **Poisson kernel** on the real line. Expressing (7.3.5) in terms of P_y we write

$$u(x+iy) = \frac{1}{\sqrt{2\pi}}\int_{-\infty}^{\infty} f(s)P_y(x-s)\,ds = \frac{1}{\sqrt{2\pi}}\int_{-\infty}^{\infty} f(x-s)P_y(s)\,ds. \tag{7.3.10}$$

This features the solution of the Dirichlet problem in the upper half-plane as a *convolution* of the boundary function f with the Poisson kernel. The Poisson kernel has a wealth of properties and links complex analysis in the upper half-plane to Fourier analysis of functions on the real line.

Example 7.3.6. (Applying Poisson's integral formula) Solve the Dirichlet problem in the upper half-plane with boundary data on the real line

$$f(x) = u(x) = \begin{cases} C & \text{if } a < x < b, \\ 0 & \text{otherwise,} \end{cases}$$

where C is a constant.

Solution. We apply the first formula in (7.3.5) and get for $-\infty < x < \infty$, $y > 0$,

$$u(x+iy) = \frac{y}{\pi}\int_{-\infty}^{\infty}\frac{f(s)}{(x-s)^2+y^2}\,ds = \frac{y}{\pi}\int_a^b\frac{C}{(x-s)^2+y^2}\,ds,$$

since $f(s)$ is 0 outside the interval $a < s < b$. We evaluate the integral in terms of the inverse tangent, using an obvious change of variables,

$$\begin{aligned}
u(x+iy) &= \frac{y}{\pi}\int_a^b\frac{C}{(x-s)^2+y^2}\,ds \\
&= \frac{y}{\pi}\frac{C}{y^2}\int_a^b\frac{1}{\left(\frac{s-x}{y}\right)^2+1}\,ds \\
&= \frac{C}{\pi}\int_{\frac{a-x}{y}}^{\frac{b-x}{y}}\frac{ds}{s^2+1} \\
&= \frac{C}{\pi}\left[\tan^{-1}\left(\frac{b-x}{y}\right)+\tan^{-1}\left(\frac{x-a}{y}\right)\right].
\end{aligned}$$

We can give a concrete geometric interpretation of this answer.

Let

$$\alpha_1 = \tan^{-1}\left(\frac{b-x}{y}\right)$$

$$\alpha_2 = \tan^{-1}\left(\frac{x-a}{y}\right)$$

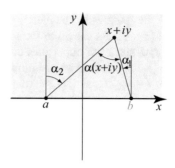

as shown in Figure 7.48. Then

$$u(x+iy) = \frac{C}{\pi}(\alpha_1 + \alpha_2) = \frac{C}{\pi}\alpha(x+iy),$$

where $\alpha(x+iy)$ is the harmonic measure of the interval (a,b); that is, α is the angle at the point $x+iy$ subtended by the interval (a,b) (Figure 7.48). □

Fig. 7.48 For a fixed interval (a,b), the angle at $x+iy$ subtended by (a,b) is a harmonic function $\alpha(x+iy)$ called the harmonic measure of (a,b).

Suppose that Ω is a region, f_1 and f_2 are two functions defined on the boundary of Ω, and u_1 and u_2 are solutions of the Dirichlet problems on Ω with boundary values f_1 and f_2, respectively. Since Laplace's equation $\Delta u = 0$ is linear, it is straightforward to check that $u = u_1 + u_2$ solves the Dirichlet problem on Ω with boundary values $u = f_1 + f_2$ on the boundary of Ω. This useful process of generating a solution by adding solutions of two or more related problems is called **superposition** of solutions. It will appear again in our study of other linear equations.

Example 7.3.7. (Superposing solutions) Let (a_1, b_1) and (a_2, b_2) be two disjoint intervals on the real line, and let T_1 and T_2 be two complex numbers. Solve the Dirichlet problem in the upper half-plane with boundary data $f(x) = T_1$ if $x \in (a_1, b_1)$, T_2 if $x \in (a_2, b_2)$ and 0 otherwise.

Solution. For $j = 1,2$, let $f_j(x) = T_j$ if x is in (a_j, b_j), 0 otherwise. Clearly, $f(x) = f_1(x) + f_2(x)$. From Example 7.3.6, the solution of the Dirichlet problem in the upper half-plane with boundary values $f_j(x)$ on the real line is

$$u_j(z) = \frac{T_j}{\pi}\left[\tan^{-1}\left(\frac{b_j - x}{y}\right) - \tan^{-1}\left(\frac{a_j - x}{y}\right)\right].$$

Thus the solution to our original problem is $u(z) = u_1(z) + u_2(z)$. □

In the next example, we use the Poisson integral formula and the Joukowski mapping introduced in Example 7.1.8.

Example 7.3.8. (Joukowski mapping) Solve the Dirichlet problem shown in Figure 7.49.

Solution. We transform the problem into a problem in the upper half-plane by using the Joukowski mapping $w = J(z) = \frac{1}{2}(z + \frac{1}{z})$. As shown in Example 7.1.8, J takes the upper unit circle onto $[-1, 1]$, and the semi-infinite intervals $(-\infty, -1]$ and $[1, \infty)$ onto themselves. Thus the boundary conditions in the transformed Dirichlet problem in the upper half-plane are $f(w) = 100$ for real $-1 < w < 1$ and $f(w) = 0$ for real w outside this interval. According to Example 7.3.6, the solution in the upper half of the w-plane is

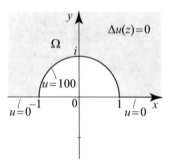

Fig. 7.49 A Dirichlet problem on the upper half space minus the unit disk.

$$\frac{100}{\pi}\left[\tan^{-1}\left(\frac{1 - \operatorname{Re} w}{\operatorname{Im} w}\right) + \tan^{-1}\left(\frac{1 + \operatorname{Re} w}{\operatorname{Im} w}\right)\right].$$

Replacing w by $J(z)$ we obtain the solution in the region Ω: For all (x, y) in Ω,

$$u(x + iy) = \frac{100}{\pi}\left[\tan^{-1}\left(\frac{1 - \operatorname{Re} J(z)}{\operatorname{Im} J(z)}\right) + \tan^{-1}\left(\frac{1 + \operatorname{Re} J(z)}{\operatorname{Im} J(z)}\right)\right]$$

With a little bit more work, the answer could be expressed in terms of x and y. □

Example 7.3.9. (A Poisson boundary distribution) Solve the Dirichlet problem in the upper half-plane (shown in Figure 7.50) with boundary data the **Poisson temperature distribution** on the real line:

$$P_a(x) = \sqrt{\frac{2}{\pi}}\frac{a}{x^2 + a^2}, \text{ where } a > 0.$$

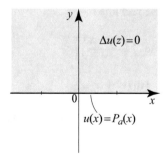

Fig. 7.50 The Dirichlet problem in the upper half-plane with boundary data on the real line.

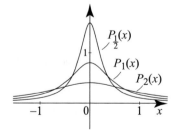

Fig. 7.51 The Poisson kernel $P_y(x)$ for various values of $y > 0$. Note that $P_y(x)$ is positive, even, and bell-shaped; also $\lim_{y \downarrow 0} P_y(0) = \infty$.

This problem models the steady-state distribution in a large sheet of metal with insulated lateral surface, whose boundary along the x-axis is kept at a fixed temperature distribution described by a Poisson kernel $P_a(x)$, where $a > 0$ is a positive constant. The temperature at the origin is $u(0) = P_a(0) = \sqrt{\frac{2}{\pi}} \frac{1}{a}$, which tends to ∞ as a tends to 0. Away from the origin, the temperature decays to 0 like $a/(x^2 + a^2)$. So smaller values of $a > 0$ correspond to temperature distributions concentrated around 0. This fact is shown in Figure 7.51.

Solution. We apply the first formula in (7.3.5) and get for $-\infty < x < \infty$, $y > 0$,

$$u(x+iy) = \frac{y}{\pi} \int_{-\infty}^{\infty} \frac{P_a(s)}{(x-s)^2 + y^2} \, ds = \frac{ay}{\pi} \sqrt{\frac{2}{\pi}} \int_{-\infty}^{\infty} \frac{ds}{(s^2 + a^2)((x-s)^2 + y^2)}.$$

We evaluate the last integral using the residue method completing the contour with a semicircle in the upper half-plane. The function $h(z) = 1/(z^2 + a^2)((x-z)^2 + y^2)$ has two simple poles in the upper half-plane, at $z = ia$ and at $z = x + iy$. We compute the residues at these points using Proposition 5.1.3(ii). We have

$$\text{Res}\,(h,\,ia) = \frac{1}{(x-ia)^2 + y^2} \text{Res}\left(\frac{1}{(z^2 + a^2)},\, ia\right) = \frac{1}{(x-ia)^2 + y^2} \frac{1}{2ia},$$

and

$$\text{Res}\,(h,\,x+iy) = \frac{1}{(x+iy)^2 + a^2} \text{Res}\left(\frac{1}{(x-z)^2 + y^2},\, x+iy\right) = \frac{1}{(x+iy)^2 + a^2} \frac{1}{2iy}.$$

Applying Proposition 5.3.4 we obtain

$$u(x+iy) = \frac{ay}{\pi} \sqrt{\frac{2}{\pi}} \, 2\pi i \Big(\text{Res}\,(h(z),\, ia) + \text{Res}\,(h(z),\, x+iy) \Big)$$

$$= 2ayi \sqrt{\frac{2}{\pi}} \left(\frac{1}{(x-ia)^2 + y^2} \frac{1}{2ia} + \frac{1}{(x+iy)^2 + a^2} \frac{1}{2iy} \right)$$

$$= \sqrt{\frac{2}{\pi}} \left(\frac{y}{(x-ia)^2 + y^2} + \frac{a}{(x+iy)^2 + a^2} \right)$$

$$= \sqrt{\frac{2}{\pi}} \frac{a+y}{x^2 + (a+y)^2},$$

where the last equality follows by elementary algebraic manipulations. This solves the problem. But note that the last expression is precisely $P_{a+y}(x)$. Thus

$$u(x+iy) = P_{a+y}(x),$$

which shows that the solution of the Dirichlet problem with a Poisson boundary data $P_a(x)$ is another Poisson distribution $P_{a+y}(x)$. This amazing fact about temper-

ature problems with Poisson boundary data can also be explained using properties
of convolutions and Fourier transforms.   □

Exercises 7.3

In Exercises 1–15, solve the Dirichlet problem described by the accompanying figure (Figures 7.52–7.66), using the methods of this section. Examples 7.3.6 and 7.3.7 and superposition of solutions are useful.

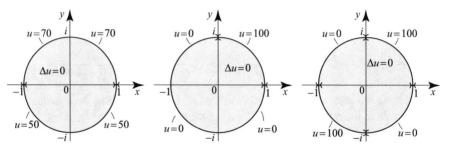

Fig. 7.52 Exercise 1.   **Fig. 7.53** Exercise 2.   **Fig. 7.54** Exercise 3.

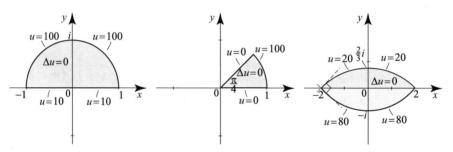

Fig. 7.55 Exercise 4.   **Fig. 7.56** Exercise 5.   **Fig. 7.57** Exercise 6.

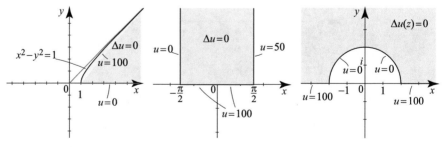

Fig. 7.58 Exercise 7.   **Fig. 7.59** Exercise 8.   **Fig. 7.60** Exercise 9.

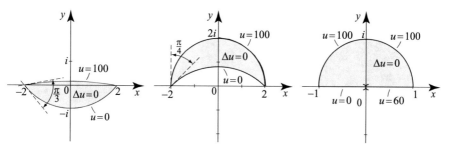

Fig. 7.61 Exercise 10.   **Fig. 7.62** Exercise 11.   **Fig. 7.63** Exercise 12.

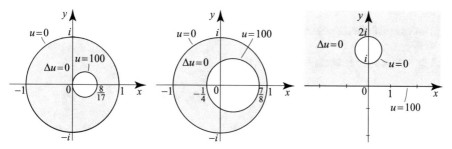

Fig. 7.64 Exercise 13.   **Fig. 7.65** Exercise 14.   **Fig. 7.66** Exercise 15.

16. Generalize Example 7.3.6 as follows. Suppose that $I_1 = (a_1, b_1)$, $I_2 = (a_2, b_2)$, ..., $I_n = (a_n, b_n)$ are n disjoint intervals on the real line, $T_1, T_2, \ldots, T_n$ are n real or complex values.
(a) Show that a solution of the Dirichlet problem in the upper half-plane with boundary data

$$f(x) = \begin{cases} T_j & \text{if } a_j < x < b_j, \\ 0 & \text{otherwise,} \end{cases}$$

is

$$u(x+iy) = \frac{1}{\pi} \sum_{j=1}^{n} T_j \left[\tan^{-1} \left(\frac{b_j - x}{y} \right) - \tan^{-1} \left(\frac{a_j - x}{y} \right) \right].$$

If any one of the a_j's is infinite, say $a_1 = -\infty$, then $\tan^{-1} \left(\frac{a_1 - x}{y} \right) = \tan^{-1}(-\infty) = -\frac{\pi}{2}$. Similarly, if one of the b_j's is infinite, say $b_n = \infty$, then $\tan^{-1} \left(\frac{b_n - x}{y} \right) = \tan^{-1}(\infty) = \frac{\pi}{2}$.
(b) Let $z = x + iy$. Show that the answer can be written as

$$u(z) = \frac{1}{\pi} \sum_{j=1}^{n} T_j \left(\text{Arg}\,(z - b_j) - \text{Arg}\,(z - a_j) \right).$$

17. (a) Solve the Dirichlet Problem in Figure 7.67 by using the Poisson integral formula.

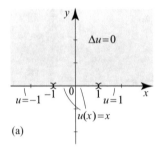

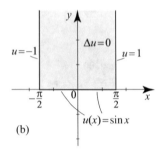

Fig. 7.67 Dirichlet problem on the upper half space.

Fig. 7.68 Dirichlet problem on a half strip.

(b) Solve the Dirichlet problem in Figure 7.68 using the conformal map $w = \sin z$ and the result in part (a).

18. Show that the solution in Example 7.3.1 is

$$u(x+iy) = 50 + \frac{100}{\pi}\left[\tan^{-1}\left(\frac{y}{1-x}\right) + \tan^{-1}\left(\frac{y}{1+x}\right)\right], \quad x^2 + y^2 < 1.$$

19. Isotherms in Example 7.3.1. Recall the solution of the corresponding Dirichlet problem in the upper half of the w-plane, $\frac{100}{\pi}(\pi - \operatorname{Arg} w)$.
(a) Show that the isotherms in the w-plane are rays emanating from the origin.
(b) Conclude that the isotherms in the z-plane are arcs of circles in the unit disk passing through the points -1 and 1. [*Hint*: Consider the pre-image of the isotherms in the w-plane by the mapping $\phi(z) = i\frac{1-z}{1+z}$.]

20. Study the isotherms of the Dirichlet problem in the w-plane in Example 7.3.3 and then determine the isotherms of the original problem in the z-plane.

21. Complete the proof of Theorem 7.3.4 by showing that (7.3.9) holds. [*Hint*: Organize your proof as follows. Show $1 - |\psi(z)|^2 = \frac{4y}{x^2+(1+y)^2}$. Then show $|e^{i\phi} - \psi(z)|^2 = |\psi(s) - \psi(z)|^2 = 4\frac{(x-s)^2+y^2}{(1+s^2)(x^2+(1+y)^2)}$.]

22. Show that $P_a(x) = \frac{1}{a}P_1\left(\frac{x}{a}\right)$ for all $a > 0$. Thus the graph of $P_a(x)$ is the graph of $P_1(x)$, scaled vertically by a factor of $\frac{1}{a}$ and scaled horizontally by a factor of a.

23. Show that for any $y > 0$, $\frac{1}{\sqrt{2\pi}}\int_{-\infty}^{+\infty} P_y(x)\,dx = 1$, where $P_y(x)$ is the Poisson kernel.

In Exercises 24–29, solve the Dirichlet problem in the upper half-plane with the boundary values the indicated $f(x)$ on the x-axis. Use residues to evaluate the Poisson integral. In Exercises 28 and 29, a is a nonzero real number.

24. $f(x) = \dfrac{x}{x^2+x+1}$ **25.** $f(x) = \dfrac{1}{x^4+1}$ **26.** $f(x) = \dfrac{\sin x}{x}$

27. $f(x) = \cos x$ **28.** $f(x) = \cos(ax)$ **29.** $f(x) = \sin(ax)$

Solve the Dirichlet problems depicted in Figures 7.69–7.73 by transforming them into a Dirichlet problem in the upper half-plane. To solve these Dirichlet problems in the upper half-plane, use the Arg function for Exercises 30–32 and the Poisson integral formula for Exercises 33–34.

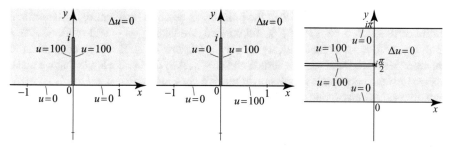

Fig. 7.69 Exercise 30.   **Fig. 7.70** Exercise 31.   **Fig. 7.71** Exercise 32.

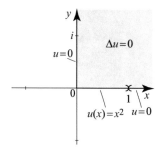

Fig. 7.72 Exercise 33.   **Fig. 7.73** Exercise 34.

35. Let S denote the region in the upper half-plane consisting of all z such that $|z| > 1$. Consider the Dirichlet problem in S with boundary conditions $u(x, y) = A$ if (x, y) is on the upper unit circle, and $u(x, 0) = B$ for all $|x| > 1$. Using the result of Exercise 7, Section 6.3, show that a solution of this Dirichlet problem is

$$u(x, y) = \frac{3B - A}{2} + \frac{A - B}{\pi} \left(\mathrm{Arg}\left((1 + i) \frac{J(z) - 1}{J(z) + i} \right) - \mathrm{Arg}\left((1 - i) \frac{J(z) + 1}{J(z) + i} \right) \right) \quad (z = x + iy),$$

where J is the Joukowski function. [*Hint*: Consider the mapping $z \mapsto \frac{J(z) - i}{J(z) + i}$ from S to D.]

7.4 The Schwarz-Christoffel Transformation

In this section we describe a method for constructing one-to-one analytic mappings of the upper half-plane onto polygonal regions. We start by setting the notation.

Suppose that Ω is a region in the w-plane, bounded by a positively oriented polygonal path P with n sides. Let $w_1, w_2, \ldots, w_n$ denote the vertices of P, counted consecutively as we trace the path through its positive orientation; see Figure 7.74

when $n = 5$. If P is bounded, then the point w_n is taken to be the initial and terminal point of the closed path P. If P is unbounded, we take $w_n = \infty$ and think of P as a polygon with $n - 1$ vertices $w_1, w_2, \ldots, w_{n-1}$ (Figure 7.75). It will be convenient to measure the exterior angle at a vertex, and so we let θ_j denote the angle that we make as we turn the corner of the polygon at w_j. We choose $0 < |\theta_j| < \pi$ ($j = 1, \ldots, n$); a positive value corresponds to a left turn, and a negative value corresponds to a right turn. In Figure 7.74, θ_2 is negative while all other θ_j are positive.

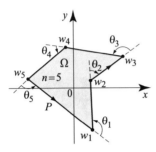

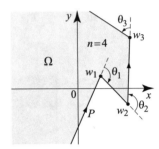

Fig. 7.74 Positively oriented polygonal boundary with corner angles measured from the outside.

Fig. 7.75 Unbounded polygonal region with n sides ($n = 4$) and $n - 1$ vertices.

Theorem 7.4.1. (Schwarz-Christoffel Transformation) *Let Ω be a region bounded by a polygonal path P with vertices at w_j (counted consecutively) and corresponding exterior angles θ_j. Then there is a one-to-one conformal mapping f of the upper half-plane onto Ω, such that*

$$f'(z) = A\,(z - x_1)^{-\frac{\theta_1}{\pi}}\,(z - x_2)^{-\frac{\theta_2}{\pi}}\cdots(z - x_{n-1})^{-\frac{\theta_{n-1}}{\pi}}, \qquad (7.4.1)$$

where A is a constant, the x_j's are real and satisfy $x_1 < x_2 < \cdots < x_{n-1}$, $f(x_j) = w_j$, $\lim_{z \to \infty} f(z) = w_n$, and the complex powers are defined by their principal branches.

The points x_j ($j = 1, \ldots, n - 1$) on the x-axis are the pre-images of the vertices of the polygonal path P in the w-plane. Two of the x_j's may be chosen arbitrarily, so long as they are arranged in ascending order. We can express the fact that $\lim_{z \to \infty} f(z) = w_n$ by writing $f(\infty) = w_n$. In the case of an unbounded polygon P, we have $f(\infty) = \infty$.

Definition 7.4.2. The mapping f whose derivative is the function in (7.4.1) is called a **Schwarz-Christoffel transformation**, after the German mathematicians Karl Herman Amandus Schwarz (1843–1921) and Elwin Bruno Christoffel (1829–1900). Since f is an antiderivative of the function in (7.4.1), we can write

$$f(z) = A \int (z - x_1)^{-\frac{\theta_1}{\pi}}\,(z - x_2)^{-\frac{\theta_2}{\pi}}\cdots(z - x_{n-1})^{-\frac{\theta_{n-1}}{\pi}}\,dz + B. \qquad (7.4.2)$$

The constants A and B depend on the size and location of the polygonal path P.

The full proof of Theorem 7.4.1 is quite complicated. We only sketch a part of the proof that illustrates the ideas behind the construction of the transformation.

Proof. (Sketch) Consider a mapping f whose derivative is given by (7.4.1), and let w_j denote the image of x_j, where $x_1 < x_2 < \cdots < x_{n-1}$ are real. To understand the effect of the mapping f on the real axis, recall from Section 7.1 that a conformal mapping f at a point z_0 acts like a rotation by an angle $\arg f'(z_0)$. Thus the mapping whose derivative is in (7.4.1) acts like a rotation by an angle

$$\arg f'(z) = \tag{7.4.3}$$
$$\arg A - \frac{\theta_1}{\pi}\arg(z - x_1) - \frac{\theta_2}{\pi}\arg(z - x_2) - \cdots - \frac{\theta_{n-1}}{\pi}\arg(z - x_{n-1}).$$

For $z = x$ on the x-axis with $x < x_1$, we have $x - x_j < 0$ for all $j = 1, 2, \ldots, n-1$, hence $\arg(x - x_j) = \pi$ for all $j = 1, 2, \ldots, n-1$, and so from (7.4.3) we get

$$-\infty < x < x_1, f(x) \in [w_n, w_1] \implies \arg f'(x) = \arg A - (\theta_1 + \theta_2 + \cdots + \theta_{n-1}). \tag{7.4.4}$$

Thus if $w_n = f(\infty)$, then all the points in the interval $(-\infty, x_1)$ are mapped onto a line segment starting with w_n and ending with $w_1 = f(x_1)$ and at an angle given by (7.4.4) (Figure 7.76). For x in the interval (x_1, x_2), we have $x - x_1 > 0$ and $x - x_j < 0$ for all $j = 2, \ldots, n-1$, hence $\arg(x - x_1) = 0$ and $\arg(x - x_j) = \pi$ for all $j = 2, \ldots, n-1$, and so from (7.4.3) we obtain

$$x_1 < x < x_2, f(x) \in [w_1, w_2] \implies \arg f'(x) = \arg A - (\theta_2 + \cdots + \theta_{n-1}). \tag{7.4.5}$$

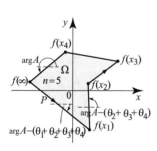

Fig. 7.76 Arguments of the line segments of the polygon P.

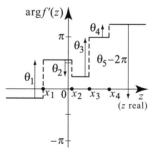

Fig. 7.77 As x crosses x_j from left to right, $\arg f'(x)$ changes abruptly by θ_j then remains constant until it crosses x_{j+1}.

Thus, at x_1 we have an abrupt change in the argument (the path turns left by angle θ_1) and then all the points in (x_1, x_2) are mapped onto the line segment with initial point w_1 and terminal point $w_2 = f(x_2)$, and at an angle given by (7.4.5) (Figure 7.77). Continuing in this fashion we obtain for $j = 2, \ldots, n-1$

$$x_{j-1} < x < x_j, f(x) \in [w_{j-1}, w_j] \implies \arg f'(x) = \arg A - (\theta_j + \cdots + \theta_{n-1}). \quad (7.4.6)$$

Finally we find that after an abrupt change in the argument, the points in the interval (x_{n-1}, ∞) are mapped onto a line segment with initial point w_{n-1}, and at angle

$$x_{n-1} < x < +\infty, f(x) \in [w_{n-1}, w_n] \implies \arg f'(x) = \arg A. \quad (7.4.7)$$

In the case of a bounded polygon, this line segment will connect back to w_n (Exercise 11). The polygon is then closed; for closed polygons, after turning the last corner our combined angle of turns is $\theta_1 + \theta_2 + \cdots + \theta_n = 2\pi$. Thus we have shown that the mapping whose derivative is given by (7.4.1) takes the real line onto a polygon with vertices $w_j = f(x_j)$ and exterior angles θ_j. Since the upper half-plane is to our left as we traverse the real line rightward, conformality ensures that the image region is to our left as we trace P in the positive sense (i.e., f maps the upper half-plane onto the interior of P). The converse of these statements is also true, although we will not prove it. That is, for every polygonal path P with vertices w_j and exterior angles θ_j, it can be shown that we can find ordered real numbers $x_1, \ldots, x_{n-1}$, and complex numbers A and B such that the mapping in (7.4.2) whose derivative is given by (7.4.1) takes the real line onto P. Moreover, two of the x_j's can be chosen arbitrarily. ∎

Example 7.4.3. (Schwarz-Christoffel transformation for a sector) Find a Schwarz-Christoffel transformation that maps the upper half-plane onto the sector of angle $0 < \alpha < \pi$ shown in Figure 7.78.

Solution. Obviously one answer is $f(z) = z^{\frac{\alpha}{\pi}}$, but let us see how this answer comes out of (7.4.2). Since the region has two sides, we have $n = 2$. From Figure 7.78, the exterior angle at $w_1 = 0$ is $\pi - \alpha$. In the z-plane, choose $x_1 = 0$, then (7.4.2) yields

$$\begin{aligned} f(z) &= A \int z^{-\frac{\pi-\alpha}{\pi}} dz + B \\ &= A \int z^{-1+\alpha/\pi} dz + B \\ &= A \frac{\pi}{\alpha} z^{\frac{\alpha}{\pi}} + B, \end{aligned}$$

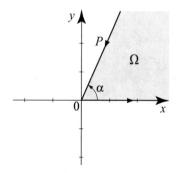

Fig. 7.78 Sector of angle α.

where all branches are principal. In order to have $f(0) = 0$, we take $B = 0$.

Obviously any positive value of A will work, so we can take $f(z) = z^{\frac{\alpha}{\pi}}$. □

For our next example, we need the following useful formula, whose proof is sketched in Exercise 12. For z in the upper half-plane ($\text{Im} z > 0$), $0 \le \alpha \le \pi$, and all real numbers a, we have

$$(z+a)^{\frac{\alpha}{\pi}}(z-a)^{\frac{\alpha}{\pi}} = (-1)^{\frac{\alpha}{\pi}}(a^2-z^2)^{\frac{\alpha}{\pi}}, \tag{7.4.8}$$

with all branches being principal. For example, if $\alpha = \frac{\pi}{2}$, $a = 1$, and $\text{Im} z > 0$, then

$$(z+1)^{\frac{1}{2}}(z-1)^{\frac{1}{2}} = i(1-z^2)^{\frac{1}{2}}. \tag{7.4.9}$$

Example 7.4.4. (The inverse sine as a Schwarz-Christoffel transformation) Find a Schwarz-Christoffel transformation that maps the upper half-plane onto the semi-infinite vertical strip in Figure 7.79.

Solution. We know that $\sin z$ maps the infinite strip in Figure 7.79 onto the upper half-plane. So the mapping that we are looking for is the inverse of $\sin z$. Let us see how this comes out of (7.4.2). In the w-plane, take $w_1 = -\frac{\pi}{2}$, $w_2 = \frac{\pi}{2}$, with exterior angles $\theta_1 = \theta_2 = \frac{\pi}{2}$. In the z-plane, take $x_1 = -1$ and $x_2 = 1$. Then (7.4.2) yields

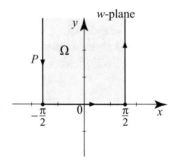

Fig. 7.79 A semi-infinite vertical strip with positively oriented boundary.

$$f(z) = A\int (z+1)^{-\frac{1}{2}}(z-1)^{-\frac{1}{2}}\,dz + B.$$

Using (7.4.9) and a well-known antiderivative, we get

$$f(z) = -Ai\int \frac{1}{\sqrt{1-z^2}}\,dz + B = -Ai\sin^{-1}z + B,$$

where $\sin^{-1}z$ is the principal branch of the inverse sine function; that is,

$$\sin^{-1}z = -i\text{Log}\left(iz + e^{\frac{1}{2}\text{Log}(1-z^2)}\right)$$

as shown in Example 1.8.8.

Setting $f(-1) = -\frac{\pi}{2}$, we find

$$-Ai\sin^{-1}(-1) + B = -\frac{\pi}{2} \quad \Rightarrow \quad Ai\frac{\pi}{2} + B = -\frac{\pi}{2}.$$

Setting $f(1) = \frac{\pi}{2}$, we find

$$-Ai\sin^{-1}(1) + B = \frac{\pi}{2} \quad \Rightarrow \quad -Ai\frac{\pi}{2} + B = \frac{\pi}{2}.$$

Solving for A and B, we find $A = i$ and $B = 0$. Hence $f(z) = \sin^{-1} z$.   □

Like many constructions involving Schwarz-Christoffel transformations, the next example gives rise to elliptic integrals (see Section 5.5). Although these integrals are very difficult to evaluate, they are extensively tabulated and can be conveniently evaluated numerically using standard functions in most computer systems.

Example 7.4.5. (Schwarz-Christoffel transformation for a triangle) Find a Schwarz-Christoffel transformation that maps the upper half-plane onto the right isosceles triangle in Figure 7.80.

Solution. It is clear that the triangle is determined by two consecutive vertices and their corresponding angles. In the w-plane, take $w_1 = -1$, $w_2 = 1$, with exterior angles $\theta_1 = \theta_2 = \frac{3\pi}{4}$. In the z-plane, we freely choose the points $x_1 = -1$ and $x_2 = 1$. Then (7.4.2) yields

$$f(z) = A \int_{[0,z]} \frac{1}{(\zeta+1)^{\frac{3}{4}}} \frac{1}{(\zeta-1)^{\frac{3}{4}}} d\zeta + B$$

$$= A \int_{[0,z]} \frac{d\zeta}{(-1)^{\frac{3}{4}}(1-\zeta^2)^{\frac{3}{4}}} + B,$$

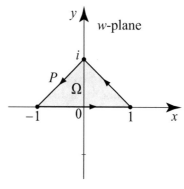

Fig. 7.80 Positively oriented isosceles triangle.

where we have used (7.4.8) with $\alpha = \frac{3\pi}{4}$. This integral cannot be expressed in terms of elementary functions for arbitrary z, but we will be able to evaluate it for $z = \pm 1$ in order to determine the constants A and B. Setting $f(-1) = -1$ and using $(-1)^{\frac{3}{4}} = e^{\frac{3\pi i}{4}}$, we get

$$-1 = e^{-\frac{3\pi i}{4}} A \int_0^{-1} \frac{d\zeta}{(1-\zeta^2)^{\frac{3}{4}}} + B = -e^{-\frac{3\pi i}{4}} A \int_0^1 \frac{d\zeta}{(1-\zeta^2)^{\frac{3}{4}}} + B,$$

or

$$-e^{-\frac{3\pi i}{4}} AI + B = -1, \quad \text{where} \quad I = \int_0^1 \frac{d\zeta}{(1-\zeta^2)^{\frac{3}{4}}}. \tag{7.4.10}$$

To evaluate the integral I, we make the change of variables $\zeta = \sin x$, $d\zeta = \cos x\, dx$, then

$$I = \int_0^{\frac{\pi}{2}} \frac{dx}{(\cos x)^{\frac{1}{2}}} = \frac{1}{2} \frac{\Gamma(1/4)\Gamma(1/2)}{\Gamma(3/4)} \approx 2.622,$$

where we have appealed to Exercise 25, Section 4.2, to evaluate the integral in terms of the gamma function, while the approximate value of the integral was obtained with the help of a computer. Setting $1 = f(1)$, we get

$$e^{-\frac{3\pi i}{4}} A I + B = 1. \tag{7.4.11}$$

Solving for A and B in (7.4.11) and (7.4.10), we find $B = 0$ and $A = \dfrac{e^{\frac{3\pi i}{4}}}{I}$, and so

$$f(z) = \frac{e^{\frac{3\pi i}{4}}}{I} e^{-\frac{3\pi i}{4}} \int_{[0,z]} \frac{d\zeta}{(1-\zeta^2)^{\frac{3}{4}}} = \frac{1}{I} \int_{[0,z]} \frac{d\zeta}{(1-\zeta^2)^{\frac{3}{4}}}. \qquad \square$$

The next two examples illustrate a limiting technique in computing Schwarz-Christoffel transformations.

Example 7.4.6. (An *L*-shaped region) Find a Schwarz-Christoffel transformation that maps the upper half-plane onto the *L*-shaped region in Figure 7.81.

Solution. To determine the orientation of the boundary in such a way that the region becomes interior to a positively oriented boundary, we think of the region as a limit of a region with vertices at $w_1 = 0$, $w_2 > 0$, and $w_3 = 1 + i$, and corresponding exterior angles $\theta_1 = \frac{\pi}{2}$, $\theta_2 = \alpha$, and $\theta_3 = \beta$ (Figure 7.82).

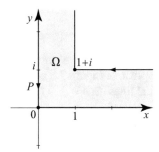

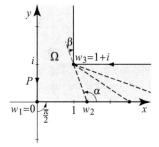

Fig. 7.81 An *L*-shaped region with positively oriented boundary. As we follow the boundary according to this orientation, the region is to our left.

Fig. 7.82 The region in Figure 7.81 is thought of as the limit of the semi-infinite region with vertices w_1, w_2, w_3 as w_2 tends to infinity along the *x*-axis.

As $w_2 \to \infty$, $\theta_2 \to \pi$, and $\theta_3 \to -\frac{\pi}{2}$. Thus, we may think of our region as having a vertex at infinity with exterior angle $\theta_2 = \pi$ and a vertex at $w_3 = 1 + i$ with exterior angle $-\frac{\pi}{2}$. In fact, setting $\theta_j = \pi$ forces $f(x_j) = \infty$. Now we may only choose two of the three x_j's arbitrarily, but in fact a solution can be found in $x_1 = -1$, $x_2 = 0$, and $x_3 = 1$. Other choices of the three x_j's will typically result in other *L*-shaped regions. From (7.4.2) and (7.4.9) (noticing that $\text{Im}\, z < 0$ in our case) we obtain

$$f(z) = A \int (z+1)^{-\frac{1}{2}} z^{-1} (z-1)^{\frac{1}{2}} \, dz + B = A \int \frac{z-1}{iz(1-z^2)^{\frac{1}{2}}} \, dz + B.$$

Observe that

$$\frac{z-1}{iz(1-z^2)^{\frac{1}{2}}} = \frac{-i}{(1-z^2)^{\frac{1}{2}}} + \frac{i}{z(1-z^2)^{\frac{1}{2}}}.$$

Letting $z = \frac{1}{\zeta}$, we note that $\mathrm{Im}\,\zeta < 0$ and $dz = -\frac{d\zeta}{\zeta^2}$, hence

$$\int \frac{i\,dz}{z(1-z^2)^{\frac{1}{2}}} = -i\int \frac{d\zeta}{\zeta(1-\frac{1}{\zeta^2})^{\frac{1}{2}}} = \int \frac{d\zeta}{(1-\zeta^2)^{\frac{1}{2}}} = \sin^{-1}\zeta + C,$$

where the justification involving the square root manipulation is left to Exercise 11. Thus

$$f(z) = \int \frac{-i}{(1-z^2)^{\frac{1}{2}}} + \frac{i}{z(1-z^2)^{\frac{1}{2}}}\,dz = A\left(-i\sin^{-1}z + \sin^{-1}\frac{1}{z}\right) + B,$$

where inverse sines are principal branches. Setting $f(-1) = 0$, we get $A(-i(-\frac{\pi}{2}) + (-\frac{\pi}{2})) + B = 0$. Setting $f(1) = 1+i$, we get $A(-i\frac{\pi}{2} + \frac{\pi}{2}) + B = 1+i$. Solving for A and B, we get $A = \frac{i}{\pi}$ and $B = \frac{1+i}{2}$, and so we obtain

$$f(z) = \frac{1}{\pi}\left(\sin^{-1}z + i\sin^{-1}\frac{1}{z}\right) + \frac{1+i}{2}. \qquad \square$$

Example 7.4.7. (A doubly slit plane) Find a Schwarz-Christoffel transformation that maps the upper half-plane onto the doubly slit plane in Figure 7.83, which consists of the w-plane minus the semi-infinite horizontal lines $\mathrm{Im}\,w = \pm i$ and $\mathrm{Re}\,w < 0$.

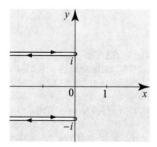

Fig. 7.83 A doubly slit plane with a positively oriented boundary consisting of four sides.

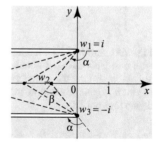

Fig. 7.84 The region in Figure 7.83 is obtained as a limit of the triangular regions as w_2 tends to $-\infty$ along the x-axis.

Solution. To define the region as the interior of a positively oriented boundary requires four sides as shown in Figure 7.83. We think of the region as a limit of a region with vertices at $w_1 = i$, $w_2 = t$ (where $t < 0$) and $w_3 = -i$, and corresponding exterior angles $\theta_1 = \alpha$, $\theta_2 = \beta$, and $\theta_3 = \alpha$ (Figure 7.84). As $t \to -\infty$, $\theta_1 \to -\pi$,

$\theta_2 \to \pi$, and $\theta_3 \to -\pi$. Thus, we may think of our region as having a vertex at $w_2 = \infty$ with exterior angle $\theta_2 = \pi$ and two vertices at $\pm i$ with each exterior angle being $-\pi$. Taking $x_1 = -1$, $x_2 = 0$, and $x_3 = 1$, we get from (7.4.2)

$$f(z) = A \int (z+1)\frac{1}{z}(z-1)\,dz + B = A \int \left(z - \frac{1}{z}\right) dz + B = \frac{A}{2}z^2 - A\,\mathrm{Log}\,z + B.$$

Setting $f(-1) = i$ and $f(1) = -i$, we get

$$\begin{cases} i & = \frac{A}{2} - A\,\mathrm{Log}\,(-1) + B \\ -i = \frac{A}{2} - A\,\mathrm{Log}\,(1) + B, \end{cases} \Rightarrow \begin{cases} i & = \frac{A}{2} - Ai\pi + B \\ -i = \frac{A}{2} + B. \end{cases}$$

Solving for A and B we get $A = -\frac{2}{\pi}$ and $B = \frac{1}{\pi} - i$, and so we obtain

$$f(z) = -\frac{1}{\pi}z^2 + \frac{2}{\pi}\mathrm{Log}\,z + \frac{1}{\pi} - i. \qquad \square$$

Image of Level Curves

Suppose that f is a Schwarz-Christoffel transformation taking the upper half-plane onto a region Ω in the w-plane. If we want to solve a Dirichlet problem $\Delta U(w) = 0$, the technique is to map the problem to a corresponding problem in the upper half of the z-plane, where a solution u of Laplace's equation $\Delta u(z) = 0$ with boundary conditions can be obtained more easily. The solution in the w-plane is then $U(w) = u(f^{-1}(w))$. However, the Schwarz-Christoffel transformation yields f, not f^{-1}, and it is not always possible or easy to invert f in closed form; try for instance the preceding examples. Nevertheless, the conformal map f is still very useful: It allows us to find isotherms of U without actually knowing it. As we now show, this is because the image under f of an isotherm $u(z) = C$ (where C is a constant) is an isotherm $U(w) = C$ in the region Ω.

Proposition 7.4.8. (Image of Level Curves) *With the preceding notation, the image under $w = f(z)$ of the level curve $u(z) = C$ is a level curve $U(w) = C$. Thus, if $\gamma(t)$, $a < t < b$, parametrizes an isotherm of u, then $f(\gamma(t))$, $a < t < b$, parametrizes a corresponding isotherm of U.*

Proof. Since $\gamma(t)$ is a level curve of u, $u(\gamma(t)) = C$. As $U(w) = u(f^{-1}(w))$, we conclude that $U(f(\gamma(t))) = u(\gamma(t)) = C$, and hence $f(\gamma(t))$ is a level curve of U. $\blacksquare$

This proposition is of course true for any one-to-one conformal map, not just for Schwarz-Christoffel transformations acting on the upper half-plane. We illustrate with an example.

Example 7.4.9. (Isotherms in the L-shaped region) Find the isotherms of the Dirichlet problem $\Delta U(w) = 0$ in the L-shaped region in Figure 7.85.

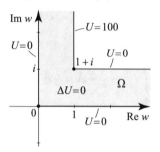

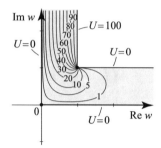

Fig. 7.85 The Dirichlet problem on an L-shaped region.

Fig. 7.86 Isotherms for the Dirichlet problem on an L-shaped region.

Solution. From Example 7.4.6, we know that the Schwarz-Christoffel transformation $f(z) = \frac{1}{\pi}\left(\sin^{-1} z + i\sin^{-1}\frac{1}{z}\right) + \frac{1+i}{2}$ maps the upper half-plane onto the L-shaped region. Since the boundary data switches from 0 to 100 at $f(1) = 1+i$, the corresponding Dirichlet problem in the upper half-plane is $\Delta u = 0$, $u(x) = 0$ for $x < 1$, and $u(x) = 100$ for $x > 1$. We immediately write down the solution using the argument function:

$$u(z) = 100 - \frac{100}{\pi}\operatorname{Arg}(z-1).$$

The isotherms $u = C$ thus satisfy $\operatorname{Arg}(z-1) = \pi - \frac{\pi C}{100}$ and are rays emanating from the point $z = 1$. Each ray is parametrized by $\gamma(t) = 1 + te^{i\left(\pi - \frac{\pi C}{100}\right)}$, where $0 < t < \infty$. By Proposition 7.4.8, the image of this ray under $w = f(z)$ is the isotherm $U = C$ in the L-shaped region, and it is parametrized by

$$f(\gamma(t)) = \frac{1}{\pi}\left(\sin^{-1}\left(1 + te^{i\left(\pi - \frac{\pi C}{100}\right)}\right) + i\sin^{-1}\left(\frac{1}{1 + te^{i\left(\pi - \frac{\pi C}{100}\right)}}\right)\right) + \frac{1+i}{2}.$$

Some isotherms are plotted in Figure 7.86.   □

Fluid Flow

We now investigate problems in two-dimensional fluid flow, using our knowledge of harmonic functions and the technique of conformal mappings. In an ideal situation where fluid is flowing over a two-dimensional surface represented by an unbounded region Ω in the z-plane, assuming that the fluid is incompressible (fixed density) and irrotational (circulation around a closed path is zero), it can be shown that there is a harmonic function ϕ such that the velocity V in Ω is described by the gradient of ϕ. That is, for (x, y) in Ω we have

$$V(x, y) = \nabla \phi(x, y) = \left(\frac{\partial \phi}{\partial x}, \frac{\partial \phi}{\partial y} \right)(x, y) = \big(\phi_x(x, y), \phi_y(x, y) \big). \qquad (7.4.12)$$

Definition 7.4.10. The function ϕ is called the **velocity potential** of the flow. The curves defined by the relation

$$\phi(x, y) = C_1 \qquad (7.4.13)$$

are called the **equipotential curves** or **equipotential lines**. The **streamlines** of the flow are the curves that are orthogonal to the equipotential curves. If the streamlines are expressed in the form

$$\psi(x, y) = C_2, \qquad (7.4.14)$$

for some function ψ, then, from Section 2.5, we know that we can take ψ to be the harmonic conjugate of ϕ in Ω. The function ψ is called the **stream function**. The fluid will flow on the level curves of ψ. If we let

$$\Phi(z) = \phi(x, y) + i \psi(x, y) \quad (z = x + iy \text{ in } \Omega), \qquad (7.4.15)$$

then Φ is analytic in Ω. It is called the **complex potential** of the flow.

Thus the real part of the complex potential is the velocity potential, and the level curves of its imaginary part are the streamlines. If Γ denotes the boundary of Ω, we would expect the fluid to flow along Γ or that Γ is one of the streamlines. Thus the points on Γ should satisfy (7.4.14).

For a simple example of a complex potential, we consider a uniform rightward flow in the upper half-plane with complex potential $\Phi(z) = z = x + iy$ (Figure 7.87). Here the velocity potential is $\phi(x, y) = x$, and the velocity at each point is the vector $(1, 0)$, which is the gradient of $\phi(x, y) = x$. The stream function is $\psi(x, y) = y$, and the streamlines are $\psi(x, y) = C$ for constants $C \geq 0$. In accordance with the properties of a flow, the boundary $y = 0$ is one of the streamlines corresponding to the value $C = 0$.

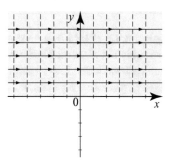

Fig. 7.87 Streamlines in a uniform rightward flow in the upper half-plane.

We find the streamlines using conformal mapping techniques. Let f be a one-to-one conformal mapping of the upper half of the z-plane onto Ω, taking the x-axis onto Γ, the boundary of Ω. Suppose that a stream function ψ is known for the upper half-plane. By the properties of conformal mappings, $\Psi = \psi \circ f^{-1}$ is harmonic in Ω. Proposition 7.4.8 tells us that the images of streamlines $\psi(z) = C$ under the mapping f are streamlines $\Psi(w) = C$. Since the real axis is a streamline for ψ and Γ is the image of the real axis, we conclude that Γ is a streamline for Ψ. Thus Ψ is a stream function for Ω.

In the following examples, we take the simple stream function $\psi(z) = y$ for the upper half-plane. Streamlines for the region Ω are found by using Proposition 7.4.8.

Example 7.4.11. (Fluid flow in a sector) Find and plot the streamlines for the sector in Figure 15, where fluid flows in along the line $\text{Arg}\,w = \frac{\pi}{4}$ and flows out along $\text{Arg}\,w = 0$.

Solution. From Example 7.4.3, the Schwarz-Christoffel transformation $f(z) = z^{\frac{1}{4}}$ maps the upper half-plane to the sector $0 < \arg z < \frac{\pi}{4}$. We use the simple stream function in the upper half-plane $\psi(z) = y$ to generate a solution. Streamlines in the z-plane are parametrized as $\gamma(x) = x + i y_0$ for fixed y_0. Streamlines in the w-plane are images of these under f; we have $f(\gamma(x)) = (x + i y_0)^{\frac{1}{4}}$. As the parameter x increases, the streamlines are traced in the manner shown in Figure 7.88. Fluid comes in along $\text{Arg}\,w = \frac{\pi}{4}$ and out along $\text{Arg}\,w = 0$.   □

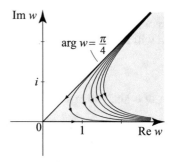

Fig. 7.88 Streamlines in a sector.

Example 7.4.12. (Fluid flow in the doubly slit plane) Find and plot the streamlines for the doubly slit plane in Figure 7.83, where fluid flows in from the upper left, past the double obstacle, and flows out to the lower left.

Solution. In Example 7.4.7, we found that $f(z) = -\frac{1}{\pi}z^2 + \frac{2}{\pi}\text{Log}\,z + \frac{1}{\pi} - i$ is a conformal mapping of the upper half-plane onto the doubly slit plane Ω. Taking $\psi(z) = y$ to be the stream function for the upper half-plane, for each $y_0 \geq 0$ we have a streamline parametrized by

$$\gamma(x) = x + i y_0, \quad -\infty < x < \infty.$$

By Proposition 7.4.8, the curves $f(\gamma(x))$, $-\infty < x < \infty$, are streamlines in the doubly slit plane. They are

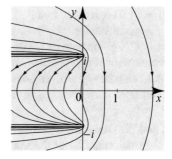

Fig. 7.89 Streamlines in a doubly slit plane.

$$f(x+iy_0) = -\frac{1}{\pi}(x+iy_0)^2 + \frac{2}{\pi}\text{Log}(x+iy_0) + \frac{1}{\pi} - i \qquad (-\infty < x < \infty).$$

The streamlines are plotted in Figure 7.89. Note that the central channel serves neither as a source of fluid nor as a final destination; fluid flows in and then flows out. The fluid far into the central channel is almost stagnant. ☐

Exercises 7.4

In Exercises 1–6, find the Schwarz-Christoffel transformation of the upper half-plane onto the region described by the following figures. Use the labeled points to set up the integral (7.4.2).

1. Figure 7.90. [*Hint*: Use (7.4.9) and integrate.]

2. Figure 7.91. [*Hint*: $\frac{z+1}{(z-1)^{\frac{1}{2}}} = \frac{z-1+2}{(z-1)^{\frac{1}{2}}} = (z-1)^{\frac{1}{2}} + \frac{2}{(z-1)^{\frac{1}{2}}}$.]

3. Figure 7.92. [*Hint*: $\frac{(z+1)^{\frac{1}{2}}}{(z-1)^{\frac{1}{2}}} = \frac{z+1}{i(1-z^2)^{\frac{1}{2}}} = \frac{z}{i(1-z^2)^{\frac{1}{2}}} + \frac{1}{i(1-z^2)^{\frac{1}{2}}}$.]

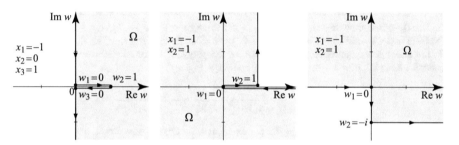

Fig. 7.90 Exercise 1.   **Fig. 7.91** Exercise 2.   **Fig. 7.92** Exercise 3.

4. Figure 7.93. [*Hint*: $\frac{(z-1)^{\frac{1}{2}}}{(z+1)^{\frac{1}{2}}} = \frac{z-1}{i(1-z^2)^{\frac{1}{2}}} = \frac{z}{i(1-z^2)^{\frac{1}{2}}} - \frac{1}{i(1-z^2)^{\frac{1}{2}}}$.]

5. Figure 7.94. [*Hint*: Let $z = \sin\zeta$, where $-\frac{\pi}{2} \leq \text{Re}\,\zeta \leq \frac{\pi}{2}$ and $\text{Im}\,\zeta \geq 0$. Then $(1-z^2)^{\frac{1}{2}} = \cos\zeta$. Use $\cos^2\zeta = \frac{1+\cos 2\zeta}{2}$, integrate, and then use $\sin 2\zeta = 2\sin\zeta\cos\zeta$.]

6. Figure 7.95. [*Hints*: In this problem $\theta_1 = \pi$, $\theta_2 = -\frac{\pi}{2}$, and $\theta_3 = -\frac{\pi}{2}$ (in reality it is $\frac{3\pi}{2}$, but this is identified with $-\frac{\pi}{2}$ as $|\frac{3\pi}{2}| > \pi$). We have

$$\frac{(z-1)^{\frac{1}{2}}}{z+1} = \frac{1}{(z-1)^{\frac{1}{2}}} \frac{z-1}{z+1} = \frac{1}{(z-1)^{\frac{1}{2}}} - \frac{2}{(z-1)^{\frac{1}{2}}} \frac{1}{z+1}.$$

In the second term, use

$$\frac{1}{z+1} = \frac{i}{2\sqrt{2}}\left(\frac{1}{\sqrt{z-1}+i\sqrt{2}} - \frac{1}{\sqrt{z-1}-i\sqrt{2}}\right),$$

and change variables $u = \sqrt{z-1} - i\sqrt{2}$ and $v = \sqrt{z-1} - i\sqrt{2}$. You cannot find A and B just from $f(1) = 0$. Instead, first argue that $\text{Arg}\,A = -\frac{\pi}{2}$ via identity (7.4.6) by looking at the angle of the negative imaginary axis, which is the image $(1,\infty)$. Thus we have $A = -i|A|$. Now get $B = i\pi\sqrt{2}|A|$ from $f(1) = 0$. To get $|A|$, use that $\text{Im}\,f(x) = 1$ for $x < -1$.]

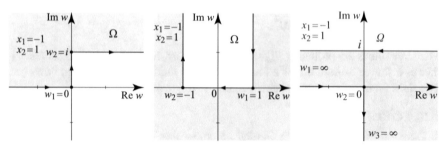

Fig. 7.93 Exercise 4.   **Fig. 7.94** Exercise 5.   **Fig. 7.95** Exercise 6.

7. Consider the Dirichlet problem in the semi-infinite strip of Example 7.4.4, where the base is kept at temperature $100°$ and the other two vertical sides at temperature $0°$. Solve this Dirichlet problem and describe the isotherms.

8. Find and plot the streamlines in the region of Example 7.4.3.

9. Project Problem: Closure of the polygon. In this exercise, we show that for a closed polygon where $\theta_1 + \cdots + \theta_n = 2\pi$, the integral formula for $f(z)$, (7.4.2), converges to w_n as $z \to \infty$.

(a) Use $\theta_n < \pi$ to show that $\theta_1 + \cdots + \theta_{n-1} > \pi$. Define $\beta_j = \frac{\theta_j}{\pi}$ for $j = 1, \ldots, n-1$.

(b) Note that the coefficients A and B in (7.4.1) will dilate, rotate, and translate the mapping and do not affect convergence of $f(z)$ or closure of the polygon, so we take $A = 1$ and $B = 0$. For concreteness, pick $x_0 = 1 + \max_{1 \le j \le n-1} x_j$, and set

$$f(z) = \int_{[x_0, z]} \frac{d\zeta}{(\zeta - x_1)^{\beta_1} (\zeta - x_2)^{\beta_2} \cdots (\zeta - x_{n-1})^{\beta_{n-1}}}.$$

(c) We show that f has a limit on the positive real axis. Restrict $z = x$ real, and use the limit comparison test for the integrand (against $\frac{1}{x^{\beta_1 + \cdots + \beta_{n-1}}}$) to show that

$$\lim_{y \to \infty} f(y) = \int_{x_0}^{\infty} \frac{dx}{(x - x_1)^{\beta_1} (x - x_2)^{\beta_2} \cdots (x - x_{n-1})^{\beta_{n-1}}}$$

is finite. Define w_n to be this number.

(d) To show that $\lim_{|z| \to \infty} f(z) = w_n$ write $z = Re^{i\theta}$, $R > 0$ and $0 \le \theta \le \pi$. Then

$$|f(Re^{i\theta}) - f(R)| = \left| \int_{[R, Re^{i\theta}]} \frac{dz}{(z - x_1)^{\beta_1} (z - x_2)^{\beta_2} \cdots (z - x_{n-1})^{\beta_{n-1}}} \right| \le R\pi M(R),$$

where $M(R)$ is the maximum of the absolute value of the integrand on the upper semicircle of radius R. Now $\lim_{R \to \infty} (M(R) R^{\beta_1 + \cdots + \beta_{n-1}}) = 1$, so $RM(R) = R^{1 - \beta_1 - \cdots - \beta_{n-1}} (M(R) R^{\beta_1 + \cdots + \beta_{n-1}}) \to 0$, and so $|f(Re^{i\theta}) - f(R)| \to 0$ uniformly in θ as $R \to \infty$. Since $f(R) \to w_n$, we conclude that $f(z) \to w_n$ as $z \to \infty$. A similar argument works if z lies in the lower half-plane, i.e., when $\pi < \theta \le 2\pi$.

10. (a) Follow the outlined steps to show that for $\text{Im}\, z > 0$, a real, and $0 \le \alpha \le \pi$,

$$(z - a)^{\frac{\alpha}{\pi}} = (-1)^{\frac{\alpha}{\pi}} (a - z)^{\frac{\alpha}{\pi}}, \quad \text{all branches principal.} \tag{7.4.16}$$

Fix z with $\text{Im}\, z > 0$. We must prove $e^{\frac{\alpha}{\pi} \text{Log}(z - a)} = e^{i\alpha} e^{\frac{\alpha}{\pi} \text{Log}(a - z)}$, and it will be sufficient to prove $\text{Log}(z - a) = \text{Log}(a - z) + i\pi$, or that $\text{Arg}(z - a) = \text{Arg}(a - z) + \pi$. We know that for each z, $\text{Arg}(z - a) = \text{Arg}(a - z) \pm \pi$. However, $0 < \text{Arg}(z - a) < \pi$ and $-\pi < \text{Arg}(a - z) < 0$, so $0 < \text{Arg}(z - a) - \text{Arg}(a - z) < 2\pi$ and we must use the plus sign. This proves (7.4.16).

(b) Follow the outlined steps to prove (7.4.8). Use part (a) to show

$$(z-a)^{\frac{\alpha}{\pi}}(z+a)^{\frac{\alpha}{\pi}} = (-1)^{\frac{\alpha}{\pi}}(a-z)^{\frac{\alpha}{\pi}}(a+z)^{\frac{\alpha}{\pi}},$$

and so we must show $(a-z)^{\frac{\alpha}{\pi}}(a+z)^{\frac{\alpha}{\pi}} = (a^2-z^2)^{\frac{\alpha}{\pi}}$. It is sufficient to show that

$$\mathrm{Arg}\,(a-z) + \mathrm{Arg}\,(a+z) = \mathrm{Arg}\,(a^2-z^2).$$

We know that for each z, $\mathrm{Arg}\,(a-z) + \mathrm{Arg}\,(a+z) = \mathrm{Arg}\,(a^2-z^2) + 2k\pi$, where k is 0, 1, or -1. However $-\pi < \mathrm{Arg}\,(a-z) < 0$ and $0 < \mathrm{Arg}\,(a+z) < \pi$, so the left side is in $(-\pi, \pi)$, and hence $k = 0$.

11. In this exercise we prove that for $\mathrm{Im}\,\zeta < 0$,

$$\zeta\left(1 - \frac{1}{\zeta^2}\right)^{\frac{1}{2}} = -i(1-\zeta^2)^{\frac{1}{2}}, \tag{7.4.17}$$

all branches being principal. Expanding the above in terms of the logarithm, it will be sufficient to show that

$$\mathrm{Arg}\,\zeta + \frac{1}{2}\mathrm{Arg}\left(1 - \frac{1}{\zeta^2}\right) = -\frac{\pi}{2} + \frac{1}{2}\mathrm{Arg}\,(1-\zeta^2),$$

or that $h(\zeta) = 2\,\mathrm{Arg}\,\zeta + \mathrm{Arg}\,(1 - \frac{1}{\zeta^2}) + \pi - \mathrm{Arg}\,(1 - \zeta^2) = 0$. We know that $h(\zeta) = 2k\pi$, where k is an integer that may depend on ζ. However, since the images under $w_1 = 1 - \frac{1}{\zeta^2}$ and $w_2 = 1 - \zeta^2$ of the lower half-plane $\mathrm{Im}\,\zeta < 0$ do not include the negative real axis, h is continuous. A continuous function which takes on discrete values must be constant (Lemma 5.7.8), and so k cannot depend on ζ. Pick $\zeta = -i$ and conclude $k = 0$.

12. Project Problem: (Channel width formula). We can write the logarithm as a Schwarz-Christoffel transformation $\mathrm{Log}\,z = \int \frac{dz}{z}$, where there is one vertex, $x_1 = 0$, $\theta_1 = \pi$, $w_1 = \infty$.

The selection of $\theta_1 = \pi$ forces a semi-infinite channel in the left half-plane (in this case the channel is also infinite into the right half-plane). For points Δx on the real axis near $x_1 = 0$, we have

$$f(\Delta x) - f(-\Delta x) = -i\pi,$$

and so the channel width is π. We now see that this type of behavior is exhibited wherever we set an angle $\theta_{j_0} = \pi$.

(a) Take a Schwarz-Christoffel transformation where a particular $\theta_{j_0} = \pi$, and other $\theta_j \leq \pi$. Then

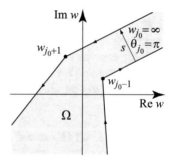

Fig. 7.96 The channel separation s.

$$f(z) = A \int \frac{dz}{(z-x_1)^{\frac{\theta_1}{\pi}} \cdots (z-x_{j_0}) \cdots (z-x_{n-1})^{\frac{\theta_{n-1}}{\pi}}} + B.$$

Show that $\lim_{z \to x_{j_0}} |f(z)| = \infty$ as z tends to x_{j_0} from the upper half-plane. [*Hint*: Write $f'(z) = \frac{g(z)}{z - x_{j_0}}$ where g is analytic near x_{j_0}).]

(b) Define the channel separation complex number s by $s = \lim_{r \downarrow 0} (f(x_{j_0} + r) - f(x_{j_0} - r))$. A typical case is shown in Figure 7.96, and its absolute value $|s|$ represents the channel width. Parametrize the upper semicircle from $x_{j_0} - r$ to $x_{j_0} + r$ and get

$$s = A \lim_{r \downarrow 0} \int_0^\pi \frac{-i r e^{i(\pi-t)}\, dt}{\left(x_{j_0} + r e^{i(\pi-t)} - x_1\right)^{\frac{\theta_1}{\pi}} \cdots \left(r e^{i(\pi-t)}\right) \cdots \left(x_{j_0} + r e^{i(\pi-t)} - x_{n-1}\right)^{\frac{\theta_{n-1}}{\pi}}}$$

$$= \frac{-iA\pi}{\left(x_{j_0} - x_1\right)^{\frac{\theta_1}{\pi}} \cdots \left(x_{j_0} - x_{n-1}\right)^{\frac{\theta_{n-1}}{\pi}}},$$

where in the denominator of the final expression, the term $x_{j_0} - x_{j_0}$ is skipped. Conclude that the channel width is

$$|s| = \frac{|A|\pi}{\left|x_{j_0} - x_1\right|^{\frac{\theta_1}{\pi}} \cdots \left|x_{j_0} - x_{n-1}\right|^{\frac{\theta_{n-1}}{\pi}}} \tag{7.4.18}$$

where again $|x_{j_0} - x_{j_0}|$ is skipped.
(c) Verify the channel width formula for the L-shaped region of Example 7.4.6.

7.5 Green's Functions

Suppose that Ω is a simply connected region bounded by a simple (closed) path Γ as shown in Figure 7.97. Let f be a piecewise continuous function on Γ and consider the Dirichlet problem

$$\Delta u(z) = 0 \qquad \text{for } z \text{ in } \Omega; \tag{7.5.1}$$

$$u(\zeta) = f(\zeta) \qquad \text{for } \zeta \text{ on } \Gamma. \tag{7.5.2}$$

Our goal in this section is to derive a formula that expresses the solution as a path integral over Γ, involving the boundary function f and the so-called **Green's function** of the region Ω, which is a function that depends only on Ω.

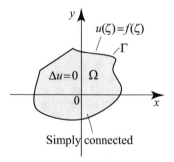

Fig. 7.97 A Dirichlet problem in a simply connected region.

Suppose that Ω and Ω' are two regions bounded by simple paths Γ and Γ'. Let ϕ be a one-to-one analytic map of Ω onto Ω'. We will further suppose that ϕ is analytic and one-to-one on Γ. It follows from Theorem 7.1.5 that ϕ maps the boundary of Ω to the boundary of Ω'. Suppose that F is a real differentiable function of two variables defined on Γ'. We think of complex numbers as points in the complex plane, and consider $F(z)$ for z on Γ'. We write $\frac{\partial F}{\partial n_{\Gamma'}}$ or simply $\frac{\partial F}{\partial n}$ to denote the directional derivative of F in the direction of the outward unit normal vector to the path Γ'. By definition, this is the dot product of the gradient of F, $\nabla F = (F_x, F_y)$, with the outward unit normal vector n_{Γ}'. Thus

$$\frac{\partial F}{\partial n_{\Gamma'}} = \nabla F \cdot n_{\Gamma'},$$

where each expression is computed at a given point on Γ'; see Figure 7.98.

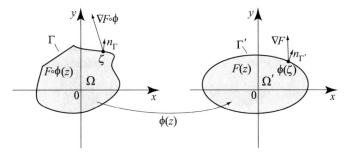

Fig. 7.98 The mapping ϕ is analytic and one-to-one on Ω. It maps boundary to boundary and preserves angles.

Although normal derivatives are tedious to compute in general, they are easy to express in some important special cases. For example, if Γ' is any circle centered at the origin, then $\frac{\partial F}{\partial n_{\Gamma'}}$ is just the radial derivative of F (see Figure 7.99):

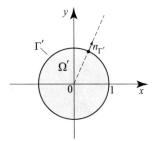

$$\frac{\partial F}{\partial n_{\Gamma'}} = \frac{\partial F}{\partial r}.$$

If $F(z) = \ln|z|$ and Γ' is the unit circle, then for all points on the unit circle

Fig. 7.99 For a circle centered at the origin, the normal derivative is the radial derivative.

$$\left.\frac{\partial F}{\partial n_{\Gamma'}}\right|_{|z|=1} = \left.\frac{\partial}{\partial r}\ln r\right|_{r=1} = \left.\frac{1}{r}\right|_{r=1} = 1.$$

Our goal is to relate the normal derivative of F on Γ' to the normal derivative of $F \circ \phi$ on Γ. Recall that if ϕ is analytic and $|\phi'(z)| \neq 0$ at some point z, then ϕ rotates a path through z by a fixed angle and scales by $|\phi'(z)|$. So ϕ maps a normal vector to Γ at ζ to a normal vector to Γ' at $\phi(\zeta)$, and it scales its modulus by $|\phi'(\zeta)|$. Since the normal derivative measures the rate of change of the function in the direction of the normal vector to the curve, thinking as we do with the chain rule, we expect the normal derivative of $F \circ \phi$ at ζ to be equal to the normal derivative of F at $\phi(\zeta)$ times $|\phi'(\zeta)|$. This expectation turns out to be correct, and we have the following change of variables formula:

$$\frac{\partial(F \circ \phi)}{\partial n_{\Gamma}}(\zeta) = |\phi'(\zeta)|\frac{\partial F}{\partial n_{\Gamma'}}(\phi(\zeta)). \tag{7.5.3}$$

The proof of (7.5.3) is presented at the end of this section. The importance of this formula is that it incorporates the effect of the conformal properties of analytic functions. Let us move a step closer to the desired formula for Green's functions by deriving a formula that uses the boundary values of u to reproduce its value at a special point inside Ω.

Lemma 7.5.1. (Change of Variables) *Suppose that $w = \phi(z)$ is a one-to-one analytic mapping of a simply connected region Ω and its boundary Γ onto the open unit disk D and its boundary C. Let u be a function harmonic on Ω and piecewise continuous on the boundary Γ. Let z_0 in Ω be the point such that $\phi(z_0) = 0$. Then*

$$u(z_0) = \frac{1}{2\pi} \int_\Gamma u(\zeta) \frac{\partial \ln |\phi(\zeta)|}{\partial n} \, ds, \tag{7.5.4}$$

where $ds = |d\zeta|$ is the element of arc length on Γ. Hence if Γ is parametrized by $\gamma(t)$, $a \le t \le b$, then $ds = |\gamma'(t)| \, dt$.

Proof. We will make use of (7.5.3), but first we note one useful result. Refer to Figure 7.100 for the explanation of the notation.

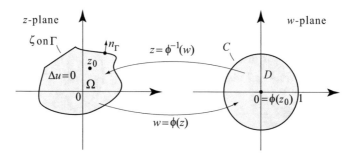

Fig. 7.100 If ϕ is analytic and one-to-one, then ϕ^{-1} is also analytic.

The function ϕ^{-1} is analytic from the closed unit disk (in the w-plane) onto Ω and its boundary (in the z-plane). So the function $u \circ \phi^{-1}$ is harmonic on the open unit disk, being the composition of a harmonic function u with an analytic function ϕ^{-1} (Theorem 6.1.10). Moreover, $u \circ \phi^{-1}$ is piecewise continuous on C. Thus, by the mean value property of harmonic functions (Corollary 6.1.19), we have

$$\frac{1}{2\pi} \int_0^{2\pi} u(\phi^{-1}(e^{it})) \, dt = u(\phi^{-1}(0)) = u(z_0). \tag{7.5.5}$$

Our goal now is to show that the integral in (7.5.4) is precisely the integral that we just evaluated in (7.5.5). Parametrize C by e^{it}, $0 \le t \le 2\pi$. Then Γ will be parametrized by $\phi^{-1}(e^{it})$, $0 \le t \le 2\pi$. The element of arc length on Γ is

$$\left| \frac{d}{dt} \phi^{-1}(e^{it}) \right| dt = \left| \frac{ie^{it}}{\phi'(\phi^{-1}(e^{it}))} \right| dt = \frac{1}{|\phi'(\phi^{-1}(e^{it}))|} \, dt.$$

Using (7.5.3) to perform the change of variables $\zeta = \phi^{-1}(e^{it})$, we transform the integral in (7.5.4) into

$$\frac{1}{2\pi} \int_0^{2\pi} u(\phi^{-1}(e^{it})) \left. \frac{\partial \ln |w|}{\partial r} \right|_{w=e^{it}} |\phi'(\phi^{-1}(e^{it}))| \frac{dt}{|\phi'(\phi^{-1}(e^{it}))|}$$

$$= \frac{1}{2\pi} \int_0^{2\pi} u(\phi^{-1}(e^{it})) dt = u(z_0),$$

in view of (7.5.5).   ∎

Let us note the following interesting property of the logarithm that we derived in the preceding proof: If ϕ is a conformal mapping of Γ and its interior onto the unit circle C and its interior, then for a point ζ on Γ we have

$$\left. \frac{\partial \ln |\phi(z)|}{\partial n_\Gamma} \right|_{z=\zeta} = |\phi'(\zeta)|. \tag{7.5.6}$$

Composing ϕ with an appropriate linear fractional transformation, we are able to reproduce the values of u at any point inside Ω, not just $z_0 = \phi^{-1}(0)$ as shown in (7.5.4). Let z be in Ω and think of $\phi(z)$ as a fixed point inside the unit disk in the w-plane. Consider the linear fractional transformation

$$\tau_z(w) = \frac{w - \phi(z)}{1 - \overline{\phi(z)}w}. \tag{7.5.7}$$

This is a one-to-one mapping from the unit disk onto itself and from the unit circle onto itself; see Proposition 4.6.2. Let us compose τ_z with ϕ and define

$$\Phi(z, \zeta) = \tau_z(\phi(\zeta)) = \frac{\phi(\zeta) - \phi(z)}{1 - \overline{\phi(z)}\phi(\zeta)}, \quad z, \zeta \text{ in } \Omega. \tag{7.5.8}$$

This is a function of two variables z and ζ, but we often think of it as a function of ζ alone for a fixed value of z. As a function of ζ, it clearly maps z to 0; that is, $\Phi(z, z) = 0$ (Figure 7.101).

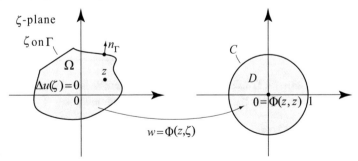

Fig. 7.101 We think of $\Phi(z, \zeta)$ as a function of one variable ζ in Ω, for fixed z in Ω. As a function of ζ, $\Phi(z, \zeta)$ is analytic and one-to-one from Ω onto the unit disk and takes z to 0; that is, $\Phi(z, z) = 0$.

Using $\Phi(z, \zeta)$ in place of $\phi(\zeta)$ in (7.5.4), we are able to reproduce the value of u at any point z in Ω.

Theorem 7.5.2. (Green's Functions) *Suppose that Ω is a simply connected region with boundary Γ, and ϕ is a one-to-one analytic function on Ω and its boundary onto the unit disk and its boundary. Let u be a function that is harmonic on Ω and piecewise continuous on Γ. For z and ζ in Ω, let $\Phi(z, \zeta)$ be as in (7.5.8). Then, for any z in Ω, we have*

$$u(z) = \frac{1}{2\pi} \int_{\Gamma} u(\zeta) \frac{\partial}{\partial n} \ln|\Phi(z, \zeta)| \, ds, \qquad (7.5.9)$$

where $ds = |d\zeta|$ is the element of arc length on Γ.

Definition 7.5.3. The function

$$G(z, \zeta) = \ln|\Phi(z, \zeta)| = \ln \left| \frac{\phi(\zeta) - \phi(z)}{1 - \overline{\phi(z)}\,\phi(\zeta)} \right|, \qquad z, \zeta \text{ in } \Omega, \qquad (7.5.10)$$

is called the **Green's function** for the region Ω. Formula (7.5.9) is a **generalized Poisson integral formula** for the simply connected region Ω.

Green's function plays a fundamental role in the solution of important partial differential equations (Laplace's equation and Poisson's equation).

Like the Poisson formulas on the disk and in the upper half-plane, formula (7.5.9) can be used to solve a general Dirichlet problem in a simply connected region Ω, where the boundary data is piecewise continuous. Of course, this solution depends on the explicit formula for the conformal mapping of Ω onto the unit disk. Once this mapping is determined, Green's functions can be used to solve the Dirichlet problem. We illustrate these ideas with several examples and show how we can recapture the Poisson formulas from Green's functions.

We often write the Green's function $G(z, \zeta)$ in terms of the real and imaginary parts of $z = x + iy$ and $\zeta = s + it$. We also write the Green's function using polar coordinates of z and ζ, where $z = re^{i\theta}$ and $\zeta = \rho e^{i\eta}$.

Example 7.5.4. (Green's function and Poisson formula for the disk) (a) Show that the Green's function for the unit disk in polar coordinates is

$$G(z, \zeta) = \ln \left| \frac{\rho e^{i\eta} - re^{i\theta}}{1 - r\rho e^{i(\eta - \theta)}} \right|, \qquad \text{for } z = re^{i\theta} \text{ and } \zeta = \rho e^{i\eta}. \qquad (7.5.11)$$

As a specific illustration, fix $z = \frac{2}{5}$ in the unit disk, and plot the function $\zeta \mapsto G(\frac{2}{5}, \zeta)$, for ζ in the unit disk. This is Green's function for the unit disk anchored at a specific point $z = \frac{2}{5}$ in the unit disk.
(b) Derive the Poisson integral formula for the unit disk.

Solution. (a) We use (7.5.10). In this case, the conformal mapping ϕ of the unit disk onto itself is simply $\phi(z) = z$, and so

$$
\begin{aligned}
G(z, \zeta) &= \ln|\Phi(z, \zeta)| \\
&= \ln\left|\frac{\phi(\zeta) - \phi(z)}{1 - \overline{\phi(z)}\,\phi(\zeta)}\right| \\
&= \ln\left|\frac{\zeta - z}{1 - \bar{z}\zeta}\right|,
\end{aligned}
$$

and (7.5.11) follows upon replacing z by $re^{i\theta}$ and ζ by $\rho e^{i\eta}$.

The plot of the function $G(\frac{2}{5}, \zeta)$ on the unit disk is shown in Figure 7.102.

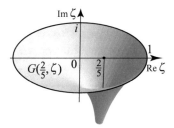

Fig. 7.102 Green's function $G(\frac{2}{5}, \zeta)$ for the unit disk anchored at $z = \frac{2}{5}$. Note that $G(\frac{2}{5}, \zeta) = 0$ for all ζ on the boundary and $G(\frac{2}{5}, \zeta)$ has a singularity at $\zeta = \frac{2}{5}$.

(b) To derive the Poisson integral formula for the unit disk, we must write out (7.5.9) when Γ is the unit circle. In this case, $ds = d\eta$, where $0 \le \eta \le 2\pi$. Using (7.5.6), we find that

$$
\frac{\partial}{\partial n} \ln\left|\frac{\zeta - z}{1 - \bar{z}\zeta}\right|_{|\zeta|=1} = \left|\frac{d}{d\zeta}\left(\frac{\zeta - z}{1 - \bar{z}\zeta}\right)\right|_{\zeta = e^{i\eta}} = \left|\frac{1 - |z|^2}{(1 - \bar{z}\zeta)^2}\right|_{\zeta = e^{i\eta}}
$$

$$
= \frac{1 - r^2}{|1 - re^{-i\theta}e^{i\eta}|^2} = \frac{1 - r^2}{1 - 2r\cos(\theta - \eta) + r^2}.
$$

Plugging into (7.5.9), we find, for $z = re^{i\theta}$ with $0 \le r < 1$,

$$
u(z) = \frac{1 - r^2}{2\pi} \int_0^{2\pi} \frac{u(e^{i\eta})}{1 - 2r\cos(\theta - \eta) + r^2}\, d\eta,
$$

which is Poisson's formula on the unit disk. $\square$

Before we move to the next example, let us understand the role of $\Phi(z, \zeta)$ in (7.5.8). Since Φ is the composition of two conformal mappings, it is itself a conformal mapping of Ω onto the unit disk, and from (7.5.8) we have $\Phi(z, z) = 0$. By the Riemann mapping theorem, $\Phi(z, \zeta)$ is uniquely determined by these properties, up to a unimodular multiplicative constant. In particular $|\Phi(z, \zeta)|$ is uniquely determined and so is the Green's function for the region. (The uniqueness part in the Riemann mapping theorem is not difficult to prove, and so we are not appealing to a deep result here.) Consider, for example, the linear fractional transformation

$$
\tau(\zeta) = \frac{z - \zeta}{\bar{z} - \zeta} \tag{7.5.12}
$$

where z is in the upper half-plane. If ζ is real so that $\bar{\zeta} = \zeta$, then

$$\left|\frac{z-\zeta}{\bar{z}-\zeta}\right| = \left|\frac{z-\zeta}{\bar{z}-\bar{\zeta}}\right| = \frac{|z-\zeta|}{|z-\bar{\zeta}|} = 1.$$

Thus τ maps the real line onto the unit circle, and since it takes z onto the origin, it follows that τ maps the upper half-plane onto the unit disk, and thus $\tau(\zeta) = \Phi(z, \zeta)$ for the upper half-plane.

Example 7.5.5. (Green's function and Poisson's formula in the upper half-plane) (a) Show that the Green's function for the upper half-plane is

$$G(z, \zeta) = \frac{1}{2} \ln \frac{(x-s)^2 + (y-t)^2}{(x-s)^2 + (y+t)^2}, \quad \text{for } z = x+iy,\ \zeta = s+it\ (y, t > 0). \quad (7.5.13)$$

Fix $z = 1 + i$ in the upper half-plane, and plot the function $\zeta \mapsto G(1+i, \zeta)$, for ζ in the upper half-plane. This is Green's function for the upper half-plane anchored at a specific point $z = 1 + i$ in the upper half-plane.
(b) Derive the Poisson integral formula for the upper half-plane.

Solution. (a) According to (7.5.10), Green's function for the upper half-plane is $\ln|\Phi(z, \zeta)|$ where $\Phi(z, \zeta)$ is in (7.5.12). Thus,

$$G(z, \zeta) = \ln \left|\frac{z-\zeta}{\bar{z}-\zeta}\right|$$

$$= \frac{1}{2} \ln \frac{|z-\zeta|^2}{|\bar{z}-\zeta|^2}$$

$$= \frac{1}{2} \ln \frac{(x-s)^2 + (y-t)^2}{(x-s)^2 + (-y-t)^2},$$

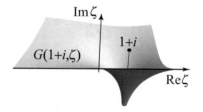

Fig. 7.103 Green's function $G(1+i, \zeta)$ for the upper half-plane anchored at $z = 1 + i$. Note that $G(1+i, \zeta) = 0$ for all ζ on the boundary and $G(1+i, \zeta)$ has a singularity at $\zeta = 1 + i$.

which is equivalent to (7.5.13).The function $G(1+i, \zeta)$ is plotted in Figure 7.103
(b) To derive Poisson's integral formula in the upper half-plane we compute the normal derivative in (7.5.9). If Γ is the real s-axis, then the normal derivative is clearly the derivative in the negative direction along the imaginary t-axis. Thus,

$$\frac{\partial}{\partial n} G(z, \zeta) = -\frac{1}{2} \frac{\partial}{\partial t} \ln \frac{(x-s)^2 + (y-t)^2}{(x-s)^2 + (y+t)^2}.$$

A straightforward calculation of the derivative, then setting $t = 0$, yields

$$\frac{\partial}{\partial n} G(z, \zeta) = \frac{2y}{(x-s)^2 + y^2}.$$

Plugging into (7.5.9) yields

$$u(z) = \frac{y}{\pi} \int_{-\infty}^{\infty} \frac{u(s)}{(x-s)^2 + y^2} \, ds \quad (z = x + iy),$$

which is Poisson's formula for the upper half-plane.   □

We give one more example of a Green's function.

Example 7.5.6. (Green's function for a semi-infinite vertical strip) Compute
Green's function for a semi-infinite vertical strip.

We can map the strip Ω in Figure 7.104
conformally onto the upper half-plane
using the mapping $w = \sin z$. Composing the function (7.5.12) with this, we
obtain a one-to-one analytic mapping
of Ω onto the unit disk, taking z in Ω
onto the origin. Thus the Green's func-
tion for Ω is

$$G(z, \zeta) = \ln \left| \frac{\sin z - \sin \zeta}{\sin z - \overline{\sin \zeta}} \right|. \quad □$$

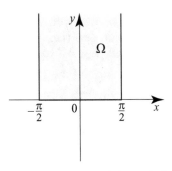

Fig. 7.104 A semi-infinite vertical
strip.

Next we prove next some interesting properties of Green's functions.

Theorem 7.5.7. (Properties of Green's Functions) *Suppose that Ω is a simply
connected region with boundary Γ, and let ϕ, $\Phi(z, \zeta)$, and $G(z, \zeta)$ be as in Theorem 7.5.2. Then the Green's function G has the following properties:*
(i) $G(z, \zeta) \leq 0$ for all z and ζ in Ω;
(ii) $G(z, \zeta) = 0$ for all z in Ω and ζ on Γ;
*(iii) $G(z, \zeta) = G(\zeta, z)$ for all z and ζ in Ω (**symmetric property**);*
*(iv) for each z in Ω, there is a function $\zeta \mapsto u_1(z, \zeta)$ such that $u_1(z, \zeta)$ is har-
monic for all ζ in Ω, $u_1(z, \zeta) = -\ln|z - \zeta|$ for all ζ on the boundary Γ, and
$G(z, \zeta) = u_1(z, \zeta) + \ln|z - \zeta|$ for all $\zeta \neq z$ in Ω.*

Properties (*i*) and (*ii*) could be verified on the graphs of the Green's functions in
Figures 7.102 and 7.103. Before we prove the theorem, we illustrate the properties
in Figure 7.105 for a typical case where Ω is the upper half-plane and Green's
function is anchored at $z = 1 + i$.

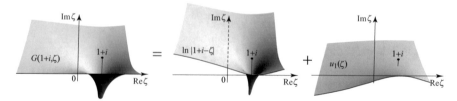

Fig. 7.105 A Green's function $G(z_0, \zeta)$ anchored at $z_0 = 1+i$ is the sum of a logarithm, $\ln|z_0 - \zeta|$, and a harmonic function, $u_1(\zeta)$, such that $u_1(\zeta) = -\ln|z_0 - \zeta|$ on the boundary. As a result, $G(z_0, \zeta)$ vanishes on the boundary and has a singularity at z_0 like $\ln|z_0 - \zeta|$.

Proof. Fix z in Ω. From the definition of ϕ and Φ (see (7.5.7) and (7.5.8)), we have that $\Phi(z, \zeta)$ is in the open unit disk D (i.e., $|\Phi(z, \zeta)| < 1$) for all ζ in Ω and $\Phi(z, \zeta)$ is on the unit circle C (i.e., $|\Phi(z, \zeta)| = 1$) for all ζ on Γ. This clearly proves (i) and (ii), because $\ln|x| < 0$ if $|x| < 1$ and $\ln|x| = 0$ if $|x| = 1$. For (iii), we have

$$G(z, \zeta) = \ln\left|\frac{\phi(\zeta) - \phi(z)}{1 - \overline{\phi(z)}\,\phi(\zeta)}\right| = \ln\left|\frac{\phi(z) - \phi(\zeta)}{1 - \overline{\phi(z)}\,\phi(\zeta)}\right| = \ln\left|\frac{\phi(\zeta) - \phi(z)}{1 - \overline{\phi(\zeta)}\,\phi(z)}\right| = G(\zeta, z).$$

To prove (iv), fix z in Ω and consider

$$\psi(z, \zeta) = \frac{\phi(\zeta) - \phi(z)}{\zeta - z} \frac{1}{1 - \overline{\phi(z)}\phi(\zeta)} \qquad (\zeta \neq z \text{ in } \Omega).$$

Clearly, $\psi(z, \zeta)$ is analytic for all $\zeta \neq z$ in Ω. What happens as ζ approaches z? We have

$$\lim_{\zeta \to z} \psi(z, \zeta) = \lim_{\zeta \to z} \frac{\phi(\zeta) - \phi(z)}{\zeta - z} \frac{1}{1 - \overline{\phi(z)}\phi(\zeta)} = \frac{\phi'(z)}{1 - |\phi(z)|^2},$$

which is finite because $|\phi(z)| < 1$ and nonzero because ϕ is one-to-one and so $\phi'(z) \neq 0$. Hence $\psi(z, \zeta)$ has a removable singularity at z (Theorem 4.5.12). By defining ψ at $\zeta = z$ to be

$$\psi(z, z) = \frac{\phi'(z)}{1 - |\phi(z)|^2},$$

$\psi(z, \zeta)$ becomes analytic and nonzero for all ζ in Ω. Set $u_1(z, \zeta) = \ln|\psi(z, \zeta)|$; then u_1 is harmonic for all ζ in Ω. But for $\zeta \neq z$

$$u_1(z, \zeta) = \ln|\psi(z, \zeta)| = \ln\left|\frac{\phi(\zeta) - \phi(z)}{1 - \overline{\phi(z)}\phi(\zeta)}\right| - \ln|\zeta - z| = G(z, \zeta) - \ln|z - \zeta|.$$

Also, $u_1(z, \zeta) = -\ln|z - \zeta|$ on the boundary because $G(z, \zeta) = 0$ on the boundary, and so (iv) holds. ∎

Because the function $\zeta \mapsto u_1(z, \zeta)$ is the solution of a Dirichlet problem in Ω with boundary values $-\ln|z - \zeta|$, if Ω is bounded, this solution is unique. Thus the representation of Green's function in Theorem 7.5.7(iv) is unique when Ω is bounded. Property (iv) in Theorem 7.5.7 can be used to define the Green's function of a domain. That is, any function $G(z, \zeta)$ that satisfies (iv) also satisfies (7.5.9).

Proof of the change of variables formula (7.5.3)

Suppose that $\gamma(t) = x(t) + iy(t)$ is a parametrization of a smooth path with $\gamma'(t) = x'(t) + iy'(t) \neq 0$, where t lies in some interval. If we assume our path has a positive orientation, then an outward unit normal may be obtained by rotating the tangent $\gamma'(t)$ clockwise by $\pi/2$ and dividing by its absolute value (Figure 7.106). Hence $n = \frac{\gamma'}{i|\gamma'|}$ or

$$n_\Gamma = \frac{1}{|\gamma'(t)|}\left(y'(t), -x'(t)\right). \quad (7.5.14)$$

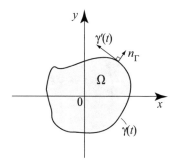

Fig. 7.106 Outward tangent and normal

Let ϕ be as in the text preceding (7.5.3). To simplify the notation, let us write $\phi(z) = u(x, y) + iv(x, y)$, with u, v real-valued, write F as $F(u, v)$, and denote partial derivatives by subscripts. So

$$(F \circ \phi)_x = \frac{\partial}{\partial x} F(u(x,y), v(x,y)) = F_u u_x + F_v v_x, \quad (7.5.15)$$

where the last equality follows from the chain rule in two dimensions. Similarly,

$$(F \circ \phi)_y = \frac{\partial}{\partial y} F(u(x,y), v(x,y)) = F_u u_y + F_v v_y. \quad (7.5.16)$$

Using the definition of the normal derivative, (7.5.14), (7.5.15), and (7.5.16), we get

$$\frac{\partial}{\partial n_\Gamma} F \circ \phi(t) = \nabla(F \circ \phi) \cdot n_\Gamma(t) = \frac{1}{|\gamma'(t)|}\left((F \circ \phi)_x, (F \circ \phi)_y\right) \cdot (y'(t), -x'(t))$$

$$= \frac{1}{|\gamma'(t)|}\left((F_u u_x + F_v v_x)y'(t) - (F_u u_y + F_v v_y)x'(t)\right). \quad (7.5.17)$$

Consider now the path Γ', which is parametrized by

$$\phi(\gamma(t)) = u(x(t), y(t)) + i v(x(t), y(t)).$$

Conformality ensures that the outward normal to $\phi(\gamma(t))$ is still turned clockwise from the tangent, so in analogy with (7.5.14) we obtain

$$
\begin{aligned}
n_{\Gamma'}(t) &= \frac{1}{\left|\frac{d}{dt}\phi(\gamma(t))\right|} \left(\frac{d}{dt} v(x(t), y(t)), -\frac{d}{dt} u(x(t), y(t)) \right) \\
&= \frac{1}{|\phi'(\gamma(t))\gamma'(t)|} \left(v_x x'(t) + v_y y'(t), -u_x x'(t) - u_y y'(t) \right).
\end{aligned}
$$

Thus, for t in the interval of parametrization, we have

$$
\begin{aligned}
\frac{\partial F}{\partial n_{\Gamma'}}(t) &= \nabla F \cdot n_{\Gamma'}(t) \\
&= \frac{(F_u, F_v) \cdot \left(v_x x'(t) + v_y y'(t), -u_x x'(t) - u_y y'(t) \right)}{|\gamma'(t)| \, |\phi'(\gamma(t))|} \\
&= \frac{F_u \left(v_x x'(t) + v_y y'(t) \right) + F_v \left(-u_x x'(t) - u_y y'(t) \right)}{|\gamma'(t)| \, |\phi'(\gamma(t))|}. \quad (7.5.18)
\end{aligned}
$$

Comparing (7.5.17) and (7.5.18) and using the Cauchy-Riemann equations, $u_x = v_y$, $u_y = -v_x$, we verify the validity of (7.5.3). ∎

Exercises 7.5

In Exercises 1–8, derive the Green's function for the region depicted in the accompanying figure (Figures 7.107–7.114).

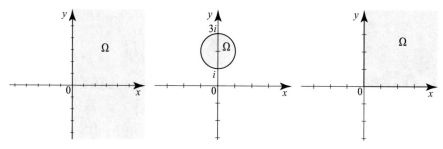

Fig. 7.107 Exercise 1.   **Fig. 7.108** Exercise 2.   **Fig. 7.109** Exercise 3.

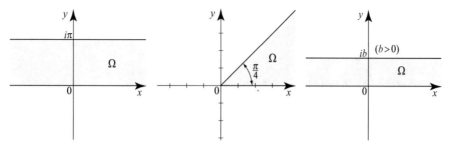

Fig. 7.110 Exercise 5.   **Fig. 7.111** Exercise 4.   **Fig. 7.112** Exercise 6.

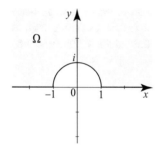

Fig. 7.113 Exercise 7.

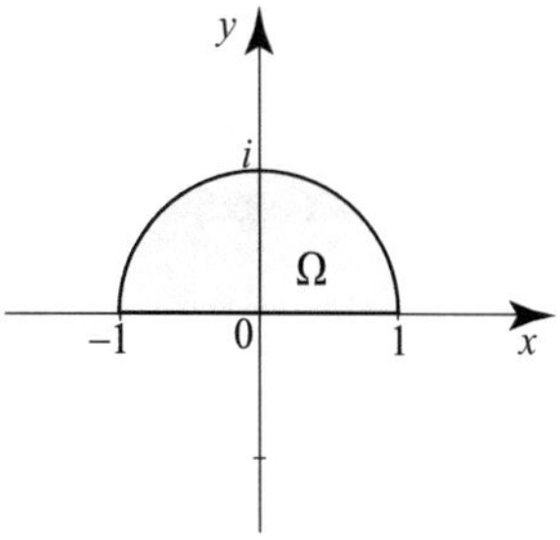

Fig. 7.114 Exercise 8.

9. Project Problem: Poisson's formula in the first quadrant. (a) Derive the following Poisson formula in the first quadrant for the Dirichlet problem in Figure 7.115, using Green's function:

$$u(x+iy) = \frac{y}{\pi} \int_0^\infty f(s) \left(\frac{1}{(x-s)^2 + y^2} - \frac{1}{(x+s)^2 + y^2} \right) ds$$
$$+ \frac{x}{\pi} \int_0^\infty g(t) \left(\frac{1}{x^2 + (y-t)^2} - \frac{1}{x^2 + (y+t)^2} \right) dt.$$

(b) Consider the special case in which $g(t) = 0$. Use a symmetry argument to show that the solution in this case is the same as the restriction to the first quadrant of the solution of the Dirichlet problem in the upper half-plane with boundary data on the real axis: $u(s) = f(s)$ if $s > 0$ and $u(s) = -f(-s)$ if $s < 0$.

(c) Consider the special case in Figure 7.115 in which $f(s) = 0$. Use a symmetry argument to show that the solution in this case is the same as the restriction to the first quadrant of the solution of the Dirichlet problem in the right half-plane with boundary data on the imaginary axis given by $u(it) = g(t)$ if $t > 0$ and $u(it) = -g(-t)$ if $t < 0$.

(d) Write your answers in (b) and (c) using the Poisson integral formula for the upper half-plane and the right half-plane. Then sum the solutions to rederive your answer in (a).

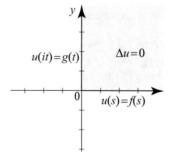

Fig. 7.115 The Dirichlet problem on a quarter plane.

10. Poisson's formula in a semi-infinite vertical strip. Derive Poisson's formula in the region of Example 7.5.6, using Green's function.

7.6 Poisson's Equation and Neumann Problems

In this section we use Green's functions to solve the Poisson boundary value problem on a region Ω, then apply similar techniques to solve another important type of problem known as a Neumann problem. Suppose h is a continuous function on the region Ω and f is a piecewise continuous function on the boundary Γ of Ω, which is a closed simple curve with positive orientation. We start with the Poisson problem, which consists of the **inhomogeneous Laplace equation** known as **Poisson's equation** along with a condition specifying the values of the function on the boundary Γ of Ω (Figure 7.116):

$$\Delta u(z) = h(z) \quad \text{for all } z \text{ in } \Omega; \tag{7.6.1}$$

$$u(\zeta) = f(\zeta) \quad \text{for all } \zeta \text{ on } \Gamma. \tag{7.6.2}$$

This equation models, for example, the time-independent (or steady-state) temperature distribution in a medium in the presence of heat sources. It also arises in the study of the velocity potential of an incompressible ideal fluid flow in the presence of sources or sinks. In this section we show that the solution of (7.6.1)–(7.6.2) has a simple expression in terms of the Green's function of Ω. Obviously, if h is identically 0 in Ω, (7.6.1) reduces to Laplace's equation.

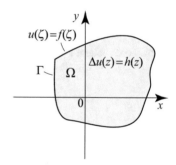

Fig. 7.116 A Poisson problem in a region Ω.

Before we solve the problem, we would like to state the following simple but important **superposition principle**.

Theorem 7.6.1. (Superposition Principle) *Suppose that u_1 satisfies (7.6.1) in Ω and $u_1(\zeta) = 0$ for all ζ on Γ. Suppose that u_2 satisfies Laplace's equation in Ω and $u_2(\zeta) = f(\zeta)$ for all ζ on Γ. Then $u = u_1 + u_2$ satisfies Poisson's equation (7.6.1) with boundary conditions (7.6.2).*

The proof is immediate and is omitted. Theorem 7.6.1 allows us to decompose the general Poisson problem (7.6.1)–(7.6.2) into the sum of a Dirichlet problem with boundary values f and a Poisson problem with zero boundary values as illustrated in Figure 7.117.

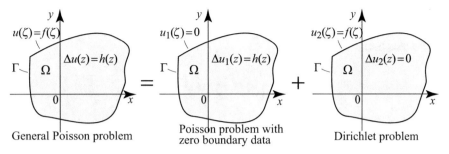

Fig. 7.117 Decomposition of a general Poisson problem into the sum of a Poisson problem with zero boundary data plus a Dirichlet problem.

Our next goal is to derive Green's identities. To do so, we make use of Green's theorem. This is stated in Theorem 3.4.2 for simply connected regions. For multiply connected regions, the version goes as follows: Let $p(x, y)$ and $q(x, y)$ have continuous first partial derivatives in Ω and on its positively oriented boundary Γ. Then we have

$$\iint_\Omega \left(p_x(x, y) + q_y(x, y)\right) dx\, dy = \int_\Gamma \left(p(x, y)\, dy - q(x, y)\, dx\right), \qquad (7.6.3)$$

where subscripts, as usual, denote partial derivatives.

Theorem 7.6.2. (Green's Identities) *Suppose that Ω is a bounded region whose boundary Γ consists of a finite number of simple closed positively oriented paths (as in Theorem 3.7.2). Let u and v have continuous second partial derivatives on Ω and Γ. Then we have* **Green's first identity**

$$\iint_\Omega \left(u\Delta v + \nabla u \cdot \nabla v\right) dA = \int_\Gamma u \frac{\partial v}{\partial n}\, ds, \qquad (7.6.4)$$

and **Green's second identity**

$$\iint_\Omega \left(u\Delta v - v\Delta u\right) dA = \int_\Gamma \left(u\frac{\partial v}{\partial n} - v\frac{\partial u}{\partial n}\right) ds. \qquad (7.6.5)$$

Proof. Applying (7.6.3) with

$$p(x, y) = u v_x \quad \text{and} \quad q(x, y) = u v_y,$$

we obtain

$$\iint_\Omega \left(u(v_{xx} + v_{yy}) + (u_x v_x + u_y v_y)\right) dx\, dy = \int_\Gamma u\left(v_x\, dy - v_y\, dx\right). \qquad (7.6.6)$$

The integrand on the left is the same as the integrand on the left of (7.6.4). To understand the integrand on the right, let us recall that for a positively oriented curve parametrized by $\gamma(t) = x(t) + i y(t)$, t in some interval, the outward unit normal may be obtained by rotating the tangent $\gamma'(t) = x'(t) + i y'(t)$ clockwise by $\frac{\pi}{2}$ and dividing by its absolute value. Hence

$$n(t) = \frac{\gamma'(t)}{i|\gamma'(t)|} = \frac{1}{|\gamma'(t)|}(y'(t), x'(t)).$$

Thus since the normal derivative $\frac{\partial v}{\partial n}$ is by definition the gradient of v dotted with the outward unit normal vector and $ds = |\gamma'(t)| dt$, we have

$$\frac{\partial v}{\partial n} ds = (v_x, v_y) \cdot (y'(t), -x'(t)) dt = v_x dy - v_y dx,$$

which shows that the right side of (7.6.6) is the same as the right side of (7.6.4), and so (7.6.4) follows. To prove (7.6.5), we reverse the roles of u and v in (7.6.4) and get

$$\iint_\Omega (v \Delta u + \nabla u \cdot \nabla v) dx dy = \int_\Gamma v \frac{\partial u}{\partial n} ds.$$

Subtracting this from (7.6.4), we deduce (7.6.5).   ∎

We now express the solution of the Poisson problem on Ω in terms of Green's functions. We use the notation of the previous section and suppose that f and h in (7.6.1)–(7.6.2) have enough smoothness and integrability properties for the formulas to hold.

Theorem 7.6.3. (Solution of Poisson's Problem) *Suppose that Ω is a region with boundary Γ, h is a continuous function on Ω and f a piecewise continuous function on Γ. Let $G(z, \zeta)$ denote the Green's function for Ω. If $u(z)$ is a solution of Poisson's problem (7.6.1)–(7.6.2), then*

$$u(z) = \frac{1}{2\pi} \iint_\Omega h(\zeta) G(z, \zeta) dA + \frac{1}{2\pi} \int_\Gamma f(\zeta) \frac{\partial}{\partial n} G(z, \zeta) ds, \qquad (7.6.7)$$

where dA is the element of area and ds is the element of arc length.

Remark 7.6.4. The proof of this theorem uses Green's identities (7.6.4) and (7.6.5). These identities do not hold in general on unbounded regions. For this result we suppose that the region Ω is bounded in the proof below. Nevertheless, formula (7.6.7) can be used on unbounded regions and its validity there can be checked on a case-by-case basis.

Proof. Fix z in Ω, and let $\overline{B_\varepsilon(z)}$ denote the closed disk of radius $\varepsilon > 0$ centered at z in Ω, and $\Omega_\varepsilon = \Omega \setminus \overline{B_\varepsilon(z)}$. We are going to apply Green's second identity in Ω_ε.

In view of Theorem 7.5.7(iv), $G(z, \zeta)$ is a harmonic function of ζ for all $\zeta \neq z$ in Ω. In particular, for ζ in Ω_ε, $\Delta G(z, \zeta) = 0$, we also have $\Delta u = h$ because u satisfies Poisson's equation (7.6.1). We apply Green's second identity on the region Ω_ε whose boundary Γ_ε consists of Γ and the circle $C_\varepsilon(z)$ (Figure 7.118), taking $v = G(z, \zeta)$ and u equal the solution of the Poisson problem, and obtain

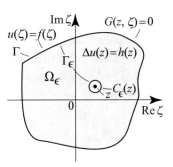

Fig. 7.118 Picture of the proof.

$$- \iint_{\Omega_\varepsilon} G(z, \zeta) h(\zeta)\, dA = \int_\Gamma \left(u(\zeta) \frac{\partial G(z, \zeta)}{\partial n} - \overbrace{G(z, \zeta)}^{0} \frac{\partial u(\zeta)}{\partial n} \right) ds$$

$$+ \int_{C_\varepsilon(z)} \left(u(\zeta) \frac{\partial G(z, \zeta)}{\partial n} - G(z, \zeta) \frac{\partial u(\zeta)}{\partial n} \right) ds,$$

because $G(z, \zeta) = 0$ for ζ on Γ. We will let $\varepsilon \downarrow 0$ and show that

$$\iint_{\Omega_\varepsilon} G(z, \zeta) h(\zeta)\, dA \rightarrow \iint_\Omega G(z, \zeta) h(\zeta)\, dA; \qquad (7.6.8)$$

$$\int_{C_\varepsilon(z)} u(\zeta) \frac{\partial G(z, \zeta)}{\partial n}\, ds \rightarrow -2\pi u(z); \qquad (7.6.9)$$

$$\int_{C_\varepsilon(z)} G(z, \zeta) \frac{\partial u(\zeta)}{\partial n}\, ds \rightarrow 0. \qquad (7.6.10)$$

This will imply (7.6.7) and complete the proof. Let us start with (7.6.10). Write $G(z, \zeta) = u_1(z, \zeta) + \ln|z - \zeta|$, where $u_1(z, \zeta)$ is harmonic, hence bounded by a constant M in some fixed disk centered at z. Also, $\frac{\partial u}{\partial n}$ is bounded in this fixed disk (say, $\left| \frac{\partial u}{\partial n} \right| \leq A$), since u has continuous partial derivatives in Ω. For ζ on $C_\varepsilon(z)$, we have $|z - \zeta| = \varepsilon$, hence $\left| G(z, \zeta) \frac{\partial u}{\partial n} \right| \leq (M + |\ln \varepsilon|) A$, and so

$$\left| \int_{C_\varepsilon(z)} G(z, \zeta) \frac{\partial u(\zeta)}{\partial n}\, ds \right| \leq (M + |\ln \varepsilon|) A \int_{C_\varepsilon(z)} ds = 2\pi\varepsilon (M + |\ln \varepsilon|) A \rightarrow 0, \text{ as } \varepsilon \rightarrow 0.$$

To prove (7.6.9), we note that on $C_\varepsilon(z)$,

$$\frac{\partial G(z, \zeta)}{\partial n} = \frac{\partial}{\partial n} (u_1(z, \zeta) + \ln|z - \zeta|) = \frac{\partial}{\partial n} u_1(z, \zeta) - \frac{1}{\varepsilon}.$$

Now

$$\int_{C_\varepsilon(z)} u(\zeta) \frac{\partial G(z, \zeta)}{\partial n} \, ds = \int_{C_\varepsilon(z)} u(\zeta) \frac{\partial}{\partial n} u_1(z, \zeta) \, ds - \frac{1}{\varepsilon} \int_{C_\varepsilon(z)} u(\zeta) \, ds.$$

The first integral on the right tends to 0 as $\varepsilon \to 0$, because $u(\zeta) \frac{\partial}{\partial n} u_1(z, \zeta)$ is bounded in $\overline{B_\varepsilon(z)}$, as in the proof of (7.6.9). To handle the second integral, note that the function $I(\varepsilon) = \frac{1}{\varepsilon} \int_{C_\varepsilon(z)} u(\zeta) \, ds = \int_0^{2\pi} u(z + \varepsilon e^{i\theta}) \, d\theta$ is continuous in ε at $\varepsilon = 0$. This follows from the continuity of u at z. So

$$\lim_{\varepsilon \to 0} \int_0^{2\pi} u(z + \varepsilon e^{i\theta}) \, d\theta = I(0) = \int_0^{2\pi} u(z) \, d\theta = 2\pi u(z),$$

which completes the proof of (7.6.9). The proof of (7.6.8) is similar. For details see Exercise 9. $\blacksquare$

Remark 7.6.5. The form (7.6.7) of the solution clearly illustrates the superposition principle. We recognize the second term on the right side of (7.6.7) as the solution of the Dirichlet problem on Ω with boundary values f (compare with Theorem 7.5.2). Looking at the other term, we also have

$$u(z) = \frac{1}{2\pi} \int\int_\Omega h(\zeta) G(z, \zeta) \, dA \qquad (7.6.11)$$

as a solution of Poisson's equation (7.6.1) with zero boundary values.

It should be mentioned that in most situations it is difficult to compute (7.6.7) in its present form. We now turn our attention to a different problem, which can be solved using an approach similar to the one we took with Green's functions.

Neumann Condition and Neumann Problems

When modeling heat problems in an insulated plate Ω, where heat exchange on the boundary is prescribed, we are led to a boundary value problem that consists of Laplace's equation along with a condition that describes the values of the normal derivative on the boundary (Figure 7.119):

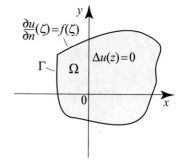

$$\Delta u(z) = 0 \qquad \text{for } z \text{ in } \Omega; \; (7.6.12)$$

$$\frac{\partial u}{\partial n}(\zeta) = f(\zeta) \qquad \text{for } \zeta \text{ on } \Gamma. \, (7.6.13)$$

Fig. 7.119 A Neumann problem in a region Ω.

Such a problem is called a **Neumann problem** (after the German mathematician Carl Gottfried Neumann (1832–1925)) and sometime referred to as a **Dirichlet**

problem of the second kind. Condition (7.6.13) is known as a **Neumann condition**. The normal derivative at the boundary describes the rate of exchange of heat
with the surrounding medium or the flux of heat across the boundary. For example,
the condition $f(\zeta) = 0$ corresponds to an insulated point where there is no exchange
of heat with the surrounding medium. Since $f(\zeta)$ represents the flux of heat across
the boundary of u and u represents a steady-state temperature distribution inside Ω,
we would expect the total flux across the boundary to be zero; that is, f cannot be
arbitrary, and it has to satisfy the **compatibility condition**

$$\int_\Gamma f(\zeta)\,ds = 0. \tag{7.6.14}$$

Indeed, (7.6.14) follows from the following useful property of harmonic functions,
by setting $f(\zeta) = \frac{\partial u}{\partial n}$.

Proposition 7.6.6. (Normal Derivative of Harmonic Functions) *Suppose that u is
harmonic in a bounded region Ω and its boundary Γ. Then*

$$\int_\Gamma \frac{\partial u}{\partial n}\,ds = 0. \tag{7.6.15}$$

Proof. Reversing the roles of u and v in Green's first identity (7.6.4) and then picking $v = 1$, we get

$$\iint_\Omega \Delta u\,dA = \int_\Gamma \frac{\partial u}{\partial n}\,ds. \tag{7.6.16}$$

Since $\Delta u = 0$, the proposition follows. ∎

In a Neumann problem, we are asked to find a harmonic function in terms of the
values of its normal derivative on the boundary. Such a solution is not unique, since
we can add an arbitrary constant to a solution of (7.6.12)–(7.6.13) and get another
solution. So now we ask: Is the solution unique up to an arbitrary constant? The
answer is affirmative. To show this, consider the difference between two solutions,
which is harmonic and has zero normal derivative on the boundary. As we now
show, a function that is harmonic and has zero normal derivative is a constant.

Theorem 7.6.7. (Uniqueness of the Neumann solution) *Suppose that u is har-
monic on a bounded region Ω such that $\frac{\partial u}{\partial n} = 0$ on the boundary Γ of Ω. Then u is
identically constant in Ω.*

Proof. Suppose that u is harmonic in Ω and take $u = v$ in Green's first formula, so
that $\Delta v = 0$ in Ω, and get

$$\iint_\Omega (\nabla u \cdot \nabla u)\,dA = \int_\Gamma u\frac{\partial u}{\partial n}\,ds = 0,$$

because $\frac{\partial u}{\partial n} = 0$ on Γ. Now the function $\nabla u \cdot \nabla u = (u_x)^2 + (u_y)^2$ is nonnegative and
continuous on Ω. The only way for a nonnegative continuous function to integrate to
zero is to vanish identically (Lemma 3.9.5(i) proves this result in dimension one, but

the argument also works in higher dimensions.) Hence $(u_x)^2 + (u_y)^2$ is identically zero in Ω, and so $u_x = 0$ and $u_y = 0$ identically in Ω. Applying Theorem 2.4.7, we see that u is constant on Ω.   ∎

In order to express the solution of the Neumann problem as an integral, motivated by Green's function and the solution of the Dirichlet problem, we make the following definition.

Definition 7.6.8. (Neumann Functions) Suppose that Ω is a simply connected region with boundary Γ. Let z, ζ in Ω. A **Neumann function** for the region Ω is a function $N(z, \zeta)$ with the following properties:
(i) for each z in Ω, $N(z, \zeta)$ is harmonic in ζ on $\Omega \setminus \{z\}$;
(ii) $\frac{\partial N}{\partial n}(z, \zeta) = C$ for all z in Ω and ζ on Γ;
(iii) for each z in Ω, there is a function $\zeta \mapsto u_1(z, \zeta)$ such that $u_1(z, \zeta)$ is harmonic in ζ on Ω, and $N(z, \zeta) = \ln|z - \zeta| + u_1(z, \zeta)$ for all ζ in $\Omega \setminus \{z\}$.

Parts (i) and (iii) state that Neumann functions, like Green's functions, are harmonic inside Ω except for a singularity at $z = \zeta$, which is similar to the singularity of $\ln|z - \zeta|$. Part (ii) tells us that the boundary values of the normal derivative of the Neumann function are constant. This is the counterpart of the boundary condition for a Green's function, which states that a Green's function must vanish identically on the boundary. As we now show, the constant C in (ii) depends on the length of the boundary Γ.

Proposition 7.6.9. *The constant C in Definition 7.6.8 (ii), which represents the boundary value of the normal derivative of the Neumann function, is equal to*

$$C = \frac{2\pi}{L}, \tag{7.6.17}$$

where $L = \int_\Gamma ds = \ell(\Gamma)$ is the length of Γ. If L is infinite, we take $C = 0$.

Proof. The proof is based on the following interesting property of the logarithm. For any fixed z inside Ω, we have

$$\int_\Gamma \frac{\partial}{\partial n} \ln|z - \zeta| \, ds = 2\pi. \tag{7.6.18}$$

To see this, let $B_\varepsilon(z)$ be a disk of radius $\varepsilon > 0$ centered at z and contained in Ω, and let $C_\varepsilon(z)$ denote the circle centered at z with radius ε. The function $\zeta \mapsto \ln|z - \zeta|$ is harmonic in $\Omega \setminus B_\varepsilon$, so according to Proposition 7.6.6, we have

$$0 = \int_{C_\varepsilon(z)} \frac{\partial}{\partial n} \ln|z - \zeta| \, ds + \int_\Gamma \frac{\partial}{\partial n} \ln|z - \zeta| \, ds.$$

Parameterizing $C_\varepsilon(z)$ by $\zeta = z + \varepsilon e^{it}$, $0 \le t \le 2\pi$, $\frac{\partial}{\partial n} \ln|z - \zeta| = -\frac{1}{\varepsilon}$ and $ds = \varepsilon \, dt$, and so

$$0 = -\int_0^{2\pi} dt + \int_\Gamma \frac{\partial}{\partial n} \ln|z - \zeta| \, ds,$$

implying (7.6.18). Now, using parts (ii) and (iii) of Definition 7.6.8, we write

$$CL = \int_\Gamma C \, ds$$

$$= \int_\Gamma \frac{\partial N}{\partial n} \, ds$$

$$= \overbrace{\int_\Gamma \frac{\partial N_1}{\partial n} \, ds}^{0} + \int_\Gamma \frac{\partial}{\partial n} \ln |z - \zeta| \, ds$$

$$= 0 + 2\pi,$$

where the term $\int_\Gamma \frac{\partial N_1}{\partial n} \, ds$ is zero in view of identity (7.6.15) in Proposition 7.6.6. This yields (7.6.17). ∎

Neumann functions are not unique for a region (two Neumann functions can differ by a function of z and not ζ; see Exercise 12). However, any Neumann function satisfying Definition 7.6.8 will work to solve a Neumann problem. Before we express the solution of the Neumann problem in terms of the Neumann function, we note that from Theorem 7.6.7 if a solution of a Neumann problem exists, then it is unique up to an additive constant.

Theorem 7.6.10. (Solution of the Neumann Problem) *Suppose that Ω is a region bounded by a simple path Γ, and let $N(z, \zeta)$ denote a Neumann function, where z and ζ are in Ω. Then, up to an additive constant, the solution $u(z)$ of the Neumann problem (7.6.12)–(7.6.14) is*

$$u(z) = -\frac{1}{2\pi} \int_\Gamma N(z, \zeta) f(\zeta) \, ds. \tag{7.6.19}$$

The proof of Theorem 7.6.10 is similar to that of Theorem 7.6.3 (see Exercise 10). Now we derive some classical Neumann functions.

Example 7.6.11. (Neumann function for the upper half-plane) Show that the Neumann function for the upper half-plane is

$$N(z, \zeta) = \ln |z - \zeta| + \ln |\bar{z} - \zeta| \tag{7.6.20}$$

$$= \frac{1}{2} \ln \left((x - s)^2 + (y - t)^2\right) + \frac{1}{2} \ln \left((x - s)^2 + (y + t)^2\right),$$

for $z = x + iy$ and $\zeta = s + it$, $y, t > 0$ (Figure 7.120).

Solution. We know that

$$N(z, \zeta) = \ln |z - \zeta| + N_1(z, \zeta).$$

Computing the normal derivative of
$\ln |z - \zeta| = \frac{1}{2} \ln \left((x-s)^2 + (y-t)^2\right)$
along the real axis, we find

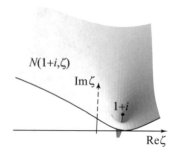

$$-\frac{d}{dt} \frac{1}{2} \ln \left((x-s)^2 + (y-t)^2\right)\Big|_{t=0}$$

$$= \frac{y-t}{(x-s)^2 + (y-t)^2}\Big|_{t=0}$$

$$= \frac{y}{(x-s)^2 + y^2}.$$

Fig. 7.120 A Neumann function
for the upper half-plane, anchored
at $z = 1 + i$.

Adding $\frac{1}{2} \ln |\bar{z} - \zeta| = \frac{1}{2} \ln \left((x-s)^2 + (y+t)^2\right)$ to $\ln |z - \zeta|$, results in adding $\frac{-y}{(x-s)^2 + y^2}$ to the normal derivative along the real axis, making the normal derivative of $\frac{1}{2} \ln |z - \zeta| + \frac{1}{2} \ln |\bar{z} - \zeta|$ zero along the real axis. This shows that $N(z, \zeta)$ as defined by (7.6.20) has 0 normal derivative along the boundary, in accordance with Proposition 7.6.9. All other properties of the Neumann function in Definition 7.6.8 are easily verified. □

The fact that the normal derivative of the Neumann function of an unbounded region is zero on the boundary allows us to construct the Neumann function for this region by using the Neumann function for the upper half-plane and a change of variables via a conformal mapping. More precisely, we have the following construction, which is valid for unbounded regions.

Proposition 7.6.12. (Neumann Function for Unbounded Regions) *Suppose that Ω is an unbounded region with boundary Γ, and ϕ is a one-to-one analytic mapping of Ω onto the upper half-plane, taking Γ onto the real axis. Then the Neumann function for Ω is*

$$N(z, \zeta) = \ln |\phi(z) - \phi(\zeta)| + \ln |\overline{\phi(z)} - \phi(\zeta)| \quad (z, \zeta \text{ in } \Omega). \qquad (7.6.21)$$

Proof. Fix z in Ω and consider the function $F(w) = \ln |\phi(z) - w| + \ln |\overline{\phi(z)} - w|$, where $\phi(z)$ and w are in the upper half-plane. By Example 7.6.11, $\frac{\partial F}{\partial n}(w) = 0$, for all w on the real axis. Applying the change of variables formula (7.5.3) it follows that $\frac{\partial F \circ \phi}{\partial n}(\zeta) = 0$ for all ζ on Γ. So $N(z, \zeta)$ has the right normal derivative on the boundary Γ. Does it have the right singularity at $\zeta = z$? Since $\overline{\phi(z)}$ is in the lower half-plane, it follows that $\ln |\overline{\phi(z)} - \phi(\zeta)|$ is harmonic for all ζ in Ω. Now let us compare $\ln |\phi(z) - \phi(\zeta)|$ to $\ln |z - \zeta|$. We have

$$\lim_{\zeta \to z} \left(\ln |\phi(z) - \phi(\zeta)| - \ln |z - \zeta|\right) = \lim_{\zeta \to z} \ln \left|\frac{\phi(z) - \phi(\zeta)}{z - \zeta}\right| = \ln |\phi'(z)|,$$

because ϕ is analytic. Since $\phi'(z) \neq 0$, $\ln|\phi(z) - \phi(\zeta)|$ and $\ln|z - \zeta|$ differ by a finite constant near z. Hence the function $N(z, \zeta)$ equals $\ln|z - \zeta|$ plus a harmonic function in Ω. ∎

Let us give an application of Proposition 7.6.12.

Example 7.6.13. (Neumann function for the first quadrant) Compute the Neumann function for the first quadrant.

Solution. Applying Proposition 7.6.12 with $\phi(z) = z^2$, we obtain that the Neumann function for the first quadrant is

$$N(z, \zeta) = \ln|z^2 - \zeta^2| + \ln|\bar{z}^2 - \zeta^2|.$$

The Neumann function for the first quadrant anchored at the point $z = 1 + i$ is shown in Figure 7.121. Using properties of the real logarithm we rewrite $N(z, \zeta)$ as

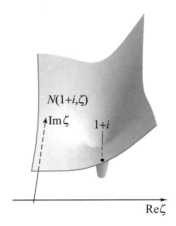

Fig. 7.121 A Neumann function.

$$N(z, \zeta) = \ln|z - \zeta| + \ln|z + \zeta| + \ln|\bar{z} - \zeta| + \ln|\bar{z} + \zeta|. \qquad \square$$

Finding Neumann functions for bounded regions is more difficult because the condition $\frac{\partial N}{\partial n} = C$ is not preserved by a conformal mapping $\phi(z)$; the normal derivative is scaled by $|\phi'(z)|$. In the exercises, we will compute the Neumann function for the unit disk.

What if we are to solve Poisson's equation (7.6.1) with Neumann boundary conditions (7.6.13)? The compatibility condition on h and f has already been handled by (7.6.16); it is

$$\iint_\Omega h(\zeta) \, dA = \int_\Gamma f(\zeta) \, ds. \qquad (7.6.22)$$

This problem is also solvable with the Neumann function, and for reference, we present an analog of Theorem 7.6.3.

Theorem 7.6.14. (Solution of Poisson-Neumann Problem) *Suppose that Ω is a region with boundary Γ, let $N(z, \zeta)$ denote a Neumann function for this region, where z and ζ are in Ω. If u is a solution to Poisson's equation (7.6.1) subject to a Neumann boundary condition (7.6.13) satisfying (7.6.22), then up to an additive constant we have*

$$u(z) = \frac{1}{2\pi} \iint_\Omega h(\zeta) N(z, \zeta) \, dA - \frac{1}{2\pi} \int_\Gamma N(z, \zeta) f(\zeta) \, ds. \qquad (7.6.23)$$

Exercises 7.6

In Exercises 1–6, derive the Neumann function for the region depicted in the accompanying figure.

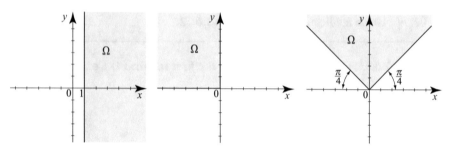

Fig. 7.122 Exercise 1.   **Fig. 7.123** Exercise 2.   **Fig. 7.124** Exercise 3.

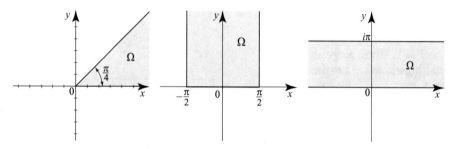

Fig. 7.125 Exercise 4.   **Fig. 7.126** Exercise 5.   **Fig. 7.127** Exercise 6.

7. Uniqueness of the solution in a Poisson problem. Show that the solution of the Poisson problem in a bounded region Ω is unique. [*Hint*: The difference between any two solutions is harmonic and has zero boundary values.]

8. Neumann function for the unit disk. Show that this function is defined for z, ζ in the unit disk by

$$N(z, \zeta) = \begin{cases} \ln|z - \zeta| + \ln\left|\frac{1}{\bar{z}} - \zeta\right| + \ln|z| & \text{if } z \neq 0, \\ \ln|\zeta| & \text{if } z = 0. \end{cases}$$

Derive this function by following the outlined steps.
(a) Write $z = re^{i\theta}$ and $\zeta = \rho e^{i\eta}$. Fix $z \neq 0$, and show that

$$\frac{\partial}{\partial n} \ln|z - \zeta|\Big|_{\rho=1} = \frac{\partial}{\partial \rho} \frac{1}{2} \ln|z - \zeta|^2\Big|_{\rho=1} = \frac{1}{2}\frac{\partial}{\partial \rho} \ln(r^2 + \rho^2 - 2r\rho\cos(\theta - \eta))\Big|_{\rho=1}$$

$$= \frac{1 - r\cos(\theta - \eta)}{1 + r^2 - 2r\cos(\theta - \eta)}.$$

(b) Write $\frac{1}{\bar{z}} = \frac{1}{r}e^{i\theta}$, use (a), and conclude that

$$\frac{\partial}{\partial n}\ln\left|\frac{1}{z}-\zeta\right|\Big|_{\rho=1} = \frac{1-\frac{1}{r}\cos(\theta-\eta)}{(\frac{1}{r})^2+1-2\cos(\theta-\eta)} = \frac{r^2-r\cos(\theta-\eta)}{1+r^2-2r\cos(\theta-\eta)}.$$

(c) Use parts (a) and (b) to show that for $z \neq 0$, $\frac{\partial}{\partial n}N(z,\zeta)\Big|_{|\zeta|=1} = 1$.

(d) Verify the remaining properties of the Neumann function for the given $N(z,\zeta)$.

9. Prove (7.6.8) by justifying the following steps:

$$\left|\iint_{\Omega_\varepsilon} G(z,\zeta)h(\zeta)\,dA - \iint_\Omega G(z,\zeta)h(\zeta)\,dA\right|$$

$$= \left|\iint_{B_\varepsilon(z)} G(z,\zeta)h(\zeta)\,dA\right|$$

$$\leq \iint_{B_\varepsilon(z)} |u_1(z,\zeta)h(\zeta)|\,dA + \iint_{B_\varepsilon(z)} |\ln|z-\zeta|h(\zeta)|\,dA$$

$$\leq C_1\iint_{B_\varepsilon(z)} dA + C_2\iint_{B_\varepsilon(z)} \ln|z-\zeta|\,dA$$

$$= C_1\varepsilon^2\pi + C_2\int_0^{2\pi}\int_0^\varepsilon r\ln|r|\,dr\,d\theta.$$

Evaluate the last integral and show that the resulting expression on the right side tends to 0 as $\varepsilon \to 0$.

10. Proof of Theorem 7.6.10. In addition to the hypothesis of the theorem, we further suppose that $\int_\Gamma u(\zeta)\,ds = A$ is finite.

(a) For fixed z in Ω, let $C_\varepsilon(z)$, $\overline{B_\varepsilon(z)}$, and Γ_ε be as in the proof of Theorem 7.6.3. Since $f(\zeta) = \frac{\partial u}{\partial n}$ in (7.6.19), we have

$$\int_\Gamma N(z,\zeta)\frac{\partial u}{\partial n}\,ds = \int_\Gamma N(z,\zeta)\frac{\partial u}{\partial n}\,ds + \int_{C_\varepsilon(z)} N(z,\zeta)\frac{\partial u}{\partial n}\,ds - \int_{C_\varepsilon(z)} N(z,\zeta)\frac{\partial u}{\partial n}\,ds$$

$$= \int_{\Gamma_\varepsilon} N(z,\zeta)\frac{\partial u}{\partial n}\,ds - \int_{C_\varepsilon(z)} N(z,\zeta)\frac{\partial u}{\partial n}\,ds$$

$$= \int_{\Gamma_\varepsilon} u(\zeta)\frac{\partial}{\partial n}N(z,\zeta)\,ds - \int_{C_\varepsilon(z)} N(z,\zeta)\frac{\partial u}{\partial n}\,ds.$$

(b) As in the proof of (7.6.10), show that $\int_{C_\varepsilon(z)} N(z,\zeta)\frac{\partial u}{\partial n}\,ds \to 0$ as $\varepsilon \to 0$.

(c) Justify the following steps:

$$\int_{\Gamma_\varepsilon} u(\zeta)\frac{\partial}{\partial n}N(z,\zeta)\,ds = \int_{C_\varepsilon(z)} u(\zeta)\frac{\partial}{\partial n}N(z,\zeta)\,ds + \int_\Gamma u(\zeta)\frac{\partial}{\partial n}N(z,\zeta)\,ds,$$

$\int_{C_\varepsilon(z)} u(\zeta)\frac{\partial}{\partial n}N(z,\zeta)\,ds \to -2\pi u(z)$ (see the proof of (7.6.9)), and

$$\int_\Gamma u(\zeta)\frac{\partial}{\partial n}N(z,\zeta)\,ds = C\int_\Gamma u(\zeta)\,ds = AC = C',$$

where C is as in Definition 7.6.8(ii).

(d) Complete the proof of Theorem 7.6.10.

11. Proof of Theorem 7.6.14. The proof mirrors the proof in the text of Theorem 7.6.3, using Green's second identity.

(a) Let $\Omega_\varepsilon = \Omega\setminus\overline{B_\varepsilon(z)}$, where $\overline{B_\varepsilon(z)}$ is the closed disk of radius $\varepsilon > 0$, centered at z. Apply Green's second identity to u and N over the region Ω_ε to get

$$-\iint_{\Omega_\varepsilon} N(z,\zeta)h(\zeta)\,dA = C\int_\Gamma u(\zeta)\,ds - \int_\Gamma N(z,\zeta)\frac{\partial u}{\partial n}\,ds$$
$$+\int_{C_\varepsilon(z)} u(\zeta)\frac{\partial N}{\partial n}\,ds - \int_{C_\varepsilon(z)} N(z,\zeta)\frac{\partial u}{\partial n}\,ds,$$

where C is the fixed value of the normal derivative of N along Γ.

(b) Argue, as in Exercise 9, that as $\varepsilon \to 0$,

$$\iint_{\Omega_\varepsilon} N(z,\zeta)h(\zeta)\,dA \to \iint_\Omega N(z,\zeta)h(\zeta)\,dA.$$

(c) Note that $C\int_\Gamma u(\zeta)\,ds$ is a constant; in fact, it is 2π times the average value of u on Γ.

(d) Just as we proved (7.6.9), show that $\int_{C_\varepsilon} u(\zeta)\frac{\partial N}{\partial n}\,ds \to -2\pi u(z)$, as $\varepsilon \to 0$.

(e) Just as we proved (7.6.10), show that $\int_{C_\varepsilon} N(z,\zeta)\frac{\partial u}{\partial n}\,ds \to 0$, as $\varepsilon \to 0$.

(f) Complete the proof of Theorem 7.6.14.

12. Project Problem: On the uniqueness of the Neumann function. Suppose we have two Neumann functions $N_1(z,\zeta)$ and $N_2(z,\zeta)$ for the same region Ω. They can be written in the form $N_1(z,\zeta) = \ln|z-\zeta| + u_1(z,\zeta)$, $N_2(z,\zeta) = \ln|z-\zeta| + u_2(z,\zeta)$, where u_1 and u_2 are harmonic functions of ζ.

(a) Show that $N_2 - N_1$ is harmonic. What is its normal derivative on Γ, the boundary of Ω?

(b) Apply Theorem 7.6.7 and conclude that N_1 and N_2 differ by a constant–that is, an expression independent of ζ. Can this expression depend on z?

(c) Conclude that $N_1(z,\zeta)$ and $N_2(z,\zeta)$ are two Neumann functions for the same region if and only if they differ by any function of z alone.

13. Project Problem: Symmetry of the Neumann function. Unlike Theorem 7.5.7 for Green's functions, Definition 7.6.8 for Neumann functions does not mention that they are symmetric, and in general it is not the case that $N(z,\zeta) = N(\zeta,z)$. In this problem we discover that symmetry can be recaptured by imposing an extra condition on the Neumann function and that this can always be done without disrupting its role in the solution of the Neumann-Poisson problem.

(a) Refer to Exercise 12. We may add a function of z to a Neumann function and get another Neumann function. Let $N(z,\zeta)$ be a Neumann function for Ω. Show that we can find a function $F(z)$ and a Neumann function $N_0(z,\zeta) = N(z,\zeta) - F(z)$ such that $\int_\Gamma N_0(z,\zeta)\,ds$ ($ds = |d\zeta|$) is independent of z. (It is crucial that Γ has finite length.)

(b) Applying Green's second identity to $N(z_1,\zeta)$ and $N(z_2,\zeta)$ over the region Ω_ε and taking the limit as $\varepsilon \to 0$ (as in the proof of Theorem 7.6.3), we get

$$N(z_1,z_2) - N(z_2,z_1) = \frac{1}{2\pi}\int_\Gamma \left(N(z_1,\zeta)\frac{\partial N(z_2,\zeta)}{\partial n} - N(z_2,\zeta)\frac{\partial N(z_1,\zeta)}{\partial n}\right)\,ds.$$

Use the fact that the normal derivative of $N(z,\zeta)$ is C (a constant) and part (a) to conclude that $N_0(z,\zeta) = N_0(\zeta,z)$.

(c) As a double-check, replace $N(z,\zeta)$ by $N_0(z,\zeta)$ in (7.6.23) and show that the equation remains unchanged. [*Hint*: Remember that $F(z)$ is a constant as far as the integration is concerned, and use the compatibility condition (7.6.22).]

14. Neumann problem with odd boundary data. Consider a Neumann problem in the unit disk in which $f(\zeta) = f(e^{i\theta})$ is an odd function of θ, where $-\pi \leq \theta \leq \pi$. Show that $u(z) = 0$ for all z on the real axis inside the unit disk. [*Hint*: The functions $\ln|z - e^{i\theta}|f(e^{i\theta})$ and $\ln|\frac{1}{z} - e^{i\theta}|f(e^{i\theta})$ are odd functions of θ if z is a real number.]

Appendix

Topology: Distance Between Sets

If K and S are two nonempty sets, the **distance** between K and S is defined to be the smallest value of $|z - w|$ where z is in K and w is in S. If a smallest value is not attained, we define the distance to be the infimum of all $|z - w|$, where $z \in K$ and $w \in S$. It is a fact that if K is a closed and bounded subset of $\mathbb{C}$ and S is a closed subset of $\mathbb{C}$, and K and S are disjoint, then the distance between K and S is positive and is attained for a pair of points (z_0, w_0) in $K \times S$.

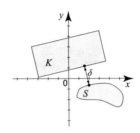

Topology: Compact Sets

A subset of the plane is compact if and only if it is bounded and closed. Compact sets are characterized by the property that every open cover has a finite subcover. Precisely, suppose that K is a compact subset of $\mathbb{C}$ and $\{U_j\}_{j \in J}$ is a (possibly uncountable) family of open subsets of the plane such that $K \subset \bigcup_{j \in J} U_j$. Then there is a finite subset J_0 of J such that

$$K \subset \bigcup_{j \in J_0} U_j,$$

i.e., K is contained in a finite subcover of the cover $\{U_j\}_{j \in J}$.

Analysis: Uniform Continuity

A function f on a set S is called **uniformly continuous** if for every $\varepsilon > 0$ there is a $\delta > 0$ such that for all z_1 and z_2 in S, $|z_1 - z_2| < \delta \Rightarrow |f(z_1) - f(z_2)| < \varepsilon$.

The key word in the definition of uniform continuity is "for all" which expresses the fact that for a given ε, the choice of δ is the same for all points in S; in the definition of continuity at a point, the choice of δ depends on the given point. Clearly, any uniformly continuous function is continuous but the converse is not true. For instance the continuous function x^2 is not uniformly continuous on $(0, \infty)$.

Theorem. *Every continuous function on a closed and bounded (compact) subset S of the plane is uniformly continuous.*

Proof. Let $\varepsilon > 0$ and let z be a point in S. Then there is a $\delta_z > 0$ such that

$$w \in S \quad \text{and} \quad |w - z| < \delta_z \implies |f(w) - f(z)| < \frac{\varepsilon}{2}.$$

The family $\{B_{\delta_z/2}(z)\}_{z \in S}$ is a open cover of S, and by compactness, it has an open subcover $\{B_{\delta_{z_1}/2}(z_1), \ldots, B_{\delta_{z_m}/2}(z_m)\}$. Let $\delta = \min(\delta_{z_1}/2, \ldots, \delta_{z_m}/2)$. Then for $|z - w| < \delta$ there is a j in

© Springer International Publishing AG, part of Springer Nature 2018
N. H. Asmar and L. Grafakos, *Complex Analysis with Applications*,
Undergraduate Texts in Mathematics, https://doi.org/10.1007/978-3-319-94063-2

$\{1,\dots,m\}$ such that $z \in B_{\delta_{z_j}/2}(z_j)$, i.e. $|z - z_j| < \delta_{z_j}/2$. This fact together with $|z - w| < \delta \le \delta_{z_j}/2$ implies $|w - z_j| < \delta_{z_j}$. Then we have

$$|z - w| < \delta \implies |f(z) - f(w)| \le |f(z) - f(z_j)| + |f(z_j) - f(w)| < \frac{\varepsilon}{2} + \frac{\varepsilon}{2} = \varepsilon,$$

since both w and z satisfy $|w - z_j|, |z - z_j| < \delta_{z_j}$. This yields the uniform continuity of f on S. ∎

Analysis: The nonempty intersection property of compact sets

Theorem. *A decreasing sequence of nonempty compact sets in the plane has a nonempty intersection.*

Proof. Let $K_1 \supset K_2 \supset \cdots \supset K_n \supset \cdots$ be a decreasing sequence of nonempty compact subsets in the plane. Suppose that $\cap_{n=1}^{\infty} K_n = \emptyset$. Then

$$K_1 = K_1 \setminus \emptyset = K_1 \setminus \left(\bigcap_{n=1}^{\infty} K_n \right) = \bigcup_{n=1}^{\infty} (K_1 \cap K_n^c) \subset \bigcup_{n=1}^{\infty} K_n^c.$$

But the sets K_n^c are open, and since their union covers K_1, by the compactness of K_1 there is a finite subcover of $\{K_n^c\}_{n=1}^{\infty}$. Then there is an m such that

$$K_1 \subset \bigcup_{n=1}^{m} K_n^c = K_m^c.$$

This implies that K_1 is disjoint from K_m which is a contradiction, since K_m is contained in K_1. ∎

Analysis: The Bolzano-Weierstrass Theorem

Lemma. *Every real sequence has a monotone subsequence.*

Proof. To prove this, let us call a positive integer n a "peak" of the sequence $\{x_n\}_{n=1}^{\infty}$ if $m > n \implies x_n > x_m$. If the sequence $\{x_n\}_{n=1}^{\infty}$ has infinitely many peaks, $n_1 < n_2 < n_3 < \cdots < n_j < \cdots$, then the subsequence $\{x_{n_j}\}_{j=1}^{\infty}$ corresponding to these peaks is monotonically decreasing. If the sequence $\{x_n\}_{n=1}^{\infty}$ has only finitely many peaks, let N be the last peak and $n_1 = N + 1$. Then n_1 is not a peak, since $n_1 > N$, which implies the existence of an $n_2 > n_1$ with $x_{n_2} \ge x_{n_1}$. Again, $n_2 > n_1 > N$ is not a peak; hence, there is $n_3 > n_2$ with $x_{n_3} \ge x_{n_2}$. Repeating this process leads to an infinite non-decreasing subsequence $x_{n_1} \le x_{n_2} \le x_{n_3} \le \dots$. In either case any real sequence $\{x_n\}_{n=1}^{\infty}$ has a monotone subsequence $\{x_{n_j}\}_{j=1}^{\infty}$. ∎

Theorem. (Bolzano-Weierstrass) *Every bounded sequence in the plane has a convergent subsequence.*

Proof. Let $z_n = x_n + iy_n$ where x_n, y_n are real numbers. If $\{z_n\}_{n=1}^{\infty}$ is bounded, then so are $\{x_n\}_{n=1}^{\infty}$ and $\{y_n\}_{n=1}^{\infty}$. By the previous lemma, the real sequence $\{x_n\}_{n=1}^{\infty}$ has a monotone subsequence $\{x_{n_j}\}_{j=1}^{\infty}$. Now consider the sequence $\{y_{n_j}\}_{j=1}^{\infty}$. By the same lemma, it also has a monotone subsequence $\{y_{n_{j_k}}\}_{k=1}^{\infty}$. By the completeness of the real number system, monotone bounded sequences converge to their supremum if they are increasing or to their infimum if they are decreasing; both limits are real numbers. It follows that both $\{x_{n_{j_k}}\}_{k=1}^{\infty}$ and $\{y_{n_{j_k}}\}_{k=1}^{\infty}$ converge to some numbers x and y, respectively. Consequently $\{z_{n_{j_k}}\}_{k=1}^{\infty}$ converges to $x + iy$ as $k \to \infty$. ∎

Index

© Springer International Publishing AG, part of Springer Nature 2018
N. H. Asmar and L. Grafakos, *Complex Analysis with Applications*,
Undergraduate Texts in Mathematics, https://doi.org/10.1007/978-3-319-94063-2

Printed in the United States
By Bookmasters